Hild

Colfax H.S.

Sierra College

Wayne
Hild
Colfax H.S.
Sierra Col.

Introduction to Partial Differential Equations and Hilbert Space Methods

Third Edition, Revised

Karl E. Gustafson
University of Colorado, Boulder

DOVER PUBLICATIONS, INC.
Mineola, New York

To my children
Amy and Garth

Bibliographical Note

This Dover edition, first published in 1999, is a combined repub-
lication of the second edition published by John Wiley & Sons,
New York, 1987 and the third edition published by International
Journal Services, Inc., Calcutta and Charleston, Ill., 1993. The text
has been corrected and updated and a new epilogue has been
added.

Library of Congress Cataloging-in-Publication Data

Gustafson, Karl E.
    Introduction to partial differential equations and Hilbert
space methods / Karl E. Gustafson. — 3rd ed., rev.
        p.   cm.
    Includes indexes.
    ISBN 0-486-61271-6 (pbk.)
    1. Differential equations, Partial. 2. Hilbert space. I.
Title.
QA374.G97   1999
515'.353—dc21   98-28587
                                                                    CIP

Manufactured in the United States of America
Dover Publications, Inc., 31 East 2nd Street, Mineola, N.Y. 11501

PREFACE TO THE THIRD EDITION, REVISED

A considerable number of typographical errors in the 2nd Edition have been corrected. A few of the more recent literature citations have been brought up to date. Some minor expository improvements have been inserted.

A symmetry of the book has been completed by the addition of Fifth and Sixth Pauses in Chapter 3. This new material supplements and follows Appendices A and B, respectively, with treatments of hyperbolic conservations laws (Riemann problem) and shock capturing methods (e.g., Godunov schemes). Both topics are of high current research interest.

This 3rd Edition conforms essentially to the Japanese Edition: Applied Partial Differential Equations, Vol. 1 (1991), Vol. 2 (1992), Kaigai Publications, Ltd., Tokyo (in Japanese).

For the present Dover Edition further typographical errors have been corrected. Some bibliographical citations have been updated or augmented. An added Epilogue treats distribution equations, semiconductor device equations, and financial derivatives equations.

A NOTE TO THE READER

We have followed a pedagogical style of . . . *once . . . twice . . . and then, again,* . . . because that is how the course evolved.

One may, if one likes, justify this in retrospect in terms of the old rules of learning: attention, association, and repetition. The first encounter must be brisk and must catch the attention; the second should better associate the connections and the more general picture; the third should emphasize and repeat important features and details.

And, as in all learning, one must have faith. One need not fall just because one stumbles over a detail or two at the beginning.

KARL GUSTAFSON
Boulder, Colorado
1999

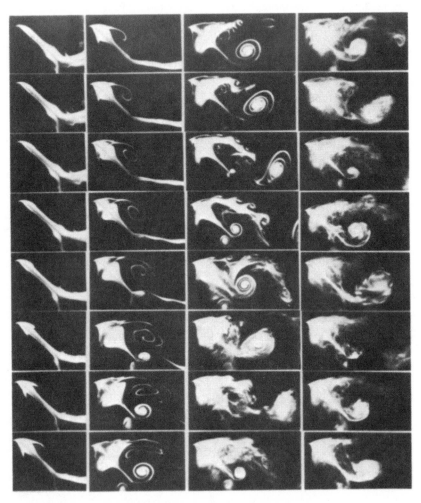

Vortex Splitting in Constant Accelerating Flow past a NACA 0015 Airfoil at
angle of attack $\alpha = 40°$ and Reynolds Number Re = 1000. Courtesy P.
Freymuth, *Prog Aerospace Sci.* **22** (1985). Numerical solution of partial differential equations can now simulate such vortex dynamics. See pp. 346–347.

PREFACE TO THE JAPANESE EDITION

This is the translation of the second edition of K.E. Gustafson's *Introduction to Partial Differential Equations and Hilbert Space Methods,* John Wiley & Sons, New York, 1987 (revised version of the 1980 First Edition). We first encountered the original English version around 1983. Its contents, structure, style, and so on appeared quite extraordinary. It was a great surprise to find a book on partial differential equations which was unlike any we had seen before. Anyone who is interested in partial differential equations and mathematical physics may be familiar with H. Hadamard (1932) and H. Bateman (1932) from the old school, I.G. Petrowski (1945, Japanese translation 1958), and R. Courant-D. Hilbert (1953, new edition 1962) from the recent past, and one or two of the current works of F. John (1971, fourth edition 1982) and R. Leis (1986) as well as other excellent literature within and outside our country. The present book gives us a unique presentation, different from any of the above-mentioned [works of] orthodox mathematical literature.

The importance of discussions concerning partial differential equations needs no emphasis in the world of science and technology. This book aims at science-engineering undergraduates, beginning level post-graduates, science technologists in general, and specialists as well as educators. As he mentions in the preface to the first edition, Professor Gustafson's motivation and purpose in writing this book was to introduce to the readers the main topics of partial differential equations, basic methods, and their application to related areas, and to develop curricula for them. . . . There may be no other work on partial differential equations which in richness of topics and multidimensional content is comparable to this book.

Critical reviews in such journals as *Mathematical Reviews* (1981, 81k: 35005; 1988, 88b: 35001), *SIAM Review* (1983, Vol. 25), *Zentralblatt fur Mathematik* (1981, Vol. 434), . . . reconfirmed our first impression and made us realize the value and the significance of publishing this book in Japanese translation (the above reviews are recommended to those who are interested). . . . We would like to emphasize what appear to us to be the three main characteristics of this book.

The topics and subjects covered herein are abundant and varied, and are not contained within the framework of mathematics but extend to the most advanced field of natural science and engineering, including physics and chemistry. The development of modern methods and rapid computational mathematics is explained from the classical through contemporary periods, using Fourier's series and basic concepts of integral calculus with many examples. Furthermore, the book is concerned with questions of essential importance to partial differential equations (existence of solutions, stability, boundary conditions, etc.), pursuing vigorously the foundation and methodological justification of separation of variables, and demonstrating at the same time close interrelationships among these methods which cannot be found in ordinary textbooks. The main theme of the original work, however, is the application of Fourier Analysis and Hilbert Space Methods to partial differential equations, as well as the application of related analytical methods and computational mathematics. Therefore, this book also serves as a basic introduction to applied mathematics. Explanations of particular themes and problems are varied in depth, but they are all stimulating, suggestive, and interesting. Historical events are not added as an accessory, but rather to illustrate historical origins and the significance of particular problems and methods, and often to raise interesting questions. It may be irrelevant, but in relation to the pure theoretical aspect of the main theme of the book we should mention that there are more comprehensive theories which are still actively studied such as those advanced by Hormander and by Sato through his theory of distributions. The purpose of this book, however, is well achieved within the theoretical framework of standard theories without reference to the theories just mentioned.

The style of this book is free and unrestrained (from the viewpoint of conventional mathematics literature) on the one hand, while the arguments are advanced analytically with a critical spirit. Thoroughly introspective, almost philosophical critical thinking underlies the book, as may be observed in comments about the solutions of partial differential equations, physics observations, supplementary discussions and so on. The style represents a cultural viewpoint which is rare among ordinary mathematics literature, and will serve as a new "intellectual" model for mathematics textbooks.

The original work, which was based on lectures delivered at the University of Colorado, the University of Minnesota, and Ecole Polytechnique Fédérale de Lausanne, clearly has an educational purpose as is stated in the preface and the guide for use. The content is structured to follow "the traditional steps of learning" reflecting the author's strategy for introducing the subject matter of the book. Creative devices are seen in the arrangement of exercises, placing Pauses between lessons, and introducing historical comments, and so on, all of which turn this book into a new, dynamic type of reading unlike traditional technical literature. Some objection may be raised to adopting the style of this book directly for undergraduate and graduate education in Japanese universities, but the underlying thinking and plan of the book have important implications. The unusual impression that this unique book may tend to give at first, to those who are accustomed to conventional technical literature, comes from the author's deep commitment to education on the subject.

We may say that this book combines and organizes three entities: abundant material from mathematics as well as related fields, solid argumentation supported by philosophical insight, and realization of the author's passionate dedication to his educational mission. Occasional rebellious feelings which we had during our translation are gone and we are humbled by this great work.

Although we were led to our task because of this book's rich content, complex illustrations, and unusual format, we often experienced the pains and difficulties of translating its lengthy sentences and penetrating written style which are uncommon in mathematics literature. Somehow we managed the entire translation of the first edition, but while we were correcting and modifying it, the second edition came out . . . and so we had to start all over again. A great deal of work, including proofreading, was involved . . . but the book's excellent content, philosophical insight and deep commitment to education encouraged us to keep going.

. . . We convey our deepest gratitude to Mr. Kuroda Hajime, the president of the Overseas Publication Trading Co., for his understanding and assistance, Mr. Mori Toshio, the head of the department of investigation, who undertook the negotiation of translation rights with Wiley, the American publisher, and Mr. Ono Tomiseki, the head of the first business division, for taking endless trouble. We are very pleased to complete our translation of this valuable book, thanks to the company's assistance. We express our sincere thanks to these three people. We would also like to give our thanks to Mr. Kamagata Masumi, the head of the business division, Mr. Sato Susumu of the same division of the Toppan Publishing Co., and to those who actually engaged in printing. We cannot thank them enough for their efforts in carrying out a great number of corrections and revisions. Furthermore, we deeply appreciate the kind responses from the author, Professor Gustafson, to our inquiries and to our request for a Preface for the Japanese translation. Despite the busy activities and schedule, Professor Gustafson was concerned with various problems which we had raised and frequently corresponded with us. We close this unusually lengthy preface with our best wishes to Professor Gustafson for his future scholarly activities.

February 1991

ABE TAKAHISA
ONDA ISAO
KOJO TOMOMI

ABOUT THE AUTHOR

Professor Gustafson, the author of the book, was born in 1935 in Manchester, Iowa. He studied engineering, physics and applied mathematics at the University of Colorado. After a research assistantship he received a Ph.D. in 1965 from the University of Maryland. His dissertation title was *A priori Bounds with Applications to Integrodifferential Boundary Problems* (dissertation director: L. E. Payne). From 1968 to the present he has been a professor at his alma mater, the University of Colorado, playing leadership roles in research and teaching. At the same time he has made great international contributions through his associations with important government committees on mathematics, applied mathematics, physics, and engineering. Recently, he has actively engaged in research development and graduate level education as a prominent leader in the fields of computational mathematics and physics. In the meantime, he has been invited to serve as visiting professor at several universities in Paris, Lausanne, Geneva, and elsewhere, and other physics and engineering research institutes where he has been involved in research and training. . . .

Following his undergraduate years, Professor Gustafson engaged in research at the Naval Research Laboratory and the University of Maryland Institute of Fluid Dynamics and Applied Mathematics. His two years as a NATO/NSF post-doctoral fellow at the Battelle Institute in Geneva and the Mathematics Institute in Rome provided experience that still exerts an important influence on his varied research and teaching activities today.

Professor Gustafson's research covers a wide range of both theoretical and applied mathematics, and his early studies of functional analysis and partial differential equations have developed into theoretical works and their applications in other areas. For the last ten years his research has included computational fluid dynamics, bifurcation theory, and multidimensional signal analysis, and with these as a base his research now includes advanced computational architectures for neurocomputing and large scale parallel scientific calculation. Up to the present, his publications now number more than two hundred papers. He has also been active in international research and conferences, and has given over two hundred lectures at numerous associations, international conferences, and symposiums in various fields of his interest. Most notably, he served as a conference chair for the VIIth International Conference on Mathematical Physics held at the University of Colorado in honor of his scholarly achievement. More recently, he directed the first IMACS Conference on Computational Physics (1990) in Boulder, chairing the IMACS Technical Committee on Computational Physics.

In addition, he has authored works on research and education, and since the publication of the first edition of this book, has published more than ten books on mathematical physics, partial differential equations, quantum dynamics, fluid dynamics. . . .

In addition, he has written nearly a hundred reviews for the four leading science research funding agencies in the United States and has served as an examiner-commentator for the Mathematical Reviews since 1969. At the same time, he has acted as a referee of more than one hundred papers for over thirty professional journals which are known to any specialist in mathematics, physics, and engineering. . . .

Professor Gustafson's enthusiasm for research guidance and education has been manifest not only on his own university campus, but throughout his long service on national-level committees . . . He has led many Ph.D.s through their graduate training, and is highly respected by his students, who receive from him a kind and thoughtful mentorship. On the other hand, his remarkable personality is easily guessed from the endless flow of foreign researchers visiting him.

CONTENTS

EPILOGUE
... Touch Ups ...

CHAPTER 1

THE USUAL TRINITIES
. . . the basics, and wherein the problems lie . . .

A partial differential equation (often referred to as a PDE)

$$F\left(u, \frac{\partial u}{\partial x_1}, \ldots, \frac{\partial^m u}{\partial x_k^s \cdots \partial x_j^r}, x_1, \ldots, x_n, t, \text{ other parameters}\right) = 0 \quad (1.1)$$

where $u = u(x_1, \ldots, x_n, t, \text{ other parameters})$ is the unknown function or relation and, where F is a function of the prescribed arguments, is a rather general entity and somewhat meaningless in the abstract. Interest usually focuses on particular equations from important applications or on generalized classes of those equations whose properties are amenable to specific description. Because the dynamics of most systems, be they physical, biological, economic, or other, usually involve at most two derivatives (e.g., velocity and acceleration), equations of second order (i.e., $m = 2$ in (1.1)) are the most important and occur the most frequently.*

One should recall from calculus that the operation of taking an ordinary or partial derivative is a linear operation, that is,

$$\partial(u + v)/\partial x = \partial u/\partial x + \partial v/\partial x.$$

Thus a partial differential equation appears as a linear operator equation (where $f(x) = $ data, presumed known)

$$Lu = f \quad (1.2)$$

if L is a combination involving only sums and compositions of derivatives, and no quadratic (or higher powers) terms such as $(\partial u/\partial x)^2$. We recall that a linear operator is a transformation L satisfying

$$L(au + bv) = aL(u) + bL(v) \quad \text{✳} \quad (1.3)$$

---

* First-order equations can often be reduced to a system of first-order ordinary differential equations and will not be treated here. See, however, Appendix A.1, and also Problem 1.9.4. Equations of order greater than two are treated in large part by the methods developed for second-order equations. See for example the Second Pause.

1

for all vectors or functions u, v in its domain $D(L)$ and all numbers or scalars a, b. The validity of linearizing assumptions when modeling physical problems varies from application to application, but linearization is usually a good first approximation and allows the use of the powerful and general linear methods (e.g., Fourier series, among others).

Thus for example the minimal surface operator

$$Lu = \left[1 + \left(\frac{\partial u}{\partial y}\right)^2\right]\frac{\partial^2 u}{\partial x^2} - 2\frac{\partial u}{\partial x}\frac{\partial u}{\partial y}\frac{\partial^2 u}{\partial x \partial y} + \left[1 + \left(\frac{\partial u}{\partial x}\right)^2\right]\frac{\partial^2 u}{\partial y^2},$$

which describes a surface stretched across a wire loop,* although nonlinear in each of its three terms, is rather well approximated by the linear Laplace operator

$$\Delta u = \frac{\partial^2 u}{\partial x^2} + \frac{\partial^2 u}{\partial y^2}$$

when the slopes are small. Moreover, by a suitable change of variable† the nonlinear minimal surface operator may be transformed into a linear operator (with variable coefficients)

$$Aw = (1 + \xi^2)\frac{\partial^2 w}{\partial \xi^2} + 2\xi\eta\frac{\partial^2 w}{\partial \xi \partial \eta} + (1 + \eta^2)\frac{\partial^2 w}{\partial \eta^2}$$

and by a further transformation into the Laplace operator

$$\Delta v = \frac{\partial^2 v}{\partial r^2} + \frac{\partial^2 v}{\partial s^2} + \frac{\partial^2 v}{\partial t^2}$$

in one higher dimension.

For these reasons, among others, in this chapter as well as in most of this book we will treat primarily second-order linear partial differential equations. Although we depart from linearity, for example, in the considerations of bifurcation theory and nonlinear waves in Section 1.8 and Problem 2.9.8, respectively, even there it will be seen that the first approximation is linear and that linear methods are used to the extent possible.

To conclude this introduction, we would like the reader to answer the following three questions.

Question 1. How do you fit a line to two given points (Diagram 1*a*)?

Question 2. Given the temperature u_0 of an object at time $t = 0$ and a law of cooling which states that the rate at which the object cools is directly propor-

---

* For a fascinating discussion of the mathematical theory of soap films see the article by F. Almgren and J. Taylor in *Scientific American* **235**, July (1976).

† The details, which would unnecessarily delay us here, may be found for example in an exercise in Problem 1.9.9.

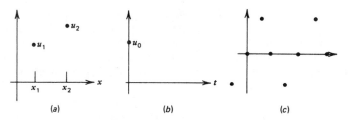

DIAGRAM 1. Elementary problems.

tional to the temperature difference between the object and its surroundings, how do you find the temperature at time t (Diagram 1b)?

Question 3. In thinking of smoothly oscillating functions, what is the first one that comes to mind (Diagram 1c)?

No doubt the reader, in his mind's eye, has already filled in the solutions to the questions. Let us do so here (Diagram 2). The answer to the first question is a line $u(x) = ax + b$, which when fit to the two points becomes

$$u(x) = \left(\frac{u_2 - u_1}{x_2 - x_1}\right)x + \left(\frac{u_1 x_2 - u_2 x_1}{x_2 - x_1}\right), \qquad x_1 \le x \le x_2.$$

The answer to the second question, if for simplicity we agree to measure temperature relative to a surrounding temperature of zero, is obtained by solving the equation $u'(t) = -ku(t)$ for $t > 0$, $u(0) = u_0$, k the given proportionality constant, which yields the solution

$$u(t) = u_0 e^{-kt}, \qquad t \ge 0.$$

We assume for the third question that the reader has imagined a sine function

$$u(x) = \sin kx.$$

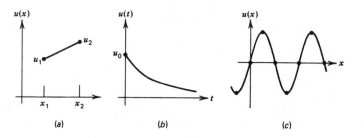

DIAGRAM 2. Elementary solutions.

Question 1 was an example of a _boundary value problem_: Stretch a function between two boundary points according to some given (e.g., physically motivated) prescription on the function over the intervening domain. Here we used the prescription that the connecting curve have minimal length, which geometrically must be a line. Mathematically, we could have expressed that prescription by

$$u''(x) = 0, \qquad x_1 < x < x_2.$$

Question 2 was an example of an _initial value problem_: Given a function initially, determine its behavior thereafter, provided that you are given a prescription (e.g., physically motivated) for how the function should evolve. The two cases in which it will decay exponentially according to $u' = -ku$ or grow exponentially according to $u' = ku$ are both included in the prescription

$$u''(x) = k^2 u(x), \qquad x \geqq 0.$$

Question 3 was an example of an _eigenvalue problem_, which is sometimes called a _characteristic value problem_. The word eigen (Ger.) means characteristic (Fr.), and perhaps a little more to the point here, it means _inherent_. What is inherent about the sine function $u(x) = \sin kx$? Its oscillatory properties, along with those of its companion the cosine function, are a result of the prescription

$$u''(x) = -k^2 u(x), \qquad -\infty < x < \infty.$$

The above three types of problems

$$\begin{bmatrix} \text{boundary value problems} \\ \text{initial value problems} \\ \text{eigenvalue problems} \end{bmatrix}$$

are among the most important encountered in partial differential equations.

1.0 Exercises

1. Integrate (by calculus):
   (a) $u''(x) = 0, \qquad -\infty < x < \infty,$
   (b) $u''(x) = 0, \qquad x > 0, \qquad u(0) = 1,$
   (c) $u''(x) = -4u(x), \qquad 0 < x < \pi.$
2. Solve:
   (a) $u''(x) = 0, \qquad 0 < x < 1, \qquad u(0) = 1, \qquad u(1) = 0,$
   (b) $u'(x) = -2u(x), \qquad x > 0, \qquad u(0) = 1,$
   (c) $u''(x) = -4u(x), \qquad 0 < x < \pi, \qquad u(0) = u(\pi) = 0.$
3. Solve:
   (a) $\dfrac{\partial u(x, y)}{\partial x} = 0, \qquad -\infty < x < \infty, \qquad -\infty < y < \infty,$
   (b) $\dfrac{\partial^2 u(x, y)}{\partial x^2} = 0, \qquad -\infty < x < \infty, \qquad -\infty < y < \infty,$

(c) $\dfrac{\partial^2 u(x, y)}{\partial x^2} = 0, \qquad x > 0, \qquad u(0, y) = 1.$

1.1 THE USUAL THREE OPERATORS AND CLASSES OF EQUATIONS

The following three operators usually form the base for a study of partial differential equations*:

(1) Laplace operator, $\Delta = \dfrac{\partial^2}{\partial x_1^2} + \cdots + \dfrac{\partial^2}{\partial x_n^2}$,

(2) Diffusion operator, $\dfrac{\partial}{\partial t} - \Delta$,

(3) D'Alembert operator, $\Box = \dfrac{\partial^2}{\partial t^2} - \Delta$.

The three general classes of partial differential operators that one encounters are: (1) elliptic operators, (2) parabolic operators, and (3) hyperbolic operators, respectively. By a change of variables (see below, and Problem 1.9.1) of particular type, one reduces an elliptic operator L to its canonical form, the Laplacian Δ in another coordinate system. Likewise, parabolic and hyperbolic operators may be

---

*A remark on notations, of which there are many. Our preference is to minimize the importance of notation while recognizing its validity. Let us, therefore, without further ado, become familiar with several in use.

Usually we will take $u(x_1, \ldots, x_n, t)$ to be scalar (e.g., real) valued; if u is vector valued, one needs to be a little more careful with some of the notations. To illustrate briefly, several equivalent notations for the Laplacian Δ:

$$\Delta u = \sum_{i=1}^{n} \partial^2 u/\partial x_i^2 \qquad \text{Calculus form}$$

$$= u_{x_1 x_1} + \cdots + u_{x_n x_n} \qquad \text{PDE form}$$

$$= u_{,ii} \qquad \text{Einstein or tensor summation convention}$$

$$= \text{div(grad } u) \qquad \text{Vector analysis form}$$

$$= \nabla^2 u \qquad \text{Electrostatic form}$$

$$= u_{xx} + u_{yy} \qquad \text{Two-dimensional form}$$

The notation for the other principal partial differential operators is similar and should cause no difficulty. One will need to recognize that $u_t = u_{,t} = \partial u/\partial t$ denotes the partial derivative with respect to t, and that t often represents time.

transformed to the diffusion and d'Alembert operators, respectively. The Laplace* operator Δ is sometimes called the potential operator the diffusion operator is often called the heat operator,† and the d'Alembert‡ operator is often called the wave operator, due to their frequent appearance in applications bearing the latter names. Although the relative importance of the various names and viewpoints could be a subject for endless debate, we may summarize them in Table 1.1.

In Problem 1.9.1 the reader will find more details concerning the classification procedure for a partial differential equation, and in Problem 1.9.2 how it relates to certain properties of the equation.§ Let it suffice here to classify the general second-order equation in two independent variables¶

$$Au_{xx} + 2Bu_{xy} + Cu_{yy} + Du_x + Eu_y + Fu + G = 0$$

according to the discriminant rule (which is verified in Problem 1.9.1):

Discriminant	Equation Type
$d > 0$	elliptic
$d = 0$	parabolic
$d < 0$	hyperbolic

where $d = AC - B^2$. The names of the three equation types are thus seen to be related to the conic sections of the same name. The lower order terms are simply ignored in this procedure. It follows that an elliptic equation may be transformed by a linear 1–1 change of variables to the two-dimensional Laplacian form

$$u_{\xi\xi} + u_{\eta\eta} + \text{(lower-order terms)} = 0,$$

a parabolic equation to the canonical form

$$u_{\xi\xi} + \text{(lower-order terms)} = 0$$

or

$$u_{\eta\eta} + \text{(lower-order terms)} = 0,$$

* After Laplace and his use of the operator in *Mecanique Celeste* in 1785.

† There seems to be no person whose name has been traditionally attached to the general diffusion operator. Although a number of diverse names spring to mind, such as Thompsonian, Joulean, Brownian, Markovian, Eddingtonian, among a multitude of others, perhaps if history had been more accurate the potential operator would have been called the Newtonian and the diffusion operator would have been called the Fourierian. For an account of the lives and scientific contributions of Newton, Fourier, d'Alembert (and Laplace), see E. Bell, *Men of Mathematics*, Simon and Schuster, New York, 1937, for example.

‡ After d'Alembert and his 1746 solution of the vibrating string problem.

§ In mathematics the student soon discovers that any "transformation to canonical form" can be an oversimplification and even a deception. Exactly what is "canonical," what are the invariants, and so on, must be clarified. On the other hand, often the canonical form itself is the one that appears most frequently and is therefore of primary importance, and that is the case in partial differential equations.

¶ Regard all coefficients as constant. Generally, A, B, C may depend on x, y, u, $u_{,i}$, but not on $u_{,ij}$.

Table 1.1

Mathematical Quantity	Surnamed	Physically Named	Classification Type
Δ_n	Laplacian	Potential operator	Elliptic
$\dfrac{\partial}{\partial t} - \Delta_{n-1}$	(Heat)	Diffusion operator	Parabolic
$\square = \dfrac{\partial^2}{\partial t^2} - \Delta_{n-1}$	d'Alembertian	Wave operator	Hyperbolic

and a hyperbolic equation to one of the canonical forms

$$u_{\xi\xi} - u_{\eta\eta} + \text{(lower-order terms)} = 0$$

or

$$u_{\xi\eta} + \text{(lower-order terms)} = 0.$$

In particular the Laplacian, heat and wave operators are thus seen to be the canonical forms for the three types of partial differential equations.*

Classification of higher-order operators (e.g., interesting fourth-order elliptic operators appear in the theory of elasticity) is similar but more complicated, as are systems of partial differential equations.

If an operator has nonconstant coefficients, the classification is only local (i.e., it may change as the point (x, y) moves from one part to another part of the domain Ω). This is the case, for example, in the Tricomi operator of gas dynamics

$$Lu = yu_{xx} + u_{yy}.$$

The difference in type can be quite crucial: the elliptic region (Fig. 1.1) corresponds

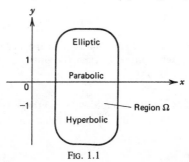

Fig. 1.1

* There is a rather harmless joke about mathematicians at the University of Minnesota, one of several centers for the study of partial differential equations. Whereas at most universities the mathematicians ask whether a mathematical visitor be topologist, algebraist, analyst, and so on, the story is that at Minnesota they ask whether he or she is elliptic, parabolic, or hyperbolic.

to subsonic and smooth flow, the parabolic boundary to a sonic barrier, and the hyperbolic region to supersonic propagation of shock waves. From the hyperbolic region behavior in this example, from the Poisson formula for the solution to the heat equation that we will see presently, and from the discussion in the introduction of the minimal surface equation, we may let our intuition be directed as follows:

Equation Type	Solution Behavior
Elliptic	a stationary and energy-minimizing surface
Parabolic	a very smoothing and spreading flow
Hyperbolic	a disturbance-preserving wave

This induction does in fact provide a reasonably correct intuitive picture for our future guidance.

Problem 1. (a) Using the discriminant rule, verify the classification stated above for the Tricomi operator L. (b) Show that the minimal surface operator L is indeed elliptic. (c) Show the same for the transformed operator A with variable coefficients given on page 2.

Problem 2. (a) The d'Alembert (or wave) operator

$$Lu = u_{tt} - u_{xx}$$

is only one of two canonical forms for a hyperbolic equation. The other canonical form is

$$u_{rs}.$$

Find a 1–1 linear change of variables $r = c_{11}t + c_{12}x$, $s = c_{21}t + c_{22}x$, which transforms the equation $u_{tt} - u_{xx} = 0$ into the equation $u_{rs} = 0$. (b) For a review exercise in elementary linear algebra, letting C denote the matrix of the transformation found in part (a), calculate the inverse transformation

$$C^{-1}: \begin{pmatrix} r \\ s \end{pmatrix} \to \begin{pmatrix} t \\ x \end{pmatrix}.$$

(c) Geometrically, what are the solutions to the one-dimensional Laplacian equation $u_{xx} = 0$? What are they when you also ask that $u(x) = 0$ at $x = 0$ and $u(x) = 1$ at $x = b \neq 0$?

Problem 3. To develop some intuition, write down by inspection (i.e., trial and error) some solutions of the basic equations mentioned above:

(a) $u_{xx} + u_{yy} = 0$

(b) $u_{xx} - u_t = 0$

(c) $u_{xx} - u_{tt} = 0$

(d) $yu_{xx} + u_{yy} = 0$

(e) $(1 + (u_y)^2)u_{xx} - 2u_y u_x u_{xy} + (1 + (u_x)^2)u_{yy} = 0.$

1.1 Exercises

1. Apply the classification rule to
   (a) $4u_{xx} + 2u_{yy} + 27u_y = 0$,
   (b) $4u_{xx} - 2u_{yy} = 0$,
   (c) $4u_{xx} - u_y = 0$.

2. Verify that
   (a) $u = \ln r$ satisfies the Laplacian $\Delta_2 u = 0$ for $r = (x^2 + y^2)^{1/2} \neq 0$.
   (b) the Gaussian $(4\pi t)^{-1/2} e^{-(x^2/4t)}$ satisfies the Heat equation $u_t - u_{xx} = 0$, $t \neq 0$.
   (c) a moving sine wave $\sin(x + t)$ satisfies the Wave equation $u_{tt} - u_{xx} = 0$.

3. (a) Show that if $Au_{xx} + 2Bu_{xy} + Cu_{yy} = 0$ is parabolic, then the change of variables

$$\begin{cases} \zeta = x \\ \eta = rx + y, \end{cases}$$

   where $r = -B/A$, transforms the equation to the canonical form

$$u_{\zeta\zeta} = 0.$$

   (b) If $Au_{xx} + 2Bu_{xy} + Cu_{yy} = 0$ is hyperbolic, then the change of variables

$$\begin{cases} \zeta = r_1 x + y \\ \eta = r_2 x + y, \end{cases}$$

   where r_1 and r_2 are the roots of $Ar^2 + 2Br + C = 0$, transforms the equation to the canonical form

$$u_{\zeta\eta} = 0.$$

   Hence $u = f(\zeta) + g(\eta) = f(r_1 x + y) + g(r_2 x + y)$, where f and g are arbitrary (twice differentiable) functions, is always a solution of the partial differential equation.
   (c) Classify $yu_{xx} + 2xu_{xy} + yu_{yy} = 0$.

1.2 THE USUAL THREE TYPES OF PROBLEMS

One thinks in terms of three types of problems occurring in applications,* although as we shall see one runs into almost all combinations of the three types of operators and the three types of problems.

---

* Further discussion of the physical derivation of such problems will be found later and, in particular, in Section 1.7.

1.2.1. Boundary Value Problems (hereafter sometimes abbreviated as BVP)

An important example is the famous Dirichlet problem*

$$\begin{cases} \Delta u = 0 \text{ "in } \Omega\text{",} & \text{i.e., } x \in \Omega, \\ u = f \text{ "on } \partial\Omega\text{",} & \text{i.e., } x \in \partial\Omega, \end{cases}$$

where Ω is a specified domain (also called a region, an open connected set in Euclidean n-space) and where f is the given boundary value defined on the boundary $\partial\Omega$. (Fig. 1.2a).

1.2.2. Initial Value Problems (hereafter sometimes abbreviated as IVP)

An illustrative example is the heat equation (in an infinite idealized rod)

$$\begin{cases} u_t - u_{xx} = 0, & -\infty < x < \infty, \quad t > 0, \\ u(x, 0) = f(x), & -\infty < x < \infty, \quad t = 0, \end{cases}$$

where f is the given initial temperature distribution and $u(x, t)$ the evolving temperature distribution for positive time t (Fig. 1.2b).

1.2.3. Eigenvalue Problems† (hereafter sometimes abbreviated as EVP)

A simple example‡ is the ordinary differential equation (hereafter sometimes abbreviated as ODE)

$$\begin{cases} -v'' = \lambda v, & 0 < x < \pi, \\ v(0) = v(\pi) = 0, \end{cases}$$

The given boundary value f

$\partial\Omega$ = boundary of Ω

FIG. 1.2a

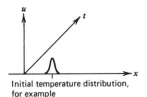

Initial temperature distribution, for example

FIG. 1.2b

* Which permeated (and caused) much of 18th- and 19th-century mathematics. In the Dirichlet problem it is desired to obtain, for each data function f continuous on the boundary $\partial\Omega$, a solution u continuous on the closure of the domain Ω. The Dirichlet problem was theoretically regarded as solved for a domain Ω when the existence of such a solution could be demonstrated. With the work of Poincaré and Hilbert, among others, this was accomplished for most reasonable domains. See Section 1.5.3 for further historical remarks on this problem.

† Also called characteristic value or fundamental frequency problems, among others.

‡ This is sometimes called the Rayleigh equation, after Lord Rayleigh.

a solution (the second eigenfunction) illustrated in Figure 1.2c. This equation arises as we shall see below after an application of the method of separation of variables to the vibrating string problem (Fig. 1.2d)

$$\begin{cases} u_{tt} - u_{xx} = 0, & 0 < x < \pi, \quad t > 0, \\ u(0, t) = u(\pi, t) = 0, & t \geq 0, \\ u(x, 0) = f_1(x), & u_t(x, 0) = f_2(x). \end{cases}$$

The latter describes a vibrating string (with small displacements only, to keep the description linear) with initial position and velocity given and held fixed at the ends. The former (EVP) always has the trivial (zero) solution and one is only interested in nonzero solutions (fundamental modes) $v_n(x)$ and in the corresponding λ_n (eigenvalues) from which the general solution to the vibration problem may be constructed by the method of separation of variables.

We note that the vibration problem just mentioned involves boundary values, initial values, eigenvalues, and both hyperbolic and elliptic operators.

For purposes of review and later use let us recall how one solves the Rayleigh eigenvalue problem (1.2.3) above by elementary considerations from the theory of ordinary differential equations. By trial of $v(x) = e^{zx}$ in the equation

$$v''(x) + \lambda v(x) = 0,$$

one is led to the so-called auxiliary equation

$$z^2 + \lambda = 0$$

and the two fundamental solutions

$$y_1(x) = e^{z_1 x}, \qquad y_2(x) = e^{z_2 x},$$

where z_1 and z_2 are the roots $\pm \sqrt{-\lambda}$ from the auxiliary equation. All other solutions are exactly the linear combinations of the two fundamental solutions. When $\lambda < 0$, this provides the two real valued fundamental solutions

$$y_1(x) = e^{rx}, \qquad y_2(x) = e^{-rx},$$

where $r = \sqrt{-\lambda}$. When $\lambda = 0$, one obtains from the above the constant functions, and in order to have two linearly independent fundamental solutions one takes

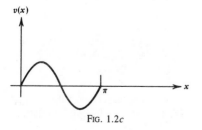

FIG. 1.2c

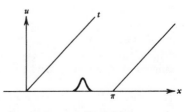

FIG. 1.2d

$$y_1(x) = 1, \qquad y_2(x) = x.$$

For $\lambda > 0$, fundamental solutions $e^{\pm i\lambda^{1/2}x}$ found above are complex valued and in order to obtain two linearly independent real valued fundamental solutions one takes the real and imaginary parts

$$y_1(x) = \cos \lambda^{1/2}x, \qquad y_2(x) = \sin \lambda^{1/2}x.$$

Upon trying $v = c_1 y_1 + c_2 y_2$ for arbitrary constants c_1 and c_2 for each of these three cases $\lambda < 0$, $\lambda = 0$, $\lambda > 0$, along with the boundary conditions, one sees quickly that the first two cases permit only the trivial solution. Thus the eigenfunctions for the problem are

$$v_n(x) = \sin \lambda_n^{1/2}x, \qquad n = 1, 2, 3, \ldots,$$

with corresponding eigenvalues

$$\lambda_n = 1, 4, 9, \ldots, n^2, \ldots$$

Note that

$$v_n(x) = c_n \sin \lambda_n^{1/2}x$$

where c_n is any arbitrary constant, is also a solution to the eigenvalue problem, with the same eigenvalue λ_n.

Problem 1. (a) Solve by inspection the Dirichlet problem, where Ω is the unit square $0 \leq x \leq 1$, $0 \leq y \leq 1$, and where the data is:

$$f(x, y) = \begin{cases} x \text{ for } 0 \leq x \leq 1, & y = 0, \\ 1 \text{ for } x = 1, & 0 \leq y \leq 1, \\ x \text{ for } 0 \leq x \leq 1, & y = 1, \\ 0 \text{ for } x = 0, & 0 \leq y \leq 1. \end{cases}$$

(b) Show that there is no polynomial solution to the problem of part (a) when the data $f(x, y) = x$ on the first and third sides is replaced by quadratic data $f(x, y) = x^2$. (c) For further exercise, either (i) keep trying to find a continuous solution to the problem of part (b), or (ii) investigate the same problem with the Laplacian replaced by the minimal surface operator.

Problem 2. Solve by inspection the heat equation given above for the initial data (a) $f(x) = 1$; (b) $f(x) = x$; (c) $f(x) = x^2$.

Problem 3. Solve by inspection the wave equation given above for the initial data (a) $f_1(x) = \sin x$ and $f_2(x) = 0$; (b) $f_1(x) = 0$ and $f_2(x) = \sin x$; (c) the same as in (a) and (b) but with $\sin x$ replaced by $\sin 2x$.

The notion of "characteristic curves" for a given partial differential equation, and how those characteristic curves relate to where you may have initial data as concerns the solvability of an initial value problem, may be found in Problem 1.9.2 and in Chapter 3, Appendix A.

1.2 Exercises

1. Boundary value problems (BVP) should be distinguished as either interior or exterior, and usually our first impressions take us to the former, on which solutions are more readily understood.

   (a) Show the interior BVP

   $$\begin{cases} u''(x) = 0 & \text{in } \Omega = (-1, 1), \\ u(x) = f(x) & \text{on } \partial\Omega = \{\text{the points } x = \mp 1\}, \end{cases}$$

   has exactly one solution.

   (b) Show the unbounded BVP

   $$\begin{cases} u''(x) = 0 & \text{in } \Omega = (0, \infty), \\ u(x) = f(x) & \text{on } \partial\Omega = \{\text{the points } x = 0, \infty\}, \end{cases}$$

   has exactly one, no, or an infinite number of solutions.

   (c) Show the exterior BVP

   $$\begin{cases} u''(x) = 0 & \text{in } \Omega = (-\infty, -1) \cup (1, \infty), \\ u(x) = f(x) & \text{on } \partial\Omega = \{\text{the points } x = \mp 1\}, \end{cases}$$

   has an infinite number of solutions.

2. Initial value problems (IVP) are usually the principal topic in a first course in ordinary differential equations. For review and some geometrical intuition, solve the following IVPs.

   (a) $\begin{cases} u_t(x, t) = 0, & -\infty < x < \infty, \quad t > 0, \\ u(x, 0) = f(x), & -\infty < x < \infty, \quad t = 0. \end{cases}$

   (b) $\begin{cases} u_t(x, t) + 3u(x, t) = 0, & -\infty < x < \infty, \quad t > 0, \\ u(x, 0) = f(x), & -\infty < x < \infty, \quad t = 0, \end{cases}$

   (c) $\begin{cases} u_{tt}(x, t) = 0, & -\infty < x < \infty, \quad t > 0, \\ u(x, 0) = f(x), & -\infty < x < \infty, \quad t = 0. \end{cases}$

3. Eigenvalue problems are concerned with fundamental modes of vibration in a given physical system. There is a great conceptual difference between linear and so-called nonlinear eigenvalue problems.

   (a) Show that for any linear eigenvalue problem

   $$Lu = \lambda u$$

   any scalar multiple of an eigenfunction u is also an eigenfunction.

   (b) Confirm that $v = 100 \sin 5x$ is a solution to the Rayleigh eigenvalue problem.

(c) Verify that a scalar multiple of an eigensolution $u_\lambda(x)$ of the nonlinear eigenvalue problem

$$u''(x) = \lambda u^3(x)$$

is not in general an eigensolution, at least not for the same "eigenvalue" λ.

1.3 THE USUAL THREE QUESTIONS (AND THE OTHER THREE)

Given problems of the above type one usually asks the following three questions:

(1) *Existence* (of at least one solution u)

(2) *Uniqueness* (i.e., either one or zero solutions)

(3) *Stability* (often called continuous dependence of solutions $\{u\}$ upon the data $\{f\}$)

There are three other questions of equal importance, especially in applications:

(1') *Construction* (of the "physical" solution)

(2') *Regularity* (e.g., how "differentiable" is the found solution)

(3') *Approximation* (especially when exact construction of the solution is not possible)

Of course in practice the best method of demonstrating (1) existence is to (1') construct* an exact solution that is sufficiently regular (2') to be substituted into the original problem, meeting all requirements originally imposed. If uniqueness (2) has been proven, then one has *the* solution. Unfortunately, and in contrast to most ordinary differential equations, solution representations in partial differential equations often involve limiting processes such as series or integrals, and the solutions are generally found to be not representable in closed form as elementary functions. Thus, for example, a truncated Fourier series solution used in practice is in fact only an approximate solution (3') using approximate data f; however, if stability (3) has been demonstrated, a small error in the data induces only a small error in the solution, and this is usually a satisfactory state of affairs.

By the common terminology a problem for which (1), (2), and (3) have been shown is said to be *well-posed* or *properly posed*. One might say instead that the mathematical model is complete and consistent. Exactly the right type of data, initial conditions, and boundary conditions are present to determine exactly the mathematical solution. This does not say how well the mathematical model describes the physical problem, which may possess different solution properties due to un-accounted-for effects. The latter is a problem in model building that the mathematics may or may not clarify.†

---

* Any method is acceptable, such as guessing, Fourier expansion, Green's function representation, integral transform, formal arguments, change of variable, or whatever.

† For example, the solution formula given in the following Problem 2 shows that the heat equation model given above is often not realistic, in that linearizing the model forces an infinite speed of heat

Let us discuss further these questions, which are at the heart of the theory of partial differential equations, in terms of the three examples given above. We consider a two-dimensional Dirichlet problem on a "nice"‡ bounded domain Ω as depicted in Figure 1.3a. For intuitive purposes it is instructive to think first of the nonlinear version of the Laplacian Δ, namely, the minimal surface operator

$$Lu = (1 + (u_y)^2)u_{xx} - 2u_xu_yu_{xy} + (1 + (u_x)^2)u_{yy}$$

and the nonlinear minimal surface problem

$$\begin{cases} Lu = 0 & \text{in } \Omega, \\ u = f & \text{on } \partial\Omega. \end{cases}$$

As discussed in the introduction, one may imagine $f(x, y)$ as describing a wire loop; the minimal surface can be thought of as the "soap film" stretching across the wire. Because the equation on Ω is homogeneous (zero data), we are disregarding gravity. If slopes are small, then L is approximately the Laplacian Δ, and the Dirichlet problem

$$\begin{cases} \Delta u = 0 & \text{in } \Omega, \\ u = f & \text{on } \partial\Omega. \end{cases}$$

can be thought of as a problem involving a membrane stretched across the given boundary data, even though the curvature requirements of L and Δ are not exactly the same.

Proceeding with this intuitive view of the problem, one immediately rules out the spurious solution (Fig. 1.3b) to the Dirichlet problem,

$$\begin{cases} u(x, y) \equiv 0, & \text{for } (x, y) \text{ in } \Omega, \\ u(x, y) \equiv f(x, y), & \text{for } (x, y) \text{ on } \partial\Omega. \end{cases}$$

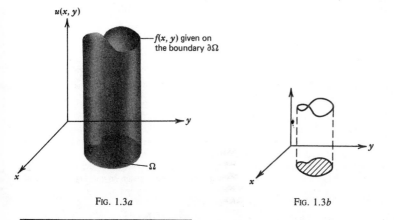

FIG. 1.3a FIG. 1.3b

propagation into the solution. The latter may be seen by considering data $f(x)$ concentrated on a very small interval of time $t = 0$. See Problem 2 at the end of this section.

‡ The question of "nice" domains is subtle (see Section 1.6.1 and Problem 1.9.5). However, it turns out that most domains Ω of practical interest are all right.

Thus we see that any interesting solution must be at least continuous at the boundary $\partial\Omega$. This is an example of the regularity requirement (2'). If we now regard the Dirichlet problem as a linear system L of two linear conditions (namely, that $\Delta u = 0$ in Ω and $Iu = f$ on $\partial\Omega$, where I just denotes the identity operator), we see that a reasonable "domain" for L is the set of functions $u(x, y)$ that are continuous on $\overline{\Omega} \equiv \Omega + \partial\Omega$, that is, on the closure of Ω. Similarly, because one wants to take two derivatives, it is reasonable to require that u be twice continuously differentiable in Ω.* For such functions one easily shows (see Problem 1.9.3) the following.

Maximum Principle. Let Ω be any bounded domain, and let $u(x, y)$ in $C^0(\overline{\Omega}) \cap C^2(\Omega)$ be a harmonic† function in Ω. Then u attains its maximum value on $\overline{\Omega}$ somewhere on $\partial\Omega$.

The following problems illustrate the three types of questions (1), (2), and (3).

Problem 1. (a) Use the Maximum Principle to show uniqueness (2) among $C^0(\overline{\Omega}) \cap C^2(\Omega)$ functions for the Dirichlet boundary value problem (1) on a bounded domain. (b) Discuss whether the C^2 requirement is appropriate. (c) Can you obtain a (3) stability statement from the Maximum Principle?

Problem 2. Given the Poisson representation formula‡ for the heat propagation initial value problem (2), namely, that

$$u(x, t) = \frac{1}{\sqrt{4\pi t}} \int_{-\infty}^{\infty} e^{[-(x-y)^2/4t]} f(y) \, dy, \qquad t > 0,$$

discuss to what extent it provides a solution to the problem. Assume that the data f is very nice.

Problem 3. Examine the following boundary value problem

$$\begin{cases} u_{xx} + 4xu = 16, & 0 < x < 2 \\ u(0) = 0 \\ u(2) = -1 \end{cases}$$

as concerns the six [(1), (2), (3), (1'), (2'), (3')] questions.

---

* The shorthand is: $D(L) = C^0(\overline{\Omega}) \cap C^2(\Omega)$, that is, functions $u(x, y)$ continuous on $\overline{\Omega}$ possessing second partial derivatives $u_{xx}, u_{xy}, u_{yx}, u_{yy}$ all of which are continuous in Ω. In general, the meaning of $C^k(S)$ for an arbitrary set S should be clear to the reader. Notice that the word "domain" is used in two different senses: the geometrical sense (for Ω) and the function sense (for L). But both senses are really the same; u is a function on the x in the domain Ω and L is a function on the $u(x)$ in the domain $D(L)$.

† That is, $\Delta u = 0$ in Ω, as in the Dirichlet problem.

‡ This is an example of a powerful solution method known as *Green's Functions*, which we shall further discuss later. This method is, however, sometimes more powerful than usable, due to the difficulty in many cases of writing down explicitly the Green's Function, even when it is known to exist.

Some remarks on these three problems. Concerning Problem 1, there is a second (for slightly less general domains Ω) general method for demonstrating uniqueness, known as the *energy method*, that we will discuss later in Section 1.6. Problem 2 involves the complications of an infinite domain and some review of differentiation under the integral. We suggest leaving it somewhat open-ended at this time. On the other hand, the student will profit by beginning its consideration as concerns the three questions (and the other three) now. Problem 3, although an ordinary differential equation, will already indicate a number of considerations that arise as one asks the six questions of this section.

For a proof of the Maximum Principle, see Problem 1.9.3. Use of maximum principles provides uniqueness results for elliptic and parabolic equations but generally speaking not for hyperbolic equations. For the latter, energy and characteristic methods may be employed.

Existence results hold rather widely for initial value problems if the initial data is good. One may recall for example the standard general existence statement (Picard theorem) for the ordinary differential equation initial value problem

$$\begin{cases} y'(x) = f(x, y(x)) \\ y(0) = y_0. \end{cases}$$

For completeness we include in Problem 1.9.4 a short discussion of the extension of this result to initial value problems for partial differential equations, the so-called Cauchy–Kowalewski theorem.

The other three questions:

    (1′) Construction of the solution

    (2′) Regularity (i.e., verification of the solution)

    (3′) Approximation of the solution

are of equal importance to those of (1) existence, (2) uniqueness, and (3) stability. In the actual use of partial differential equations in science and other applications, no doubt (1′), and if not (1′) then (3′), must be given the greatest attention. In this book we have attempted to maintain some balance concerning the relative merits of the Questions (1), (2), (3), (1′), (2′), and (3′), but overall have placed the major emphasis on (1′) the construction of the solution. A rigorous treatment of Question (2′) for certain types of problems will be found in Section 2.2. Numerical methods for Question (3′) will be found in Section 2.6 and Chapter 3, Appendix B.

1.3 Exercises

  1. For the BVP

$$\begin{cases} u''(x) = 0, & 0 < x < 1, \\ u(0) = b, & u'(1) = a, \end{cases}$$

    discuss (a) existence, (b) uniqueness, and (c) stability.

2. For the BVP of Exercise 1, discuss (a) construction, (b) regularity, and (c) approximation.

3. For the BVP

$$\begin{cases} a_0(x)u''(x) + a_1(x)u'(x) + a_2(x)u(x) = 0, & 0 < x < 1, \\ u(0) = a, \quad u(1) = b, \end{cases}$$

discuss (a) existence, (b) uniqueness, and (c) construction.

1.4   THE USUAL THREE TYPES OF "BOUNDARY CONDITIONS"

We put "boundary conditions" in quotation marks because initial values as well as boundary values often appear in one of the following three ways, and in combinations thereof. If we consider the Dirichlet problem (1), but interpret it as describing a membrane hanging under a force F (e.g., of gravity), one has the so-called Poisson equation

$$\Delta u = F \text{ in } \Omega$$

with the following types of boundary conditions:

(1) $u = f$ on $\partial\Omega$, Dirichlet boundary condition

(2) $\partial u/\partial n = f$ on $\partial\Omega$, Neumann boundary condition*

(3) $\partial u/\partial n + ku = f$ on $\partial\Omega$, $k > 0$, Robin boundary condition

corresponding to a (1) fixed-boundary membrane, (2) free membrane, and (3) elastically supported membrane. A combination, a so-called "mixed" problem, might be one with the membrane fixed on part of $\partial\Omega$ and free on the other part, for example. We will usually assume that Ω is a nice domain, as in Figure 1.4.

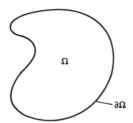

Ω

$\partial\Omega$

FIG. 1.4 A typical nice domain.

---

* $\partial u/\partial n$ denotes the outer normal derivative on $\partial\Omega$, that is, $\partial u/\partial n = \text{grad } u \cdot n$, where n is the unit vector pointing outward from and normal to the boundary $\partial\Omega$. Unless stated to the contrary, both the gradient grad u and the (unique) unit outer normal n are presumed to exist.

All three types of boundary conditions may also occur, for example, in steady-state heat conduction problems, plate and membrane problems, in problems involving electrostatic or gravitational potentials, and in fluid dynamics and elsewhere (see Section 1.7). Note that the vibrating string problem above employs fixed end, that is, Dirichlet (1) boundary conditions and both Dirichlet (1) and Neumann (2) initial data. If one is involved in a problem on an unbounded domain Ω, one usually imposes some kind of decay rate "boundary condition at infinity" to make the problem properly posed. This may be as simple as $u \to 0$ as $x \to \infty$, or more complicated specifications may be required.

Thus, for example, we have the Dirichlet problem,

$$\begin{cases} \Delta u = 0 \text{ in } \Omega, \\ u = f \text{ on } \partial\Omega, \end{cases}$$

the so-called Neumann problem,

$$\begin{cases} \Delta u = 0 \text{ in } \Omega, \\ \dfrac{\partial u}{\partial n} = f \text{ on } \partial\Omega, \end{cases}$$

and the Robin problem,*

$$\begin{cases} \Delta u = 0 \text{ in } \Omega, \\ \dfrac{\partial u}{\partial n} + ku = f \text{ on } \partial\Omega, \qquad k > 0. \end{cases}$$

In each of these problems it was assumed that no forcing data F were present within the domain Ω. When such data F are present, we have for example the Dirichlet–Poisson problem,

$$\begin{cases} \Delta u = F \text{ in } \Omega \\ u = f \text{ on } \partial\Omega \end{cases}$$

and similarly the Neumann–Poisson problem and the Robin–Poisson problem.

As mentioned above, the vibrating string problem given in Section 1.2, namely,

$$\begin{cases} \text{(i)} & u_{tt} - u_{xx} = 0, & 0 < x < \pi, & t > 0, \\ \text{(ii)} & u(0, t) = u(\pi, t) = 0, & t \geqq 0, \\ \text{(iii)} & u(x, 0) = f_1(x), & 0 < x < \pi, & t = 0, \\ \text{(iv)} & u_t(x, 0) = f_2(x), & 0 < x < \pi, & t = 0, \end{cases}$$

possesses (i) zero "Poisson" data on the (x, t) domain for $t > 0$; (ii) Dirichlet boundary conditions at the ends of the string at $x = 0$ and $x = \pi$ for all time t; (iii) a "Dirichlet" initial condition; and (iv) a "Neumann" initial condition. Normally one does not belabor the naming of all these conditions but rather builds up some intuition and experience as to what type of solution behavior one can expect

---

* The coefficient k may be variable or constant, depending on the context.

in terms of the various conditions. A great aid in the latter is a willingness to think physically about the various conditions.

For example, in the vibrating string problem one can imagine indeed a vibrating string. For the linear model to be reasonably correct one should think in terms of small vibrations only. The "Dirichlet" boundary condition should be thought of as a *fixed-end* condition in which the ends of the string are kept fixed and at the same horizontal level. If you imagine yourself as thus holding the string rather tautly, along comes a friend who (iii) lifts the string to a plucked position and then at time $t = 0$ instantaneously releases it into vibration. The degree to which he fails to release it "straight-down," owing to an imbalance between thumb and forefinger, will show up in the initial velocity (iv) that he accidentally has thus imparted to its motion.

A Neumann end condition for the vibrating string problem would be of the form

$$\text{(ii)} \quad u'(0, t) = u'(\pi, t) = 0, \qquad t \geq 0,$$

the prime denoting derivative with respect to x. Here one visualizes a "flapping" string, where the vibrations level out near the ends, so that even though the ends of the string may be going up and down, the string remains horizontal at the very ends. This is what happens at the end of a snapped whip, for example. What happens at the other end?

A Robin end condition for the vibrating string problem would take the form

$$\text{(ii)} \quad \begin{cases} u'(0, t) - k_1 u(0, t) = 0, \\ u'(\pi, t) + k_2 u(\pi, t) = 0, \end{cases}$$

where k_1 and k_2 are positive constants. This may be thought of, for example, as a string constrained by springs at each end, the restoring force being proportional to the amount of distention of the string. Another situation of this type would be that of a vibrating cord held by rings on poles at each end.

In the actual modeling of any of these situations one would need of course to think about the degree of linearity in the actual physical situation, friction terms, and so on. Nevertheless, a physical intuition, even though only approximate, of what the partial differential equation may be describing is a valid asset in its further mathematical consideration.

Problem 1. Solve the equation $u''(x) = 0$, $0 < x < 1$, along with the following boundary conditions: (a) $u(0) = 0$, $u(1) = 1$; (b) $u'(0) = 0$, $u'(1) = 1$; (c) $-u'(0) + u(0) = 1$, $u'(1) + u(1) = 1$. The prime here denotes the ordinary derivative d/dx.

Problem 2. For the equation $u''(x) = 0$, $0 < x < 1$, (a) solve with "mixed" boundary conditions (Neumann at one end, Robin at the other end): $u'(0) = 0$, $u(1) + 2u'(1) = 1$. (b) Given that $u(0) = 0$, what types of the three boundary conditions can you entertain at $x = 1$? (c) Synthesize a general statement from (a) and (b), and prove it.

Problem 3. For the equation $u''(x) = 0$, $0 < x < 1$, and for constant $k > 0$, (a) solve with "boundary condition" $-u'(0) - ku(0) = 0$, $u'(1) - ku(1) = 0$. (b) Solve with boundary condition $-u'(0) - ku(0) = 1$, $u'(1) - ku(1) = 1$. (c) Speculate on some type of general statement for the problem

$$\Delta u = 0 \text{ in } \Omega,$$

$$\frac{\partial u}{\partial n} - ku = f \text{ on } \partial\Omega.$$

This type of boundary condition should be contrasted with the Robin boundary condition and is sometimes called a Steklov boundary condition.

As mentioned in Section 1.3, there is a general theorem guaranteeing the existence of a unique analytic solution in a neighborhood of the initial data, provided that the latter is analytic, the coefficients of the equation are analytic, and that certain other conditions are satisfied. This theorem (Picard–Cauchy–Kowaleski) will be found in Problem 1.9.4. In this way, for any of the "boundary conditions" considered in the present section, for example, Dirichlet, Neumann, Robin, elastic, mixed, Steklov, and so on, the partial differential equations along with those "boundary conditions" regarded as "local initial conditions" can usually be shown, if necessary by adding more data, to possess local solutions near the boundary.

Thus, for a partial differential equation, especially a boundary value problem, the task as concerns the existence question is that of establishing a global, rather than a local, solution. And the best way to do this, when possible, is to actually construct the solution.

1.4 Exercises

1. Identify the following boundary conditions:
   (a) $u(r) = 0$ for $r = (x_2 + y_2)^{1/2} = 1$,
   (b) $\partial u(r)/\partial r = 0$ for $r = (x_2 + y_2)^{1/2} = 1$,
   (c) $(\partial u(r)/\partial r + 3u = 0$ for $r = (x_2 + y_2)^{1/2} = 1$.
2. (a) Show that the exterior Dirichlet problem

$$\begin{cases} \Delta u = 0 & \text{in } r > 1, \\ u = c & \text{on } r = 1, \end{cases}$$

   has two solutions, where $x = (x_1, x_2, x_3)$ and where $r = (x_1^2 + x_2^2 + x_3^2)^{1/2}$.
   (b) What other boundary condition could you add to obtain just one solution?
   (c) Obtain the other solution by another boundary condition at $r = 1$.

3. Without solving, prove or disprove uniqueness of solutions of the following general Dirichlet–Poisson problem

$$\begin{cases} u''(x) = F(x), & 0 < x < 1, \\ u(0) = a, & u(1) = b, \end{cases}$$

by means of (a) geometry, (b) maximum principle, and (c) calculus.

1.5 THE USUAL THREE SOLUTION METHODS (WITH HISTORICAL REMARKS)

Roughly speaking there are three principal analytical methods for solving partial differential equations:

(1) *Separation of Variables* (also called the *Fourier method*, or solution by *eigenfunction expansion*)

(2) *Green's Function* (also called *fundamental singularities*, or solution by *integral equations*)

(3) *Variational Formulation* (also called the *energy method*, or solution by the *calculus of variations*)

We will illustrate these three methods in terms of the examples of problems already given above, and of course return to them in more detail in the later sections.

1.5.1. Separation of Variables Method

To illustrate the separation of variables method* let us consider again the vibrating string problem (Fig. 1.5a) with fixed ends as considered earlier:

$$\begin{cases} \text{(PDE)} & u_{tt} - u_{xx} = 0, & 0 < x < \pi, \quad t > 0, \\ \text{(BV)} & u(0, t) = u(\pi, t) = 0, & t \geq 0, \\ \text{(IV 1)} & u(x, 0) = f(x), & 0 < x < \pi, \\ \text{(IV 2)} & u_t(x, 0) = 0, & 0 < x < \pi. \end{cases}$$

For simplicity we have assumed in the second initial condition (IV 2) that the initial

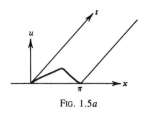

FIG. 1.5a

---

* Historically, the first use of this method is usually credited to Daniel Bernoulli in 1755, and for the vibrating string problem.

t-velocity has been taken to equal zero.* One hopes for a solution $u(x, t) = v(x)w(t)$ with the variables "separated" (it is not) and substitutes into the PDE so that

$$v(x)w''(t) - v''(x)w(t) = 0.$$

One then hopes that the solution u is never zero (it is, at times) and divides by it, arriving at

$$\frac{w''(t)}{w(t)} = \frac{v''(x)}{v(x)} = \text{some constant, say } -\lambda,$$

since a function of t equal to a function of x must be constant. For $u(x, t) = v(x)w(t)$ to satisfy the BV $u(0, t) = u(\pi, t) = 0$, it suffices and is reasonable to require that $v(0) = v(\pi) = 0$, and thus we have arrived at the Rayleigh eigenvalue problem given previously, namely

$$\begin{cases} -v''(x) = \lambda v(x), & 0 < x < \pi, \\ v(0) = v(\pi) = 0. \end{cases}$$

From ODE (see Section 1.3) this problem has an infinite number of solutions $v_n(x) = $ (arbitrary constant) $\cdot \sin nx$ and corresponding eigenvalues $\lambda_n = n^2 = 1, 4, 9, \ldots$, where $n = 1, 2, 3, \ldots$. For $u(x, t) = v(x)w(t)$ to satisfy the IV 2 it is reasonable to require that $w'(0) = 0$ and thus for each now-determined λ_n we have from the above relation the IVP

$$\begin{cases} -w''(t) = \lambda_n w(t), & t > 0, \\ w'(0) = 0. \end{cases}$$

From ODE this problem has solution $w_n(t) = ($ arbitrary constant $) \cdot \cos nt$, and we have thus arrived at an infinite number of solution candidates

$$u_n(x, t) = \text{(arb. const.)} v_n(x)w_n(t) = c_n \sin nx \cos nt.$$

It is easily verified that each $u_n(x, t)$ satisfies the PDE, BV, and IV 2 of the original problems.

It remains then to satisfy the initial condition (IV 1): $u(x, 0) = f(x)$. This clearly will not be satisfied by any one of the $u_n(x, t)$ candidates unless f is itself a multiple of a fundamental mode $\sin nx$.†

---

* Physically, this means that you are able to release the string from its initial position $f(x)$ exactly vertically. Mechanically, it means that you can do so without any imbalance between the thumb and forefinger. But one can treat in the same way the other case $f_1 = 0$, f_2 given, where the string is held straight across, taut, with no initial displacement, but with a push in the t-direction. One then finds the combined solution for both data (IV 1 and IV 2) present, by adding. Mathematically, this is allowed by the linearity of the problem.

† Bernoulli's controversial idea (1755) was that any initial displacement $f(x)$ was itself a linear combination of such sine functions and that thus a number of modes $u_n(x, t)$ will suffice when properly combined for the solution. The calculation of the coefficients was finally resolved by Euler in 1777. It should be mentioned that Lagrange in 1760 independently employed separation of variables, for the vibrating string problem, and that d'Alembert in 1746 had previously solved the vibrating string problem by another method (see Section 2.5).

If we go ahead and try the most general "linear combination" of the $u_n(x, t)$, that is,

$$u(x, t) = \sum_{n=1}^{\infty} c_n u_n(x, t) = \sum_{n=1}^{\infty} c_n \sin nx \cos nt$$

we find ourselves facing three mathematical questions:

1. First, can we apply the PDE across the infinite series, that is, can we differentiate term by term so that (with $\Box = \partial^2/\partial t^2 - \partial^2/\partial x^2$ here)

$$\Box u(x, t) \stackrel{?}{=} \sum_{n=1}^{\infty} c_n \Box u_n(x, t) = 0.$$

If so,[†] then $u(x, t)$ satisfies the PDE, BV, and IV 2 requirements of the problem, and then the second question presents itself.

2. Is

$$u(x, 0) = \sum_{n=1}^{\infty} c_n \sin nx \stackrel{?}{=} f(x);$$

that is, do we obtain with our solution candidate the initial displacement (IV 1)? Restated, is "every" function $f(x)$ representable in a Fourier sine series? If so,[*] we have, formally and employing the Wallis formulas of calculus, that for each $m = 1, 2, 3, \ldots,$

$$\int_0^{\pi} f(x) \sin mx \, dx = \sum_{n=1}^{\infty} c_n \int_0^{\pi} \sin mx \sin nx \, dx = c_m \cdot \frac{\pi}{2}.$$

Thus the (Fourier) coefficients c_m, and thereby the separation of variables solution, are determined.

3. Is $u(x, t)$ the "physical" solution? For this it would suffice to give a uniqueness proof among a class of functions u (in the domain $D(L)$ of the problem) possessing enough continuity and regularity (2') so as to rule out other spurious solutions such as $u(x, t) \equiv 0$ for $t > 0$, $u(x, 0) = f(x)$, analogous to our previous discussion in Section 1.3 of such pathological and uninteresting solutions. There are several ways to argue this, but we defer them to later.

In like manner other PDEs may be solved by separation of variables, as will be seen later. When the geometrical setting is different (e.g., spherical rather than

---

[†] And this should bring to mind that by advanced calculus one knows that what is needed here is enough uniform convergence, as will be made precise later (Sec. 2.2).

[*] That it is the case for any square-integrable f is usually called the Riesz–Fisher[††] Theorem in the Lebesgue integration theory. Although Euler first found the Fourier coefficents, the name Fourier derives from the use of such expansions in the theory of heat conduction problems published by Fourier in the book *Théorie analytique de la chaleur* in 1822.

[††] After F. Riesz and E. Fischer both in the *Compt. Rend. Acad. Sci. Paris* **144** (1907), who showed l_2 equivalent to L^2, and as a corollary, the cited result.

rectangular) one sometimes encounters Fourier expansions most naturally in terms of eigenfunctions other than sines and cosines.‡ If, however, the domain Ω is unbounded (for example), one encounters situations§ where the eigenfunctions correspond to eigenvalues such as the $\lambda_n = 1, 4, 9, \ldots$ above are not complete enough for expansion of arbitrary functions; some aspects of these problems will be discussed later in Section 2.7 and may be resolved in terms of eigenfunctions ranging over a continuous set of λ and solution u given by integrals in place of series.

Problem 1. (a) Solve (formally) by separation of variables the Dirichlet problem (1) shown in Figure 1.5b with data zero on three sides. (b) Argue how to solve that problem with data present on all four sides by breaking the problem into four problems of type (a) by linearity and change of variable.

Problem 2. (a) Convert the Laplacian Δ to polar coordinates (as a review exercise in calculus). (b) Solve formally by separation of variables the Dirichlet problem (1) shown in Figure 1.5c, where Ω is the unit open disc.

Problem 3. (a) Solve explicitly, that is, find the Fourier coefficients, Problems 1 and 2, with $f(x) = x$, $f(\theta) = \theta$, respectively. (b) Do (a) with data e^x, e^{θ} respectively.

Before leaving (for now) the separation of variables method, and as it concerns Problems 1–3 above, let us mention a few conventions found in the literature. We will use c_n to denote as above a general Fourier coefficient; that is, if a function f is expanded in terms of a set of eigenfunctions $\{\varphi_n\}$, we will write

$$f = \sum_{n=1}^{\infty} c_n \varphi_n$$

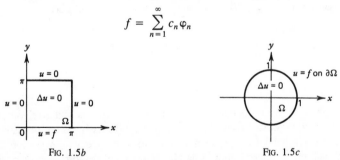

FIG. 1.5b FIG. 1.5c

‡ For example, Legendre in 1784–85, interested in attraction properties of solids of revolution, solved the problem via separation of variables employing an eigenfunction expansion in terms of the polynomials now bearing his name. Motivated to some extent by problems in the theory of sound propagation (with experiments performed by Sturm and Colladon in Lake Geneva in 1826), Sturm and Liouville in 1836 developed a general theory of eigenfunction expansions for second-order ordinary differential equations (see Section 2.4).

§ The λ_n encountered thus far are called the "point-spectrum" of the differential operator L, but in these situations just mentioned one has present also a "continuous-spectrum," as seen in spectroscopy for example. Further amplification of these concepts will be found in Chapter 2.

where $c_n = (f, \varphi_n)$ in the appropriate inner* product for the problem. Above we had for example (but here "normalizing" the φ_n)

$$c_n = \int_0^\pi f(x)\varphi_n(x)\, dx$$

where $\varphi_n(x) = (2/\pi)^{1/2} \sin nx$, the $(2/\pi)^{1/2}$ being a normalizing factor. Here the φ_n are prenormalized so that the "length" $(\varphi_n, \varphi_n) = 1$. For example, one has tacitly prenormalized in this way in the case of Euclidean 3-space, where the $\{\varphi_n\}$ are the base vectors $\mathbf{i}$, $\mathbf{j}$, and $\mathbf{k}$; the need for later normalizations then disappears. On the other hand, the classical treatments often just used the coefficients $c_n = (f, \varphi_n)/(\varphi_n, \varphi_n)$ for nonnormalized basis vectors such as the $\varphi_n(x) = \sin nx$, which is equivalent to the above procedure of normalizing first.

With reference to Problem 2 above and in general to the question of expansion of a function $f(\theta)$ in terms of sines and cosines on the interval $-\pi < \theta < \pi$, one needs both the sines and cosines for general functions because the former are all odd functions and the latter are all even functions.† If the data f is even about zero the sine terms will vanish, whereas if f is odd about zero, the cosine terms will vanish, but in general all must be present. Classical notation is to write $f = \Sigma c_n\varphi_n$ on $-\pi < \theta < \pi$ in the form

$$f = \tfrac{1}{2}a_0 + \sum_{n=1}^{\infty} (a_n \cos n\theta + b_n \sin n\theta)$$

where $a_n = \pi^{-1} \int_{-\pi}^{\pi} f(\varphi) \cos n\varphi\, d\varphi$, $n = 0, 1, 2, \ldots$, and where $b_n = \pi^{-1} \int_{-\pi}^{\pi} f(\varphi) \sin n\varphi\, d\varphi$, $n = 1, 2, 3, \ldots$. Thus the Fourier coefficients have been split into the cosine coefficients $\{a_n\}$ and the sine coefficients $\{b_n\}$ as have been the eigenfunctions $\{\varphi_n\}$. The normalizing factor here would be

$$(\varphi_n, \varphi_n)^{-1/2} = \left(\int_{-\pi}^{\pi} \sin^2 n\varphi\, d\varphi\right)^{-1/2}$$

$$= \left(\int_{-\pi}^{\pi} \cos^2 n\varphi\, d\varphi\right)^{-1/2} = \pi^{-1/2}.$$

The expression just written above is commonly referred to as the Trigonometric or Trigonometric Fourier Series of f to distinguish it from expansions in terms of other types of eigenfunctions $\{\varphi_n\}$, such as those to be enumerated in Section 2.1.

In solving Problem 2 above one tries solutions of the form $u(r, \theta) = R(r)\Theta(\theta)$ and arrives at the separated ODEs

$$-\Theta''(\theta) = \lambda\,\Theta(\theta), \qquad -\pi < \theta < \pi,\ddagger$$

---

* Other words used are "scalar," "dot," and so forth. See Section 2.3 for a more precise statement.
† A function f is odd about the point zero if $f(-\theta) = -f(\theta)$ and even about the point zero if $f(-\theta) = f(\theta)$; that is, "origin" and "y-axis" symmetry, respectively.
‡ We could have used $0 < \theta < 2\pi$ or any other interval of length 2π to parameterize the unit circle. The above is most common.

and

$$r^2 R''(r) + r R'(r) - \lambda R(r) = 0, \qquad 0 \leqq r < 1.$$

What are the boundary conditions to determine the eigenvalues λ_n and eigenfunctions Θ_n? Here one remembers the prescription of $u \in C^2(\Omega)$ for the Dirichlet problem, and in order to guarantee at least C^1 interior continuity* one specifies that

$$\Theta(-\pi) = \Theta(\pi) \quad \text{and} \quad \Theta'(-\pi) = \Theta'(\pi).$$

It then follows that $\Theta''(-\pi) = \Theta''(\pi)$ due to the continuity at the end points of the resulting solutions of this ODE eigenvalue problem.

From the above eigenvalue problem for $\Theta(\theta)$ one thus arrives at $\lambda_n = 0, 1, 4,$ $\ldots, n^2, \ldots$ and the corresponding trigonometric eigenfunctions Θ_n. The resulting equation for $R(r)$ with λ_n in it has general solutions $R_n(r) = c_1 r^n + c_2 r^{-n}$, $n \neq 0$, $R_0(r) = c_1 + c_2 \ln r$.† Here one then imposes a boundary condition that $R(r)$ be bounded near $r = 0$ so that the c_2 terms drop out, so that again one has taken note of the desired $C^2(\Omega)$ interior regularity requirement of the original PDE. The rest of the details in Problem 2 are left to the student.

The boundary data in Problem 3 above suffer discontinuities but, nonetheless, as will be seen later, the separation of variables method still provides solutions u to the PDEs in the interiors of the domains. In general the Fourier series can be expected to converge to the average value of data which have jumps.§

A related question is solution regularity at "corners" of a domain such as in Problem 1, whether or not the data f are continuous. This is an interesting and more complicated question that also arises, for example, in fourth-order (e.g., with operator $L = \Delta^2$) elliptic problems describing plate stress and deformation, and we do not go into these matters here.

To see the effect of discontinuities in the boundary $\partial\Omega$ itself, one needs only to take a soapfilm with a loop of square configuration having very slight raggedness at the corner. You will find yourself making a smoother solder joint if the original square loop was not smooth enough. Most treatments of boundary value problems more or less assume a nice smooth boundary even when it is not so. Proper treatment of corners needs more care.

With these remarks we leave the separation of variables method in its formal status. The three mathematical questions—of existence, uniqueness, and stability— will be treated further in terms of the separation of variables solution in the next chapter by some considerations from advanced calculus, Fourier series, and the theory of partial differential equations. It should be emphasized once more perhaps

---

  * Many textbooks are surprisingly fuzzy on this point, stating a desire for periodic solutions, and so forth, which although somewhat equivalent, nonetheless obscures the fact that regularity is the real concept entering here.

  † As may be seen by recognizing the equation as an Euler equation from ODE, for example, or alternately by using the power series method for ODEs with a regular singular point. See any elementary ODE book, or for example the references given in Problem 2.9.4.

  § See the Dini Tests of Section 1.9.6(2).

that upon encountering a partial differential equation one is always encouraged to try the separation of variables method inasmuch as it provides the most useful concrete way of solving PDEs and thereby answering explicitly the questions — construction, regularity, approximation—of the solution.

1.5(1) Exercises

1. Calculate the Fourier sine coefficients

$$c_n = \frac{2}{\pi} \int_0^\pi f(x) \sin nx \, dx, \qquad n = 1, 2, \ldots$$

for the following functions
(a) $f(x) = \sin x$,
(b) $f(x) = \sin^3 x$,
(c) $f(x) = x(\pi - x)$.

2. Using the general solution formula

$$u(x, t) = \sum_{n=1}^\infty c_n \sin nx \cos nt$$

derived in the text, find the separation of variables solutions of the vibrating string problem with initial position
(a) $f(x) = \sin x$, $\quad 0 < x < \pi$,
(b) $f(x) = \sin^3 x$, $\quad 0 < x < \pi$,
(c) $f(x) = x(\pi - x)$, $\quad 0 < x < \pi$.

3. Solve the heat equation problem

(PDE) $\quad u_t - u_{xx} = 0$, $\quad 0 < x < \pi$, $\quad t > 0$,
(BVP) $\quad u(0, t) = u(\pi, t) = 0$, $\quad t \geq 0$,
(IVP) $\quad u(x, 0) = f(x)$, $\quad 0 < x < \pi$,
with (a) $f(x) = \sin x$, (b) $f(x) = \sin^3 x$, and (c) $f(x) = x(\pi - x)$.

1.5.2. Green's Function Method

The most general* theoretical way of "inverting" a PDE $Lu = f$ is by use of Green's† functions. And "inverting" is the correct word here since one obtains an "inverse" operator L^{-1}, and thus the solution $u = L^{-1}f$, in much the same spirit as when solving a matrix equation $Ax = y$ by calculating the inverse matrix A^{-1}.

---

* For example, the separation of variables method breaks down if the domain Ω is not of some special geometry such as rectangular, spherical, cylindrical, and so forth; one need only try it on a general two-dimensional domain to see this, the λ_n no longer being independent of one coordinate's position. Generally speaking, the Green's function method works for arbitrary domains but with the corresponding disadvantage that it yields the solution in a "pointwise" rather than "global" form. Also there may be difficulty in explicitly obtaining the Green's function for the problem.

† As also the case with the Green's identities to be discussed later, Green's functions are named after George Green, see his "Essay on the application of mathematical analysis to the theory of electricity and magnetism," 1828.

As it usually turns out in this method one obtains a representation of L^{-1} by means of an integral operator‡ whose kernel $G(P, Q)$ is the Green's function for the problem in question:

$$u(P) = \int_{\Omega} G(P, Q)f(Q)\, dV_Q, \qquad P \in \Omega.$$

One may interpret this solution representation as an "adding up and simultaneous weighting" of the data f, much as A^{-1} can be regarded as a "weighting" of the data $\mathbf{y}$ in the matrix case.

The ramifications of the theory of Green's functions are many and sometimes complicated. We restrict ourselves here to just three interpretations, both mathematical and physical. For partial differential equations most Green's functions are singular as the point Q approaches P. Our first illustration of a Green's function is not singular, however, and is in fact no more than the variation of constants formula from ODE.

Let L denote the one-dimensional Dirichlet–Poisson problem

$$\begin{cases} -v''(x) = f(x), & 0 < x < \pi, \\ v(0) = v(\pi) = 0, \end{cases}$$

which we denote loosely by $Lv = f$ (understanding as usual that L also connotes, in addition to the operator $-\Delta$ on the domain, that the identity operator I maps v onto zero data on the boundary). From ODE we recall that the homogeneous equation $-v'' = 0$ has two fundamental solutions $\varphi_1 = 1$ and $\varphi_2 = x$, and, by the variation of constants formula (see Problem 1), the inhomogeneous equation $-v''(x) = f(x)$ has the general solution

$$v(x) = c_1 + c_2 x + \int_0^x (s - x)f(s)\, ds.$$

Substitution of the boundary conditions yields

$$c_1 = 0 \quad \text{and} \quad c_2 = \pi^{-1} \int_0^\pi (\pi - s)f(s)\, ds,$$

and thus

$$\begin{aligned} v(x) &= \int_0^\pi \frac{x(\pi - s)}{\pi} f(s)\, ds + \int_0^x (s - x)f(s)\, ds \\ &= \int_0^x \left[\frac{x(\pi - s)}{\pi} + (s - x) \right] f(s)\, ds + \int_x^\pi \frac{x(\pi - s)}{\pi} f(s)\, ds \\ &= \int_0^\pi G(x, s)f(s)\, ds, \end{aligned}$$

‡ The use of such integral operators was pioneered by Liouville in 1832. His method was used by Neumann in 1877 and improved by Fredholm and Hilbert in 1899 in finally resolving the Dirichlet problem for general domains.

where the Green's function $G(x, s)$ is given by

$$G(x, s) = \begin{cases} s(\pi - x)/\pi, & s \leqq x, \\ x(\pi - s)/\pi, & s \geqq x. \end{cases}$$

This example illustrates a number of properties most Green's functions enjoy, including the symmetry $G(x, s) = G(s, x)$ and the dependence (π here) on the particular domain and on the particular boundary conditions.*

A second interpretation of the Green's function is as a fundamental singularity yielding a point source in a physical problem such as heat conduction. As will be seen in Section 1.7, under linearizing assumptions the flow of heat in a cooling body is described by the heat equation, with $u(x, y, z, t)$ denoting the temperature,

$$u_t - \Delta u = 0$$

and in the case of steady, that is, time-independent, flow, wherein $u = u(x, y, z)$, one has the equation

$$\Delta u = 0$$

for the temperature distribution. In imagining such a steady outward heat flow (Fig. 1.5d) one needs a source at P and this is provided by the singular function

$$S(P, Q) = \frac{1}{4\pi r_{PQ}}$$

where P (which denotes (x, y, z)) is regarded as fixed within Ω and where Q (which denotes another point in the body or on its boundary) varies over $\overline{\Omega}$. Here r_{PQ} denotes the Euclidean distance between P and Q. The interpretation of the singular function $S(P, Q)$ as a source function for the heat flow then proceeds as follows, first noting that $S(P, Q)$ is harmonic in $\Omega - \{P\}$:

$$\Delta_Q S(P, Q) = 0, \qquad P \neq Q.$$

Putting an ε-sphere about P (see Fig. 1.5e), one has the temperature rate of change on $|P - Q| = \varepsilon$ given by the heat flow

$$-k \, \text{grad}_Q S(P, Q) = \frac{-k\mathbf{n}}{4\pi\varepsilon^2}$$

where $\mathbf{n}$ is the unit outer normal on $|P - Q| = \varepsilon$ and where k† denotes the thermal conductivity of the material there. Normalizing the coefficient k to 1 and integrating over the sphere $|P - Q| = \varepsilon$, one has 1 calorie of heat entering the rest of Ω from

---

* In general, the Green's function for a PDE should be regarded as a nontrivial entity, perhaps not so easy to find, inasmuch as it contains all information about the problem, such as the operators, boundary conditions, and domain geometry.

† Units of k are calories per second here. See Section 1.7 for a further discussion of heat conduction.

FIG. 1.5d

FIG. 1.5e

the sphere $|P - Q| \leq \varepsilon$ through the sphere $|P - Q| = \varepsilon$ in 1 second, as seen by calculating the heat flux through the sphere $|P - Q| = \varepsilon$:

$$\int_{|P-Q|=\varepsilon} -k \, \text{grad}_Q S(P, Q) \cdot \mathbf{n} \, ds = 1.$$

Since this argument does not depend on ε, ε may be taken arbitrarily small and $S(P, Q)$ may be interpreted as the temperature distribution in Ω due to a unit heat source located at P.

This brings us to a third interpretation of Green's functions in terms of what we shall loosely call the (Dirac) delta function property. More precisely the arguments depend on the calculation of improper integrals, the details of which we will postpone for the time being (see, however, Problems 2 and 3). In this third interpretation, we may write (remembering from the second interpretation that $S(P, Q)$ is harmonic for $P \neq Q$) the "fundamental solution" or "delta function" relation

$$-\Delta_Q S(P, Q) = \delta(P, Q),$$

where the delta "function" $\delta(P, Q)$ is defined to be the quantity* given by the properties that $\delta(P, Q) = 0$ when $P \neq Q$ and

$$\int_\Omega w(Q)\delta(P, Q) \, dV_Q = w(P)$$

for arbitrary smooth functions w on Ω. Let us use the latter "point evaluator" property to find the solution $u(P)$ of the Poisson-Dirichlet problem

$$\begin{cases} -\Delta u = F \text{ in } \Omega, \\ u = 0 \text{ on } \partial\Omega. \end{cases}$$

---

* It exists, but as a "measure" rather than a function. We do not need any measure-theoretic aspects of the delta function here.

We assume that the solution u exists and is sufficiently regular so that we may multiply by an arbitrary function v also vanishing on $\partial\Omega$ and integrate as follows:

$$\int_\Omega vF = -\int_\Omega v\,\Delta u = -\int_\Omega u\,\Delta v.*$$

Now supposing that there exists a function $v = v(P, Q) \equiv G(P, Q) \equiv S(P, Q) + g(P, Q)$, which is the fundamental singularity $S(P, Q)$ given above plus an adjustment term $g(P, Q)$ satisfying $\Delta_Q g(P, Q) = 0$ in Ω and "adjusting" $S(P, Q)$ so that the sum $G(P, Q) = 0$ for Q on $\partial\Omega$, on substitution into the last equation and using the delta function property $-\Delta_Q G(P, Q) = \delta(P, Q)$, one has

$$u(P) = \int_\Omega G(P, Q)F(Q)dV_Q$$

and hence the solution to the Poisson–Dirichlet problem.

Problem 1. (Review of ODE). (a) Derive the variation of constants formula used in this section for the equation $-v'' = f$. (b) Find the Green's function for the following mixed problem:

$$\begin{cases} -v''(x) = f(x), & 0 < x < \pi, \\ v(0) = 0, \\ v'(\pi) = 0. \end{cases}$$

Problem 2. (a) Show that the Dirichlet–Poisson Green's function $G(x, s)$ given above by variation of constants does in fact have the fundamental singularity and "delta function" properties in the sense that it satisfies:

$$\begin{cases} \text{D.E.;} & -\Delta_s G(x, s) = 0,\ s \neq x, \\ \delta \text{ property;} & \dfrac{dG(x, s)}{ds} = H(x, s), \text{ where } H(x, s)^\dagger \text{ is the step function shown in Figure 1.5}f,\ s \neq x, \\ \text{B.C.;} & G(x, 0) = G(x, \pi) = 0. \end{cases}$$

(b) Graph carefully the $G(x, s)$. (c) Repeat (a) and (b) for (b) of Problem 1.

Problem 3. Show that the Poisson kernel $(4\pi t)^{-1/2} e^{[-(x-y)^2/4t]}$ is the Green's function for the heat propagation problem (2) by noting (a) for fixed y, it

---

* The second equality is Green's second identity (see Section 1.6.1), which may be justified directly by an integration by parts using the divergence theorem. The student may wish, at this point, to verify it in the case of one dimension, by the usual integration by parts formula from calculus.

Often in the following pages for brevity we will omit the expressions dV, ds, dx, and so forth, when the meaning of the integral is clear. Also in the above a notation such as Δ_Q or dV_Q means that Q is the variable, P being regarded as arbitrary but fixed.

† $H(x, s)$ for "Heaviside function," as they are commonly called.

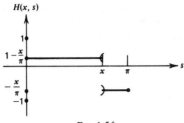

FIG. 1.5f

satisfies the heat equation $u_{xx} - u_t = 0$ in $-\infty < x < \infty$, $t > 0$, and (b) it has a "delta function" property for reproducing the initial values at $t = 0$.

1.5(2)   Exercises

1. For the ordinary differential equation

$$ay'' + by' + cy = 0$$

   with constant coefficients, review the three solution cases (a) roots real and unequal, (b) roots real and equal, and (c) roots complex conjugate.

2. The Green's function for ordinary differential equations results from applying boundary conditions to the general solution

$$y(x) = y_H(x) + y_P(x)$$

   of the equation

$$ay'' + by' + cy = f(x).$$

   Here $y_H(x)$ denotes the solution $c_1 y_1(x) + c_2 y_2(x)$ of the homogeneous equation of Exercise 1, and $y_P(x)$ denotes any particular solution of the nonhomogeneous equation.

   (a) Review the three cases for $y_1(x)$ and $y_2(x)$.
   (b) Recall the Variation of Parameters formula for $y_P(x)$.
   (c) Why does the Green's function turn out to involve integrals of products of the form $y_1(x)y_2(s)$?

3. The "delta function" has behind it now a large theory of generalized functions, measures, distributions, and indeed its full calculus warrants such structures. However, much of the confusion it can cause can be avoided by starting with the notion of "delta functional." The latter is defined simply as the transformation δ_P which maps a function $u(Q)$ to its value at a designated point $Q = P$.

   (a) The term "functional" is usually reserved in mathematics for mappings from functions to scalars, e.g., real or complex numbers. Is that the case with the delta functional δ_P?

(b) "Functionals" are also tacitly assumed unless stated to the contrary to be linear. Is that the case for δ_P?

(c) What then is the meaning of the Green's function solution

$$u(x) = \int_0^\pi G(x, s) f(s)\, ds$$

for the example worked out in the text, in terms of the "delta function" and the "delta functional"?

1.5.3. Variational Methods†

A discussion of variational methods leads to a number of questions in the general calculus of variations; in later sections we will consider some further aspects of these questions as they pertain to the study of partial differential equations. In these introductory remarks we first restrict attention to the famous Dirichlet principle and its historical significance in partial differential equations and to mathematics generally.‡

Let us consider the Dirichlet problem (specifying explicitly the regularity conditions this time):

$$\begin{cases} \Delta u = 0 \text{ in } \Omega, & u \in C^0(\overline{\Omega}) \cap C^2(\Omega), \\ u = f \text{ on } \partial\Omega, & f \in C^0(\partial\Omega). \end{cases}$$

The Dirichlet principle (stated here without any regularity assumptions on the functions or domains involved) asserts that the solution u is realized by obtaining from all functions v that are equal to f on $\partial\Omega$ the one that minimizes the energy integral

$$D(v) \equiv \int_\Omega |\text{grad } v|^2.$$

† Later we will also discuss certain *numerical methods* from this point of view (see Section 2.6).

‡ Poisson solved the Dirichlet problem for the sphere in 1820. In 1839–40 Gauss attempted a solution for arbitrary domains and concluded that some restriction might be necessary on the domain Ω. Green in his treatise of 1828 and again in 1833 asserted the existence of solutions to the Dirichlet problem according to the physical evidence, but with reservations as to the existence of a Green's function for arbitrary domains; Green's proof depended on the unproven Dirichlet principle. Thompson (Lord Kelvin) in 1847, Riemann in 1851, and Dirichlet in the early 1850s all assumed the validity of the Dirichlet principle in their work in potential theory and complex variable analytic function theory. A major crisis arose in 1870 when Weierstrass successfully placed the Dirichlet principle in question by giving examples of minimization problems in the calculus of variations wherein there are no continuous functions solving the problem, and asserted correctly that although the energy integral $D(v)$ has among the $C^0(\overline{\Omega}) \cap C^1(\Omega)$ trial functions a uniform lower bound, "whether there is a function u in the class of continuously differentiable functions that *furnishes* the lower bound was not established." The Dirichlet principle fell into disrepute for 30 years, and during the period C. Neumann (1870), H. Schwarz (1870), Poincaré (1887), and Picard (1888), among others, all obtained existence proofs for the Dirichlet problem by other methods. However, in 1899 Hilbert rescued the Dirichlet principle by showing the existence of a minimizing sequence with $C^1(\Omega) \cap C^0(\overline{\Omega})$ limit u that moreover satisfied the boundary value condition *and* the differential equation for a wide class of domains Ω. This then resolved the Dirichlet problem for all such domains Ω.

Let us illustrate what is involved here by considering the easiest Dirichlet problem, namely,

$$\begin{cases} u''(x) = 0, & 0 < x < 1, \\ u(0) = 0, & u(1) = 1, \end{cases}$$

which has solution $u(x) = x$. If we consider *a priori* as a possible choice of minimizing functions $u_n(x) = x^n$, $n = 1, 2, 3, \ldots$, (see Fig. 1.5g), we find that $D(u_n) = 1, \frac{4}{3}, \frac{9}{5}, \ldots, n^2/(2n - 1)$. Thus the minimizing trial function from among this class of functions is the harmonic one, as asserted by the Dirichlet principle.

More generally, if we consider any function v that is continuously differentiable on the interval $0 \leqq x \leqq 1$, we see by integration by parts[+] that for any $u \in C^2[0, 1]$

$$D(v) = \int_0^1 (v')^2 \, dx$$

$$= \int_0^1 (u')^2 \, dx + \int_0^1 ((v - u)')^2 \, dx + 2 \int_0^1 u'(v - u)' \, dx$$

$$= D(u) + D(v - u) + 2\left((v - u)u'\big|_0^1 - \int_0^1 (v - u)u'' \, dx \right).$$

The last term vanishes if v and u have the same given boundary values and if u is harmonic, so that by the positivity of the (energy) integral $D(v - u)$ one has

$$D(u) \leqq D(v).$$

The difficulties in the Dirichlet principle are in the other direction: in arguing the existence of a function u that actually attains the minimum energy value and in then demonstrating that u is in fact harmonic throughout the domain Ω and sufficiently regular on its closure $\bar{\Omega}$ so as to solve the differential equation completely.

Variational methods are perhaps even more important in eigenvalue problems, especially since there are very important eigenvalue problems that have not been solved* and for which one must approximate, by variational considerations, the true eigenvalues λ_n by upper and lower bounds for them. To illustrate the variational

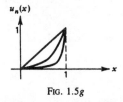

$u_n(x)$

FIG. 1.5g

[+]The key identity is the bilinearity $D(v - u) = D(v) + D(u) - 2D(v, u)$, a general property of quadratic forms and inner products, see Section 2.3. This holds for general domain Ω and will be used again in Section 2.6. The bilinearity plus $D(v, u) = D(u)$ when $v = u$ on $\partial\Omega$ provides on alternate proof of the Dirichlet variational inequality above.

* For example, in quantum mechanics, as will be discussed later. See Section 1.7.

characterization of eigenvalues let us consider the previous example (Rayleigh problem):

$$\begin{cases} -v'' = \lambda v, & 0 < x < \pi, \\ v(0) = v(\pi) = 0, \end{cases}$$

for which the eigenvalues were $\lambda_n = n^2$ and the corresponding eigenfunctions were $v_n(x) = \sin nx$, $n = 1, 2, 3, \ldots$. For an arbitrary real function $v \in C^2(0, \pi) \cap C^1[0, \pi]$ vanishing on the boundary, that is, $v(0) = v(\pi) = 0$, one has by integration by parts that

$$\int_0^\pi |v'(x)|^2 \, dx = -\int_0^\pi v(x)v''(x) \, dx$$

and in particular for an eigenfunction v_n, that

$$\lambda_n = \int_0^\pi |v_n'|^2 \, dx \Big/ \int_0^\pi v_n^2 \, dx > 0.$$

Of course from ODE we previously deduced the λ_n exactly and therefore their positiveness, but the above argument shows the latter fact more quickly and, as we shall see later, more generally and for PDE problems as well.

The quotient on the right is called Rayleigh's* quotient and, as in the case of the variational characterization of the solution of the Dirichlet boundary value problem just discussed, the energy integral $\int |v'|^2$ plays a vital role. In this EVP the variational characterization of the lowest eigenvalue† λ_1 is given by

$$\lambda_1 = \inf_v \int_0^\pi |v'|^2 \, dx \Big/ \int_0^\pi v^2 \, dx, \qquad v(0) = v(\pi) = 0,$$

and, as in the Dirichlet principle, the questions are: over which class of v is the infimum actually attained by some v_1, and is such v_1 sufficiently regular so as to satisfy the original PDE eigenvalue problem?

To illustrate the variational characterization in this problem, note that for $v_n = \sin nx$ one has

$$\frac{\int_0^\pi |v_n'|^2 \, dx}{\int_0^\pi (v_n)^2 \, dx} = \frac{n^2 \int_0^\pi \cos^2 nx \, dx}{\int_0^\pi \sin^2 nx \, dx} = n^2$$

and that (by Wallis' formula again) from calculus

$$\int_0^\pi v_n(x)v_m(x) \, dx = \begin{cases} \pi/2, & m = n \\ 0, & m \neq n \end{cases}$$

---

* After Lord Rayleigh, who did much work in diverse areas of classical spectral analysis of light, sound, color, electromagnetism, resulting in his book *Theory of Sound* in 1877 (see also the discussion in Section 2.8).

† The higher λ_n are similarly characterized but with the infimum taken only over v which are "orthogonal" to the lower eigenfunctions $v_1, \ldots, v_{n-1}$.

which, as will be made more precise later, means that the v_n form an "orthogonal" set of functions.

Problem 1. (a) Prove that $\int_0^\pi \sin nx \sin mx \, dx = \pi/2$ or 0 depending on whether $n = m$ or not; n, $m = 1, 2, 3, \ldots$. (b) Prove that $\int_{-\pi}^\pi \varphi_n(x)\varphi_m(x) \, dx$ has the same property (but different constant) when $\varphi_n(x)$ and $\varphi_m(x)$ range over the functions $\sin nx$, $n = 1, 2, 3, \ldots$, and $\cos nx$, $n = 0, 1, 2, 3, \ldots$. Such a set of functions $\{\varphi_n(x)\}$ is then said to be a mutually orthogonal set with respect to the inner product $(\varphi_n, \varphi_m) = \int \varphi_n\varphi_m \, dx$. The Fourier theorem (as will be seen later) states that any square integrable function f‡ can be written in terms of mutually orthogonal functions such as these, as we have assumed in the separation of variables method, applied formally so far.

Problem 2. Let $R(v) = \int_0^\pi (v')^2 \, dx / \int_0^\pi v^2 \, dx$ be the Rayleigh quotient discussed in the eigenvalue problem above, for functions $v \in C^1[0, \pi]$, $v(0) = v(\pi) = 0$. Consider as a "trial function" for estimating the true first eigenfunction* $v_1(x) = 2^{1/2}\pi^{-1/2} \sin x$ the function $v(x) = cx(\pi - x)$, where c is an arbitrary constant. (a) Calculate $R(v)$ and observe how close it turns out to be to the eigenvalue $\lambda_1 = 1$. (b) Calculate the root-mean-square error† $(\int_0^\pi |v - v_1|^2 \, dx)^{1/2}$ for the best value of c. This is typical of variational trial functions: they yield better eigenvalue estimates than one would expect.

Problem 3. Consider the minimization problem

$$\lambda = \text{minimum} \int_0^1 v^2 \, dx,$$

$$v \in C^0[0, 1],$$

$$v(0) = 0, \qquad v(1) = 1,$$

and show, using the $u_n(x) = x^n$ used previously above, that there is no solution.

In Problem 1.9.5(1) some further historical considerations are mentioned; there are also certain historical exercises for the student. For example, the interesting question "How did the Robin boundary condition come to be so named?" may eventually alert the student to some of the more human aspects of the subject.

In Problem 1.9.5(2) we anticipate some technical details such as: Which domains Ω are "sufficiently nice" so that one may work on them without undue worry? In practice most domains Ω cause no problems as concerns the existence of a continuous solution to the Dirichlet problem, the yielding of a Green's function as discussed in this section above, the allowing of integration by parts and use of

---

‡ That is, one of finite "length" $(f, f)^{1/2} = \left[\int\!\int f^2 \, dx\right]^{1/2} < \infty$
* Here normalized to have "length" equal to one as mentioned.
† That is, the "distance" between v and v_1 in the sense just mentioned. Other names are "metric" or "norm," names that we will employ from time to time in the following.

the divergence theorem as discussed in Section 1.6, and so on. We do not wish to become too involved with "nonnice domains" Ω, which turn out to be somewhat pathological even though mathematically interesting. On the other hand, we do not wish to discourage a curious student from reading further about them.

In Problem 1.9.5(3) we relate the variational formulation of a problem to the corresponding (Euler) partial differential equation for the problem. To do so we use the divergence theorem and Green's identities for integration by parts for partial differential equations as put forth in the following section.

1.5(3) Exercise

1. (a) Write the one-dimensional Dirichlet Principle for an interval (a, b).
   (b) Prove that the functional $\int_a^b |v'(x)|^2 \, dx$ has a finite infimum over all v being considered.
   (c) What is the difference between infimum and minimum, in this context?
2. (a) Observe a good hint for starting the path from a differential equation formulation to a variational formulation.
   (b) Do (a) for the one-dimensional Dirichlet problem.
   (c) Where is the "variation" of v in the variational formulation?
3. Give three reasons (a), (b), (c) why the regularity question becomes paramount in understanding a variational formulation of a differential equation.

FIRST PAUSE: EXAMPLES, EXPLANATIONS, EXERCISES*

Most of the fundamental concepts have now been introduced. This is a good place to pause to reinforce some of the main points by means of additional examples and exercises.

Normally, we write down the equations with convenient coefficients (e.g., equal to one) and on convenient intervals (e.g., on $(0, 1)$ or $(0, \pi)$). There is no difficulty in then treating more general situations.

Example

Solve, where k and l denote fixed physical constants,

$$\begin{cases} u_t - k u_{xx} = 0, & 0 < x < l, \quad t > 0, \\ u(0, t) = u(l, t) = 0, & t > 0, \\ u(x, 0) = x. \end{cases}$$

---

* The four Pauses in the book are intended as helpful and hopefully interesting auxiliary material to aid in practice and understanding. They should not, however, be an excuse to get bogged down, so it is recommended that one simultaneously push on.

Solution. To separate variables we assume $u(x, t) = X(x)T(t)$, which upon substitution into the equation requires

$$\frac{T'(t)}{T(t)} = k\frac{X''(x)}{X(x)} = -\lambda$$

where $-\lambda$ denotes some as yet arbitrary constant. The minus sign is there just for convenience, and the two functions in the equation must be constant due to the independence of the variables x and t. Normally, one chooses the spatial ODE first because its eigenfunctions will be more fundamental in the ensuing separation of variables solution. It is also natural to incorporate the boundary conditions into this contribution to the solution. The resulting ODE boundary-eigenvalue problem

$$\begin{cases} -kX''(x) = \lambda X(x), & 0 < x < l, \\ X(0) = X(l) = 0 \end{cases}$$

has an infinite number of solutions; namely, the eigenfunctions $X_n(x) = c_n \sin(n\pi x/l)$ corresponding to the eigenvalues $\lambda_n = kn^2\pi^2/l^2$, $n = 1, 2, 3, \ldots$. These acquired values λ_n determine the time factors $T_n(t) = d_n e^{-\lambda_n t}$ which we immediately combine with the corresponding $X_n(x)$ factors to obtain the solutions

$$u_n(x, t) = c_n X_n(x)T_n(t) = c_n \sin(n\pi x/l)e^{-\frac{kn^2\pi^2}{l^2}t}.$$

Note that we have just absorbed the arbitrary constant d_n into the arbitrary constant c_n. Forming the linear superposition and going to the limit $u(x, t) = \sum_{n=1}^{\infty} c_n u_n(x, t)$ provides our formal separation of variables solution. It vanishes on the boundary, it (formally) satisfies the partial differential equation, and will (formally) equal the initial data x if we take c_n to be the Fourier sine coefficients of x, namely

$$c_n = \frac{2}{l}\int_0^l x \sin\frac{n\pi}{l}x\,dx$$

$$= \frac{2}{l}\left[\frac{l^2}{n^2\pi^2}\sin\frac{n\pi x}{l} - \frac{xl}{n\pi}\cos\frac{n\pi x}{l}\right]\Bigg|_0^l$$

$$= \frac{2}{l}\left[-\frac{l^2}{n\pi}\cos n\pi\right] = (-1)^{n+1}\frac{2l}{n\pi}.$$

Thus the solution is

$$u(x, t) = \frac{2l}{\pi}\sum_{n=1}^{\infty}\frac{(-1)^{n+1}}{n}\sin\frac{n\pi x}{l}e^{-\frac{kn^2\pi^2}{l^2}t}.$$

Note that the physical scales are reflected in the eigenfunctions, decay rates, and Fourier coefficients of the solution.

Exercises

1. In the Example let $u(x, 0) = A$, a fixed constant, and solve.
2. Modify Example (1) to

$$\begin{cases} u_t - 2u_{xx} = u, & 0 < x < 10, \quad t > 0, \\ u(0, t) = u(10, t) = 0, & t > 0, \\ u(x, 0) = a & \text{for } 0 < x < 5, \\ u(x, 0) = -a & \text{for } 5 < x < 10. \end{cases}$$

Hint: This equation has a zeroth-order term. Try to eliminate it by the substitution $u(x, t) = e^t v(x, t)$

3. Convert the Laplacian to spherical coordinates in 3 dimensions and perform the separation of variables arguments (without solving the resulting ODEs).
4. Solve by separation of variables the plucked string problem

$$\begin{cases} u_{tt} - k^2 u_{xx} = 0, & 0 < x < l, \quad t > 0, \\ u(0, t) = u(l, t) = 0, & t > 0, \\ u(x, 0) = 2Ax/l & \text{for } 0 \leq x \leq l/2, \\ u(x, 0) = 2A(1 - x/l) & \text{for } l/2 \leq x \leq l, \\ u_t(x, 0) = 0, & 0 < x < l. \end{cases}$$

5. Solve the inhomogeneous problem

$$\begin{cases} u_t - u_{xx} = \psi(x), & 0 < x < 1, \quad t > 0, \\ u(0, t) = u(1, t) = 0, & t \geq 0, \\ u(x, 0) = f(x) & 0 < x < 1 \end{cases}$$

by separation of variables. Consider $\psi(x) = x$, $-e^{-x}$, and other easily (Fourier) integrated forcing functions, and various similarly chosen $f(x)$.
Hint: Let $u(x, t) = v(x, t) + g(x)$.

Green's functions for partial differential equations can be hard to come by, but for ordinary differential equations they come out of the variation of parameters (i.e., constants) procedure.

Example

Find the Green's function for

$$\begin{cases} u''(x) + \lambda u(x) = f(x), & 0 < x < l, \\ u(0) = u(l) = 0 \end{cases}$$

where the parameter λ is any real number.

Solution. First solving the homogeneous equation we consider three cases. For $\lambda > 0$

$$u_H(x) = c_1 \cos \lambda^{1/2} x + c_2 \sin \lambda^{1/2} x$$

and we try a particular solution

$$u_p(x) = c_1(x) \cos \lambda^{1/2}x + c_2(x)\sin \lambda^{1/2}x$$

from which the variation of constants equations

$$c_1' \cos \lambda^{1/2}x + c_2' \sin \lambda^{1/2}x = 0$$

$$c_1' \sin \lambda^{1/2}x - c_2' \cos \lambda^{1/2}x = -f(x)/\lambda^{1/2}$$

yield

$$c_1'(x) = -f(x) \sin \lambda^{1/2}x/\lambda^{1/2}$$

$$c_2'(x) = f(x) \cos \lambda^{1/2}x/\lambda^{1/2}$$

so that

$$u_p(x) = \left(-\lambda^{-1/2} \int_0^x f(s) \sin \lambda^{1/2} s \, ds \right) \cos \lambda^{1/2}x$$

$$+ \left(\lambda^{-1/2} \int_0^x f(s) \cos \lambda^{1/2} s \, ds \right) \sin \lambda^{1/2} x.$$

The general solution being $u(x) = u_H(x) + u_P(x)$, now satisfying the boundary conditions requires

$$0 = u(0) = c_1$$

$$0 = u(l)$$

$$= c_2 \sin \lambda^{1/2} l$$

$$- \lambda^{-1/2} \cos \lambda^{1/2} l \int_0^l f(s) \sin \lambda^{1/2} s \, ds$$

$$+ \lambda^{-1/2} \sin \lambda^{1/2}l \int_0^l f(s) \cos \lambda^{1/2} s \, ds$$

or

$$c_2 = - \frac{1}{\lambda^{1/2} \sin \lambda^{1/2}l} \int_0^l \sin \lambda^{1/2}(l - s)f(s) \, ds.$$

Thus the general solution for arbitrary forcing data $f(x)$ may be written

$$u(x) = \sin \lambda^{1/2}x \int_0^l \frac{\sin \lambda^{1/2}(l - s)}{-\lambda^{1/2} \sin \lambda^{1/2}l} f(s) \, ds + \int_0^x \frac{\sin \lambda^{1/2} (x - s)}{\lambda^{1/2}} f(s) \, ds$$

$$= \int_0^x \left[\frac{\sin \lambda^{1/2}x \sin \lambda^{1/2} (s - l)}{\lambda^{1/2} \sin \lambda^{1/2}l} + \frac{\sin \lambda^{1/2}(x - s)}{\lambda^{1/2}} \right] f(s) \, ds$$

$$+ \int_x^l \frac{\sin \lambda^{1/2} x \sin \lambda^{1/2} (s - l)}{\lambda^{1/2} \sin \lambda^{1/2}l} f(s) \, ds.$$

The quantity in the brackets with common denominator reduces to

$$[\cdots] = \frac{\sin \lambda^{\frac{1}{2}}s \sin \lambda^{\frac{1}{2}} (x - l)}{\lambda^{\frac{1}{2}} \sin \lambda^{\frac{1}{2}}l}$$

so that we may write

$$u(x) = \int_0^l G(x, s)f(s) \, ds$$

where

$$G(x, s) = \begin{cases} \dfrac{\sin \lambda^{\frac{1}{2}}s \sin \lambda^{\frac{1}{2}}(x - l)}{\lambda^{\frac{1}{2}} \sin \lambda^{\frac{1}{2}} l}, & 0 \leqq s \leqq x, \\[3mm] \dfrac{\sin \lambda^{\frac{1}{2}}x \sin \lambda^{\frac{1}{2}}(s - l)}{\lambda^{\frac{1}{2}} \sin \lambda^{\frac{1}{2}} l}, & x \leqq s \leqq l. \end{cases}$$

Note that the Green's function becomes singular exactly at the eigenvalues of the problem, $\lambda = n^2\pi^2/l^2$. A more advanced theory (called either Integral Equations or Fredholm Theory) studies the Green's functions at these singularities. When the data f is orthogonal to the eigenfunctions, the solution often may still be represented by a reduced Green's function.

As the dimension of the space domain increases (i.e., as we move from ordinary differential equations to partial differential equations), the singularity of the Green's function must increase to accommodate the desired delta function reproducing property mentioned earlier. These singularities in $n = 2$ and 3 space dimensions are best understood in terms of the Green's Identities of the next section.

Exercises

6. Show for the Example that for $\lambda = 0$

$$G(x, s) = \begin{cases} s(x - l)/l, & 0 \leqq s \leqq x, \\ x(s - l)/l, & x \leqq s \leqq l, \end{cases}$$

and for $\lambda < 0$

$$G(x, s) = \begin{cases} \dfrac{\sinh ((-\lambda)^{\frac{1}{2}}s) \sinh ((-\lambda)^{\frac{1}{2}} (x - l))}{(-\lambda)^{\frac{1}{2}} \sinh ((-\lambda)^{\frac{1}{2}}l)}, & 0 \leqq s \leqq x, \\[3mm] \dfrac{\sinh ((-\lambda)^{\frac{1}{2}}x) \sinh ((-\lambda)^{\frac{1}{2}} (s - l))}{(-\lambda)^{\frac{1}{2}} \sinh ((-\lambda)^{\frac{1}{2}}l)}, & x \leqq s \leqq l, \end{cases}$$

7. Find the Green's function for

$$\begin{cases} u''(x) + u(x) = f(x), & 0 < x < \pi/2, \\ u(0) = u(\pi/2) = 0. \end{cases}$$

Then find the solution when $f(x) = x$.

8. Find the Green's function for

$$\begin{cases} u''(x) + u(x) = f(x), & 0 < x < \pi/2, \\ u(0) + u'(0) = 0, \\ u(\pi/2) + u'(\pi/2) = 0. \end{cases}$$

Then find the solution when $f(x) = x$.

After *separation of variables* and *Green's function representation*, we have categorized the third principal method for solving partial differential equations as the *variational method*. This is the classical viewpoint. A modern viewpoint might emphasize instead the *numerical method*. In compromise, we have accommodated them both in a unified manner in Section 2.6 of Chapter 2. For those who wish at this point to learn some of the essentials of numerical techniques, see also Appendix B in Chapter 3.

The variational formulation of a given partial differential equation is often obtained merely by multiplying through by u and integrating by parts. This leads to the Green's Identities of the next section. Some of the main ideas can often be illustrated in one space dimension.

Example

Show variationally that the first eigenvalue Λ_1 of the clamped (elastic) rod

$$\begin{cases} -u'''' = \Lambda u'', & a < x < b, \\ u(a) = u(b) = 0, \\ u'(a) = u'(b) = 0, \end{cases}$$

is greater than or equal to the second eigenvalue λ_2 of the fixed string

$$\begin{cases} -w'' = \lambda w, & a < x < b, \\ w(a) = w(b) = 0 \end{cases}$$

on the same interval.

Solution. Of course we could in this ODE version* solve the two problems exactly for all eigenvalues and eigenfunctions. Proceeding here variationally, we recall an important integration by parts formula

$$(uv' - vu')\bigg|_a^b = \int_a^b (uv'' - vu'') \, dx$$

for any twice continuously differentiable functions u and v. Letting $v = u''$ we have, for all u satisfying the clamped rod boundary conditions,

$$0 = \int_a^b uu'''' \, dx - \int_a^b (u'')^2 \, dx.$$

---

* The PDE version was a conjecture of A. Weinstein and was shown to be generally true by L. Payne, *Arch. Rat. Mech. Anal.* **4** (1955). See the Second Pause.

Upon multiplying the rod eigenvalue equation by u, integrating, and substituting into the above relation, we have for any eigenfunction u

$$\int_a^b (u'')^2 \, dx \; = \; -\Lambda \int_a^b uu'' \, dx \; = \; \Lambda \int_a^b (u')^2 \, dx.$$

Thus the eigenvalues Λ_n of the rod are given by the Rayleigh quotients

$$\Lambda_n \; = \; R_{\substack{\text{clamped} \\ \text{rod}}}(u_n) \equiv \int_a^b (u_n''(x))^2 \, dx \Big/ \int_a^b (u_n'(x))^2 \, dx$$

and moreover, as in the Dirichlet Principle of Section 1.5.3, they are characterized variationally by

$$\Lambda_n \; = \; \underbrace{\inf R(u)}_{\substack{\text{all } u \perp u_1, \ldots, u_{n-1}, \\ u(a) = u(b) = u'(a) = u'(b) = 0}}$$

Here we use the $\perp$ symbol $u \perp v$ to denote that $\int_a^b uv \, dx = 0$. In particular, Λ_1 is the infimum of the Rayleigh quotient over all admissible u (e.g., those satisfying the prescribed boundary conditions and possessing enough derivatives to do the integrations by parts). For the Rayleigh fixed end string problem we already know that the variational characterization of its eigenvalues is similarly given by

$$\lambda_n \; = \; R_{\substack{\text{fixed} \\ \text{string}}}(w_n) \; = \; \underbrace{\inf \int_a^b (w')^2 \, dx \Big/ \int_a^b w^2 \, dx.}_{\substack{\text{all } w \perp w_1, \ldots, w_{n-1}, \\ w(a) = w(b) = 0}}$$

We now observe that u_1', where u_1 denotes the first eigenfunction of the rod problem, satisfies the boundary conditions for the string problem. Should it also be orthogonal to w_1, the first string eigenfunction, we would have the desired result:

$$\Lambda_1 \; = \; \int_a^b (u_1'')^2 \Big/ \int_a^b (u_1')^2 \geqq \lambda_2.$$

Assuming $\int_a^b u_1' w_1 \neq 0$, we may modify the above trial function u_1' by forming the new trial function

$$v \; = \; \alpha u_1' + u_1$$

in which we choose $\alpha = -\int_a^b u_1 \, w_1 / \int_a^b u_1' w_1$ so that

$$0 \; = \; \int v w_1 \; = \; \alpha \int u_1' \, w_1 + \int u_1 w_1$$

This trial function yields the desired inequality

$$\lambda_2 \leqq \frac{D(v)}{\int_a^b v^2} = \frac{\alpha^2 \int_a^b (u_1'')^2 + \int_a^b (u_1')^2}{\alpha^2 \int_a^b (u_1')^2 + \int_a^b u_1^2} = \frac{\alpha^2 \Lambda_1 + 1}{\alpha^2 + \left(\int_a^b u_1^2 \Big/ \int_a^b (u_a')^2\right)} \leqq \Lambda_1$$

The cross terms vanished because the boundary conditions gave

$$\int_a^b u_1'' u_1' = \frac{1}{2} \int_a^b ((u_1')^2)' = 0,$$

$$\int_a^b u_1' u_1 = \frac{1}{2} \int_a^b (u_1^2)' = 0.$$

The last inequality used $\int_a^b (u_1')^2 / \int_a^b u_1^2 \leqq \Lambda_1 \equiv \int_a^b (u_1'')^2 / \int_a^b (u_1')^2$. This follows from integration by parts and Schwarz's Inequality*

$$\int_a^b (u_1')^2 = -\int_a^b u_1 u_1'' \leqq \left(\int_a^b (u_1)^2\right)^{1/2} \left(\int_a^b (u_1'')^2\right)^{1/2}.$$

Exercises

9. (a) Solve the clamped rod and fixed string problems exactly. (b) Verify that the eigenfunctions of the rod are orthogonal. (c) Show variationally that as the length of the interval (a, b) is increased all eigenvalues in both problems decrease.

1.6   THREE IMPORTANT MATHEMATICAL TOOLS

Among the mathematical tools that we employ we single out in this preliminary chapter three of the most important, which we designate as:

(1) *Divergence Theorem*

(2) *Inequalities*

(3) *Convergence Theorems*

One could have labeled them instead as:

(1') *Green's Identities*

(2') *Schwarz's Inequality*

(3') *Differentiation term by term,*

---

* See the next section.

to be overly specific; or to be overly general, as

(1″) *Exterior Differential Forms*

(2″) *A priori Estimates*

(3″) *Advanced Calculus*

Here we will supply only some of the basic facts,* more general theory being available in books on the just listed double-primed subjects.

1.6.1. Divergence Theorem

The divergence theorem† is the higher dimensional form of the fundamental theorem of calculus and can be stated rather simply as

$$\int_\Omega \frac{\partial f}{\partial x_i} = \oint_{\partial\Omega} fn_i$$

where n_i is the ith component of the unit outer normal to the surface $\partial\Omega$ (Fig. 1.6a; **n** is the unit outer normal).

We write this important theorem in several forms. The general assumption, as in the fundamental theorem of calculus, is that the function being differentiated is C^1. Later we discuss the regularity assumptions more precisely, especially as they concern the domain Ω and its boundary $\partial\Omega$. For now we tacitly assume that the outer normal derivative $\partial u/\partial n \equiv \text{grad } u \cdot \mathbf{n}$ is well defined. As it turns out, edges and corners can be tolerated as they make no contribution to a higher dimensional surface integral.

The Divergence Theorem:

$$\int_\Omega \frac{\partial f}{\partial x_i} = \oint_{\partial\Omega} fn_i \qquad\qquad \text{component form, most general}$$

$$\int_a^b f'(t)\, dt = f(b) - f(a) \qquad\qquad \text{calculus ``one-dim.'' form}$$

$$\int_\Omega \text{div } \mathbf{f} = \oint_{\partial\Omega} \mathbf{f} \cdot \mathbf{n} \qquad\qquad \text{vector form}$$

* A little time spent here in learning to manage these three tools is worthwhile. The student who has had advanced calculus has probably seen the divergence theorem (but probably not enough of it), almost certainly not enough inequalities, and not the applications of convergence theorems in the uniform norm. On the other hand, for those who wish to push quickly ahead, they may do so, keeping well in mind from calculus (1) integration by parts, (2) Schwarz's inequality, and (3) that most nice things converge.

† Also often called the Gauss integral theorem, and sometimes Green's theorem. The latter is usually reserved for the special two-dimensional form given in the listing below.

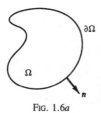

FIG. 1.6a

$$\int_\Omega (f_i)_{,i} = \oint_{\partial\Omega} f_i n_i$$

summation notation form

$$\iint_\Omega [Q_x - P_y] \, dx \, dy = \oint_{\partial\Omega} P \, dx + Q \, dy$$

$\begin{cases} \text{``Green's Theorem'' form,} \\ \text{``Stoke's Theorem'' form} \end{cases}$

$$\int_\Omega \operatorname{grad} u \cdot \operatorname{grad} v = \oint_{\partial\Omega} u \frac{\partial v}{\partial n} - \int_\Omega u \, \Delta v$$

Green's identity I

$$\oint_{\partial\Omega} u \frac{\partial v}{\partial n} - \oint_{\partial\Omega} v \frac{\partial u}{\partial n} = \int_\Omega u \, \Delta v - \int_\Omega v \, \Delta u$$

Green's identity II

$$u(P) = -\frac{1}{4\pi} \iiint_\Omega \frac{1}{r_{PQ}} \Delta_Q u + \frac{1}{4\pi} \oiint_{\partial\Omega} \frac{1}{r_{PQ}} \frac{\partial u}{\partial n_Q}$$

$$-\frac{1}{4\pi} \oiint_{\partial\Omega} u \frac{\partial}{\partial n_Q} \left(\frac{1}{r_{PQ}} \right)$$

Green's identity III
($n = 3$ case)

A proof of the divergence theorem can get quite complicated and we defer to the references for a more complete discussion.* We consider here only a "simple" region Ω or a region Ω that can be decomposed into a finite number of simple regions, and this way we will see that a proof of the divergence theorem reduces in fact to the fundamental theorem of calculus. A simple region here will be taken to mean one cylindrical in one coordinate axis direction with disjoint top and bottom surfaces that are C^1 surfaces themselves. See Fig. 1.6b for some simple regions. Cubes are simple regions and a given large domain Ω can often be decomposed into a grid of cubes and a remainder of simple regions as shown in the figure. The

* The best general reference is perhaps O. D. Kellogg, *Foundations of Potential Theory* (Dover, New York, 1953), although the proof for regular regions is scattered throughout the book. A modern proof, focusing principally on Ω a cube, can be found in M. Spivak, *Calculus on Manifolds* (Benjamin, New York, 1965). In this vein see also H. Flanders, *Differential Forms* (Academic Press, New York, 1963). The most general conditions from an integration theory viewpoint may be found in Hassler Whitney's book, *Geometric Integration Theory* (Princeton Univ. Press, Princeton, N.J., 1957).

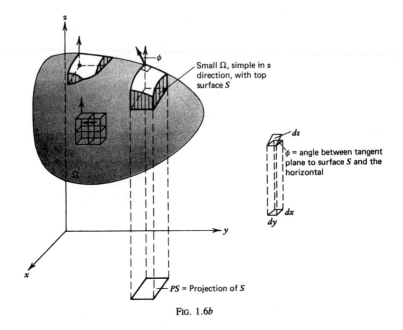

FIG. 1.6b

small region closest to the z axis in Fig. 1.6b is not simple but the same proof will work for it; alternately, it may be further decomposed into two simple subregions. Assuming that the divergence theorem holds for simple regions, the contributions from opposing surfaces on the cubes cancel if the function f is continuous and one is left with simple regions along the boundary of the domain to evaluate.

Looking at the simple region Ω as shown in the figure, with top surface S and bottom surface parallel to the xy plane, the other sides being cylindrical, and assuming the given function f is C^1 over the whole of this subregion and its boundary, one has

$$\int_\Omega f_z \equiv \iiint_\Omega \frac{\partial f}{\partial z}\, dV = \iint_{PS} dx\, dy \int_{\text{"bottom"}}^{\text{"top"}} \frac{\partial f}{\partial z}\, dz$$

$$= \iint_{PS} [f(x, y, \text{top } z) - f(x, y, \text{bottom } z)]\, dx\, dy$$

by the fundamental theorem of calculus and the definition of the iterated integral. Having evaluated the volume integral, we next look at the surface integral, noting that $n_z = 0$ on the cylindrical sides, $n_z = -1$ on the bottom:

$$\oint_{\partial\Omega} f n_z \equiv \oiint_S f \cos\varphi\, ds - \oiint_{PS} f(x, y, \text{bottom } z)\, dx\, dy.$$

Since φ is also the angle* between the horizontal and the infinitesimal tangent plane (see Fig. 1.6b) at any point on the surface S,

$$\iint_S f \cos \varphi \, ds = \iint_{PS} f(x, y, \text{top } z) \, dx \, dy$$

and the theorem is proved.

Looking now at the vector form of the divergence theorem, we see that one need only add up the component terms and use the just shown component form; that is, if $\mathbf{f} = (f_1, \ldots, f_n)$, then

$$\int_\Omega \text{div } \mathbf{f} \, dV = \sum_{i=1}^n \int_\Omega \frac{\partial f_i}{\partial x_i} \, dV = \sum_{i=1}^n \oint_{\partial\Omega} f_i n_i \, ds = \oint_{\partial\Omega} \mathbf{f} \cdot \mathbf{n} \, ds,$$

since the integral† of a finite sum is the finite sum of the integrals. Likewise with the summation notation form.

The Green's theorem form of the divergence theorem will be considered next; we have also mentioned that it is of the "Stokes Theorem" form but the latter is usually reserved for the three-dimensional version,‡ the proof there (and in higher dimensions) being similar. Let Ω be a two-dimensional domain with $\partial\Omega$ described as the $\text{arc } x = x(s)$, $y = y(s)$, in terms of arc length s (Fig. 1.6c). Let $\mathbf{f}$ be the vector function

$$\mathbf{f}(x, y) \equiv (f_1(x, y), f_2(x, y)) = (Q(x, y), - P(x, y)).$$

Then

$$\iint_\Omega [Q_x - P_y] \, dx \, dy = \iint_\Omega ((f_1)_x + (f_2)_y) = \iint_\Omega \text{div } f \, dV = \oint_{\partial\Omega} f_i n_i \, ds$$

$$= \oint_{\partial\Omega} \left(f_1 \frac{dy}{ds} - f_2 \frac{dx}{ds} \right) ds = \oint_{\partial\Omega} Q \, dy + P \, dx.$$

---

* It is helpful to recall from calculus the formula for surface area,

$$A(S) = \iint_{PS} (1 + u_x^2 + u_y^2)^{1/2} \, dx \, dy$$

for a surface $u(x, y)$ over the xy plane. The tangent plane to the surface at a point is given by $z - z_i = u_x \cdot (x - x_i) + u_y(y - y_i)$ and has attitude numbers $(u_x, u_y, -1)$, the xy plane has attitude numbers $(0, 0, -1)$, $\cos \varphi = (0, 0, -1) \cdot (u_x, u_y, -1)/1 \cdot (1 + u_x^2 + u_y^2)^{1/2}$, and

$$A(S) = \lim \sum \Delta S = \lim \sum dx_i \, dy_i/\cos \varphi_i \iint_{PS} (1 + u_x^2 + u_y^2)^{1/2} \, dx \, dy$$

by the usual limiting procedure in defining the integral. The argument in proving the divergence theorem is thus essentially the same one.

† Being a linear operator.

‡ $\oint_S [(R_y - Q_z) \cos \alpha + (P_z - R_x) \cos \beta + (Q_x - P_y) \cos \gamma] \, dS = \oint_{\partial S} [P \, dx + Q \, dy + R \, dz]$

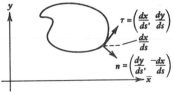

FIG. 1.6c

We turn next to the Green's identities I, II, and III that are vital in a study of PDEs. The first two may be regarded as just integration by parts, (e.g., as a higher dimensional version of integration by parts with the divergence theorem replacing the fundamental theorem of calculus). Actually, the key is in reality just *differentiation by parts,** i.e.*, the product rule,

$$(uv_{,i})_{,i} = uv_{,ii} + u_{,i}v_{,i}$$

following by an integration over Ω

$$\int_\Omega (uv_{,i})_{,i} \, dV = \int_\Omega u \, \Delta v \, dV + \int_\Omega \text{grad } u \cdot \text{grad } v \, dV$$

and application of the divergence theorem

$$\int_\Omega (uv_{,i})_{,i} \, dV = \oint_{\partial\Omega} uv_{,i}n_i \, ds \equiv \oint_{\partial\Omega} u \frac{\partial v}{\partial n} \, ds,$$

which yields the first identity

$$\int_\Omega \text{grad } u \cdot \text{grad } v = \oint_{\partial\Omega} u \frac{\partial v}{\partial n} - \int_\Omega u \, \Delta v.$$

We have assumed that u and v are in $C^1(\overline{\Omega}) \cap C^2(\Omega)$ in order to differentiate and integrate. A special case of Green's first identity is that with $u = v$,

$$\int_\Omega |\text{grad } u|^2 = \oint_{\partial\Omega} u \frac{\partial u}{\partial n} - \int_\Omega u \, \Delta u$$

---

* In one dimension this is the elementary calculus formula $(uv)' = uv' + vu'$, and in the same way the given summation notation formula just adds up the corresponding partial derivatives. Written out for $n = 2$ and $n = 3$ this becomes, respectively,

$$\frac{\partial}{\partial x}\left(u \frac{\partial v}{\partial x}\right) + \frac{\partial}{\partial y}\left(u \frac{\partial v}{\partial y}\right) = \left(u \frac{\partial^2 v}{\partial x^2} + u \frac{\partial^2 v}{\partial y^2}\right) + \left(\frac{\partial u}{\partial x}\frac{\partial v}{\partial x} + \frac{\partial u}{\partial y}\frac{\partial v}{\partial y}\right)$$

and

$$\frac{\partial}{\partial x_1}\left(u \frac{\partial v}{\partial x_1}\right) + \frac{\partial}{\partial x_2}\left(u \frac{\partial v}{\partial x_2}\right) + \frac{\partial}{\partial x_3}\left(u \frac{\partial v}{\partial x_3}\right)$$
$$= u\left(\frac{\partial^2 v}{\partial x_1^2} + \frac{\partial^2 v}{\partial x_2^2} + \frac{\partial^2 v}{\partial x_3^2}\right) + \left(\frac{\partial u}{\partial x_1}\frac{\partial v}{\partial x_1} + \frac{\partial u}{\partial x_2}\frac{\partial v}{\partial x_2} + \frac{\partial u}{\partial x_3}\frac{\partial v}{\partial x_3}\right).$$

where the Dirichlet integral $\int_\Omega |\text{grad } u|^2$ represents energy, virtual mass, capacity, torsional rigidity, and other important quantities.

Letting $D(u, v) = \int \text{grad } u \cdot \text{grad } v$, as in the ODE case considered in the discussion of variational methods, and noting that

$$D(u, v) = D(v, u)$$

by the symmetry of the ordinary dot product for vectors in n-space, we have Green's second identity,

$$\oint_{\partial\Omega} u \frac{\partial v}{\partial n} - \oint_{\partial\Omega} v \frac{\partial u}{\partial n} = \int_\Omega u \, \Delta v - \int_\Omega v \, \Delta u.$$

The validity of the third Green's identity follows from Green's second identity with v taken to be $v_{PQ} = S(P, Q) = (4 \pi r_{PQ})^{-1}$, the singular source function discussed in Section 1.5.2 along with the delta function property that

$$\int_\Omega u \, \Delta v \equiv \int_\Omega u(Q) \, \Delta_Q v_{PQ} = -u(P)$$

at interior points P in Ω. The verification involves principally the observation that the singular function v_{PQ} in the integrands in Green's second identity necessitates a consideration of the integrals there as singular (or improper) integrals. For example, by calculus the meaning of the above integral is (see Fig. 1.6d)

$$\int_\Omega u \, \Delta_Q \left(\frac{1}{4\pi r_{PQ}}\right) \equiv \lim_{\varepsilon \to 0} \int_{\Omega_\varepsilon} u \, \Delta_Q \left(\frac{1}{4\pi r_{PQ}}\right)$$

provided that the limit exists.*

Writing then Green's second identity for the region Ω_ε as shown in Figure 1.6d with P an interior point of the original region Ω and with a small three-

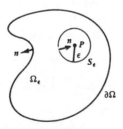

FIG. 1.6d

* Which it does of course in this case, since

$$\int_{\Omega_\varepsilon} u \, \Delta_Q \left(\frac{1}{4\pi r_{PQ}}\right) = 0$$

due to the fact that $1/r_{PQ}$ is harmonic in Ω_ε.

dimensional ball (solid sphere) B_ε centered at P, so that $\partial\Omega_\varepsilon$ is composed of the two disjoint parts $\partial\Omega$ and $S_\varepsilon \equiv \partial B_\varepsilon$, one has

$$0 = \int_{\Omega_\varepsilon} u\, \Delta_Q \frac{1}{4\pi r_{PQ}}$$

$$= \int_{\Omega_\varepsilon} v_{PQ}\, \Delta_Q u - \oint_{\partial\Omega} v_{PQ} \frac{\partial u}{\partial n_Q} + \oint_{\partial\Omega} u \frac{\partial v_{PQ}}{\partial n_Q} + \oint_{S_\varepsilon} u \frac{\partial v_{PQ}}{\partial n_Q} - \oint_{S_\varepsilon} v \frac{\partial u}{\partial n_Q}$$

where

$$v = \frac{1}{4\pi r_{PQ}} \quad \text{and} \quad \frac{\partial}{\partial n_Q}\left(\frac{1}{r_{PQ}}\right)\bigg|_{S_\varepsilon} = -\frac{\partial}{\partial r}\left(\frac{1}{r_{PQ}}\right)\bigg|_{S_\varepsilon} = \frac{1}{\varepsilon^2}.$$

The Green's third identity then follows from the above second identity upon taking the limit of both sides as $\varepsilon \to 0$, provided that the first term on the right makes sense and the last two terms on the right yield a contribution $u(P)$. The first integral

$$\int_{\Omega_\varepsilon} \left(\frac{1}{4\pi r_{PQ}}\right) \Delta_Q u$$

is regular and converges for $u \in C^2(\Omega)$ because the integration element dV is of the form

$$dV = r^2\, dr d \text{ (solid angle)}$$

in three dimensions. The next to the last integral is seen to converge to $u(P)$ by use of the usual calculus mean value theorem for integrals for continuous functions u, as follows:

$$\oint_{S_\varepsilon} u(Q) \frac{\partial}{\partial n_Q}\left(\frac{1}{4\pi r_{PQ}}\right) = \frac{1}{4\pi\varepsilon^2} \oint_{S_\varepsilon} u(Q)\, ds_Q = \frac{4\pi\varepsilon^2}{4\pi\varepsilon^2} u(Q_\varepsilon) = u(Q_\varepsilon) \to u(P),$$

where Q_ε is a guaranteed but unspecified point on S_ε; by $u \in C^0(\Omega)$ one has necessarily that $Q_\varepsilon \to P$ and $u(Q_\varepsilon) \to u(P)$ as $\varepsilon \to 0$. The last integral on the right is similarly seen to converge to zero as $\varepsilon \to 0$, for $u \, \varepsilon \, C^1(\Omega)$, by the existence of a Q'_ε on S_ε converging to P,

$$\oint_{S_\varepsilon} \frac{1}{4\pi r_{PQ}} \frac{\partial u}{\partial n_Q} = \frac{1}{4\pi\varepsilon} \oint_{S_\varepsilon} \frac{\partial u}{\partial n_Q} = \frac{4\pi\varepsilon^2}{4\pi\varepsilon} \frac{\partial u}{\partial n_Q}(Q'_\varepsilon) = \varepsilon \frac{\partial u}{\partial n_Q}(Q'_\varepsilon) \to 0.$$

We wish to make three supplemental remarks at this point concerning the divergence theorem and Green's identities before leaving them.

Remark 1

The constant 4π that occurred here for three dimensions is in general replaced by $\omega_n = 2\pi^{n/2}/\Gamma(n/2)$, the area of the unit sphere in n dimensions, and the source function $V_{PQ} \equiv S_{PQ}$ is replaced by

$$S(P, Q) = \frac{1}{(n - 2)\omega_n r_{PQ}^{n-2}}, \qquad n > 2,$$

and

$$S(P, Q) = \frac{1}{2\pi} \ln\left(\frac{1}{r_{PQ}}\right), \qquad n = 2.$$

Thus in particular the two-dimensional Green's third identity is

$$u(P) = -\frac{1}{2\pi} \iint_\Omega \ln\left(\frac{1}{r_{PQ}}\right) \Delta u + \frac{1}{2\pi} \oint_{\partial\Omega} \ln\left(\frac{1}{r_{PQ}}\right) \frac{\partial u}{\partial n_Q}$$

$$- \frac{1}{2\pi} \oint_{\partial\Omega} u \frac{\partial}{\partial n_Q} \ln\left(\frac{1}{r_{PQ}}\right).$$

Remark 2

Replacing the fundamental singularity $S(P, Q)$, in Green's third identity by the Green's function

$$G(P, Q) = S(P, Q) + g(P, Q),$$

where $g(P, Q)$ is a smooth adjusting function that adds or subtracts just the right amount to the fundamental singularity, for each fixed P, so that $G(P, Q)$ satisfies the boundary conditions for a given problem (this was also discussed earlier in Section 1.5), yields interesting results. For example, for the Dirichlet problem, $g(P, Q)$ is required to satisfy (Fig. 1.6e)

$$\begin{cases} \Delta_Q g(P, Q) = 0, & Q \text{ in } \Omega, \\ g(P, Q) = -S(P, Q), & Q \in \partial\Omega; \end{cases}$$

since its boundary value $-S(P, Q)$ is continuous on $\partial\Omega$ the standard existence theory for the Dirichlet problem guarantees* a harmonic solution $g_P \in C^2(\Omega) \cap C^1(\overline{\Omega})$ for every P. Upon insertion† of the resulting Green's function $G(P, Q)$ into Green's III we get

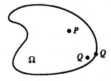

FIG. 1.6e

* See for example the book by Kellogg cited previously in this section.

† This may be checked in either of two ways. One may repeat the argument given above in going from Green's second to third identities but with $S(P, Q)$ replaced by $S(P, Q) + g(P, Q)$, or one may use Green's second identity for $g(P, Q)$ and then add those terms to the previous result.

$$u(P) = \int_\Omega G(P, Q)F(Q) - \oint_{\partial\Omega} \frac{\partial G}{\partial n_Q}(P, Q)f(Q)$$

as a representation formula for the solution of the general Dirichlet–Poisson problem for general domains‡;

$$\begin{cases} -\Delta u = F \text{ in } \Omega, \\ u = f \text{ on } \partial\Omega. \end{cases}$$

For example, for the Poisson problem

$$\begin{cases} -\Delta u = F \text{ in } \Omega \\ u = 0 \text{ on } \partial\Omega \end{cases} \quad \text{i.e., } Lu = \begin{pmatrix} -\Delta \\ I \end{pmatrix} u = \begin{pmatrix} F \\ 0 \end{pmatrix} \text{ on } \begin{pmatrix} \Omega \\ \partial\Omega \end{pmatrix}$$

regarded as a linear operator equation with two components, the above representation formula is just the expression for the inverse operator L^{-1},

$$u = L^{-1} \begin{pmatrix} F \\ 0 \end{pmatrix} = \int_\Omega G(P, Q)F(Q)\, dV_Q.$$

One proceeds similarly for the composite Dirichlet–Poisson problem. It is intuitively reasonable that the inverse of a differential operator should be an integral operator as in elementary calculus. Thus Green's third identity interpreted as L^{-1} completes the circle of ideas

Fund. Thm. of Calculus $\Rightarrow$ divergence theorem $\Rightarrow$ Green's identity

$$\Uparrow \qquad\qquad\qquad\qquad\qquad\qquad\qquad\qquad \Downarrow$$
$$\int \text{ as } \left(\frac{d}{dx}\right)^{-1} \Leftarrow \begin{matrix} \text{Integral ops. to} \\ \text{Invert. Diff. ops.} \end{matrix} \Leftarrow \int_\Omega G(P, Q) \text{ as } \Delta_{D.B.C.}^{-1}$$

although it should be mentioned in qualification that there are many other ramifications and consequences of the notion of integral operators.

Remark 3

One of the corollaries of the above is the *mean value theorem* for harmonic functions,

$$u(P) = \frac{1}{\omega_n r^{n-1}} \oint_{S_{PQ}=r} u(Q)\, dS_Q$$

for $n \geq 2$, which states that a harmonic function is everywhere equal to its own spherical averages. Indeed, this property may be taken to be the defining property for harmonic functions and then shown to be equivalent to the requirement that $\Delta u = 0$. A similar "solid mean value* theorem" holds for harmonic functions u, namely,

‡ As usual we do not define the most general domain. Some authors dodge this by defining a regular domain to be one on which the divergence theorem holds, a tautological procedure at best. See Problem 1.9.5.

* This is a mean value theorem by noting that the volume of the unit ball is ω_n/n.

$$u(P) = \frac{n}{\omega_n r^n} \int_{r_{PQ} \leq r} u\,(Q)\,dV_Q.$$

To see how the above mean value theorem results from Green's third identity we need the Green's function $G(P, Q)$ for the unit sphere. For the case $n = 3$ it may be seen† that $G(P, Q)$ is given by

$$G(P, Q) = \frac{1}{4\pi r_{PQ}} - \frac{1}{4\pi r_{OQ} r_{PQ'}}$$

where $Q' = r_{OQ}^{-2}\,Q$ (see Fig. 1.6f). By symmetry and trigonometry one has

$$G(Q, P) = G(P, Q) = \frac{1}{4\pi r_{PQ}} - \frac{1}{4\pi r_{OP} r_{QP'}}\ .$$

Holding P fixed and differentiating $G(Q, P)$ with respect to Q, one has by a calculation

$$\frac{\partial G}{\partial n_Q}\ (P\ \text{fixed inside},\ Q\ \text{on}\ \partial) = \left(\frac{-1}{4\pi r_{PQ}^2} + \frac{1}{4\pi r_{QP'}^2}\right)\frac{1}{r_{PQ}}.$$

Noting by similar triangles that then $r_{QP'} = r_{QP}/r_{OP}$, one thus has from the representation formula of Remark 2, for the Dirichlet problem for the unit sphere (Fig. 1.6g)

$$\begin{cases} \Delta u = 0\ \text{in}\ r < 1, \\ u = f\ \text{on}\ r = 1, \end{cases}$$

that

$$u(P) = -\oint_{r=1} \frac{\partial G}{\partial n} f\,ds = \frac{1}{4\pi} \oint_{r_{OQ}=1} \left(\frac{1 - r_{OP}^2}{r_{PQ}^3}\right) f(Q)\,ds_Q, \qquad n = 3.$$

Note that the data is weighted more heavily as Q approaches P along the boundary. This formula is sometimes called the Poisson integral representation

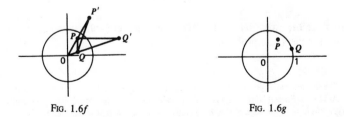

FIG. 1.6f   FIG. 1.6g

† This may be verified directly. One way to derive it is by the "physical" method of electrostatic images, which is essentially what we have done here, by use of the image points P' and Q'. For each Q on the boundary, a "counterbalancing" point source is placed at P' so that the ratio of the potentials $\frac{1}{4}\pi r_{PQ}$ and $\frac{1}{4}\pi r_{OP'}$ is the constant r_{OP}.

of the solution. In particular, when P is taken at the origin, the boundary points Q are all equidistant, the averaging is perfect, and one obtains the mean value theorem

$$u(0) = \frac{1}{\omega_3} \int_{|Q|=1} u(Q) \, ds_Q$$

given at the beginning of this remark; by considering spheres of radius r rather than $r = 1$, one obtains similarly the other versions given there.

To oversimplify: This all works because the fundamental singularity was taken to be

$$S(P, Q) = S(r) = \frac{1}{(n-2)\omega_n r^{n-2}}$$

so that upon differentiation one has

$$S'(r) = \frac{\partial S}{\partial r} = \frac{1}{\omega_n r^{n-1}} = \text{the reciprocal of the surface area of the } r \text{ sphere in } n \text{ space.}^*$$

Problem 1.† By use of the divergence theorem, show uniqueness for the Dirichlet–Poisson problem

$$\begin{cases} \Delta u = f \text{ in } \Omega, \\ u = g \text{ on } \partial\Omega, \end{cases}$$

on a bounded nice domain Ω (Fig. 1.6h). State some regularity assumptions on f, g, u, and Ω.

Problem 2. Given nice Ω, $\partial\Omega$, and g, do you think that a solution always exists for the problem

$$\begin{cases} \Delta u = 0 \text{ in } \Omega, \\ \dfrac{\partial u}{\partial n} = g \text{ on } \partial\Omega? \end{cases}$$

Analyze this question via the divergence theorem and via the uniqueness question for the same problem.

---

* For further variations on this theme see, for example, F. John, *Plane Waves and Spherical Means Applied to Partial Differential Equations* (Wiley-Interscience, New York, 1955) and J.L. Synge, *The Hypercircle in Mathematical Physics* (Cambridge Univ. Press, Cambridge, 1957).

† These problems are illustrative of the second main method for proving uniqueness, the so-called energy method. Viewing $D(u, u)$ as the energy, one shows $D(u, u) = 0$ for the case of zero data, as one would expect physically. Then one concludes by sufficient regularity and the zero boundary data that the solution must be zero for zero data. Recall that uniqueness for $Lu = f$ is equivalent to showing that $Lu = 0$ has only the trivial solution, for any linear operator L.

Fig. 1.6*h*

Problem 3. Having considered the so-called first (Dirichlet) and second (Neumann) boundary value problems above, consider now the third boundary value problem,

$$\begin{cases} \Delta u = f \text{ in } \Omega, \\ \dfrac{\partial u}{\partial n} + ku = g \text{ on } \partial\Omega, \end{cases}$$

and make a uniqueness decision for both the cases $k > 0$ (Robin condition) and $k < 0$ (Steklov condition: eigenvalues in boundary operator).

1.6 (1)   Exercises

1. (a) Let Ω be a bounded region in R^3 for which the divergence theorem holds, and let $x = (x_1, x_2, x_3)$. Show that the volume of Ω is given by

$$\text{vol } (\Omega) = \frac{1}{3} \oint_{\partial\Omega} x \cdot n.$$

   (b) Extend (a) to the case of R^m, $x = (x_1, \ldots, x_m)$, namely

$$\text{vol } (\Omega) = \frac{1}{m} \oint_{\partial\Omega} x \cdot n.$$

   (c) If $\Omega = B_m(r) = \{x \in R^m : |x| \leq r\}$ and $\partial\Omega = S_m(r) = \{x \in R^m : |x| = r\}$, show that

$$\text{vol } (\Omega) = \frac{r}{m} \cdot \text{Area } (\partial\Omega).$$

   Check the validity of this formula when $m = 2, 3$.

2. (a) Check the validity for $m = 2$ of the formula

$$\text{Vol } (B_m(r)) = 2r \, \text{Vol } (B_{m-1}(r)) \int_0^{\pi/2} \cos^m\theta \, d\theta.$$

   (b) Repeat (a) for the case $m = 3$ dimensions.
   (c) Compute the volume and surface area of $B_4(r)$ in R^4.

3. (a) Check that the mean value theorem $u(P) = \dfrac{1}{\omega_n r^{n-1}} \displaystyle\oint_{S_{PQ}=r} u(Q)dS_Q$ holds
   for solutions to $u''(x) = 0$, i.e., in the one-dimensional case.

   (b) Check that solutions to $\Delta_2 u = 0$ of the form $u(x, y) = ax + by + c$
   also satisfy the mean value theorem.

   (c) Show conversely that a C^2 function that satisfies the mean value theorem
   in $\mathbb{R}^1$ is harmonic, i.e., satisfies $u''(x) = 0$.

1.6.2. Inequalities

Inequalities are the basis of analysis and this is especially true for the theory of
partial differential equations.[*] That this is the case can be described roughly by
the statement that one cannot in general solve partial differential equations, one
can only approximate them.

As the second basic tool let us therefore briefly consider some inequalities that
are useful throughout mathematics and not just for partial differential equations.
Proofs of these inequalities will be found at the end of the section.

The most important inequality is Schwarz's[†] inequality, which is essential,
among others, for an understanding of Fourier series.

1. Schwarz's Inequality: $|(u, v)| \leq \|u\| \, \|v\|$. Other notations are: $\langle u, v \rangle$, $u \cdot v$,
$\langle u|v \rangle$, $\int u(x)v(x) \, dx$, $\int u(x)\overline{v}(x) \, dx$, $\sum_{i=1}^{n} a_i b_i$, $\sum_{i=1}^{\infty} a_i \overline{b}_i$, $D(u, v)$ and so on, depending on the context in which the so-called inner product[‡] (u, v) is being used.
In all cases it is presupposed that one is working with an inner product space of
functions or vectors, that is, one in which there is a natural inner product defined
that possesses all the properties of the calculus dot product. This is made more
precise in the next chapter when the notion of a Hilbert space is discussed.

The geometrical way to understand Schwarz's inequality[§] is as in calculus,

$$\mathbf{v}_1 \cdot \mathbf{v}_2 = |\mathbf{v}_1| \, |\mathbf{v}_2| \cos \theta.$$

There $|\mathbf{v}|$ denotes the Euclidean length of the vector $\mathbf{v}$. In the same way for any
inner product

$$(u, v) = \|u\| \cdot \|v\| \cos \theta$$

defines an angle (Fig. 1.6i) between u and v, and the norm $\|u\|$ is defined by

$$(u, u) = \|u\|^2.$$

A related inequality to the Schwarz inequality is the following (any $\varepsilon > 0$):

---

[*] Some of the basic differential inequalities are usually now called Sobolev inequalities, after the
Russian mathematician, Sobolev. An important one will be discussed in this section.

[†] Other names are Cauchy's inequality (by the French) and Buniakowsky's inequality (by the
Russians).

[‡] Other terminology is: dot product, scalar product, ket. See Section 2.3 for the properties thereof.

[§] Proofs of these inequalities follow this discussion of them.

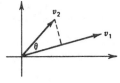

FIG. 1.6i

1′. Arithmetic–Geometric Mean Inequality: $|(u, v)| \leq \varepsilon\|u\|^2 + (4\varepsilon)^{-1}\|v\|^2$. This inequality (for short, the A–G Inequality) is useful from time to time when it is desired to give different weights to the vectors u and v in bounding the inner product. Let us illustrate this point by obtaining a stability result for the Dirichlet–Poisson problem on a bounded domain, using a combination of Schwarz and A–G Inequalities.

1″. Stability and Approximation for the Dirichlet–Poisson Problem

$$\begin{cases} \Delta u = F \text{ in } \Omega, \\ u = f \text{ on } \partial\Omega. \end{cases}$$

We seek an inequality* of the form

$$\int_\Omega u^2 \leq c_1 \oint_{\partial\Omega} u^2 + c_2 \int_\Omega (\Delta u)^2 \tag{EST}$$

for all nice (e.g., $u \in C^2(\Omega) \cap C^1(\overline{\Omega})$) functions u defined on $\overline{\Omega}$. Once obtained, we have a *bound* on the particular solution u of the given problem in terms of the given data,

$$\int_\Omega u^2 \leq c_1 \oint_{\partial\Omega} f^2 + c_2 \int_\Omega F^2.$$

Moreover, since the problem is linear, for $u = u_1 - u_2$ the difference of two solutions, one has the *stability* statement

$$\int_\Omega |u_1 - u_2|^2 \leq c_1 \oint_{\partial\Omega} |f_1 - f_2|^2 + c_2 \int_\Omega |F_1 - F_2|^2.$$

The latter becomes a *method of approximation* by considering any approximation φ to the true solution u, giving error bound ($c_i' \cong c_i$, see Problem 3a)

$$\|u - \varphi\|_{L^2(\Omega)} \leq c_1'\|f - \varphi\|_{L^2(\partial\Omega)} + c_2'\|F - \Delta\varphi\|_{L^2(\Omega)}.$$

* Such inequalities are called *a priori estimates* in the PDE literature because they are shown *a priori* (i.e., before the fact), that is, without talking about data F and f, for sufficiently nice but otherwise arbitrary functions u. As such they not only serve in *stability* and *approximation* considerations as shown here but also are basic to general *existence* proofs in the theory of PDE. If you must see more on the latter, see the references given in Problem 2.9.5.

Let us then proceed to derive the desired estimate (EST) above.

The problem by linearity can be written as two separate problems†

$$\begin{cases} \Delta h = 0 \text{ in } \Omega \\ h = f \equiv u \text{ on } \partial\Omega \end{cases} \text{ and } \begin{cases} \Delta v = F \equiv \Delta u \text{ in } \Omega \\ v = 0 \text{ on } \partial\Omega \end{cases} \quad \text{where } u = h + v.$$

Then by the A–G inequality

$$\int_\Omega u^2 = \int_\Omega (h^2 + v^2 + 2hv) \leq (1 + \varepsilon) \int_\Omega h^2 + (1 + \varepsilon^{-1}) \int_\Omega v^2.$$

By a variational approach we may bound v in terms of F, as was done in Section 1.5.3 for the ODE case. The (see Section 1.9.5(3)) variational characterization of the first eigenvalue λ_1 for the membrane problem

$$\begin{cases} -\Delta w = \lambda w \text{ in } \Omega, \\ w = 0 \text{ on } \partial\Omega, \end{cases}$$

is

$$0 < \lambda_1 = \inf_{\substack{w = 0 \\ \text{on } \partial\Omega}} \left[\frac{-\int_\Omega w \, \Delta w}{\int_\Omega w^2} \right],$$

as may be seen formally by multiplying the equation by w and integrating. Of course we have not specified the regularity class of w in the variational characterization, but it includes $C^2(\Omega) \cap C^1(\overline{\Omega})$. From the variational characterization we have then for an arbitrary nice function w vanishing on $\partial\Omega$ that, using Schwarz's inequality,

$$\int_\Omega w^2 \leq \frac{1}{\lambda_1} \left(-\int_\Omega w \, \Delta w \right) \leq \frac{1}{\lambda_1} \left(\int_\Omega w^2 \right)^{1/2} \left(\int_\Omega (\Delta w)^2 \right)^{1/2}$$

and thus for the v in the second problem above,

$$\int_\Omega v^2 \leq \frac{1}{\lambda_1^2} \int_\Omega F^2 \equiv \frac{1}{\lambda_1^2} \int_\Omega (\Delta u)^2.$$

For h as defined above in terms of u we use what is sometimes called the auxiliary problem method, or, from a more mathematical viewpoint, an *adjoint problem*; namely, let ψ be defined by

$$\begin{cases} \Delta\psi = h \text{ in } \Omega \\ \psi = 0 \text{ on } \partial\Omega \end{cases} \text{ recalling that } \begin{cases} \Delta h = 0 \text{ in } \Omega \\ h = u = f \text{ on } \partial\Omega \end{cases}.$$

---

† We write them in both the *a priori* (i.e., data is just u and Δu for any given u) and the *a posteriori* (i.e., assume the data f and F are given for the given Dirichlet–Poisson problem) ways for "both audiences."

Then by Green's second identity we have

$$\oint_{\partial\Omega} h \frac{\partial \psi}{\partial n} - \oint_{\partial\Omega} \psi \frac{\partial h}{\partial n} = \int_{\Omega} h \, \Delta\psi - \int_{\Omega} \psi \, \Delta h$$

and upon substitution of the defined quantities above, using Schwarz's inequality,

$$\int_{\Omega} h^2 = \oint_{\partial\Omega} h \frac{\partial \psi}{\partial n} \leq \left(\oint_{\partial\Omega} u^2 \right)^{1/2} \left[\oint_{\partial\Omega} \left(\frac{\partial \psi}{\partial n} \right)^2 \right]^{1/2}.$$

By means of an *auxiliary problem* (see the discussion and Exercise 5 in the next Pause) one can obtain an inequality

$$\oint_{\partial\Omega} \left(\frac{\partial \psi}{\partial n} \right)^2 \leq k(\Omega) \int_{\Omega} (\Delta\psi)^2$$

for functions ψ vanishing on the boundary $\partial\Omega$, where $k = k(\Omega)$ is a constant that depends on the geometry of Ω and $\partial\Omega$. Accepting the latter fact we have

$$\int_{\Omega} h^2 \leq k \oint_{\partial\Omega} u^2$$

and thus, upon substitution into the original decomposition, the desired *a priori* estimate

$$\int_{\Omega} u^2 \leq c_1 \oint_{\partial\Omega} u^2 + c_2 \int_{\Omega} (\Delta u)^2$$

with $c_1 = k(1 + \varepsilon)$ and $c_2 = (1 + (4\varepsilon)^{-1})\lambda_1^{-2}$, for arbitrary $\varepsilon > 0$. An optimal ε could then be chosen in terms of the two domain constants λ_1 and k.

2. The Triangle Inequality: $\|u + v\| \leq \|u\| + \|v\|$. The meaning of this inequality lies in its name. See Figure 1.6*j* in two-space, interpreting the norm $\|\cdot\|$ as length. As shown for example in advanced calculus or elementary topology, the triangle inequality* is typical of any metric space† and is a property thereof usually expressed as

$$d(x, y) \leq d(x, z) + d(z, y)$$

for any x, y, z in the metric space. Since any set S can be made into a metric space by defining $d(x_1, x_2) = 0$ or 1 as x_1 and x_2 are the same point or not, respectively, this would seem to make the triangle inequality *a priori* very general. However, in partial differential equations the set S is usually a rather carefully chosen vector space and there is required a certain art (or experience) in picking the right metric.

---

* Another name is Minkowsky's inequality.

† A metric space is a set S equipped with a "length" $d : S \rightarrow$ reals that satisfies (i) $d(x_1, x_2) \geq 0$ and $d(x_1, x_2) = 0$ iff $x_1 = x_2$, (ii) $d(x_1, x_2) = d(x_2, x_1)$, and most importantly the triangle inequality, (iii) $d(x_1, x_2) \leq d(x_1, x_3) + d(x_3, x_2)$ A norm $\|u\| = d(u, 0)$ has all these properties and moreover the property that $\|cu\| = |c| \, \|u\|$ so that, for example, $\|-u\| = \|u\|$.

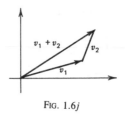

Fig. 1.6*j*

As a metric the triangle inequality above can be expressed as $d(u, -v) \leqq d(u, 0) + d(0, -v)$ and likewise one has in metric notation $d(u, v) \leqq d(u, 0) + d(0, v)$ and in norm notation

$$\|u - v\| \leqq \|u\| + \|v\|.$$

A useful lower bound for $\|u - v\|$ obtained easily (see Problem 1) by addition and subtraction is the following:

2'. Inverse Triangle Inequality ($\Delta_{\leqq}^{-1}$ for short).

$$\|u - v\| \geqq |\,\|u\| - \|v\|\,|.$$

The triangle (and therefore the inverse triangle inequality also) inequality holds more generally than the Schwarz inequality because for the formula all one needs is some vector space with a metric on it, whereas for the latter one needs an inner product. That is, Schwarz's inequality is intrinsically related to the notion of orthogonality. One may describe this situation roughly by saying that the triangle inequality holds for all *Banach* spaces of vectors or functions, whereas the Schwarz inequality is available only in the subclass of Banach spaces called *Hilbert* space.* We will describe the very important concept of Hilbert space in Chapter 2 when discussing Fourier series; but for the moment let us just mention the most important examples; namely, the so-called $L^p(\Omega)$ spaces, whose definition is as follows. Let Ω be an otherwise arbitrary set or domain in n space with the usual (dx) Lebesgue measure on it; then

$$L^p(\Omega) = \left\{ \text{functions } f(x) \;\middle|\; \int_\Omega |f(x)|^p \, dx < \infty \right\}, \qquad 1 \leqq p \leqq \infty.$$

Here f may be a real valued or complex valued function.† For $p = 2$ the space $L^2(\Omega)$ will be seen to be a Hilbert space. The others are Banach spaces. The special

---

* One can introduce a ''semi'' -inner product and a resulting type of Schwarz inequality even in a Banach space. The resulting structure is very useful in describing energy dissipation of evolving systems. See Bonsall and Duncan, *Numerical Ranges I, II* (Cambridge Univ. Press, Cambridge, 1971, 1973) and Antoine and Gustafson, *Adv. in Math.* **41**(1981). See also Gustafson and Rao, *Numerical Range* (Springer, Berlin, 1997) for further information.

† In many books on functional analysis and in some applications it must be complex valued. Here for simplicity we will deal mostly with $f(x)$ real valued. The latter suffices, for example, for many applications in continuum mechanics.

case of $p = \infty$ is in fact the case of a "maximum norm" and is taken to mean $\sup_{\Omega_{a.e.}} |f(x)| < \infty$, where Ω a.e. means "almost everywhere," that is, except on a subset of Lebesgue measure zero.

We will not get into the technicalities of Lebesgue measure and Lebesgue integrable functions here,‡ but will instead make statements such as "think of a sequence of piecewise C^1 functions and define the integral of the limit as the limit of the integrals of the sequence." Most of our functions and data will in fact be the usual smooth Riemann integrable functions familiar from calculus, but it will be important to realize the principal merit of the larger class of Lebesgue integrable functions of being closed under limits.§

Important special cases are $L^2(-\infty, \infty)$ and $L^2(0, \pi)$, for example. Our discussion of Fourier series will be seen to emphasize $L^2(0, \pi)$, $L^2(-\pi, \pi)$, and $L^1(-\pi, \pi)$. For all p the norm is given by

$$\|f\|_p = \left[\int_\Omega |f(x)|^p \, dx \right]^{1/p} .$$

A very important variation on Schwarz's inequality is the Hölder inequality.

2″. Hölder Inequality (for L^p spaces).

$$\int_\Omega |fg| \leq \|f\|_p \|g\|_q, \qquad p^{-1} + q^{-1} = 1$$

Note that for the case $p = 2$, then the so-called conjugate index q has value $q = 2$ also, and the Hölder inequality reduces to the Schwarz inequality

$$\int_\Omega |fg| \leq \|f\|_2 \|g\|_2.$$

Stated another way, the Hölder inequality asserts that if $f \in L^p$ and $g \in L^q$, then $fg \in L^1$, that is, fg is an integrable function, that is, its integral is finite.

3. The Sobolev Inequality.

$$\|u(P)S(P, Q)\|_{L^2(R^3)} \leq (1/2\pi)\|\text{grad } u\|_{L^2(R^3)}$$

Stated another way, for any smooth function u of bounded support, that is, a function $u(x)$ that is zero outside of some sufficiently large sphere, one has for any fixed $y \in R^3$ that

$$\int \frac{u^2(x)}{|x - y|^2} \, dx \leq 4 \int |\text{grad } u|^2 \, dx.$$

---

‡ Among the many excellent references is the book by E. Asplund and L. Bungart, *A First Course in Integration* (Holt, Reinhart, and Winston, New York, 1966). See also the Lebesgue dominated convergence theorem in 1.9.6(3).

§ The instructor may wish to give his or her own qualifications of these remarks at this point.

Here, and for later convenience we have used the notation $x \equiv P = (x_1, x_2, x_3)$, $y \equiv Q = (y_1, y_2, y_3)$ for two points in 3-space.

3'. Why the Atom Does Not Collapse.
The Schwarz and Sobolev inequalities may be used to show now why the atom is stable, that is, why the electron does not just give up and allow itself to be pulled into the nucleus. With complications the argument given below for the hydrogen atom extends to all atoms.

What is to be shown is that the kinetic energy due to the motion of an electron about the nucleus is sufficiently large so as to counterbalance the potential energy pulling the electron toward the nucleus. The kinetic energy is represented by $\frac{1}{2}D(u, u)$ and the potential energy due to the Coulomb potential $1/r$ by $\int (1/r)u \cdot u$. The net energy level is given by λ in the eigenvalue problem for the Schrödinger equation

$$-\tfrac{1}{2}\Delta u - \frac{Z}{r}u = \lambda u,$$

where $u \in L^2(R^3)$ and vanishes sufficiently rapidly at infinity. One may in fact take $u \in C_0^\infty(R^3)$, the infinitely differentiable functions of compact support,* that is, functions u vanishing outside some sufficiently large sphere, since technical considerations allow the passage by approximation to all required u in the domain of the operator. We have written the equation in atomic, or Hartree, units, and Z represents the charge.

Considering first the case of charge $Z = 1$, we obtain, as is our custom by now with elliptic operators, the variational characterization of the energy spectrum values of λ, namely,

$$\lambda = \frac{\frac{1}{2}D(u, u) - \int (1/r)u^2}{\int u^2},$$

by multiplying the equation by u, integrating over $\Omega = R^3$, and employing Green's first identity. If it can be shown that there exists a lower bound for all eigenvalues λ, then the atom is stable. Recall that in the Bohr model (Fig. 1.6k) of the atom the electron is restricted to certain orbits, releasing energy $\lambda_i - \lambda_j$ when dropping down to a closer ith orbit from a jth orbit. The existence of a lower bound for all energies λ thus guarantees a minimal or "ground-state" orbit below which the electron cannot drop.

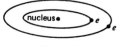

FIG. 1.6k

---

* Also called the "test functions." See Problem 2.9.3 for a proof of the denseness of the test functions C_0^∞ in $L^2(\Omega)$.

Now by the Schwarz, A–G, and Sobolev inequalities, we have for the potential energy term, using the $L^2(R^3)$ norm notation $\|u\| = [\int_{R^3} |u|^2 \, dx]^{1/2}$, that

$$\int \frac{1}{r} u^2 \leq \left\| \frac{1}{r} u \right\| \|u\| \leq 2D(u, u)^{1/2} \|u\| \leq \varepsilon D(u, u) + \frac{1}{\varepsilon} \|u\|^2$$

so that

$$\lambda \geq \frac{\left(\frac{1}{2} - \varepsilon \right) D(u, u) - \frac{1}{\varepsilon} \|u\|^2}{\|u\|^2}.$$

Taking $\varepsilon = \frac{1}{2}$ yields the lower bound for the energies λ,

$$\lambda \geq -2.$$

In the sequence of inequalities used above there was naturally some precision loss. The spectrum for the hydrogen operator may be found exactly (Fig. 1.6l) and consists of the negative eigenvalues $\lambda_n = -1/2n^2$, $n = 1, 2, 3, \ldots$, plus the positive "continuous" spectrum* $[0, \infty)$. The eigenfunctions $u_n(x)$ for the eigenvalues λ_n correspond to "bound states" and to orbits in which the electron is bound about the nucleus, tending eventually to the lowest energy λ_1 and the corresponding "ground state." The positive energies λ in $[0, \infty)$ correspond to "scattering states" in which the kinetic energy wins out and the electrons eventually drift (or speed) away.

We have considered above the case $Z = 1$, but the same result holds replacing $1/r$ by Z/r for higher charges.† In that case one takes $\varepsilon = 1/2Z$ and obtains lower energy bound $-2Z^2$. The true negative eigenvalues are $\lambda_n = -Z^2/2n^2$, $n = 1, 2, 3, \ldots$.

Proofs of the Inequalities

A simple proof of the Schwarz inequality in the real case is provided by the fact that a real quadratic polynomial has real zeros if and only if the discriminant is positive or zero. Letting

$$Q(t) \equiv (u + tv, u + tv) = \|u\|^2 + t(2(u, v)) + t^2 \|v\|^2,$$

$$-\frac{1}{2} \qquad -\frac{1}{8} \quad 0$$

FIG. 1.6l The spectrum of the hydrogen operator.

---

* See Section 2.7 for a further discussion of the continuous spectrum.

† Even allowing artificial elements, the upper bound on realistic charges is 137. See the discussion in *Physics Today*, August (1976), concerning new evidence for the existence of superheavy elements corresponding to charges 116, 124, 126, and 127. Later work leaves this question unsettled.

since $Q(t) = \|u + tv\|^2 \geqq 0$, necessarily the discriminant $B^2 - 4AC \leqq 0$, that is,

$$4(u, v)^2 - 4\|u\|^2\|v\|^2 \leqq 0,$$

yielding the inequality. Clearly u is a multiple of v if and only if equality holds (see Problem 1). The complex case is handled by noting that $(u, v) + (v, u) = 2Re(u, v)$ since $(v, u) = (\overline{u, v})$.

The A–G inequality follows from the simple fact that $(a - b)^2 \geqq 0$, since then

$$2ab \leqq a^2 + b^2.$$

This contains the classical arithmetic mean–geometric mean comparison

$$\sqrt{ab} \leqq \frac{a + b}{2},$$

and by letting $\sqrt{a} = (\sqrt{2\varepsilon})u$ and $\sqrt{b} = (1/\sqrt{2\varepsilon})v$ in the above one has

$$uv \leqq \varepsilon u^2 + (4\varepsilon)^{-1}v^2.$$

Likewise from Schwarz's inequality one has the A–G inequality

$$|(u, v)| \leqq \|u\|\|v\| \leqq \varepsilon\|u\|^2 + (4\varepsilon)^{-1}\|v\|^2.$$

In the Hilbert space case the triangle inequality follows easily from the Schwarz inequality since

$$\|u + v\|^2 = \|u\|^2 + \|v\|^2 + 2Re(u, v)$$
$$\leqq (\|u\| + \|v\|)^2.$$

The inverse triangle inequality verification is left for Problem 1.

There are several ways to prove Hölders inequality. It follows for example from Young's inequality* for two positive numbers a and b

$$ab \leqq \frac{a^p}{p} + \frac{b^q}{q}, \qquad p^{-1} + q^{-1} = 1,$$

which is a generalization of the A–G inequality to the case $p \neq 2$. Putting $a = |f|/\|f\|_p$ and $b = |g|/\|g\|_q$ into Young's inequality, we have upon integrating that

$$\frac{\int |fg|}{\|f\|_p\|g\|_q} \leqq \frac{1}{p}\frac{\int |f|^p}{\|f\|_p^p} + \frac{1}{q}\frac{\int |g|^q}{\|g\|_q^q} = 1$$

and thus Hölders inequality.

Young's inequality can be shown from the rational function

---

* There are inequalities more general than Young's inequality also, roughly describable as corresponding to more general functions than the $Q(t)$ and $R(t)$ above.

$$R(t) = p^{-1}t^p + q^{-1}t^{-q}, \qquad p^{-1} + q^{-1} = 1, \qquad t > 0.$$

$R(t)$ is seen to be strictly convex with a minimum at $t = 1$ by taking two derivatives:

$$R'(t) = t^{p-1} - t^{-(q+1)}, \quad R'(t_0) = 0 \Rightarrow t_0^{p+q} = 1 \Rightarrow t_0 = 1,$$

$$R''(t) = (p - 1)t^{p-2} + (q + 1)t^{-(q+2)}$$
$$= q^{-1}pt^{p-2} + (q + 1)t^{-(q+2)} > 0.$$

Young's inequality then follows from $R(a^{1/q}b^{-1/p}) \geqq R(1) = 1$.

We may show the Sobolev inequality by use of Green's first identity and Schwarz's inequality. Recall that we have stated the Sobolev inequality

$$\left\| \frac{u}{|x - y|} \right\|_2 \leqq 2D(u, u)^{1/2}$$

for any smooth function u that vanishes for sufficiently large x, y being fixed, using here the notation $D(u, u) = \int |\text{grad } u|^2$ as before, the integrals thus being taken over all of 3-space R^3 but in fact only over a bounded domain Ω that supports the given function u (Fig. 1.6m). The calculation is as follows: Let all volume integrals in the following be taken over the interior of an R-sphere containing the support set of the function $u(x)$, and let

$$r = ((x_1 - y_1)^2 + (x_2 - y_2)^2 + (x_3 - y_3)^2)^{1/2} = |x - y|.$$

Then by Green's first identity we have

$$D(u^2, \ln r) = \oint u^2 \frac{\partial}{\partial n}(\ln r) - \int u^2 \, \Delta(\ln r) = \int \frac{u^2(x)}{|x - y|^2},$$

the surface integral vanishing and* $\Delta(\ln r) = -1/r^2$. On the other hand, by (two applications of) Schwarz's inequality, and observing that

$$\sum_{i=1}^{3} \frac{(x_i - y_i)^2}{|x - y|^2} = 1,$$

---

* Some intermediate calculations were the following:

$$\left(\frac{1}{|x - y|}\right)_{x_i} = -\frac{(x_i - y_i)}{|x - y|^3}, \quad (\ln r)_{x_i} = \frac{x_i - y_i}{r^2} = \frac{1}{r}r_{x_i},$$

$$r_{x_i} = \frac{x_i - y_i}{r} \text{ from } r^2 = \sum_{i=1}^{3} (x_i - y_i)^2,$$

$$r_{x_i x_i} = \frac{r^2 - (x_i - y_i)^2}{r^3},$$

$$(\ln r)_{x_i x_i} = \frac{1}{r}r_{x_i x_i} + \left(\frac{1}{r}\right)_{x_i} r_{x_i} = \frac{r^2 - (x_i - y_i)^2 - (x_i - y_i)^2}{r^4}.$$

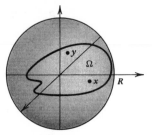

FIG. 1.6m

we have

$$D(u^2, \ln r) = \int \text{grad}(u^2) \cdot \text{grad}(\ln r)$$

$$= 2 \int u \sum_{i=1}^{3} u_{x_i} \frac{(x_i - y_i)}{|x - y|^2}$$

$$\leq 2 \left[\int \int \frac{u^2(x)}{|x - y|^2} \right]^{1/2} D(u, u)^{1/2}$$

which yields the Sobolev inequality.

Problem 1. (a) Prove that one obtains equality in the Schwarz's inequality if and only if u is a multiple of v. (b) Verify the inverse triangle inequality from the triangle inequality and the discussed properties of the norm $\|u\|$.

Problem 2. (a) Consider the EVP

$$\begin{cases} -\Delta u = \lambda u \text{ in } \Omega, \\ u = 0 \text{ on } \partial\Omega, \end{cases}$$

on an arbitrary bounded smooth domain Ω in n-dimensional space, and derive formally the variational characterization of the eigenvalues λ_n. (b) Solve for the λ_n and the corresponding eigenfunctions $u_n(x, y)$ by the method of separation of variables for the case of Ω the square (Fig. 1.6n).

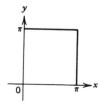

FIG. 1.6n

Problem 3. With respect to the stability result for the Dirichlet–Poisson problem

$$\|u_1 - u_2\|^2 \leqq c_1\|f_1 - f_2\|^2 + c_2\|F_1 - F_2\|^2$$

obtained in the text in terms of L^2 norms, consider the following questions:
(a) How does one go back and forth between inequalities such as

$$a \leqslant c_1'b_1 + c_2'b_2 \quad \text{and} \quad a^2 \leqslant c_1b_1^2 + c_2b_2^2$$

as was done in the discussion there? (b) Obtain a similar stability result in the maximum norms for the same problem. (c) What are your chances of getting directly an *a priori* estimate of the form

$$\int_\Omega u^2 \leqq (\text{const}) \oint_{\partial\Omega} u^2$$

for a given bounded domain Ω and functions u satisfying $\Delta u = 0$ in Ω, $u \in C^2(\Omega) \cap C^1(\overline{\Omega})$?

1.6(2)   Exercises

1. Write down Schwarz's Inequality for the cases (a) Euclidean three space, (b) the Dirichlet integral $D(v) = \int_\Omega |\text{grad } v|^2$, and (c) infinite vectors $x = (x_1, x_2, x_3, \ldots)$.
2. Write down the Triangle Inequality for the cases (a), (b), (c) of Exercise 1.
3. (a) Prove b from (a) or (c) in Exercise 1.
   (b) Verify Schwarz's inequality for

$$x = (4, 5, 6), \quad y = (7, 8, 9).$$

   (c) Are these vectors close or far from pointing in the same direction?

1.6.3.   Convergence Theorems

It will be convenient to later call upon the following convergence theorems when discussing solutions or proposed solutions of certain partial differential equations. Let us first recall three common types of convergence that one encounters not only here but in numerical analysis, signal analysis, and elsewhere in mathematics and applications. We will describe the results on a finite interval $[a, b]$, but they hold equally as well on any more general reasonable bounded closed domain $\overline{\Omega}$ in n-space. Let $u_n(x)$, $n = 1, 2, 3, \ldots$, be a sequence of functions defined on $a \leqq x \leqq b$. Let $u(x)$ be the candidate for limit function.

1. Pointwise Convergence of a Sequence of Functions. For every x in $[a, b]$, for every given $\varepsilon > 0$, there exists $N = N(x, \varepsilon)$ such that $|u_n(x) - u(x)| < \varepsilon$ is guaranteed for all $n > N$. Then $u_n(x) \to u(x)$ pointwise on $[a, b]$. The meaning as

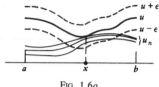

FIG. 1.6*o*

illustrated in Figure 1.6*o* is that the $u_n(x)$ eventually all are contained within the ε-tube about $u(x)$, at least at the point x.

2. Uniform (Pointwise) Convergence of a Sequence of Functions. For every given ε > 0, there exists $N = N(\varepsilon)$ such that $|u_n(x) - u(x)| < \varepsilon$ is guaranteed for all $n > N$. Then $u_n(x) \to u(x)$ uniformly (pointwise) on $[a, b]$. The meaning as illustrated in Figure 1.6*p* is that the u_n eventually all are contained within the ε-tube about u, over the whole interval. In terms of the previous definition of just pointwise convergence, the meaning is that N can be found independent of x. In terms of norms, if we let

$$\|v\| = \max_{a \leq x \leq b} |v(x)|$$

for a continuous function (i.e., $v \in C^0[a, b]$), then uniform convergence of u_n to u can be written as

$$\|u_n - u\| \to 0 \quad \text{as} \quad n \to \infty.$$

3. (Root) Mean Square (RMS) Convergence of a Sequence of Functions. For every given ε > 0, there exists $N = N(\varepsilon)$ such that

$$\left[\int_a^b |u_n(x) - u(x)|^2 \, dx \right]^{1/2} < \varepsilon$$

is guaranteed for all $n > N$. Then $u_n \to u$ in (root) mean square. The meaning as illustrated in Figure 1.6*q* is that the area squared between $u_n(x)$ and $u(x)$ eventually becomes arbitrarily small, although the relative pointwise errors on decreasing subintervals can be rather out of control. In terms of norms, if we let $\|v\|^2 = \int_a^b |v(x)|^2 \, dx$ for any square integrable function (i.e., $v \in L^2(a, b)$), then RMS convergence of u_n to u can be written as

$$\|u_n - u\| \to 0 \quad \text{as} \quad n \to \infty.$$

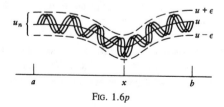

FIG. 1.6*p*

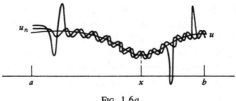

FIG. 1.6q

This is more commonly called L^2 convergence and represents convergence in the Hilbert space $L^2(a, b)$.

The same defintions apply with the u_N (we use capital N rather than lower case n when denoting partial sums, following a standard convention) taken to be the partial sums of an infinite series, namely,

$$u_N(x) \equiv s_N(x) = \sum_{n=1}^{N} u_n(x)$$

and

$$u(x) = \sum_{n=1}^{\infty} u_n(x) = \lim_{N \to \infty} s_N(x)$$

since an (infinite) series is always defined as the limit of the partial sum (finite) series s_N. Usually in an infinite series the "lim" has been tacitly assumed to mean the existence of the pointwise limit for each x in the sense of pointwise convergence. However, it could just as well have been taken to mean "lim" in the third sense above, and that is exactly what we do when discussing general Fourier series.

Nature of the Limit Function. Given a sequence $u_n(x)$ but possibly not given in advance a candidate limit function $u(x)$, the so-called Cauchy criteria for convergence are very useful. In this criteria one guarantees an arbitrarily small $u_n - u_m$ for sufficiently large n and m. It then usually follows* that there exists a limit function u to which the sequence u_n converges.

Thus (1) pointwise convergence becomes $|u_n(x) - u_m(x)| < \varepsilon$ for all n, m $\geq N$; (2) uniform (pointwise) convergence becomes $|u_n(x) - u_m(x)| < \varepsilon$ for all

---

* To mathematicians this is an almost religious point. In fact, one could describe the situation somewhat irreverently by saying that if the desired limit $u(x)$ does not explicitly exist, the mathematician will define it to exist. The resulting system will then be said to be "complete," although it may really be complete only because it has been made so by the addition of idealized elements to "fill up any pinholes." The meaning of the word "complete" should therefore always be "pinned down" when encountered. Completeness for pointwise convergence (1) above follows from the axioms in the construction of the real number system; roughly, that the limit of a Cauchy sequence of irrational numbers will be nothing new. Completeness for uniform convergence (2) above is described by mathematicians by saying that $C^0[a, b]$ in the maximum norm is a Banach space. Completeness for RMS convergence (3) becomes the statement that $L^2(a, b)$ is a Hilbert space, and requires the notion of Lebesgue integral. These notions will be further discussed in Section 2.3.

$n, m \geq N$ and for all x in $[a, b]$; and (3) RMS (L^2) convergence becomes $\int_a^b |u_n(x) - u_m(x)|^2 \, dx < \varepsilon$ for all $n, m \geq N$. Simple examples (e.g., the $u_n(x) = x^n$ of Problem 3, Section 1.5.3) show that the pointwise limit of a sequence of continuous functions need not be continuous. However, a sequence $u_n(x)$ that satisfies the Cauchy criterion for uniform convergence will have a continuous limit function, as is shown in advanced calculus. Likewise, the Lebesgue integration theory allows sufficiently general integrable functions so as to guarantee an L^2 limit function $u(x)$ for every sequence $u_n(x)$ converging in L^2 in the Cauchy criterion sense.

Implications between the Three Types of Convergences. In brief, one has the diagram:

The two heavy implications follow easily. By definition uniform pointwise convergence (2) $\Rightarrow$ pointwise convergence (1), and by use of Schwarz's inequality one has (2) $\Rightarrow$ (3). The two light implications (called Dini tests) hold in the special case in which $u_N \equiv s_N$ is the Nth partial sum of the Fourier (trigonometric) series for a function $u \equiv f$ possessing the indicated continuity properties. For a further discussion of such Dini tests, which guarantee pointwise convergence, see Problem 1.9.6(2).

The point here is that a Fourier series representation of a given function f will always converge in the third (L^2) sense (as will be shown in Chapter 2), but will converge pointwise to f only in special cases and under additional hypotheses such as those in the Dini tests (Problem 1.9.6(2)).

By way of further illustration and emphasis of this last-mentioned point, recall as discussed in Section 1.5.1 that when using the separation of variables method for solving a partial differential equation, we wrote the given data function f as a Fourier series

$$f \cong \sum c_n \varphi_n,$$

where c_n was (f, φ_n) in the appropriate inner product for the application. For example, and to be specific, on the interval $0 \leq x \leq \pi$ and for $f \in L^2(0, \pi)$ we employed the Fourier sine series expansion of f, in which we had

$$\varphi_n(x) = \left(\frac{2}{\pi}\right)^{1/2} \sin nx$$

and

$$c_n = \int_0^\pi f(s) \varphi_n(s) \, ds,$$

$n = 1, 2, 3, \ldots$. Letting $u_N(x) \equiv s_N(x) = \Sigma_{n=1}^{N} c_n \varphi_n(x)$ be the Fourier series truncated to the partial sums consisting of the first N terms therein, the question then becomes, in what senses is $f = \Sigma_{n=1}^{\infty} c_n \varphi_n$? The Fourier theory to be discussed in Section 2.3 will establish that one always has $s_N \to f$ in the L^2 sense, that is, in convergence sense (3) above. If, moreover, $f(x)$ is Hölder continuous at x_0,* then (by the Dini test, see Problem 1.9.6) one has $s_N \to f$ pointwise at x_0, that is, in convergence sense (1) above. If, moreover, f is continuously differentiable on the interval, that is, $f \in C^1[0, \pi]$, and also satisfies the (necessary) condition that $f(0) = f(\pi)$, then $s_N \to f$ uniformly pointwise on the interval.

Problem 1. Verify the implication (2) $\Rightarrow$ (3) in the above diagram.

Problem 2. (a) Determine the Fourier sine series for $f(x) = x$ on $[0, \pi]$. (b) Discuss the three convergences for $u_n(x) \equiv s_n(x)$ the partial Fourier sums.

Problem 3. Discuss the three types of convergence for the sequence $u_n(x) = x^n$ on $[0, 1]$, $n = 1, 2, 3, \ldots$.

Sufficient Conditions for the Three Types of Convergences. As concerns pointwise convergence (1) of a series $\Sigma \, u_n$ of constant terms u_n (or of a series $\Sigma \, u_n(x)$ with x fixed), one obtains in calculus a number of tests. Let us recall here for our later use the d'Alembert or *ratio test*: if there exist $\rho < 1$, N such that $|u_{n+1}/u_n| < \rho$ for all $n \ge N$, then the given constant term series $\Sigma \, u_n$ converges and moreover converges absolutely. Belaboring the point, that means there are numbers A and B so that $|A - s_N| < \varepsilon$ and $|B - r_N| < \varepsilon$ for any given $\varepsilon > 0$ whenever $n \ge N_A$ and $n \ge N_B$, respectively, where $s_N = \Sigma_{n=1}^{N} u_n$ and $r_N = \Sigma_{n=1}^{N} |u_n|$ denote the converging partial sums of the series.

The most important test for uniform convergence (2) is the *Weierstrass M-test*: Given the terms $u_n(x)$ of a series defined on an interval $a \le x \le b$, if one can find a dominating series of constant terms $\Sigma \, M_n$ in the sense that $|u_n(x)| \le M_n$ on $[a, b]$ for each $n = 1, 2, 3, \ldots$ and such that $\Sigma \, M_n$ converges,† then $\Sigma \, u_n$ converges uniformly pointwise on $[a, b]$. It should be remarked that one needs only to dominate a tail-end of the series $\Sigma \, u_n(x)$ since any finite number of first terms do not matter in convergence questions. Also one need not be on an interval, any point set will do.

---

* A function f is said to be Hölder continuous of exponent α (or satisfy a Lipschitz condition of order α) at point x_0 in an interval $[a, b]$ if $|f(x) - f(x_0)| \le M_\alpha |x - x_0|^\alpha$ for all x nearby and in $[a, b]$, for some $\alpha > 0$ and $M_\alpha > 0$. Special cases are $\alpha = 1$ in which f is called Lipschitz continuous, and $\alpha = M_\alpha = 1$ in which f is called contractive (or better, nonexpansive). The Hölder continuity condition is seen to be a restriction on the slope of f at x_0, and places f somewhere between continuous and differentiable (provided it is not already the latter) there.

† Thus, it is convenient to be familiar with a few common convergent series of constant terms to try for comparison with the u_n, in employing this test.

A full study of the L^2 type of convergence (3) requires a knowledge of the Lebesgue integral and elementary measure theory. For our purposes the most important fact, as will be further clarified in Section 2.3, is that a square integrable function $f \in L^2[a, b]$ possesses the property that its Fourier series partial sums $s_n(x) = \sum_{k=1}^{n} c_k \varphi_k(x)^*$ will always converge in the L^2 sense (root mean square sense (3) above) to f, even if the function $\sum_{k=1}^{\infty} c_k \varphi_k(x)$ defined by the series is not equal to f at every point. Thus when treating a PDE, and in assuming Fourier expansions for any given data f and for a solution u, one is on rather safe ground since all reasonable functions will be square integrable. Only afterward does one need to see how the "pointwise" behavior goes.

Differentiation Term by Term.† The following criterion for differentiating a series term by term will be of fundamental importance to us in verifying the validity of the separation of variables method.

Theorem D

$(\sum u_n(x))' = \sum u_n'(x)$ provided that the functions $u_n(x)$ are continuously differentiable (i.e., $C^1[a, b]$), that the series $\sum u_n(x)$ converges pointwise, and that the series $\sum u_n'(x)$ of derivatives converges uniformly pointwise.

The prime in Theorem D can be an ordinary or a partial derivative, since the proof is essentially the same.

Differentiation under the integral may be regarded as a generalization of Theorem D from sums to integrals. For example, we recall the so-called Leibnitz formula

$$\frac{d}{dy} \int_{a(y)}^{b(y)} f(x, y) \, dx = f(b(y), y)b'(y) - f(a(y), y)a'(y)$$

$$+ \int_{a(y)}^{b(y)} \frac{\partial}{\partial y} f(x, y) \, dx$$

where it is sufficient, for example, that a', b', and $\partial f/\partial y$ be continuous.

To provide a little feeling for how one may generalize things to more general functions by means of the Lebesgue integral let us mention the following generalization of Theorem D. One has $(\sum u_n)' = \sum u_n'$ "almost everywhere" provided that the u_n have "locally bounded variation" and that $\sum$ (total variation of u_n) converges. For example, this is the case if the $u_n(x)$ are continuous increasing functions with pointwise convergent series $\sum u_n(x)$.

Similarly, generalizations of differentiation under the integral may be obtained by use of the *Fundamental Lebesgue Dominated Convergence Theorem*,‡ which

---

* Provided that the φ_k, $k = 1, 2, 3, \ldots$ form what is called a maximal orthonormal set, as will be discussed further in Section 2.3.

† See any advanced calculus book for a proof of Theorem D and other related facts concerning uniform pointwise convergence and Leibnitz formulae.

‡ See Problem 1.9.6(3) for a few more details about this.

states that if the $u_n(x)$ converges pointwise (a.e.) to $u(x)$, are integrable and uniformly bounded $|u_n(x)| \leqq M(x)$ by some integrable function $M(x)$, then $\int u_n(x)\,dx \to \int u(x)\,dx$. Letting $u_n(y) = f(x, y + \Delta_n y) - f(x, y)$, one then can obtain $(d/dy)\int f(x, y)\,dx$ by the LDC Thm. via

$$\frac{d}{dy}\int f(x, y)\,dx = \lim_{n\to\infty}\int \frac{u_n(y)}{\Delta_n y}\,dx.$$

Integration Term by Term. Integration term by term is no problem. As noted above the Lebesgue dominated convergence theorem states that

$$\int u_n \to \int u$$

under very general pointwise convergence assumptions about $u_n \to u$. An advanced calculus version commonly used is the following: If $\Sigma\, u_n(x)$ converges uniformly on an interval $[a, b]$ and the u_n are continuous there (so that they are certainly integrable), then

$$\int_a^b \left(\sum_{n=1}^\infty u_n(x)\right) dx = \sum_{n=1}^\infty \left(\int_a^b u_n(x)\,dx\right).$$

The sequence version is: If $s_N(x)$ are continuous and converge uniformly to $f(x)$ on $[a, b]$, then $\int_b^a s_N(x)\,dx$ converges to $\int_a^b f(x)$.

We have considered in this section three important mathematical tools used in studying partial differential equations: the divergence theorem, inequalities, and convergence theorems.

The divergence theorem as discussed above is just one aspect of a mathematical subject called potential theory that now stretches from delicate problems in probability theory to a rather general theory of singular integral operators. In Problem 1.9.5(2) one will find a brief discussion of the question of exceptional domains on which the use of the divergence theorem may not be justified. For most of the rest of this book we will not be concerned with such domains.

The study of inequalities is a subject in its own right.*

Problem 1.9.6 contains a few additional historical remarks about the pointwise convergence of Fourier series, a proof of a Dini test for pointwise convergence, a mentioning of the so-called Gibb's effect in which the partial Fourier sums tend to pile up above and below a discontinuity in the function being so represented, and a few more remarks about the Lebesgue convergence theorems.

---

* See, for example, the books G. H. Hardy, J. E. Littlewood, and G. Polya, *Inequalities*, 2nd ed. (Cambridge Univ. Press, London, 1952), E. F. Beckenbach and R. Bellman, *Inequalities* (Springer, New York, 1971), and D. S. Mitrinovich, *Analytic Inequalities* (Springer, New York, 1970).

1.6(3)   Exercise

1. (a) Find a solution to the equation

$$\begin{cases} u'(x) = u(x), & x > 0, \\ u(0) = 1, \end{cases}$$

as a power series $u(x) = \sum_{n=0}^{\infty} a_n x^n$ using termwise differentiation.

   (b) Find a solution to the heat equation problem

$$\begin{cases} u_t - u_{xx} = 0, & -\infty < x < \infty, \quad t > 0, \\ u(x, 0) = x^2, & -\infty < x < \infty, \end{cases}$$

as a series $u(x, t) = \sum_{n=0}^{\infty} \dfrac{1}{n!} \dfrac{\partial^n u}{\partial t^n} (x, 0) t^n.$

   (c) Find a solution to the heat problem

$$\begin{cases} u_t - u_{xx} = 0, & -\infty < x < \infty, \quad t > 0, \\ u(x, 0) = \sin x, & -\infty < x < \infty, \end{cases}$$

as a series $u(x, t) = \sum_{n=0}^{\infty} \dfrac{1}{n!} \dfrac{\partial^n u}{\partial t^n} (x, 0) t^n.$

2. (a) Integrate $e^x = 1 + x + x^2/2! + \cdots$ termwise and compare to $\int_0^x e^s \, ds$

   (b) Use the ratio test to show $u(x) = \sum \dfrac{1}{n} x^n$ converges absolutely for $|x| < 1$.

   (c) Conclude that it is possible to differentiate $\sum_{n=1}^{\infty} 1/n \, x^n$ term by term and therefore

$$u(x) = -\ln(1 - x) \qquad \text{for } |x| < 1.$$

3. (a) Compute the gradient in polar ($n = 2$) coordinates.

   (b) Recalling that the Laplacian Δ_2 in polar coordinates is given by

$$\Delta_2 u = u_{rr} + r^{-1} u_r + r^{-2} u_{\theta\theta},$$

verify that $\Delta_2(r^m) = m^2 r^{m-2}$ and that $\text{grad}(r^m) = m r^{m-1}(\cos \theta, \sin \theta)$.

   (c) Let $u = 1/(1 - r) = 1 + r + r^2 + \cdots + r^m + \cdots$ for $r < 1$ in 2 dimensions. Verify the Neumann compatability relation

$$\oint_{r=R<1} \frac{\partial u}{\partial n} \, ds = \int_{r \leq R < 1} \Delta_2 u \, dx$$

by integrating the series expansions term by term.

SECOND PAUSE: EXAMPLES, EXPLANATIONS, EXERCISES

The interplay between the divergence theorem and inequalities is a fascinating and essential part of the theory and practice of partial differential equations. Let us return to the last example of the First Pause, the plate problem now considered for partial differential equations.

Example

Show variationally that the first buckling eigenvalue Λ_1 of a clamped (elastic) plate of arbitrary shape is bounded below by the second vibration eigenvalue λ_2 of the fixed membrane of the same original shape.

Solution. By original shape (or perhaps better, configuration) we mean here the domain Ω. The only smoothness we need is that Ω admit the use of the divergence theorem on functions defined over it. We recall the two problems of interest

plate:
$$\begin{cases} -\Delta^2 u = \Lambda \Delta u & \text{in } \Omega, \\ u = 0 & \text{on } \partial\Omega, \\ \partial u/\partial n = 0 & \text{on } \partial\Omega, \end{cases}$$

and

membrane:
$$\begin{cases} -\Delta w = \lambda w & \text{in } \Omega, \\ w = 0 & \text{on } \partial\Omega. \end{cases}$$

We also recall that at the end of the proof for the one-dimensional case in the previous Pause, we needed the one-dimensional version of the following inequality:

$$\int_\Omega |\text{grad } u|^2 \Big/ \int_\Omega u^2 \leq \int_\Omega |\Delta u|^2 \Big/ \int_\Omega |\text{grad } u|^2.$$

This more general multidimensional version follows as before from a combination of Green's first identity and Schwarz's inequality:

$$\int_\Omega |\text{grad } u|^2 = -\int_\Omega u\Delta u \leq \left(\int_\Omega u^2\right)^{1/2} \left(\int_\Omega (\Delta u)^2\right)^{1/2}$$

Note that we used only that u vanished on $\partial\Omega$.

Turning to the problem at hand, it is easier and physically appropriate to consider first the case of two space dimensions. The generalization to an arbitrary number of space dimensions then follows easily. To proceed, we mimic the one-dimensional proof.

First, we see that the eigenvalues are characterized by

$$\Lambda_n = \inf \int_\Omega |\Delta u|^2 \Big/ \int_\Omega |\text{grad } u|^2,$$

$$\overbrace{\text{all } u \perp u_1, \ldots, u_{n-1},}$$
$$u = \partial u/\partial n = 0 \text{ on } \partial\Omega.$$

and

$$\lambda_n = \inf \int_\Omega |\text{grad } w|^2 \Big/ \int_\Omega w^2$$

$$\overbrace{\text{all } w \perp w_1, \ldots, w_{n-1},}$$
$$w = 0 \text{ on } \partial\Omega.$$

These variational characterizations follow immediately by multiplying the differential equations by u and w, respectively, integrating over Ω and using Green's identities.

Dropping the subscripts, let $u \equiv u_1$ and $w \equiv w_1$ denote the first eigenfunctions, respectively, and consider the trial function

$$v = au_x + u$$

choosing a so that $v \perp w$. The special cases in which $u_x \perp w$ or $u \perp w$ already can be handled separately. Clamped plate eigenfunctions have the property that their partial derivatives all vanish on $\partial\Omega$. Leaving the verification of this as an exercise, we see that $v = 0$ on $\partial\Omega$ and hence we arrive at the inequality

$$\lambda_2 \leqq \frac{D(v)}{\int_\Omega v^2} = \frac{a^2 D(u_x) + D(u)}{a^2 \int_\Omega (u_x)^2 + \int_\Omega u^2} \leqq \frac{D(u_x) + a^{-2}D(u)}{\int_\Omega (u_x)^2 + a^{-2} \int_\Omega u^2}.$$

The cross terms have vanished as before because

$$D(u_x, u) = \int_\Omega (u_{xx}u_x + u_{xy}u_y) = \frac{1}{2}\int_\Omega ((u_x)^2)_x + ((u_y)_x^2) = 0$$

by the divergence theorem and the boundary conditions.

Clearly, with a second trial function $v = bu_y + u$ with b chosen so that $v \perp w$, one arrives at the same inequality with a replaced by b and u_x by u_y. Adding*
produces

$$\lambda_2 \leqq \frac{D(u_x) + D(u_y) + (a^{-2} + b^{-2})D(u)}{\int_\Omega ((u_x)^2 + (u_y)^2) + (a^{-2} + b^{-2})\int_\Omega u^2}$$

$$= \frac{\int_\Omega (\Delta u)^2 \bigg/ D(u) + (a^{-2} + b^{-2})}{1 + (a^{-2} + b^{-2})\int_\Omega u^2 \bigg/ D(u)} \leqq \Lambda_1$$

where in the last inequality we use the inequality given at the beginning of this example.

Exercises

1. Verify the variational characterizations of Λ_n and λ_n given above.
2. Verify that the eigenfunctions of a clamped plate have vanishing gradient on $\partial\Omega$. Draw a picture of a square clamped plate.
3. First for the clamped rod, then for the clamped plate, consider the special cases in which $u_x \perp w$ or $u \perp w$. For the general case rewrite the proof with trial vectors of the form $v = u_x + au$, $v = u_y + bu$. Discuss the pros and cons about which is preferred. Write proofs for the general n-dimensional case, using $v_i = a_i u_{x_i} + u$, $v_i = u_{x_i} + b_i u$, using if you like the summation convention mentioned in Section 1.1.
4. Compare the problem

$$\begin{cases} -\Delta^2 u = \Lambda \Delta u & \text{in } \Omega, \\ u = 0 & \text{on } \partial\Omega, \\ \Delta u = 0 & \text{on } \Omega. \end{cases}$$

   Draw a picture of this kind of plate.
5. Derive the inequality for functions $\psi = 0$ on $\partial\Omega$,

$$\int_{\partial\Omega} \left(\frac{\partial\psi}{\partial n}\right)^2 \leqq k(\Omega) \int_\Omega (\Delta\psi)^2,$$

---

* That $\lambda \leqq \alpha_i/\beta_i$ for $i = 1, \ldots, m$ and β_i all positive implies the inequality $\lambda \leqq (\alpha_1 + \cdots + \alpha_m)/(\beta_1 + \cdots + \beta_m)$ is immediate from adding the $\lambda\beta_i \leqq \alpha_i$.

used in the preceding Section 1.6.2 to establish a stability bound for the Dirichlet–Poisson Problem. *Hint:* consider the Steklov eigenvalue problem

$$\begin{cases} \Delta^2\psi = 0 & \text{in } \Omega, \\ \psi = 0 & \text{on } \partial\Omega, \\ \Delta\psi = \mu\partial\psi/\partial n & \text{on } \partial\Omega. \end{cases}$$

The stability constant k is related to μ_1.

Just as the combined use of the divergence theorem and inequalities is an essential part of the study of partial differential equations, so is the use of inequalities in convergence arguments. Our main interest in the tools of convergence for partial differential equations is to know that the separation of variables solutions indeed converge. A full discussion of this depends on a knowledge of the convergence of general Fourier series and is postponed until later, Section 1.9.6 and in several sections of Chapter 2.

An important* tool in demonstrating the pointwise convergence of general Fourier series expansions is the following theorem or variations of it.

Example

The Riemann–Lebesgue Lemma states that the Fourier coefficients $c_N = (h, \phi_N)$ $= \int_a^b h(x)\phi_N(x)\,dx$ tend to zero in the limit as $N \to \infty$, provided that the ϕ_N are an orthonormal set, $N = 1, 2, 3, \ldots$, with the additional property of being uniformly bounded. It is important for applications that it be demonstrated for all integrable functions h, i.e., $h \in L^1(a, b)$.

Solution. We recall the hypotheses:

$$\{\phi_N\}_{N=1}^\infty \text{ orthonormal, i.e., } \int_a^b \phi_N\phi_M\,dx = \begin{cases} 1 & N = M \\ 0 & N \neq M \end{cases},$$

$$\phi_N \text{ uniformly bounded, i.e., } |\phi_N(x)| \leq B < \infty, \text{ all } x, \text{ all } N,$$

$$h \in L^1(a, b), \text{ i.e., } \int_a^b |h(x)|\,dx < \infty.$$

The interval (a, b) is assumed bounded.

All Lebesgue integrable functions $h \in L^1(a, b)$ have the property that they are the limits of their range-truncated values. This property follows easily from their definition as limits in $L^1(a, b)$ of characteristic functions.†

All Lebesgue square integrable functions $h \in L^2(a, b)$ already have the desired property that their Fourier coefficients $c_N \to 0$ as $N \to \infty$. The proof of this is given as an Exercise.

---

* Named for the two fathers of integration theory.

†A characteristic function $\chi_S(x)$ of a set S has the values $\chi_S(x) = 1$ when $\chi \in S$, $\chi_S(x) = 0$ otherwise.

That $c_N \to 0$ as $N \to \infty$ follows easily from these two facts. Let χ_C denote the characteristic function of the subset of (a, b) on which $|h(x)|$ does not exceed the value C. Then the desired result follows immediately from

$$\underbrace{\left| \int_a^b h\phi_N \, dx - \int_a^b h\chi_C\phi_N \, dx \right|}_{c_N} \underbrace{}_{\substack{\to 0 \\ \text{as } N \to \infty}} \leq B \underbrace{\int_a^b |h - h\chi_C| \, dx,}_{\substack{\to 0 \\ \text{as } C \to \infty}}$$

noting that $h\chi_C$ is square integrable.

Exercises

6. Show that $L^2(a, b) \subset L^1(a, b)$ when (a, b) is a bounded interval.
7. Show that for $h \in L^2(a, b)$ and $\{\phi_N\}_1^\infty$ any orthonormal set in $L^2(a, b)$, $c_N = \int_a^b h\phi_N \, dx \to 0$ as $N \to \infty$.
8. Show that for $\{\phi_N\}$ the classical Fourier trigonometric expansion functions $\sin Nx$, $\cos Nx$, $N = 1, 2, \ldots$, in $L^2(-\pi, \pi)$, the hypotheses for the Riemann–Lebesgue Lemma are satisfied. Also show it for the functions $\phi_N(x) = \sin((N + \frac{1}{2})\theta)$ when normalized in $L^2(-\pi, \pi)$. The latter sequence is used importantly in the Dini convergence criteria of Section 1.9.6.

The last example and exercise in this second Pause illustrate some other important techniques useful in conjunction with the divergence theorem, inequalities, and Fourier series, in treating questions of partial differential equations.

Example

There is always a positive lower bound λ_0 for the first membrane eigenvalue λ_1 of any bounded domain Ω. This fact is sometimes called Friedrich's Theorem or Poincare's Lemma. Show how to obtain it in terms of an enclosing n-cube.

Solution. As the question is coordinate free, we may for convenience place Ω within the n-cube $(0, l) \times \cdots \times (0, l)$. For any admissible function u vanishing on $\partial\Omega$ we let $u = 0$ outside Ω. We have

$$u^2(x, \ldots) = \left(\int_0^x u_x(s, \ldots) \, ds \right)^2$$

$$\leq \left(\int_0^x ds \right) \left(\int_0^x u_x^2(s, \ldots) \, ds \right) \leq l \int_0^l u_x^2(s, \ldots) \, ds$$

by the fundamental theorem of calculus and Schwarz's inequality. Integrating over the first coordinate x,

$$\int_0^l u^2(x, \ldots)\, dx \leq l^2 \int_0^l u_x^2(x, \ldots)\, dx$$

followed by the integration over the remaining coordinates,

$$\int_0^l \cdots \int_0^l u^2\, dx_1 \cdots dx_n \leq l^2 \int_0^l \cdots \int_0^l u_{x_1}^2\, dx_1 \cdots dx_n$$

$$\leq l^2 \int_0^l \cdots \int_0^l (u_{x_1}^2 + \cdots + u_{x_n}^2)\, dx_1 \cdots dx_n$$

and hence

$$\lambda_1 = \inf \frac{D(u)}{\int_\Omega u^2} \geq \lambda_0 = l^{-2} > 0.$$

This is an example of a nonsharp bound. For one dimension and $\Omega = (0, \pi)$ we have $\lambda_1 = 1$ and $\lambda_0 = \pi^{-1}$. For Ω the square $\lambda_1 = 2$ whereas $\lambda_0 = \pi^{-2}$ and the bound gets worse as dimension goes up. For existence proofs one can sometimes get by with the existence of such a bound no matter how poor. For explicit applications one tries to sharpen these bounds.

Exercises

9. (a) The isoperimetric inequality, known already to the Greeks, states that of all plane curves of given perimeter the circle encloses the largest area. Thus for any such closed curve one has the lower bound

$$L^2 \geq 4\pi A$$

in terms of the enclosed area. As equality is obtained for the circle, $A = \pi r^2$ and $L = 2\pi r$, this bound is sharp. Prove this inequality by use of Fourier series.

(b) The following type of inequality can sometimes be useful when working on arcs in the plane. Show that

$$\sqrt{\left(\int_a^b \dot{x}(t)\, dt\right)^2 + \left(\int_a^b \dot{y}(t)\, dt\right)^2} \leq \int_a^b \sqrt{(\dot{x}(t))^2 + (\dot{y}(t))^2}\, dt$$

for any continuously differentiable $x(t)$ and $y(t)$ parametrically describing a curve in the plane.

(c) Sharpen the Friedrich's lower bound above. Also obtain it in terms of an enclosing sphere. Show why (b) won't work in (a).

1.7 SOME PHYSICAL CONSIDERATIONS (AND EXAMPLES)

Partial differential equations are created by scientists through mental perception of physical reality. Thus most important partial differential equations, although mathematical quantities in their own right, are intimately linked to physical considerations.

It follows that an understanding of the latter can aid in an understanding of the former, and that a better understanding of the latter is often a necessary ingredient for a better understanding of the former. Many successes in the many and varied applications of mathematics have required and finally have been based upon a full understanding of the intricacies of the particular application under consideration.

In this section we wish to touch briefly on this aspect of the study of partial differential equations. In (1) *Three Physical Techniques* we illustrate the use of conservation principles, linearization assumptions, and perturbation techniques, in a derivation and consideration of the heat equation. In (2) *Three Physical Settings (and Examples)* we present and describe a number of important partial differential equations from continuum (and classical) mechanics, statistical mechanics, and quantum mechanics. In (3) *Unsolved Problems* it is our aim to make it clear, albeit briefly, that a number of important questions involving partial differential equations describing fundamental physical problems remain unresolved.

1.7.1. Three Physical Techniques

There are of course many techniques used in the formulation and analysis of physical problems. Full competence in the use of just a few of them requires experience over a significant part of a lifetime. Here we wish to introduce, in the context of our study of partial differential equations, three important physically motivated techniques, namely:

Conservation (e.g., of energy) principles

Linearization (e.g., of the equation) assumptions

Perturbation (e.g., of a simpler operator) methods

To illustrate them we restrict attention to the heat equation.

It should, however, be emphasized that conservation principles apply as well to conservation of mass, momentum, and many other physical quantities. Also linearization can be imposed in a variety of ways, such as the dropping of higher order nonlinear terms, in the form of quantization of a field, and so forth. Perturbations can be singular or regular, linear or nonlinear; one may need to consider perturbations of the domain Ω; and so on. In each use of any of these techniques one must become familiar with the physical application in order to ascertain what is needed.

The usual derivation of the heat equation goes as follows. One assumes the *principle of conservation of energy* in writing a heat-balance equation (recall that heat is energy in transit) in the form

$$\frac{d}{dt} \int_{\Omega_\varepsilon} U(x, y, z, t, u(x, y, z, t)) \, dx \, dy \, dz = \int_{\partial\Omega_\varepsilon} k(x, y, z, t, u(x, y, z, t)) \frac{\partial u}{\partial n} \, ds$$

(1)

where Ω_ε is any infinitesimal portion (subdomain) of a given body Ω (domain) in question (Fig. 1.7a), U denotes the energy per unit volume,* u the temperature at the point (x, y, z) at time t, and k the coefficient of thermal conductivity for the material in the body. Equation (1) states that the rate of change of the energy in Ω_ε equals the net energy flux through the boundary $\partial\Omega_\varepsilon$, whether heat be flowing in or out or both. The units of the coefficient of thermal conductivity k are (cal)/(sec)(cm)(deg) and Equation (1) thus consists of two equivalent characterizations of the number of calories† per second flowing in or out of Ω_ε.

Assuming that the total energy function U is C^1 in all of its arguments, the left-hand side of Equation (1) may be written with the derivative under the integral. The heat flux on the right-hand side of (1) is transformable‡ by the divergence theorem to a volume integral provided that $\partial\Omega_\varepsilon$ is not too irregular and that the coefficient of thermal conductivity k is at least piecewise continuously differentiable. Thus (1) becomes

$$\int_{\Omega_\varepsilon} [U_t + U_u u_t - ((ku_x)_x + (ku_y)_y + (ku_z)_z)] \, dx \, dy \, dz = 0.$$

Assuming the temperature function u to be C^2 so that the integrand is continuous, by calculus then in fact the integrand must be identically zero, since otherwise one obtains a nonzero integral about any point where the integrand might be nonzero by taking Ω_ε (which was arbitrary) sufficiently small about that point. Hence one arrives at the partial differential equation

$$U_t + U_u u_t - (ku_{,i})_{,i} = 0 \text{ in } \Omega,$$

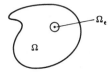

FIG. 1.7a

---

* The postulating of an internal energy function U depends upon a presumption in the first law of thermodynamics of path independence in changing temperatures along adiabatic (i.e., roughly, "noninteracting") paths, a presumption supported by but still lacking complete experimental verification.

† Recall that a calorie is defined to be the amount of heat whose absorption by one gram of water at constant atmospheric pressure gives a temperature rise from 14.5 to 15.5 degrees centigrade.

‡ This is in fact a physical statement of the divergence theorem, namely, that a surface (temperature) flux measures the (heat) quantity diverging from a (cooling) body. Similar interpretations of the divergence may be found in fluid dynamics and elsewhere.

which we have written in the summation convention for convenience.*

Assuming that the basic composition of the body is not going to change over the time interval in question, it is very reasonable to assume that the internal energy function of the body will also be time invariant, so that the equation becomes

$$U_u u_t - (k u_{,i})_{,i} = 0 \text{ in } \Omega.$$

Now one customarily imposes a *linearizing assumption* (2) on the equation by assuming that U_u and k are constant, which brings one (after normalizing the space variable units by the constant $k^{-1}U_u$) to the usual linear heat equation

$$u_t - \Delta u = 0 \text{ in } \Omega.$$

We wish however to belabor† this point and to confuse ourselves a bit with some facts. The skeptical reader should sit in a cold bath and then run some hot water in; the linear equation predicts (see Problem 1) an infinitely rapid speed of heat propagation.

The coefficient of thermal conductivity k is greatly affected in some materials by impurities therein. Although in some applications (e.g., metals formed via catalysts) this may be very important, let us go on and suppose that the material in question is pure and homogeneous so that k has no dependence on the point (x, y, z) in Ω. Moreover, we assume no fatigue factors, changes in crystalline make up, and so forth, that k has no time dependence.

Let us make the same assumptions for U_u. However, to understand better the content of such assumptions for U_u, let us pause to note that in the absence of any external work being done on the system in question, from the first law of thermodynamics‡

$$\Delta(\text{heat}) = \Delta(\text{energy}) + \Delta(\text{work})$$

and the definition of specific heat (using volume here rather than mass)

$$c = \frac{\Delta(\text{heat})}{\Delta(\text{temperature})} \text{ per unit volume}$$

we see that the quantity U_u may be regarded as the coefficient of specific heat c for the material in question. Thus the assumptions for U_u are that the specific heat

---

* And clearly the above argument holds in any dimension, although the case $n = 3$ is the most interesting one physically.

† This seems educational. The heat equation is, roughly speaking and especially in the extreme or critical temperature ranges, the "least linearizable" of the three basic types of equations.

‡ One of the earliest to comprehend that heat is related to energy and work and is not a substance was Benjamin Thompson, an American later to become Count Rumford of Bavaria, who in 1798 attributed the increased temperatures of brass residues from cannon borings to the work being done on the system. Not until Joule's work 50 years later was there any really convincing experimental evidence for the first law of thermodynamics.

c, which in general is the measure of the heat capacity of a given system, has no space or time dependence for the body Ω.

Under these assumptions on thermal conductivity and specific heat for the material, the (nonlinear) heat equation as derived above may be written as

$$u_t = \frac{1}{c(u)}\,(k(u)u_{,i})_{,i} \text{ in } \Omega. \tag{2}$$

The degree of nonlinearity of this equation depends on how much the thermal conductivity $k(u)$ and specific heat $c(u)$ vary with the temperature u. Experiments over the years have determined this dependence quite accurately, as summarized in Figures 1.7b–1.7e.

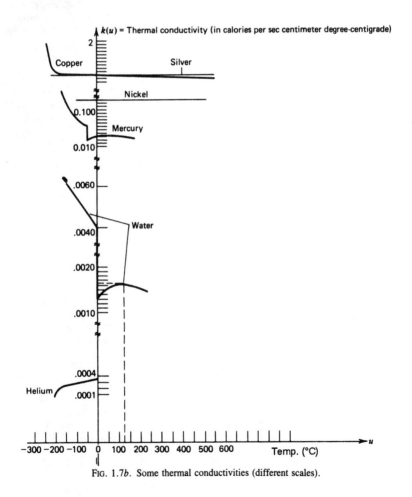

FIG. 1.7b. Some thermal conductivities (different scales).

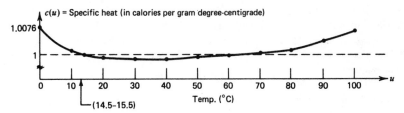

FIG. 1.7c. Specific heats of water.

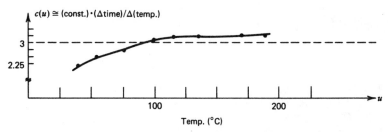

FIG. 1.7d. Specific heats of cadmium.

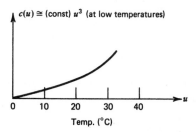

FIG. 1.7e. Specific heats of sodium chloride.

Returning to the previously recommended bathtub experience, for example, the heat propagation seems to take place in somewhat spherical outward spreading waves, suggesting* a damped wave equation (ignoring units)

$$u_t = u_{xx} - \varepsilon u_{tt}. \tag{3}$$

The damping term is u_t and the effect of the nonlinearities (i.e., the temperature dependence of k and c) is presumably carried in the (positive) constant ε. This would still amount to a forced *linearization* of the heat conduction model but now in terms of a (3) *perturbation term* εu_{tt} in the equation. This is therefore also an illustration of the third technique (3) that we wished to expose in this section. Such

---

* This has in fact been derived rather recently from a physically more rigorous point of view; see C. Catteneo, "Sur une forme de l'equation de la chaleur eliminant le paradoxe d'une propagation instantanée," *Compte Rendus Acad. Sci. Paris* 247 (1958).

perturbation terms possessing derivatives of as high order as in the unperturbed equation are commonly called singular perturbations.

A further example of the *perturbation method* (3) will be found in Problem 3 below, taken from quantum mechanics where perturbation methods are nothing less than rampant.

Problem 1. By using the Poisson representation formula (see Problem 2, Section 1.3), demonstrate that the linear heat equation propagates an initial positive unit heat impulse localized about some point x_0 with infinite speed.

Problem 2. Assure yourself of units consistency, in terms of calories, degrees, centimeters, and seconds, in the heat equation

$$cu_t - ku_{xx} = 0.$$

Problem 3. An important property of the equation for the hydrogen atom (see Section 1.6.2) is that it be self-adjoint.* The Laplacian term $H_0 \equiv -\frac{1}{2}\Delta$ is a self-adjoint operator in $L^2(R^3)$ and a general theorem† states that the same will be true for the full hydrogen operator

$$H = -\frac{1}{2}\Delta - Z/r$$

provided that the perturbation $V \equiv -Z/r$ is small in the sense that

$$\|Vu\| \leq a\|u\| + b\|H_0 u\|, \qquad b < 1,$$

for some constants a and b, independent of u. Show this, using the Schwarz and Sobolev inequalities, for nice u (e.g., $u \in C_0^\infty(R^3)$).

1.7.2. Three Physical Settings (and Examples)

We have previously described problems involving partial differential equations in the physical settings of vibrating strings, stretched membranes, heat conduction, and atoms. Let us mention the three general physical settings of:

Continuum mechanics;

Statistical mechanics;

Quantum mechanics.

Many partial differential equations arise from one of these three settings.

---

* An operator H is self-adjoint if $H = H^*$. For a precise meaning of H^*, see Section 1.9.7. The point of this problem is to establish the self-adjointness of H from that of H_0 by proving the stated inequality. For a further discussion of self-adjointness see Problem 1.9.7(1).

† Due to F. Rellich, B. Sz. Nagy, and T. Kato. See T. Kato, *Perturbation Theory for Linear Operators*, 2nd ed. (Springer, New York, 1976).

It would be an interesting but endless and hopeless task to categorize all of the settings from which partial differential equations come. Instead we present this brief discussion and some examples.

1. Continuum (and Classical) Mechanics. Many second-order PDEs are found in continuum mechanics (and, more generally, in classical mechanics). We have already seen the following examples:

Heat equation (parabolic),

Vibrating string (hyperbolic),

Sagging membrane (elliptic).

Some others follow.

Stationary irrotational incompressible fluid flow.

$$\begin{cases} \Delta u = 0 \text{ in } \Omega, \\ \dfrac{\partial u}{\partial n} = f \text{ on } \partial\Omega, \end{cases}$$

where u = velocity potential (i.e., the components v_1, v_2, v_3 of grad u are the component velocities of the fluid).

Electrostatic potential.

$$\begin{cases} \Delta\varphi = -4\pi\rho \text{ in } \Omega, \\ \varphi = f \text{ on } \partial\Omega, \end{cases}$$

where φ = the electrostatic potential (i.e., the components E_1, E_2, E_3 of grad φ are the components of the electric field strength) and ρ is the charge density in Ω.

Buckling plate (elastostatics).

$$\begin{cases} -\Delta^2 u = \lambda\Delta u \text{ in } \Omega, \\ u = \dfrac{\partial u}{\partial n} = 0 \text{ on } \partial\Omega, \end{cases}$$

where u is the vertical displacement of the plate and where the boundary conditions are those for a clamped plate Ω under edge loads that are increased until buckling occurs. The fourth-order operator $\Delta^2 \equiv \Delta\Delta$ is called the biharmonic operator. Fourth-order operators such as these occur in the theory of elasticity, and the methods for treating them resemble those used for second-order operators, to the extent possible.*

Torsional rigidity (e.g., of a beam).

$$\begin{cases} -\Delta v = 2 \text{ in } \Omega, \\ v = 0 \text{ on } \partial\Omega, \end{cases}$$

---

* As seen, for example, in the preceding Second Pause. Fourth-order operators also arise in fluid dynamics; see Appendix C.

where the quantity of interest is the torsional rigidity $T = D(v, v) = \int_\Omega |\text{grad } v|^2$, which measures resistance to twisting.

Electrostatic capacity.

$$\begin{cases} \Delta w = 0 \text{ in exterior of } \Omega, \\ w = 1 \text{ on } \partial\Omega, \end{cases}$$

where the quantity of interest is the capacity

$$C = D(w, w) = \int_{\text{exterior of } \Omega} |\text{grad } w|^2$$

which measures the charge required on $\partial\Omega$ to raise the potential one unit.

Navier–Stokes equations (viscous incompressible flow).

$$\begin{cases} \dfrac{\partial}{\partial t} u_i + u_j(u_i)_{,j} - \dfrac{1}{R} \Delta u_i = -p_{,i} \text{ in } \Omega, \qquad i = 1, 2, 3, \qquad t > 0, \\[2mm] (u_i)_{,i} = 0 \text{ in } \Omega, \\ u_i(x, 0) = f_i(x), \qquad i = 1, 2, 3, \qquad x = (x_1, x_2, x_3) \text{ in } \Omega \\ u_i(x, t) = 0, \qquad x \in \partial\Omega, \qquad t > 0, \end{cases}$$

where we have used the summation convention for convenience. Here $u = (u_1, u_2, u_3)$ are the velocity components of a fluid motion in a vessel (bounded three-dimensional Ω). The fluid is put into motion at time $t = 0$ with initial velocity (f_1, f_2, f_3) and is thereafter governed by the first (Navier–Stokes) equations. The second equation states that the motion is incompressible. The last (boundary) condition is one of adherence of the fluid at the boundary.

The above examples involving a steady-state (nontime-varying) problem, are elliptic. In like manner, many time-evolving problems yield examples of parabolic PDEs, and vibrating or oscillating problems examples of hyperbolic PDEs.

2. Statistical Mechanics. We use the term loosely. Roughly speaking one finds here more often first-order integrodifferential equations rather than second-order equations. We list only a few.

Linear Boltzmann equation (transport equation).

$$\begin{cases} \dfrac{\partial n}{\partial t} = -v \text{ grad}_x n + \int_{R^3} k(x, v, v')n(x, v', t) \, dv' - \sigma(x, v)n, \qquad t > 0 \\ n(x, v, 0) = f(x, v), \end{cases}$$

where $n = n(x, v, t)$ is a particle density function for a bounded domain $\Omega \subset R^3$, $x = (x_1, x_2, x_3)$, $v = (v_1, v_2, v_3)$. The six-dimensional space (x, v) is called the μ-space or phase space in statistical mechanics. The bounded domain may be, for example, a nuclear reactor (neutron transport problem) or a star (radiative transfer problem). The equation describes the motion of a cloud of noninteracting particles, the first term on the right-hand side describing the free motion, the second the input at (x, v) due to scattering from other regions in momentum space and production by other particles, the third a loss due to absorption or scattering. The integral operator $\int k(x, v, v') \cdot dv'$ produces its value of particles when applied to a single particle.

Carleman system (kinetic theory of gas).

$$\begin{cases} \dfrac{\partial u}{\partial t} + \dfrac{\partial u}{\partial x} + a(u^2 - v^2) = 0, & 0 < x < 1, \quad t > 0 \\[2mm] \dfrac{\partial v}{\partial t} - \dfrac{\partial v}{\partial x} + a(v^2 - u^2) = 0, & 0 < x < 1, \quad t > 0 \\[2mm] u(0, t) = v(1, t) = 0, & t > 0 \\[1mm] u(x, 0) = f(x) \geqq 0, & 0 < x < 1 \\[1mm] v(x, 0) = g(x) \geqq 0, & 0 < x < 1, \end{cases}$$

where a is a positive constant. The unknowns $u(x, t)$ and $v(x, t)$ represent densities of particles of two types that together comprise the gas, and the equations may be regarded as simplified coupled Boltzmann equations.

Nonlinear Debye–Hückel equation (plasma).

$$\left\{ \Delta\varphi(x) = 2\lambda \left[1 - \frac{e^{-\varphi(x + 1/2)}}{\int_\Omega e^{-\varphi(x)} \, dx} \right], \right.$$

where φ possesses certain periodic boundary conditions. Without going into further details,* φ represents the potential for a jellium model in a self-consistent field approximation corresponding to certain assumptions about the existence of long-range ordering in Coulomb systems.

There are a great many nonlinear equations of physical importance and of recent formulation. Some will be found in Section 1.8. We have included this particular one here because it exhibits three properties of interest: (a) the $x + \frac{1}{2}$ represents a spatial translation and this equation is thus an example of a whole class of equations with displaced (advanced, retarded, delayed) arguments; (b) as an ODE one can play with it in a number of interesting ways (see Problem 2); and (c) as a PDE it, like many others, resists solution.

3. Quantum Mechanics.

3. Quantum Mechanics. Again we will be brief. The unknown function $u(x, t)$ represents a probability of finding a particle at x at time t, in the sense that

$$\int_\Omega |u(x, t)|^2 \, dx, \qquad x = (x_1, x_2, x_3),$$

---

* See P. Choquard, "Long-range ordering in one-component Coulomb systems," *Physical Reality and Mathematical Description*, Enz and Mehra (eds.), D. Reidel (1974), Dordrecht-Holland.

is the probability that the particle is in Ω at time t. Thus one wants the normalization $\int_{R^3}|u(x, t)|^2\, dx = 1$.

Schrödinger equation (single free particle).

$$\begin{cases} \dfrac{\partial u}{\partial t} = -i\Delta u, & x \in R^3, \quad t > 0 \\[2mm] u(x, 0) = f(x), \end{cases}$$

where we have normalized the units. Because of the i this is not parabolic and solutions have a somewhat different behavior than in the heat equation.

Klein–Gordon equation (relativistic free particle).

$$\begin{cases} \dfrac{\partial^2 u}{\partial t^2} - \Delta u = \lambda u, & x \in R^3, \quad \lambda > 0, \end{cases}$$

where λ corresponds to a positive continuous spectrum* of eigenvalues, under physically appropriate normalization boundary conditions on the eigenfunctions u, the latter called de Broglie waves. This is thus a hyperbolic eigenvalue problem.

Dirac equation (relativistic free particle with spin).

$$\begin{cases} \dfrac{\partial u_1}{\partial t} = -\dfrac{\partial u_4}{\partial x_1} + i\dfrac{\partial u_4}{\partial x_2} - \dfrac{\partial u_3}{\partial x_3} - imu_1, \\[3mm] \dfrac{\partial u_2}{\partial t} = -\dfrac{\partial u_3}{\partial x_1} - i\dfrac{\partial u_3}{\partial x_2} + \dfrac{\partial u_4}{\partial x_3} - imu_2, \\[3mm] \dfrac{\partial u_3}{\partial t} = -\dfrac{\partial u_2}{\partial x_1} + i\dfrac{\partial u_2}{\partial x_2} - \dfrac{\partial u_1}{\partial x_3} - imu_3, \\[3mm] \dfrac{\partial u_4}{\partial t} = -\dfrac{\partial u_1}{\partial x_1} - i\dfrac{\partial u_1}{\partial x_2} + \dfrac{\partial u_2}{\partial x_3} + imu_4. \end{cases}$$

One can write the equation in more compact form using matrices, but we wrote it out here to show the exact coupling. The Dirac equation is a first-order system for $u = (u_1, u_2, u_3, u_4)$ a four-component vector, $u = u(x, t)$, $x = (x_1, x_2, x_3)$, $-\infty < t < \infty$. The Dirac equation is related to the Klein–Gordon equation and is approachable mathematically in some ways via Schrödinger equation techniques.

Helium atom equation (nonrelativistic).

$$\left\{ -\tfrac{1}{2}\Delta_1 u - \tfrac{1}{2}\Delta_2 u - \dfrac{2u}{r_1} - \dfrac{2u}{r_2} + \dfrac{1}{r_{12}}\, u = \lambda u, \right.$$

where $u = u(x_1, x_2, x_3, x_1', x_2', x_3')$, Δ_1 is the Laplacian for x, Δ_2 is the Laplacian for x', $r_1 = (x_1^2 + x_2^2 + x_3^2)^{1/2}$, r_2 the same for x', and $r_{12} = ((x_1 - x_1')^2 + (x_2$

---

* A further discussion and precise definition of continuous spectrum will be found in Section 2.7. For our purposes here it means that the equation represents a continuous range of physically realizable energy values all the way from 0 to ∞.

$- x_2')^2 + (x_3 - x_3')^2)^{1/2}$. The normalizing boundary condition is that u must be square integrable, and since it is an eigenvalue problem, one may take

$$\int |u(x, x')|^2 \, dx \, dx' = 1.$$

The spectrum is known to consist of a lower spectrum of distinct eigenvalues plus a continuous spectrum $[-1, \infty)$.

Problem 1. For the vibrating clamped plate problem

$$\begin{cases} \Delta^2 u = \mu u \text{ in } \Omega \\ u = 0 \text{ on } \partial\Omega \\ \partial u/\partial n = 0 \text{ on } \partial\Omega \end{cases}$$

(a) show the eigenvalues to be positive, and (b) obtain a positive lower bound for them.

Problem 2. Consider the one-dimensional version of the nonlinear Debye–Hückel equation stated above,

$$u''(x) = 2\lambda \left[1 - \frac{e^{-u(x + 1/2)}}{N(u)} \right], \qquad 0 \leqq x \leqq 1, \qquad \lambda > 0,$$

where $N(u) = \int_0^1 e^{-u(x)} \, dx$, with periodicity condition

$$u(x) = u(x + 1) = u(1 - x), \qquad -\infty < x < \infty,$$

and normalization condition

$$\int_0^1 u(x) \, dx = 0.$$

(a) Set $u(0) = 1$ and try by graphical inspection to determine an appropriate solution. (b) Try the substitution $v = u'$, differentiate to obtain v'', and see what results.

Problem 3. Try separation of variables for the helium atom equation given above.

1.7.3. Some Unsolved Problems

For example, it is unknown when the first eigenfunction in the clamped plate problem is always of one sign over the domain. The Navier–Stokes equation initial value problem for data not necessarily small has not yet been completely resolved. The Boltzmann equations have been solved only for certain geometries Ω. Equations such as the Debye–Hückel equation are only beginning to be investigated. The nonsolvability of the helium eigenvalue problem has been a longstanding one in quantum mechanics, and its nonsolvability causes the consequent nonsolvability of Schrödinger equations for lithium and all more complicated atoms.

There are many other important partial differential equations that we have not discussed that remain unresolved.

Problem 1. Could the geometry or topology of the plate Ω influence whether the first eigenfunction of the clamped plate changes sign?

Problem 2. Which of the main questions (1), (2), (3), (1'), (2'), (3') for the Navier–Stokes equation initial value problem are not resolved?

Problem 3. Why, intuitively, should you expect to have trouble solving the Helium equation?

1.7 Exercises

1. Show that the Laplacian term $H_0 \equiv \frac{1}{2}\Delta$ is "formally self-adjoint" in the sense that

$$\int_{R^3} (H_0 u(x))v(x)\, dx \;=\; \int_{R^3} u(x)(H_0 v(x))dx$$

for all sufficiently smooth functions u and v vanishing outside some bounded subdomain Ω in R^3.

2. Show that you may solve exactly the following plate eigenvalue problem.

$$\begin{cases} -\Delta^2 u = \Lambda\Delta u & \text{in } \Omega, \\ \quad u = 0 & \text{on } \partial\Omega, \\ \quad \Delta u = 0 & \text{on } \partial\Omega, \end{cases}$$

when Ω is a rectangular plate.

3. Can you separate variables in the other clamped plate problems?

1.8 ELEMENTS OF BIFURCATION THEORY

The word "bifurcation" means a "splitting," and in the context of nonlinear differential equations it connotes a situation wherein at some critical (e.g., physical) parameter λ the number of solutions to the equation changes. The critical value of λ is often called a bifurcation point.*

This concept of a change in the number of solutions of a differential equation is not, in and of itself, a new one to us. Let us recall the basic eigenvalue problem

$$\begin{cases} -u''(x) = \lambda u(x), & 0 < x < \pi, \\ u(0) = u(\pi) = 0. \end{cases}$$

---

* A more appropriate word would be "emanation" point, since the splitting can be into any number of new solutions.

The solutions were the eigenfunctions $u_n(x) = c_n \sin nx$ at the eigenvalues $\lambda_n = 1, 4, 9, \ldots, n^2, \ldots$, where c_n was an arbitrary constant. If we plot the eigenvalue parameter λ along the horizontal axis and the solution norms† vertically, we have the solution diagram (Fig. 1.8a). For all other values of λ the only solution was the trivial solution $u = 0$, which is interesting only as the solution branch from which the others emanate. In bifurcation problems especially in continuum mechanics one usually works with real valued solutions only, and often one wishes to distinguish between a positive and negative solution (e.g., thinking physically, the buckling up of a rod, as contrasted with the buckling down of it). This we can easily do in the above example by taking the vertical axis to be c_n rather than $\|u_n\|$, the "signed magnitude" of the solutions. See Figure 1.8b for the resulting diagram.

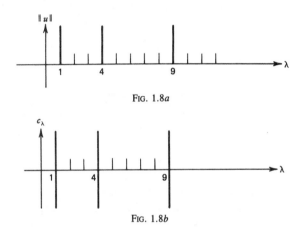

FIG. 1.8a

FIG. 1.8b

Figures 1.8a and 1.8b, although for a linear problem, are examples of what are called the *bifurcation diagram* for a nonlinear problem. The eigenvalues λ_n are the bifurcation points. At each of the λ_n the number of solutions changes from one solution (the trivial one) to an infinite number of them.

In general, and as we shall see, the effect of a nonlinearity in a problem is to bend the solution branches. Let us consider a simple example to see how that comes about.

Consider first the mildly* nonlinear problem

---

† In this diagram it does not matter whether we use the L^2 norm $\|u\| = (\int_0^\pi u^2)^{1/2}$ or the maximum norm $\|u\| = \max_{0 \le x \le \pi} |u(x)|$, although it will change the vertical scale. Unless stated otherwise, we will use the L^2 norm for $\|u\|$.

* This type of nonlinearity consists only of a "rescaling" and in that sense is not a genuine nonlinearity. Nonetheless, they occur in a few physical problems and are sometimes treated in the literature. A *genuine nonlinearity* is one such as u^3 or more generally $f(u)u$, wherein $f(u)$ is a vector that actually "turns" the vector u to a new direction in the space. *Mild* nonlinearities such as $\|u\|^2 u$ could also be called *pseudo-nonlinearities*.

$$\begin{cases} -u''(x) = \lambda\|u\|^2 u(x), & 0 < x < \pi, \\ u(0) = u(\pi) = 0. \end{cases}$$

As before the trivial solution $u = 0$ is a solution branch for all λ. Since for any nontrivial solution u the quantity $\lambda\|u\|^2$ is a constant bearing the same sign as λ, as above in the linear problem we are led by the boundary conditions to the solution candidates $u_n(x) = c_n \sin nx$. Substituting the latter into the equation then yields the solution branches $\lambda\|u\|^2 = n^2$, which in terms of the coefficient amplitudes is

$$c_n = \pm\left(\frac{2}{\pi}\right)^{1/2} \frac{n}{\lambda^{1/2}}.$$

The bifurcation diagram for this problem is thus Figure 1.8c. The only "bifurcation point" in this example is at $\lambda = \infty$.

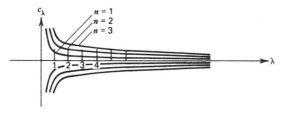

FIG. 1.8c

To obtain a nonlinear example that more closely resembles the linear one, let us put the linear term back in. Hence let us consider the example

$$\begin{cases} -u''(x) = \lambda u(x) + \|u\|^2 u(x), & 0 < x < \pi, \\ u(0) = u(\pi) = 0. \end{cases}$$

Again, if $u(x)$ is a solution, so is $-u(x)$, so that we can expect as above a bifurcation diagram with solution branches occurring symmetrically above and below the horizontal λ axis. Also as above we see from the boundary conditions that the solution candidates are $u_n(x) = c_n \sin nx$, which upon substitution into the equation yields the relation $\lambda + \|u\|^2 = n^2$. Thus the solution branches are

$$c_n = \pm\left(\frac{2}{\pi}\right)^{1/2} (n^2 - \lambda)^{1/2}$$

and the bifurcation diagram is (see Fig. 1.8d) the branches emanating from the bifurcation points $\lambda_n = 1, 4, \ldots, n^2, \ldots$, which are the eigenvalues of the original linear eigenvalue problem. If we change the nonlinearity from $\|u\|^2 u$ to $-\|u\|^2 u$, the branches become

$$c_n = \pm\left(\frac{2}{\pi}\right)^{1/2} (\lambda - n^2)^{1/2}$$

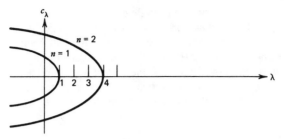

FIG. 1.8*d*

and the bifurcation diagram becomes Figure 1.8*e*. The latter two diagrams (Figs. 1.8*d* and 1.8*e*) are quite typical of those found for a large number of nonlinear problems.

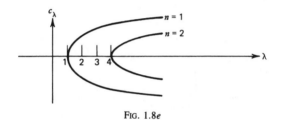

FIG. 1.8*e*

We turn now to a consideration of some physical settings in which such problems arise: (a) buckling of a yard (or meter) stick; (b) a rotating string; and (c) chemical kinetics.

(a) Buckling of a Yardstick

Suppose you push a yardstick against the wall. A picture of the physical situation then is Figure 1.8*f*. It can be shown that a model equation for this situation is

$$\begin{cases} -\varphi''(x) = \lambda \sin \varphi(x), & 0 < x < l, \\ \varphi'(0) = \varphi'(l) = 0. \end{cases}$$

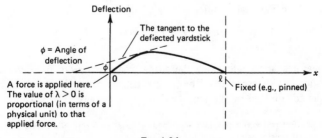

FIG. 1.8*f*

Note that the maximum value of the angle of deflection is expected at the ends. Let $\varphi(0) = \varphi_0$ be that maximum value for each given "force" $\lambda \geqq 0$. The model equations can be solved (see Problem 1) by ordinary differential equation techniques and the bifurcation diagram is Figure 1.8g. The first branch corresponds to a buckling mode as drawn in Figure 1.8f, the second branch to the second buckling mode (Fig. 1.8h), and so on. The actual attainability of the higher modes will depend on the flexibility of the yardstick (eventually it will break).

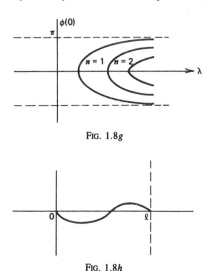

FIG. 1.8g

FIG. 1.8h

(b) A Rotating String

A string, held at the top, of length l, and rotated faster and faster from the top, will assume different shapes. Given that the string is sufficiently flexible and that it is subject only to the force of gravity pulling it straight down and to the centrifugal force caused by a rotation of constant angular velocity, the string satisfies a nonlinear partial differential wave equation in space and time. For simplicity let us look only for solutions that represent a string whose shape does not change as it is rotated at a constant angular velocity around the vertical axis.* One then arrives at the nonlinear ordinary differential equation

$$\begin{cases} -u''(s) = \lambda(u^2(s) + s^2)^{-1/2} u(s), & 0 < s < l, \\ u(0) = u'(l) = 0. \end{cases}$$

Here s measures arc length along the string, measured from the free end, and differentiation is with respect to s. The eigenvalue parameter λ is the angular rotation speed (squared) divided by the (constant) force of gravity. The unknown $u(s)$ is

---

* This in effect performs a nonlinear separation of variables for us, eliminating both time and conceivable spatial nonsymmetrics from the problem.

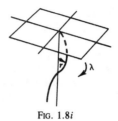

FIG. 1.8*i*

(modulo scale multiplication by a physical constant) the derivative with respect to arc length of the radial distance r from the vertical axis, as shown in Figure 1.8*i*. Thus the boundary conditions for the equation given above mean that at the bottom end, $r'(0) = 0$ and the string is parallel to the vertical axis, as in the free end condition for the linear vibrating string problems we considered in earlier sections, and at the top end $r''(l) = 0$ so that the string looks linear there.

We will not solve this problem here.[†] Note that the trivial solution is $u(s) = r'(s) = 0$, the case in which the string hangs vertically downward under the force of gravity, rotating within that axis. When the ratio λ of angular velocity to force of gravity increases to an adequately large value (which turns out to be the first eigenvalue of the linearized equation; see Problem 2) the string departs from the trivial mode and is found in the mode drawn in figure 1.8*i*. More complicated modes prevail as centrifugal force increases its advantage over gravity.

(c) Chemical Kinetics

In chemical kinetics[*] one finds systems of partial differential equations governing chemical reactions between various agents, such as

$$\begin{cases} (C_i)_t = k_i \Delta C_i + f_i(C_j, T) \\ T_t = k \Delta T + f(C_j, T) \end{cases}$$

along with appropriate initial and boundary values. The C_j are concentrations and T is the temperature in appropriate units. A special case that arises in combustion theory and under the assumption that concentrations remain constant over the short time interval in question leads to the single nonlinear parabolic equation

$$T_t = k \Delta T + q_0 c_0 e^{-a_0/T},$$

where q_0 is the heat of reaction, a_0 is a known gas energy constant, and where c_0 is the initial concentration. Letting $u(x, t)$ denote the temperature $T(x, t)$ in accordance with our usual notation, and regarding $\lambda = k^{-1}q_0 c_0$ as the eigenvalue parameter, we may consider the equation

[†] See I. Kolodner, "Heavy rotating string," *Comm. Pure Appl. Math.* **8** (1955).

[*] See, for example, the books by Gavalas, *Nonlinear Differential Equations of Chemically Reacting Systems* (Springer, New York, 1968) and Prigogine and Nicolis, *Self-organization in Nonequilibrium Systems: From Dissipative Structures to Order Through Fluctuations* (Wiley, New York, 1977).

$$\begin{cases} -\Delta u = \lambda e^{-a_0/u} & \text{in } \Omega \\ u = u_0 & \text{on } \partial\Omega, \end{cases}$$

for the steady-state equilibrium temperature $u(x)$, given that a constant temperature u_0 is to be maintained on the boundary.

For physical reasons let us consider only the case $\lambda \geqq 0$. For ease of visualization let us take Ω to be the unit disk in two dimensions. It may then be seen that the bifurcation diagram for solutions is of the form shown in Figure 1.8j.

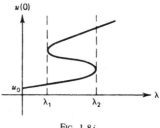

FIG. 1.8j

Although such equations arise in a variety of nonlinear heat source problems, one interesting interpretation is that in which the unit disk is a solid fuel rocket cross section with u_0 representing a constant burn on the outside and with λ representing the concentration of the propellant in the cross section. Higher concentrations λ produce a higher center burn temperature $u(0)$. Between critical concentrations λ_1 and λ_2 there are three possible center burn temperatures. No doubt in certain applications one would like to achieve the highest one.

For a more accessible version of this third type of problem, let us consider the ordinary differential equation

$$\begin{cases} -u''(x) = \lambda e^{u(x)}, & -1 < x < 1, \\ u(-1) = u(1) = 0. \end{cases}$$

For example, model equations of this type arise in the theory of gravitational equilibrium of polytropic stars. This particular equation may be integrated exactly (see Problem 3) and has a bifurcation diagram as shown in Figure 1.8k.

For the three physical examples given above the model equations were finally formulated for ease of solution in reduced form as ordinary differential equations. The first could then be solved by means of an associated initial value problem (see Problem 1). The second can also be solved by that method. Both the second and third problems may be solved by another approach often useful for nonlinear problems, which we now describe for the case of the third problem. This method carries several names such as the method of Hammerstein integral equations, the method of monotone iterations, application of the theory of positive cones in ordered Banach spaces. It is highly dependent upon the fact that the inverse L^{-1} of the differential operator $Lu = -u''$ with Dirichlet boundary conditions is both positive and com-

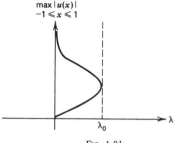

FIG. 1.8k

pact. These are strong conditions, but often prevail for elliptic operators on bounded domains. Further discussion of some of these points will be found in Chapter 2. This method also has the property of being quite amenable to numerical approximation procedures.

Considering then for illustration the last problem above, with nonlinearity e^u, we employ our Green's function method to rewrite the problem as the integral equation

$$u(x) = \lambda \int_{-1}^{1} G(x, s) e^{u(s)} ds,$$

where $G(x, s)$ is the Green's function for the operator $Lu = -u''$ with Dirichlet boundary conditions. $G(x, s)$ is easily seen to be

$$G(x, s) = \begin{cases} \frac{1}{2}(1 + s)(1 - x), & s \leq x, \\ \frac{1}{2}(1 - s)(1 + x), & s \geq x. \end{cases}$$

It is known* that if there can be demonstrated the existence of both an "upper solution" $v(x) \geq \lambda \int G(x, s) e^{v(s)} ds$ and a "lower solution" $w(x) \leq \lambda \int G(x, s) e^{w(s)} ds$, then a solution exists. Moreover, this solution may be found as the limit of the iteration

$$u_{n+1}(x) = \lambda \int_{-1}^{1} G(x, s) e^{u_n(s)} ds$$

where the initial approximation point u_0 to begin the iteration may be taken to be either an upper or a lower solution. In practice it is usually the latter.

In particular, let us take

$$u_0(x) \equiv 0,$$

* See, for example, the survey paper by H. Amann, "Fixed point equations and nonlinear eigenvalue problems in ordered Banach spaces," *SIAM Review* **18** (1976).

an easily seen lower solution to the problem for any $\lambda \geq 0$. We then have

$$u_1(x) = \tfrac{1}{2}\lambda(1 - x^2),$$

$$u_2(x) = 1 - e^{(1/2)\lambda(1-x^2)} + \tfrac{1}{2}\lambda e^{(1/2)\lambda}\left[\int_{-1}^{1} e^{-(1/2)\lambda s^2}\,ds + x\left(-\int_{-1}^{x} e^{-(1/2)\lambda s^2}\,ds\right.\right.$$

$$\left.\left.\vdots \qquad + \int_{x}^{1} e^{-(1/2)\lambda s^2}\,ds\right)\right].$$

It is left to the student (see Problem 3), as both a mathematical and philosophical exercise, to consider further these iterations.

We close this section with a short discussion of the concepts of *secondary bifurcation* and *turbulence*.

The initial or *primary bifurcation* usually occurs as a splitting off of a nontrivial solution from the trivial (or a constant) solution $u \equiv 0$. This was the case in the examples above. The nontrivial solution branch may then bifurcate again. This is called a *secondary bifurcation*. On a bifurcation diagram this second bifurcation would look like Figure 1.8*l*.

In certain problems in fluid dynamics, experiments have revealed a number of secondary bifurcations, seemingly arranged in a hierarchy of increasing numbers of bifurcations. This led to a famous conjecture by Landau that the state of turbulence could be viewed as the end result of such an infinite chain of increasingly complicated bifurcations. There is on the other hand countervailing evidence that indicates that turbulence may result after only a few secondary bifurcations. Thus our understanding of the nature of turbulence in a number of important physical applications remains unclear.†

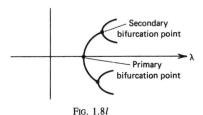

FIG. 1.8*l*

Problem 1. (a) Solve the buckling yardstick problem by considering the corresponding initial value problem

$$\begin{cases} \varphi''(x) + \lambda \sin \varphi(x) = 0, & x > 0, \\ \varphi(0) = \varphi_0, \\ \varphi'(0) = 0, \end{cases}$$

† For some further aspects of turbulence, see Appendix C.2.

where $\lambda > 0$ and $0 < \varphi_0 < \pi$, accepting only the solutions that satisfy $\varphi'(l) = 0$. (b) Determine to what extent your solution can be given explicitly or only implicitly. (c) Deduce the bifurcation diagram for this problem as drawn in this section (Fig. 1.8g).

Problem 2. One "linearizes" a nonlinear equation by dropping the higher order terms.* Rather general theorems and "metatheorems" in bifurcation theory† then assert that the solution branches of the nonlinear equation emanate from eigenvalues of the linearized equation. (a) Find the linearizations for the six examples considered above. (b) Investigate how well the branches emanate according to the stated assertion.

Problem 3. (a) Determine when the two nonlinear temperature source problems given above have no positive solutions for $\lambda < 0$. (b) Integrate exactly the nonlinear boundary value problem

$$\begin{cases} -u''(x) = \lambda e^{u(x)}, & -1 < x < 1, \\ u(-1) = u(1) = 0. \end{cases}$$

Hint: Make a change of variable. (c) Investigate, analytically or by computer, further iterations of the problem of part (b) as begun in the text above.

1.9 SUPPLEMENTARY DISCUSSIONS AND PROBLEMS

This section contains eight subsections, Problem 1.9.1 through 1.9.8, each with material and exercises supplementing the same-numbered sections in the text. Then there is a final subsection, Problem 1.9.9, containing three "confirmation" exercises and further suggested exercises and problems.

Problem 1.9.1 contains some further details for the classification of partial differential equations as discussed in Section 1.1. It is our feeling that in an introduction to the subject one should not spend too much time on these classification details. Similarly, in most of the supplementing material to follow, much has been left to the instructor's discretion and student's curiosity as to how far to go in pursuing those matters. For example, in Problem 1.9.2 we have given only a minimal treatment of characteristics. That subject can on the other hand be studied in great

---

* There are other, more rigorous ways to describe the linearization of such equations, but this rule applies well in practice. We are speaking here of "linearization about zero," that is, "about the zero solution in u." The idea is that, for small solutions, the higher-order terms are smaller and have only a secondary effect.

† For the large and rapidly growing literature on this important subject let us suggest the review article by I. Stakgold, *SIAM Rev.* **13** (1971), the recent book by M. M. Vainberg and V. A. Trenogin, *Theory of Branching of Solutions of Nonlinear Equations* (Noordhoff, Leyden, 1974), and the treatises of D. Joseph, *Stability of Fluid Motions I, II,* (Springer, Berlin, 1976). There are so many other important works on this subject that we cannot include a full bibliography here. For more recent approaches see M. Golubitsky and D. Schaeffer, *Singularities and Groups in Bifurcation Theory* (Springer, New York, 1985).

detail, especially as it concerns the behavior of general hyperbolic partial differential operators, solutions of equations of mixed type, and nonlinear problems. See Appendix A.

Problem 1.9.3 outlines a proof of the maximum principles for elliptic and parabolic equations. The basic Picard–Cauchy–Kowalewski existence theorem for initial value problems is given in Problem 1.9.4.

Problem 1.9.5 contains three parts that to some extent augment the three parts of Section 1.5, but that also provide additional background material for the subject as a whole. Because we develop rather fully the method of separation of variables in this book, in Problem 1.9.5(1) we instead have chosen to stimulate a historical perspective for the student. In Problem 1.9.5(2) we attempt to take care of, in an admittedly summarily way, the questions about domain pathologies. In Problem 1.9.5(3) we give the derivation of a partial differential equation from its variational formulation. Parts of Problem 1.9.5, especially (2) and (3), will require that the reader begin the sixth section of the chapter, specifically, 1.6.1(1).

In an arrangement similar to that of Problem 1.9.5, Problem 1.9.6 contains three parts: (1) a short history about mathematical developments concerning the pointwise convergence of Fourier series; (2) a brief introduction to the methods of proof of Dini tests, and to the Gibbs effect; and (3) a very brief discussion of the Lebesgue dominated convergence theorem. We have no intention of getting into the details of the Lebesgue theory per se, but we feel that the dominated convergence theorem deserves emphasis here as a fundamental technique for "differentiating under the integral." We in fact employ it in that capacity in Problem 1.9.7(3) in an illustration there of a rigorous investigation of the ways in which the heat equation attains its initial values. The first two parts of Problem 1.9.7 are (1) a brief exposition of self-adjoint operators and (2) the related notion of distributional (or weak) derivative. The self-adjointness of an operator is an important concept in physical applications.

Problem 1.9.8 very briefly illustrates via the Van de Pol equation the phase portrait method for analyzing nonlinear ordinary differential equations. Our viewpoints there are that this is an important method that should not be overlooked for the special cases in which a nonlinear problem has been reduced to a second-order ordinary differential equation, and that it should be viewed as an important method in the perturbation and bifurcation theories.

Problem 1.9.1 Classification

Study some of the following conventions and facts concerning the classification procedure for second-order partial differential equations. Also feel free to come back to this problem later, since we suggest not getting bogged down at this point. A specific exercise is suggested near the end of the discussion.

Let us consider a *PDE of second order* in n variables and one unknown u on a domain Ω of points $x = (x_1, \ldots, x_n)$, $n > 1$:

$$F(x_i, u, u_{,i}, u_{,ij}) = 0.$$

A subclass is

$$Lu = Au + N(u) = 0,$$

where $Au = a^{ij}u_{,ij}$, the second-order part, can be separated from the other terms.* This equation is called *quasilinear* if $a^{ij} = a^{ij}(x_i, u, u_{,i})$, $N(u) = N(x_i, u, u_{,i})$, i.e., if whatever nonlinearity that is present in the equation does not involve second derivatives of u. It is called *almost linear or semilinear* if $a^{ij} = a^{ij}(x_i)$ and $N(u) = N(x_i, u, u_{,i})$, i.e., if it is linear in the second-order terms. It is *linear* if $a^{ij} = a^{ij}(x_i)$, $N(u) = a_iu_{,i} + au + f$, where $a_i = a_i(x)$, $a = a(x)$, $f = f(x)$. In that case L is a linear expression with possibly nonconstant coefficients. The $f(x)$ are commonly called the "data" and presumed to be known along with the coefficients.

Examples.
$(u_{x_1x_1})^2 + (u_{x_2x_2})^2$ is not quasilinear.
$(u_{x_1})^2 u_{x_1x_1} + (u_{x_2})^2 u_{x_2x_2} + u^2$ is quasilinear.
$x_1 u_{x_1x_1} + x_2 u_{x_2x_2} + u^2$ is almost linear.
$x_1 u_{x_1x_1} + x_2 u_{x_2x_2} + x_1x_2u$ is linear.
$u_{x_1x_1} + u_{x_2x_2} + u$ is linear with constant coefficients.

Usually we will be dealing with the latter case of a linear partial differential equation with constant coefficients. We will consider in the following paragraphs principally the case $n = 2$, the classification for higher dimensions being similar. Classification depends only on the highest order (here, the second-order) derivative terms, and the lower-order terms may therefore be omitted.

The main idea is quite simple. Given

$$Au = a^{ij}u_{,ij}$$

one considers an arbitrary n-vector $y = (y_1, \ldots, y_n)$ and assigns the quadratic form

$$Q_1(y) = a^{ij}y_iy_j = (Ay, y) \equiv \sum_{i=1}^{n} \sum_{j=1}^{n} a^{ij}y_iy_j.$$

---

* It should be mentioned here that often it is convenient to have an equation in the so-called divergence form

$$Lu = (a^{ij}u_{,i})_{,j} + \text{lower-order terms,}$$

rather than in the form given above,

$$Lu = a^{ij}u_{,ij} + \text{lower-order terms.}$$

For classification one sticks with the latter. The former is advantageous for such things as integrating by parts and uniqueness proofs, generally possesses better properties, and may be converted to the latter by taking the jth derivatives.

Since $u_{,ij} = u_{,ji}$ for twice continuously differentiable functions u, one may assume that $a^{ij} = a^{ji}$ so that the matrix $A = [a^{ij}]$ is an $n \times n$ symmetric matrix. By linear algebra we know that we can transform $Q_1(y)$ to principal axes form by a change of variable from $y = (y_1, \ldots, y_n)$ to $z = (z_1, \ldots, z_n)$, where the z_i are a suitable basis such that one now has the canonical form

$$Q_2(z) = k_i(z_i)^2 = \sum_{i=1}^{n} k_i(z_i)^2.$$

The number of positive, negative, and zero k_i are invariant under any (nonsingular) linear change of variable $(y_1, \ldots, y_n) \to (z_1, \ldots, z_n)$, and serve to classify the operator. For example, if all the k_i are of the same sign, the partial differential equation is elliptic; if all but one are of the same sign, the partial differential equation is hyperbolic.

The above classification is motivated by the $n = 2$ case and the canonical conic section names of elliptic, hyperbolic, and parabolic, from elementary geometry. Also, there is the discriminant test (see also Section 1.1), as in the just-mentioned setting. Let $n = 2$ and consider the second-order partial differential expression

$$Au = a^{ij}u_{,ij} = a^{11}u_{,11} + a^{12}u_{,12} + a^{21}u_{,21} + a^{22}u_{,22}$$

$$\equiv a^{11}\frac{\partial^2 u}{\partial x_1^2} + 2a^{12}\frac{\partial^2 u}{\partial x_1 \partial x_2} + a^{22}\frac{\partial^2 u}{\partial x_2^2},$$

where it may be assumed that $a^{12} = a^{21}$, that is, the coefficients of the mixed derivatives may be initially equalized.

Let the following linear change of basis be assigned*:

$$\begin{pmatrix} z_1 \\ z_2 \end{pmatrix} = \begin{pmatrix} 1 & a^{12}/a^{11} \\ 0 & 1 \end{pmatrix}\begin{pmatrix} y_1 \\ y_2 \end{pmatrix}.$$

Then (by completing the square) one has

$$Q_1(y) = a^{11}y_1 y_1 + a^{12}y_1 y_2 + a^{21}y_2 y_1 + a^{22}y_2 y_2$$

$$= a^{11}\left[\left(y_1 + \frac{a^{12}}{a^{11}}y_2\right)^2 + \frac{a^{11}a^{22} - (a^{12})^2}{(a^{11})^2}y_2^2\right]$$

$$= a^{11}(z_1)^2 + \frac{a^{11}a^{22} - (a^{12})^2}{(a^{11})}(z_2)^2 = Q_2(z).$$

---

* If $a^{11} = 0$, use a^{22} instead. If $a^{11} = a^{22} = 0$, the equation is hyperbolic, corresponding to the alternate conic section canonical form for hyperbolas $y_1 y_2 = 1$, which by (easy) change of variable goes over to the usual canonical form $z_1^2 - z_2^2 = 1$ for hyperbolas.

Classification. Let $d = a^{11}a^{22} - (a^{12})^2$ be called the discriminant. Then (definition):

$$d > 0 \Leftrightarrow Au \text{ is elliptic}$$
$$d < 0 \Leftrightarrow Au \text{ is hyperbolic}$$
$$d = 0 \Leftrightarrow Au \text{ is parabolic}$$

This is the same as saying that the k_i have equal or different signs as mentioned above. In analytical geometry the corresponding canonical forms are

$$k_1(z_1)^2 + k_2(z_2)^2 = 1$$

the three cases of ellipse, hyperbola, or (degenerate) parabola depending on whether $k_2 > 0$, $k_2 < 0$, or $k_2 = 0$, respectively, given $k_1 > 0$.

It must still, however, be shown that the above classification corresponds to and permits, by an acceptable change of variable, the transformation of the given partial differential expression Au into one of the three canonical forms:

elliptic	Laplace operator	$\dfrac{\partial^2 u}{\partial x_1^2} + \dfrac{\partial^2 u}{\partial x_2^2}$
hyperbolic	d'Alembert operator	$\dfrac{\partial^2 u}{\partial x_1^2} - \dfrac{\partial^2 u}{\partial x_2^2}$
parabolic	Diffusion operator	$\dfrac{\partial u}{\partial x_1} - \dfrac{\partial^2 u}{\partial x_2^2}$

This may be verified as follows. For the given differential expression

$$Au = a^{11}u_{x_1 x_1} + 2a^{12}u_{x_1 x_2} + a^{22}u_{x_2 x_2} + \text{(lower-order terms)}$$

with constant coefficients (the variable coefficient case in which one seeks a 1–1 but not necessarily linear change of variable is similar but involves a little more care via the implicit function theorem), we seek a linear change of variable

$$\begin{pmatrix} \zeta_1 \\ \zeta_2 \end{pmatrix} = \begin{pmatrix} c_{11} & c_{12} \\ c_{21} & c_{22} \end{pmatrix} \begin{pmatrix} x_1 \\ x_2 \end{pmatrix}, \quad \text{i.e., } \boldsymbol{\zeta} = C\mathbf{x},$$

which will change the expression A into one of the above three canonical forms. By the calculus chain rule one has

$$u_{x_1} = u_{\zeta_1}(\zeta_1)_{x_1} + u_{\zeta_2}(\zeta_2)_{x_1}$$

and so on, arriving by direct substitution at the expression

$$Au = b^{11}u_{\zeta_1 \zeta_1} + 2b^{12}u_{\zeta_1 \zeta_2} + b^{22}u_{\zeta_2 \zeta_2}$$

plus lower-order terms, where the new coefficients are

$$b^{11} = a^{11}((\zeta_1)_{x_1})^2 + 2a^{12}(\zeta_1)_{x_1}(\zeta_1)_{x_2} + a^{22}((\zeta_1)_{x_2})^2,$$

$$b^{12} = a^{11}(\zeta_1)_{x_1}(\zeta_2)_{x_1} + a^{12}((\zeta_1)_{x_1}(\zeta_2)_{x_2} + (\zeta_2)_{x_1}(\zeta_1)_{x_2}) + a^{22}(\zeta_1)_{x_2}(\zeta_2)_{x_2},$$

$$b^{22} = a^{11}((\zeta_2)_{x_1})^2 + 2a^{12}(\zeta_2)_{x_1}(\zeta_2)_{x_2} + a^{22}((\zeta_2)_{x_2})^2.$$

(The above expression, depending only on the chain rule, holds for arbitrary change of variable, not necessarily just the linear one we are considering for the case of constant coefficients.) In terms of the matrix C one may write then

$$b^{11} = a^{11}c_{11}^2 + 2a^{12}c_{11}c_{12} + a^{22}c_{12}^2,$$

$$b^{12} = a^{11}c_{11}c_{21} + a^{12}(c_{11}c_{22} + c_{21}c_{12}) + a^{22}c_{12}c_{22},$$

$$b^{22} = a^{11}c_{21}^2 + 2a^{12}c_{21}c_{22} + a^{22}c_{22}^2.$$

One may now verify the discriminant relation

$$b^{11}b^{22} - (b^{12})^2 = (a^{11}a^{22} - (a^{12})^2)[(\zeta_1)_{x_1}(\zeta_2)_{x_2} - (\zeta_1)_{x_2}(\zeta_2)_{x_1}]^2,$$

and for the present constant ceofficient case the latter factor is just

$$J = (c_{11}c_{22} - c_{12}c_{21})^2.$$

For 1–1 transformations (linear or not) this Jacobian is nonzero and hence the classification of A remains the same in the new variables ζ_1 and ζ_2.

The point is now of course to show (in the constant coefficient case this is just an exercise, and we leave this as a specific exercise for this problem) that one may in fact pick a 1–1 transformation so that one arrives at, in each of the three cases:

elliptic	$b^{12} = 0,$	$b^{11} > 0,$	$b^{22} > 0,$
hyperbolic	$b^{11} = b^{22} = 0,$	$b^{12} \neq 0,$	
parabolic	$b^{11} = b^{12} = 0,$	$b^{22} \neq 0.$	

The other hyperbolic canonical form $b^{11} > 0$, $b^{22} < 0$, $b^{12} = 0$, is available by a further change of variable. The heat equation parabolic canonical form $u_{\zeta_1} - u_{\zeta_2\zeta_2}$ with the first derivative term present will come out (if it is present) by keeping track of the lower-order terms under the change of variable.

Let us consider for example the parabolic case. Here we have the discriminant vanishing, from which we know that

$$(b^{12})^2 = b^{11}b^{22}$$

Thus when we make $b^{11} = 0$ we know also that $b^{12} = 0$, thereby leaving the equation in the desired canonical form after just solving the $b^{11} = 0$ equation. Using also the fact that $a^{12} = (a^{22}a^{11})^{1/2}$, we thus have for the C transformation the conditions for $b^{22} = 1$ and $b^{11} = b^{12} = 0$,

$$a_{11}^{1/2}c_{11} + a_{22}^{1/2}c_{12} = 0,$$

$$a_{11}^{1/2}c_{21} + a_{22}^{1/2}c_{22} = 1.$$

These clearly have solutions.

It is instructive to mention some explicit parallels with the classification and transformation procedures for the conic sections of analytical geometry. In so doing one may also get a feel for the complications of classification for more dimensions, which we will not treat here.

Consider the curve

$$9x^2 - 24xy + 16y^2 - 18x - 101y + 19 = 0.$$

This represents a parabola with focus in the second quadrant. The usual procedures of analytic geometry would first rotate away the cross term by introduction of the new coordinates (x_1, y_1) according to an angle rule

$$\begin{bmatrix} \frac{4}{5} & -\frac{3}{5} \\ \frac{3}{5} & \frac{4}{5} \end{bmatrix} \begin{bmatrix} x_1 \\ y_1 \end{bmatrix} = \begin{bmatrix} x \\ y \end{bmatrix}$$

from which the new equation is

$$25(y_1)^2 - 75x_1 - 70y_1 + 19 = 0.$$

Then the equation is in a standard form $(y_1 - \frac{7}{5})^2 = 3(x_1 + \frac{2}{5})$ from which its graph may be constructed. A further transformation to a form $y_2^2 = x_2$ can be effected by a translation of axes if desired.

Thus if asked to classify and transform to canonical form the partial differential equation

$$9u_{xx} - 24u_{xy} + 16u_{yy} - 18u_x - 101u_y + 19 = 0,$$

one could first be assured of a rotation change of variables to

$$25u_{\eta\eta} - 75u_\zeta - 70u_\eta + 19 = 0$$

with a further change of variable then taking you to

$$u_{rr} - u_t = 0.$$

Problem. (a) Classify and transform the following:

(i) $u_{xx} - u_{xt} - u_{tt} - u_x - u_t = 0$

(ii) $2u_{xx} + u_{xt} + 2u_{tt} + u = 0$

(iii) $-3u_{xx} + 10u_{xt} - 3u_{tt} + 7u_x - u_t + u = 0$

(b) Verify the discriminant relation above, at least in the linear change of variable. (c) Show for the hyperbolic and elliptic case that the linear change of variable to canonical form always exists. Comment on its degree of uniqueness. (d) Carry out both the geometrical and the differential transformations for the conic parabolic example above.

Problem 1.9.2 Characteristics

A subject related to classification (see problem 1.9.1 above) is that of characteristics of the expression Au. Characteristics for linear second-order equations are principally of interest in the hyperbolic case inasmuch as they describe the trajectories of outward spreading waves. It can be shown from the above classification procedure that the classification change of variables transformations yield the *characteristics equation*

$$\frac{dx_2}{dx_1} = \frac{a^{12} \mp \sqrt{(a^{12})^2 - a^{22}a^{11}}}{a^{11}}$$

which, due to the sign of the discriminant, has no real solutions in the elliptic case, one solution in the parabolic case, and two solutions in the hyperbolic case. These solutions are called the *characteristics or characteristic curves* for the equation.

This connection between the classification change of variables and the characteristics of an equation is contained in the following *lemma*: for $\zeta = \zeta(x, y)$ a "classification" change of variable, one obtains a vanishing b^{11} (or b^{22}) coefficient

$$a_{11}(\zeta_x)^2 + 2a_{12}\zeta_x\zeta_y + a_{22}(\zeta_y)^2 = 0$$

if and only if $\zeta(x, y) = $ constant is the general solution of the characteristic equation

$$a_{11}\left(\frac{dy}{dx}\right)^2 - 2a_{12}\left(\frac{dy}{dx}\right) + a_{22} = 0.$$

Let us prove it. Recall the implicit differentiation rule from calculus for a function $z = \zeta(x, y)$ for the special case when $y = y(x)$: from the chain rule

$$\frac{dz}{dx} = \frac{\partial z}{\partial x} \cdot 1 + \frac{\partial z}{\partial y}\frac{dy}{dx}.$$

Letting $F(x, y) = z - c$, if $F(x, y(x)) \equiv 0$ in a neighborhood of a point x_0, then the total derivative $dz/dx = 0$ there, from which emerges the important rule

$$y'(x) = -\frac{F_x}{F_y}.$$

Thus $\zeta(x, y)$ constant and a solution of the characteristics equation renders the b^{11} classification coefficient zero by substituting into it the implicit derivative relation

$$\frac{dy}{dx} = -\frac{\zeta_x}{\zeta_y}.$$

Conversely, given the relation b^{12} (or b^{22}) equal to zero, we have from the quadratic equation

$$a_{11}\left(\frac{\zeta_x}{\zeta_y}\right)^2 + 2a_{12}\left(\frac{\zeta_x}{\zeta_y}\right) + a_{22} = 0$$

the solution $(\zeta_x/\zeta_y) = (-a_{12} \pm \sqrt{a_{12}^2 - a_{11}a_{22}})/a_{11}$ which from the implicit differentiation rule gives us the characteristic equation. Notice the importance of the sign change, to which care must be paid.

Let us illustrate this classification-characteristics connection for the wave equation

$$u_{tt} - u_{xx} = 0.$$

Following the classification procedure let us transform it to the form

$$u_{\zeta\eta} = 0.$$

To do so we want

$$b^{11} = a_{11}(\zeta_x)^2 + 2a_{12}\zeta_x\zeta_t + a_{22}\zeta_t^2 = 0$$

and

$$b^{22} = a_{11}(\eta_x)^2 + 2a_{12}\eta_x\eta_t + a_{22}\eta_t^2 = 0.$$

A number of such transformations $\zeta = r(x, t)$, $\eta = s(x, t)$ were worked out in Problem 2 of Section 1.1. Even within the class of linear transformations they need not be unique. No matter how they are chosen, however, the solutions $t = t(x)$ to the characteristics equation for the wave equation will be the same. From it, namely,

$$\left(\frac{dt}{dx}\right)^2 - 1 = 0$$

we have the wave equation characteristics

$$t = x + c,$$
$$t = -x + c.$$

These are 45° lines in the (x, t) plane along which initial data impulses will propagate, a subject we will take up later (e.g., see Section 2.5). On one ζ is constant, on the other η is constant. Note that in the 45° rotated coordinates found in Problem 2 of Section 1.1, the characteristics equation for the $u_{rs} = 0$ canonical form for the wave equation gives characteristics equation $dr/ds = 0$ and characteristic curves $r = $ constant and $s = $ constant.

Problem. (a) Classify and find the characteristics of the equation

$$u_{tt} + 2u_{xt} + (\sin^2 x)u_{xx} + u_x = 0$$

(b) Find the characteristics of the heat equation $u_t - u_{xx} = 0$ and relate them to the infinite speed of heat propagation that it implies, see Problem 1 of Section 1.7. (c) Examine the characteristics of the Tricomi equation $yu_{xx} + u_{yy} = 0$ mentioned in Section 1.1.

Problem 1.9.3 Maximum Principles

The (strong) maximum principle asserts that if $\Delta u > 0$ in Ω and u is continuous on the closure $\overline{\Omega}$ where $\overline{\Omega}$ is a closed bounded domain in n-space, then u attains its maximum only on $\partial\Omega$. This follows simply from the calculus conditions for a maximum. Because $u \in C^0(\overline{\Omega})$ there is a maximum somewhere on $\overline{\Omega}$. If this obtains at an interior point $(x_1, \ldots, x_n)$ then necessarily the $u_{xi} = 0$ there and the $u_{x_i x_i} \leq 0$ so that $\Delta u \leq 0$, a contradiction. Such functions u are called strictly subharmonic because the condition $\Delta u > 0$ implies that they sag below their maximum and average values. An easily remembered example is $u(x) = x^2$.

The (weak) maximum principle states that if the inequality is not necessarily strict, i.e., $\Delta u \geq 0$ in Ω, then u attains the maximum somewhere on $\partial\Omega$, and also possibly within Ω. This follows easily from the (strong) maximum principle by letting

$$\upsilon = u + \varepsilon\, r^2$$

where $r^2 = x_1^2 + \ldots + x_n^2$. Then $\Delta\upsilon = \Delta u + 2n\varepsilon > 0$ for any positive ε so that υ attains its maximum only on $\partial\Omega$. If the bounded domain Ω now is put within an enclosing ball of radius R we have $\upsilon \leq \max_{\partial\Omega} u + \varepsilon R^2$ and hence

$$u = \upsilon - \varepsilon r^2 \leq \upsilon \leq \max_{\partial\Omega} u + \varepsilon R^2.$$

As ε was arbitrary we may let it tend to zero to guarantee that u is bounded by its maximum on $\partial\Omega$.

The (usual) maximum principle states that for harmonic functions $\Delta u = 0$,

$$\max_{\overline{\Omega}} |\, u(x)\, | \leq \max_{\partial\Omega} |\, u(x)\, |.$$

Sometimes this is stated as

$$\min_{\partial\Omega} u \leq u \text{ in } \Omega \leq \max_{\partial\Omega} u.$$

It follows easily from the (weak) maximum principle applied to u and $-u$.

Problem. (a) Verify the last statement. (b) By modifying the above arguments establish the stability bound

$$|u(x, y)| \leq \max_{\partial\Omega} |u| + (R^2/4) \max_{\overline{\Omega}} |\Delta u|$$

for any $u \in C^2(\Omega) \cap C^0(\overline{\Omega})$. Why is this a stability statement?

(c) State and prove a similar maximum principle for the heat equation,

$$\begin{cases} u_t - \Delta u = 0 \text{ in } \Omega, \quad t > 0, \\ u \text{ given at } t = 0, \\ u \text{ maintained (given) on } \partial\Omega \text{ for } t > 0, \end{cases}$$

asserting that a body Ω with no interior heat sources will attain its maximum temperature either initially (if it is being cooled) or else later (if it is being extermally heated) on its sides.

(d) If you wish, read further in the literature about maximum principles for partial differential equations and in analytic complex function theory. Also, consider why hyperbolic equations do not have natural maximum principles. As an additional exercise, consider a fourth-order elliptic operator (e.g., $\Delta\Delta u = 0$).

Problem 1.9.4 Picard–Cauchy–Kowalewski Theorem

In almost every book on ordinary differential equations one will find the following fundamental existence theorem for the initial value problem

$$\begin{cases} y'(x) = f(x, y(x)), & x > 0, \\ y(0) = y_0. \end{cases}$$

Picard Theorem. If $f(x, y)$ and $\partial f(x, y)/\partial y$ are continuous in a neighborhood of the point $x = 0$, $y = y_0$, then there exists a unique solution to the above initial value problem.

The proof usually consists of converting the ODE initial value problem to the equivalent integral equation

$$y(x) = y_0 + \int_0^x f(s, y(s)) \, ds$$

and then showing that the method of successive approximations converges.

Problem. (a) Consult a book on ordinary differential equations for this proof. (b) Rewrite the proof in terms of the contraction mapping theorem (applied to the integral equation) which guarantees a fixed point for any contracting map in a complete metric space. (c) Demonstrate that the integral equation and the differential equation initial value problem are indeed equivalent.

Not as well known is the fact that the Picard theorem has been generalized to partial differential equations. Before stating that result let us recall the following facts.

1. A function $f(x_1, \ldots, x_n)$ is analytic at a point $(x_1^0, \ldots, x_n^0)$ if there exists a neighborhood $N(x^0)$ wherein f can be represented by a convergent power series

$$f(x_1, \ldots, x_n) = \sum_{k_1, \ldots, k_n \geq 0} a_{k_1 \ldots k_n} (x_1 - x_1^0)^{k_1} \ldots (x_n - x_n^0)^{k_n}$$

where

$$a_{k_1 \ldots k_n} = \frac{1}{k_1! \cdots k_n!} \frac{\partial^{k_1 + \cdots + k_n} f(x^0)}{\partial x_1^{k_1} \cdots \partial x_n^{k_n}}$$

2. Any partial differential equation initial value problem can be reduced to a system of quasilinear (see Problem 1.9.1) equations of first order. This is done by letting each $\partial u_i / \partial x_k$ be a new variable. Thus for the equation

$$\frac{\partial^2 u}{\partial t^2} = f\left(t, x, u, \frac{\partial u}{\partial t}, \frac{\partial u}{\partial x}, \frac{\partial^2 u}{\partial t \, \partial x}, \frac{\partial^2 u}{\partial x^2}\right)$$

one lets $r = f(t, x, u, q, p, s, v)$, differentiates with respect to t to get $\partial r / \partial t = \cdots$ plus a list of the other relations, and then repeats this process after letting $\rho = \partial r / \partial t$. See (c), (d), and (e) of the problem below.

3. The initial value problem for a general quasilinear system of first order, for n independent variables plus a t variable, in N unknowns u_i, is the following (for simplicity we have translated the initial point to the origin):

$$\frac{\partial u_i}{\partial t} = \sum_{j=1}^{N} \sum_{k=1}^{n} a_{ij}^{(k)} \frac{\partial u_j}{\partial x_k} + c_i, \qquad i = 1, \ldots, N,$$

$$u_i(0, x_1, \ldots, x_n) = 0, \qquad i = 1, \ldots, N,$$

where

$$a_{ij}^{(k)} = a_{ij}^{(k)}(t, x_1, \ldots, x_n, u_1, \ldots, u_N),$$

$$c_i = c_i(t, x_1, \ldots, x_n, u_1, \ldots, u_N).$$

Cauchy–Kowalewski Theorem. If the $a_{ij}^{(k)}$ and c_i are analytic at $(0, 0, \ldots, 0, 0, \ldots, 0)$, then there exists in a neighborhood $N(0, 0, \ldots, 0)$ a unique solution to the above initial value problem and passing through the initial point 0.

The proof of this theorem is interesting and goes roughly as follows. Since all functions involved are analytic, one would expect to solve the problem by expanding everything in power series and equating the coefficients. To prove that the resulting formal power series for the solution converges, one needs to dominate it by a majorizing convergent series. For the latter one uses the series expansion of

$$M\left(1 - \frac{x_1}{R_1}\right)^{-1} \cdots \left(1 - \frac{x_n}{R_n}\right)^{-1}$$

where M and R_k are chosen appropriately. Thus the essence of the proof is proper use of a geometric series.

Problem. (a) Investigate the proof of the theroem. (b) Research the literature for more general Cauchy–Kowalewski theorems on (somewhat) arbitrary initial data strips.

(c) Let us elaborate 2 above for the case given there, namely, the second-order initial value problem

$$\begin{cases} u_{tt} = f(t, x, u, u_t, u_x, u_{tx}, u_{xx}), \\ u(0, x) = u_0(x), \\ u_t(0, x) = u_1(x), \end{cases}$$

and show that the resulting quasilinear system is equivalent. The student should verify each of the following steps.

From

$$r = u_{tt} = f(t, x, u, q, p, s, v),$$

we differentiate with respect to t to arrive at

$$r_t = f_t + f_u q + f_q r + f_p s + f_s r_x + f_v s_x$$

where the absence of some terms corresponding to x follows from the independence of x and t. The resulting quasilinear first-order system is

$$\begin{cases} u_t = q \\ q_t = r \\ p_t = q_x \\ s_t = r_x \\ v_t = s_x \\ r_t = f_t(t, x, u, q, p, s, v) + f_u q + \text{the other terms above,} \end{cases}$$

with initial conditions

$$\begin{cases} u(0, x) = u_0(x) \\ q(0, x) = u_1(x) \\ p(0, x) = u_0'(x) \\ s(0, x) = u_1'(x) \\ v(0, x) = u_0''(x) \\ r(0, x) = f(0, x, u_0(x), u_1(x), u_0'(x), u_1'(x), u_0''(x)). \end{cases}$$

Suppose we have functions (u, q, p, r, s, v) that satisfy the quasilinear problem. We claim they satisfy the change of variable equations. If so, the equivalence with the original second-order problem has been established.

From the quasilinear system, $u_t = q$. Also we have $u_{xt} = q_x = p_t$ from which $(u_x - p)_t = 0$. Hence $u_x - p = $ some function $\alpha(x)$. But at $t = 0$, we have $\alpha(x) = u_x(0, x) - p(0, x) = u_0'(x) - u_0'(x) = 0$, and hence $u_x = p$.

Because $u_{tt} = q_t = r$, the variable r is justified immediately from the definitions.

From $(u_{xt} - s)_t = 0$, we have $u_{xt} - s = $ some function $\beta(x)$, which is $u_1'(x) - u_1'(x) = 0$ at $t = 0$; similarly, $u_{xx} - v = $ some function $\gamma(x)$, which is $u_0''(x) - u_0''(x) = 0$ at $t = 0$, and hence $u_{xt} = s$ and $u_{xx} = v$.

Clearly u from the quasilinear system satisfies the original two initial conditions, and one may check in the same way as above that $(u_{tt} - f)_t = 0$ so that $u_{tt} - f = $ some function $\delta(x)$, which is $u_{tt}(0, x) - f(0, x, u_0(x), u_1(x), u_0'(x), u_1'(x), u_0''(x)) = r(0, x) - r(0, x) = 0$ at $t = 0$.

(d) Reduce as in (c) the second-order wave equation problem

$$\begin{cases} u_{tt} - u_{xx} = 0, \\ u(x, 0) = f(x), \\ u_t(x, 0) = g(x), \end{cases}$$

to an equivalent first-order system.

(e) Do it for $f(x) = x$ and $g(s) = 1$ and solve.

Problem 1.9.5

(1) Some Historical Considerations. Almost everyone who has studied any advanced mathematics has heard of Dirichlet, Dirichlet's principle, and the Dirichlet problem. Brief historical remarks concerning this problem and its central role in the development of mathematics in the latter half of the nineteenth century may be found in Section 1.5. Let us go a bit further here.

Problem. (a) Find the exact papers in which Dirichlet himself addressed the problem, principle, and boundary condition that now bear his name.

The Neumann boundary condition arises in applications in which it is desired that there be no flux of the quantity in question through the boundary. There have been several Neumanns in mathematics and more than one who have treated boundary value problems.

Problem. (b) How many Neumanns are there who have made prominent contributions to the study of partial differential equations? In particular, find the particular Neumann and the exact papers in which he addresses the boundary condition of his name.

But almost no one knows about Robin, whose name is (sometimes) attached to the third boundary condition. Our curiosity got the best of us here and we obtained a little information.* There is no doubt an interesting story here, and (although we know more) we do not go to the end of it, instead leaving it as a historical adventure for each student to pursue as he wishes.

Problem. (c) Find out more. In particular, find exactly where, if at all, Robin studied the boundary condition that now carries his name. (d) What relates Debye and Schrödinger? (e) Which Hückel, and why?

(2) Nice and Nonnice Domains. In our discussions thus far we have avoided, beyond brief statements recognizing their existence, pathologies in domains Ω that

---

* The *Oeuvres Scientifiques* of G. Robin were compiled by Louis Raffy with the help of students' lecture notes from Robin's courses given in Paris between 1892 and 1897. There are three volumes: (1) *Théorie nouvelle des fonctions, exclusivement fondée sur l'idée de nombre*; (2a) *Physique mathématique.* (2b) *Thermodynamique générale.* Robin's favorite research subject was thermodynamics, which he thought about from the age of 20 until the last months of his life. He died in his early forties (1855–1897), after having apparently burned most of his papers.

More information may be found in T. Abe and K. Gustafson, *GITOH*41 (1995) (in Japanese) and in T. Abe, K. Gustafson, "The third boundary condition—was it Robin's?" *The Mathematical Intelligencer* **20** (1998).

might cause technical difficulties. As was mentioned in Section 1.5, these questions were finally successfully resolved by Hilbert and Lebesgue, among others. Almost all domains Ω encountered in applications cause no trouble although it is not hard to imagine the physical situation in which an electrostatic charge can build up on certain spikes on the boundary $\partial\Omega$ of a domain, and not on others. Indeed, Lebesgue's example† of a domain Ω on which the Dirichlet problem is not solvable was a three-dimensional sphere punctured by the spine $y = e^{-1/x}$ of revolution (Fig. 1.9a).

FIG. 1.9a

Problem. (a) Read through the references (especially those mentioned in Section 1.6.1) to develop an understanding of nice domains in two senses: (1) divergence theorem domains, those on which one can apply the divergence theorem (called regular, normal, Gaussian domains, among others), and (2) Dirichlet problem domains, those on which one can solve the Dirichlet problem for any continuous boundary data f (called Dirichlet regions or domains, among others). (b) Write down some contributions of Poincaré and Lebesgue in particular. (c) Think about what a "cusp" or "punctured" condition might mean physically in the case, for example, (i) electrostatics or (ii) a Brownian motion inside Ω, as relating to the borderline cases between Ω nice and Ω not nice. (d) What effect does the connectivity of a domain have on whether it is a (1) divergence theorem domain or (2) Dirichlet problem domain? In particular, how about the (i) ring domain $\frac{1}{2} < x^2 + y^2 < 1$ and (ii) the punctured domain $0 < x^2 + y^2 < 1$? (e) What is the effect of dimension on these two classes of domains Ω? (f) In Section 1.5.2, the existence of the Green's function for a partial differential equation on the domain Ω in question was assumed. A third class of nice domains Ω would be those in the sense (3) for which the Green's function $G(P, Q)$ exists. Make this more precise and relate the class (3) to the Dirichlet problem domains (2). In particular, do this for two-dimensional domains by consulting the literature on analytic function theory.

(3) Variational Principles and Euler Equations. In Section 1.5.3 we mentioned the general relationship between a variationally formulated problem and the partial

---

† H. Lebesgue, "Sur des cas d'impossibilité du probleme de Dirichlet," *Comptes Rendus Soc. Math. France* (1913).

differential equation that can be derived from the variational formulation. This differential equation is then called the Euler equation for the problem, and may be linear or nonlinear. Conversely, from a given differential equation one can often profitably go to a variational formulation of the problem. Illustrations of ordinary differential equations exhibiting this duality between variational and differential problems were given in Section 1.5.3.

In this subsection we wish to illustrate for partial differential equations how one obtains the Euler equation from the variational formulation. Further ramifications will be found in Section 2.6. A full treatment would lead to a course in the Calculus of Variations.

Consider first the Dirichlet principle discussed in Section 1.5.3, wherein the variational formulation of the Dirichlet problem

$$\begin{cases} \Delta u = 0 \text{ in } \Omega, \\ u = f \text{ on } \partial\Omega, \end{cases}$$

was asserted to be

$$\begin{cases} D(u) = \min_{\substack{v=f \\ \text{on } \partial\Omega}} D(v), \\[2mm] \text{where } D(v) = \int_\Omega |\text{grad } v(x)|^2 \, dx. \end{cases}$$

Let us derive the former as the Euler equation of the latter. In doing so we shall not worry about any pathological domains Ω, we will assume that we may freely use the divergence theorem and integration by parts (see Section 1.6) on the functions defined on $\overline{\Omega}$, and we will assume that the minimizing solution u in the variational characterization exists and is not a worrisome function.

Thus, as in elementary calculus, we may regard the functional* $D(v)$ as a quantity to be minimized by setting its derivative equal to zero. Since the underlying vector space of functions v is a function space and infinite dimensional, we may attempt to perform this minimization of the functional $D(v)$ by setting all of its directional derivatives to zero:

$$0 = \frac{\partial D(u)}{\partial w}\bigg|_{\varepsilon=0} \equiv \lim_{\varepsilon \to 0} \frac{D(u + \varepsilon w) - D(u)}{\varepsilon}.$$

Since the minimization is to be taken over all reasonably nice functions v that equal f on the boundary $\partial\Omega$, the w in the above expression are taken to be any functions equal to 0 on $\partial\Omega$ so that the $v = u + \varepsilon w$ then range over all of the specified v.

The difference quotient in the above expression is easily calculated from the identity (just expand $D(u + \varepsilon w)$)

$$D(u + \varepsilon w) = D(u) + 2\varepsilon D(u, w) + \varepsilon^2 D(w),$$

---

* Generally, a functional is a mapping from vectors to scalars. Here, it assigns a real number $D(v)$ to every reasonably nice function v defined over Ω.

where

$$D(u, w) = D(w, u) = \int_\Omega \text{grad } u(x) \cdot \text{grad } w(x) \, dx,$$

from which

$$D(w, u) = 0.$$

By Green's first identity

$$D(w, u) = \oint_{\partial\Omega} w \frac{\partial u}{\partial n} \, ds - \int_\Omega w \, \Delta u \, dx$$

(see Section 1.6) and the vanishing of w on $\partial\Omega$, we thus may conclude that

$$\int_\Omega w(x) \, \Delta u(x) \, dx = 0$$

for all functions $w(x)$. This is the so-called "weak" or "dual" form of the desired statement that $\Delta u = 0$ pointwise in Ω. The latter may then be concluded for functions u that are $C^2(\Omega)$ by supposing $\Delta u(x_0)$ to be not equal to zero at some point x_0 and taking w to be a suitable continuous function of like sign and zero outside of a suitably small neighborhood of x_0, a contradiction.

As a second illustration of the calculation of the Euler equation from a variational formulation, let us consider the eigenvalue problem

$$\begin{cases} -\Delta u = \lambda u \text{ in } \Omega, \\ u = 0 \text{ on } \partial\Omega, \end{cases}$$

variationally formulated for the lowest eigenvalue λ as

$$\begin{cases} \lambda = \min_{\substack{v=0 \\ \text{on } \partial\Omega}} \frac{D(v)}{\|v\|^2}, \\ \text{where } \|v\|^2 = \int_\Omega v^2(x) \, dx. \end{cases}$$

As in the above example we may set equal to zero all directional derivatives of the functional in order to obtain the minimum, that is,

$$0 = \frac{d\lambda(u)}{dw}\bigg|_{\varepsilon=0} = \lim_{\varepsilon \to 0} \frac{\dfrac{D(u + \varepsilon w)}{\|u + \varepsilon w\|^2} - \dfrac{D(u)}{\|u\|^2}}{\varepsilon}$$

for all reasonably nice functions w that vanish on $\partial\Omega$. Using again the identity $D(u + \varepsilon w) = D(u) + 2\varepsilon D(u, w) + \varepsilon^2 D(w)$ and the similar identity $\|u + \varepsilon w\|^2 = \|u\|^2 + 2\varepsilon \int_\Omega uw \, dx + \varepsilon^2 \|w\|^2$, the difference quotient, after forming the common denominator, becomes

$$\frac{[2D(u, w) + \varepsilon D(w)]\|u\|^2 - D(u)[2 \int_\Omega uw \, dx + \varepsilon\|w\|^2]}{\|u\|^2[\|u\|^2 + 2\varepsilon \int_\Omega uw \, dx + \varepsilon^2\|w\|^2]}.$$

Taking the limit $\varepsilon \to 0$ and using Green's first identity as above then yields in order

$$D(u, w) - \frac{D(u)}{\|u\|^2} \int_\Omega uw \, dx = 0$$

and

$$\int_\Omega w(x)(-\Delta u(x) - \lambda u(x)) \, dx = 0,$$

from which $\Delta u = \lambda u$ in Ω.

Problem. (a) Do the same as was done above for the Dirichlet problem for the minimal surface equation (see the introduction to this chapter). That is, obtain the Euler equation

$$\begin{cases} (1 + (u_y)^2)u_{xx} - 2u_xu_yu_{xy} + (1 + (u_x)^2)u_{yy} = 0 \text{ in } \Omega, \\ u = f \text{ on } \partial\Omega \end{cases}$$

from the variational formulation

$$\begin{cases} E(u) = \min_{\substack{v=f \\ \text{on } \partial\Omega}} E(v), \\ \\ \text{where } E(v) = \int_\Omega [1 + |\text{grad } v(x)|^2]^{1/2} \, dx. \end{cases}$$

(b) Study the general relationships of Euler equations to their Lagrangian and Hamiltonian formulations.

(c) Give an example of a Lagrangian for the wave equation $u_{tt} - u_{xx} - u_{yy} = 0$.

Problem 1.9.6

(1) Some Historical Remarks. As in Problem 1.9.5(1) and elsewhere we have tried to bring the student's attention at least for a few moments to the historical and/or human point of view. However, time and other needs limit this endeavor. Here we restrict ourselves to a few brief remarks concerning the development of the understanding of pointwise convergence of Fourier series. Thus we are elaborating only one aspect of Section 1.6 and in fact only a part of Section 1.6.3. On the other hand, Fourier series and their pointwise convergence are, as we shall see more and more, a most essential ingredient in a study of partial differential equations.

Because we are omitting similar important historical aspects of the topics discussed in Sections 1.6.1 and 2, perhaps we should at least assign to the interested reader the following problem.

Problem. (a) Briefly trace some historical and/or human aspects of the development of (1) the divergence theorem and/or (2) inequalities.

The historical comments below pertain, in general, only to pointwise convergence for the trigonometric Fourier series and not for general Fourier series (see Chapter 2). As will be seen in the proof of the weak Dini test in 1.9.6(2) below, the establishing of pointwise convergence may depend heavily on the particular expansion functions being used.*

As mentioned previously (Section 1.5), we attribute to D. Bernoulli the idea of expanding an arbitrary function in terms of sine finctions, to Euler the calculation of the Fourier coefficients, and to F. Riesz, E. Fischer, and Lebesgue the final statement that the Fourier series for an arbitrary square integrable function on an interval will converge in the mean to the given function.

Dubois–Reymond showed in 1876 that a continuous function f can have Fourier series partial sums s_N^f that diverge at a point x_0. Fejer and Lebesgue in 1904–1905 showed that on the Lebesgue set of an integrable function f (for these concepts see 1.9.6(3) and 1.9.7(3)), s_N^f is Cesaro summable (roughly; averaged partial sums converge). Kolmogorov in 1926 found an integrable (L^1) function f such that the partial sums s_N^f diverged badly pointwise (almost everywhere in the interval). Carleson in 1966[†] showed that all square integrable functions f have the property that their Fourier partial sums s_N^f converge pointwise to f at almost all x.

Problem. (b) Add to this just-given brief history.

(2) Dini Tests and the Gibbs Effect. Again we restrict attention to the pointwise convergence of Fourier trigonometric series. Given a Fourier series solution to a partial differential equation via the method of separation of variables, we will want to know its pointwise convergence properties. The so-called Dini tests introduced in Section 1.6.3 give sufficient conditions for this. The Gibbs effect is a refinement showing the tendency of the partial sums to pile up to the left and right of a discontinuity in the given function f.

We shall state three "Dini"[‡] tests, as follows:

The *Dini Test* states that the Fourier series of a piecewise continuously differentiable function $f(x)$ of period 2π will converge pointwise to the function $f(x)$ at points x at which the function $f(x)$ is continuous, and to the average value of $f(x)$ at points x at which $f(x)$ has a jump discontinuity (Fig. 1.9b).

If (*Strong Dini Test*), moreover, $f(x)$ is continuous with piecewise continuous derivative, then the convergence is uniform on the interval (Fig. 1.9c).

---

* On the other hand it can be shown in the Sturm–Liouville Theory (see Section 2.4), but is beyond the scope of this book, that in many cases a general expansion and the trigonometric expansion will converge or diverge pointwise together.

† Carleson, "On convergence and growth of partial sums of Fourier series," *Acta Math.* **116** (1966).

‡ Named for U. Dini, an Italian mathematician, for his work on Fourier series and the theory of real variables, at Pisa in the 1870s.

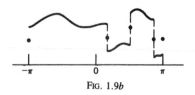

FIG. 1.9*b*

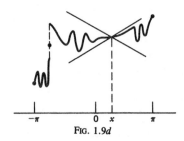

FIG. 1.9*c*

If (*Weak Dini Test*) $f(x)$ is integrable on the interval (i.e., $\int_{-\pi}^{\pi} |f(x)|\, dx < \infty$) and Hölder continuous at x, then its Fourier series converges to $f(x)$ there (Fig. 1.9*d*).

Problem. (a) Follow the proof sketch below for the (weak) Dini test. (b) Consult the literature for other versions, and prove one of them.

Let $f(x)$ satisfy $\int_{-\pi}^{\pi} |f(s)|\, ds < \infty$ on the interval $-\pi \leqq s \leqq \pi$ and be Hölder continuous at x. We recall that this is a condition between those of continuity and differentiability of a function, and means that there are constants M and $\alpha > 0$ such that for all $-\pi \leqq s \leqq \pi$

$$|f(x) - f(s)| \leqq M|x - s|^{\alpha}.$$

It has been assumed that f has been periodically extended beyond the interval $[-\pi, \pi]$.

The classical trigonometric Fourier series for $f(x)$ is

$$\tfrac{1}{2}a_0 + \sum_{n=1}^{\infty} (a_n \cos nx + b_n \sin nx)$$

where $a_n = \pi^{-1} \int_{-\pi}^{\pi} f(s) \cos ns\, ds$, $b_n = \pi^{-1} \int_{-\pi}^{\pi} f(s) \sin ns\, ds$. Let $S_N(x)$ denote the Nth partial sum, that is,

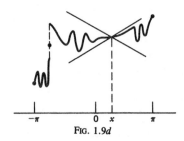

FIG. 1.9*d*

$$S_N(x) = \tfrac{1}{2}a_0 + \sum_{n=1}^{N} (a_n \cos nx + b_n \sin nx).$$

It is to be shown that $S_N(x) \to f(x)$.

Proof.

(Sketch.) (1) Establish the trigonometric identity

$$\tfrac{1}{2} + \cos \theta + \cos 2\theta + \cdots + \cos N\theta = \frac{\sin[(N + \tfrac{1}{2})\theta]}{2 \sin (\tfrac{1}{2}\theta)}$$

as follows:

$$2 \sin (\tfrac{1}{2}\theta)\left(\tfrac{1}{2} + \sum_{n=1}^{N} \cos n\theta \right) = \sin (\tfrac{1}{2}\theta) + \sum_{n=1}^{N} 2 \sin (\tfrac{1}{2}\theta) \cos n\theta$$

$$= \sin (\tfrac{1}{2}\theta) + \sum_{n=1}^{N} (\sin[(n + \tfrac{1}{2})\theta]$$

$$- \sin[(n - \tfrac{1}{2})\theta])$$

$$= \sin[(N + \tfrac{1}{2})\theta].$$

The right-hand side of the identity is sometimes called $D_N(\theta)$, the Dirichlet kernel.

(2) Substitute a_n and b_n into the partial Fourier sums at x as follows:

$$S_N(x) = \pi^{-1} \int_{-\pi}^{\pi} f(s) \left(\tfrac{1}{2} + \sum_{n=1}^{N} (\cos nx \cos ns + \sin nx \sin ns) \right) ds$$

$$= \pi^{-1} \int_{-\pi}^{\pi} f(s) \left(\tfrac{1}{2} + \sum_{n=1}^{N} \cos n(x - s) \right) ds$$

$$= (2\pi)^{-1} \int_{-\pi}^{\pi} f(s) \frac{\sin[(N + \tfrac{1}{2})(x - s)]}{\sin[\tfrac{1}{2}(x - s)]} \, ds$$

$$= (2\pi)^{-1} \int_{-\pi}^{\pi} f(x + \theta) \frac{\sin(N + \tfrac{1}{2})\theta}{\sin(\tfrac{1}{2}\theta)} \, d\theta.$$

The identity in (1), a change of variable, and periodicity of the integrand have been used.

(3) Note that $D_N(\theta)$ has the property (by use of (1))

$$\int_{-\pi}^{\pi} D_N(\theta) = \pi$$

so that upon subtraction

$$S_N(x) - f(x) = (2\pi)^{-1} \int_{-\pi}^{\pi} \left(\frac{f(x + \theta) - f(x)}{\sin(\frac{1}{2}\theta)} \right) \sin[(N + \tfrac{1}{2})\theta] \, d\theta.$$

(4) Call upon the Riemann–Lebesgue lemma (see Second Pause) that asserts: If a function $h(\theta)$ is integrable on the interval $(-\pi, \pi)$, then

$$(h, \sin[(N + \tfrac{1}{2})\theta]) \equiv \int_{-\pi}^{\pi} h(\theta) \sin[(N + \tfrac{1}{2})\theta] \, d\theta \to 0 \qquad \text{as } N \to \infty.$$

Here one requires that the functions $\varphi_N = \sin[(N + \tfrac{1}{2})\theta]$ are orthogonal in the sense

$$\int_{-\pi}^{\pi} \sin[(N + \tfrac{1}{2})\theta] \sin[(M + \tfrac{1}{2})\theta] \, d\theta = 0, \qquad N \neq M,$$

easily seen from 2 sinA sinB = cos (A-B) - cos (A+B), and that the φ_N are uniformly bounded on the interval.

(5) Letting

$$h(\theta) = \frac{f(x + \theta) - f(x)}{\sin(\frac{1}{2}\theta)} \quad \text{on } -\pi \leq \theta \leq \pi$$

and noting that $\lim \sin(\tfrac{1}{2}\theta)/\theta = \tfrac{1}{2}$ as $\theta \to 0$, one sees from (4) that

$$S_N(x) - f(x) \to 0 \qquad \text{as } N \to \infty$$

provided that any of the following hold:

$$\int_{-\pi}^{\pi} |h(\theta)| \, d\theta < \infty, \qquad \int_{-\pi}^{\pi} \left| \frac{f(x + \theta) - f(x)}{\theta} \right| d\theta < \infty,$$

f Hölder continuous or differentiable at x.

The Gibbs effect describes how the Fourier series of a function accommodates itself to the function's discontinuities. It was first pointed out for sawtooth waves by J. W. Gibbs.* Plotting the Nth partial sum of the Fourier series for such functions yields a picture like Figure 1.9e. The graph of s_N^f approaches (uniformly in the plane) the slanted sawtooth slopes plus the vertical lines shown. The amount of vertical "overshoot" can be determined (say, for a piecewise differentiable function of period 2π); it is, both at the top and the bottom, the magnitude of the jump of f at the point x_0 times the numerical factor

$$\frac{1}{\pi} \int_{\pi}^{\infty} \frac{\sin s}{s} \, ds \cong \frac{0.281}{\pi} \cong 0.0895.$$

---

* J. W. Gibbs, in a letter to *Nature* **59** (1899). For this and other contributions Gibbs has been honored by a prize in his name given each year by the American Mathematical Society for notable work in applied mathematics.

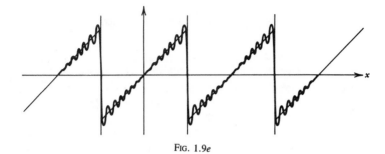

FIG. 1.9e

Problem. (c) Investigate the Gibb's effect for the square wave

$$f(x) = \begin{cases} -1, & (2n - 1)\pi < x \le 2n\pi, \\ +1, & 2n\pi < x \le (2n + 1)\pi. \end{cases}$$

Plot a few partial sums of the Fourier series for f. Then prove that the overshoot is of the magnitude given above.

(3) Lebesgue Dominated Convergence Theorem. In Section 1.6.2 we introduced the $L^p(\Omega)$ spaces. In 1.6.3 we discussed convergence in $L^2(a, b)$ essential to an understanding of Fourier series. In Section 2.4 we will treat "weighted" L^2 spaces $L^2(a, b, r)$.

On the other hand and as has been mentioned elsewhere, the majority of specific functions encountered will be integrable in the usual (Riemann) sense and there will be no need of the more extended (Lebesgue) sense of integral. The student may thus work in most cases with the L^2 convergence without worry of pathology (i.e., one may call it R^2 convergence, when it converges).

Yet there is one aspect of the Lebesgue theory extremely useful in many cases and even meaningful in the Riemann context. That is, the Lebesgue dominated convergence theorem, perhaps the most important convergence theorem in the Lebesgue theory. In particular, it is useful "differentiating under the integral." We will employ it in this way in a rigorous treatment of the attainment of initial values by solutions of the heat equation (Problem 1.9.7(3)).

The Lebesgue dominated convergence theorem states that a sequence of integrable (i.e., $L^1(\Omega)$) functions $u_n(x)$ that are uniformly (i.e., $|u_n(x)| \le M(x)$) bounded by some integrable (i.e., $L^1(\Omega)$) function $M(x)$ and which converge pointwise to a function $u(x)$ (i.e., $u_n(x) \to u(x)$ for all, or for "almost all," x in Ω, as $n \to \infty$) possess the property that then u_n converge to $u(x)$ in the mean (i.e., in $L^1(\Omega)$). Specifically, this guarantees that the limit function $u(x)$ is also integrable and that $\int_\Omega |u_n(x) - u(x)| \, dx \to 0$ as $n \to \infty$.

For differentiating under the integral, one can then for example obtain $(d/dy)\int_a^b f(x, y) \, dx$ by the Lebesgue dominated convergence theorem by letting $u_n(y) = f(x, y + \Delta_n y) - f(x, y)$, where $\Delta_n y$ denotes a small increment in y decreasing to zero as n increases, and considering the expression

$$\frac{d}{dy} \int_a^b f(x, y) \, dx = \lim_{n \to \infty} \int_a^b \frac{u_n(y)}{\Delta_n y} \, dx.$$

Problem. (a) Accepting the Lebesgue dominated convergence theorem for $L^1[a, b]$ functions, justify the above differentiation under the integral. (b) Scan any good book on Lebesgue integration theory and write down two or three other important convergence theorems found there. (c) In the dominated convergence theorem we encountered the concept that $u_n \to u$ for "almost all" x in Ω. Equivalently one says $u_n \to u$ "almost everywhere." While doing part (b), gain some understanding of "almost everywhere" convergence in terms of a few simple examples of what it means.

Problem 1.9.7

The concept of *self-adjointness* is important for partial differential equations and even more so in applications, where it, for example, confirms that all boundary conditions and eigenvalues have been accounted for in a given problem. For differential operators, the establishing of self-adjointness is rather delicate and is connected to the notion of *distributional derivative*. Let us illustrate these two concepts and their connection, in (1) and (2) below.

(1) Self-Adjoint Operators. First recall the definition of the adjoint operator A^* for an $n \times n$ matrix operator $A = [a^{ij}]$ on real Euclidean n-space, given by

$$\langle Ay, x \rangle = \sum_{j=1}^n \sum_{i=1}^n a^{ij} y_i x_j = \sum_{j=1}^n \sum_{i=1}^n a^{ji} x_i y_j = \langle y, A^* x \rangle.$$

Here A^* is just the (conjugate transpose in the complex case) transpose $[a^{ji}]$. For a differential operator let us consider specifically $L = d^2/dx^2$ in the $L^2[0, \pi]$ space of real valued (the complex valued case is similar) square integrable functions $u(x)$ defined on the interval $0 \leq x \leq \pi$. We reach a similar relation to that just given for the matrix A if we consider functions u and v that are $C^2(0, \pi) \cap C^1[0, \pi]$ and vanish at the end points, that is, $0 = u(0) = u(\pi) = v(0) = v(\pi)$. Then (by integration by parts)

$$\langle Lv, u \rangle = \int_0^\pi v''(x) u(x) \, dx = \int_0^\pi u''(x) v(x) \, dx = \langle v, Lu \rangle.$$

In such a case L is called *symmetric*. Thus the operator L, defined by the operation $Lu = u''$ and operating on a domain $D(L)$ consisting of all functions $u \in C^2(0, \pi) \cap C^1[0, \pi]$ satisfying the boundary condition $u(0) = u(\pi) = 0$, is symmetric. The general definitions of symmetric and self-adjoint operators are as follows.†

---

† We have tacitly assumed that $D(L)$ is dense in $L^2[0, \pi]$, a technicality always met in these examples. See Problem 2.9.3, where it is shown that C_0^∞ functions u are dense in $L^2[0, \pi]$.

Definition (adjoint operator L^*).   Given L, $D(L^*) = \{$all u such that there exists a w (depending on u) such that

$$\langle Lv, u \rangle = \langle v, w \rangle$$

for all v in the domain $D(L)$ of $L\}$; and then (define) $L^*u = w$. Thus

$$\langle Lv, u \rangle = \langle v, L^*u \rangle$$

for all v in $D(L)$ and all u in $D(L^*)$.

Definition (symmetric operator). L is called symmetric if $L \subset L^*$, that is, if $D(L)$ is contained in $D(L^*)$ and if $Lv = L^*v$ for all v in $D(L)$.

In other words, a symmetric operator L is symmetric in the sense that

$$\langle Lv, u \rangle = \langle v, Lu \rangle$$

for all v and u in $D(L)$.

Definition (self-adjoint operator). L is called self-adjoint if $L = L^*$, that is, L is symmetric and $D(L) = D(L^*)$.

In the matrix case above, since $D(A) = D(A^*) = $ all n-vectors, the notions of symmetric and self-adjoint coincide, and A will be symmetric if and only if the matrix $[a^{ij}]$ is symmetric, that is, $[a^{ij}] = [a^{ji}]$. The differential operator L defined above with domain $D(L) = \{u \in C^2(0, \pi) \cap C^1[0, \pi], u(0) = u(\pi) = 0\}$ is, by the integration by parts argument done above, symmetric. As it turns out, one can take second derivatives in the Lebesgue sense (equivalently: distributional sense) of functions u that are not necessarily C^2 (example: a function that is C^1 but only piecewise C^2), and for this reason $D(L)$ as given is not sufficiently large in order for L to be self-adjoint. One can define distributional derivatives of all orders, but in the following subsection we restrict attention to the second-order case.

Problem. (a) Show that the operator $Lu = u''$ with domain $D(L) = \{u \in C^2(0, \pi) \cap C^1[0, \pi], u(0) = u(\pi) = u'(0) = u'(\pi) = 0\}$ is symmetric but not self-adjoint, due to too many boundary conditions. (b) Show that the operator $Lu = u''$ with domain $D(L) = \{u \in C^2(0, \pi) \cap C^1[0, \pi]\}$ is not symmetric, due to too few boundary conditions. (c) Verify some of the statements in (1) above.

(2) Distributional Derivatives. If for all "test functions" φ in $C_0^\infty(0, \pi)$, that is, infinitely differentiable functions whose support is strictly contained within the interval $(0, \pi)$, one has the "integration by parts" relation

$$\langle L\varphi, u \rangle \equiv \int_0^\pi \varphi''(x)u(x)\,dx = \int_0^\pi \varphi(x)w(x)\,dx \equiv \langle \varphi, w \rangle$$

for a given function $u(x)$, then u is said to have distributional second derivative, defined by $u'' = w$.

Note the resemblance to the definition of the adjoint operator L^* given above. If we had restricted $D(L)$ originally to $D(L) = C_0^\infty(0, \pi)$, then the functions u possessing distributional second derivatives would have comprised exactly the larger domain $D(L^*)$.

By taking into account the Lebesgue theory, one can show that the right domain $D(L)$ to make the operator L above self-adjoint is $D(L) = \{$all square integrable u possessing distributional first- and second-order derivatives that are also square integrable, and also satisfying (in an appropriate weak sense) the boundary condition $u(0) = u(\pi) = 0\}$.

The terms distributional derivative, weak derivative, generalized derivative, and strong derivative will be found in the literature. It can be shown that in most cases (provided that they are appropriately integrable) they are the same, at least locally, and correspond to "differentiation almost everywhere," in the Lebesgue sense. Of course for all test functions and most other smooth functions encountered in practice, they are simply the usual derivative in the Riemann sense. The same statement applies to distributional partial derivatives.

> **Problem.** (a) Assuming that integration by parts calculus works for distributional derivatives just as it does for usual derivatives (it does), show the above operator L to be self-adjoint. (b) A function u is said to be harmonic in the *weak* or distribution sense if $\int_\Omega u \, \Delta\varphi \, dx = 0$ for all test functions $\varphi \in C_0^\infty(\Omega)$. Show in the one-dimensional case that if u is weakly harmonic on the interval $(0, \pi)$ and is also a $C^2(0, \pi)$ function, then u is harmonic in the usual sense on $(0, \pi)$.

(3) Heat Equation Rigorously. We wish to close this problem by a further examination of the heat equation initial value problem. In so doing we have the following three aims in mind.

(i) The use of the Lebesgue theory (and, in particular, the Lebesgue dominated convergence theorem and the Lebesgue set of a function, in a further mathematical study of partial differential equations) will be illustrated.

(ii) The physical understanding of the heat equation will be enhanced; in particular, its extremely smoothing behavior.

(iii) The power of the Green's function, once known, in providing a number of existence, uniqueness, stability, construction, regularity, and approximation conclusions, should become quite clear.

The linear heat equation

$$\begin{cases} u_t - u_{xx} = 0, & -\infty < x < \infty, \quad t > 0, \\ u(x, 0) = f(x), & -\infty < x < \infty, \end{cases}$$

was considered briefly in Problem 2 of Section 1.3. As shown there the Green's function

$$G(x, y, t) = \frac{e^{-(x-y)^2/4t}}{\sqrt{4\pi t}} \begin{cases} -\infty < x < \infty \\ -\infty < y < \infty \\ t > 0 \end{cases}$$

satisfies the heat equation for $t > 0$. Hence verifying the *existence* of the solution

$$u(x, t) = \int_{-\infty}^{\infty} G(x, y, t) f(y)\, dy, \qquad t > 0,$$

reduced to giving conditions under which one can differentiate under the integral.

For the *uniqueness* question one needs to look rather carefully at *regularity* conditions to be placed on the class $\{u\}$ of solution candidates. All of this could be done for example by reference to standard advanced calculus by restricting attention to functions $\{u\}$ and data $\{f\}$ that are continuous with enough continuous derivatives dying off rapidly enough at infinity in both x and t. See also Section 2.2 and Problem 1 therein.

We will now outline how one can obtain further results by use of elements of Lebesgue integration theory. It will be left as a problem at the end of this section that the student should verify some of the calculations. For a discussion of the L^p spaces and theory see Section 1.6.3 and Problem 1.9.6(3).

Theorem

Let $f \in L^p(-\infty, \infty)$ for some $1 \leqq p \leqq \infty$. Then the Green's function representation formula

$$u(x, t) = \int_{-\infty}^{\infty} G(x, y, t) f(y)\, dy$$

provides an infinitely differentiable solution $u(x, t)$ of the heat equation for $-\infty < x < \infty$, $t > 0$. Moreover, $u(x, t)$ converges pointwise to $f(x)$ as $t \to 0$ for "almost all" x. If f is continuous with compact support, the pointwise convergence as $t \to 0$ is uniform in x and the solution $u(x, t)$ is continuous for $-\infty < x < \infty$, $t \geqq 0$. Stability with respect to initial data holds in the sense

$$\max_{\substack{-\infty < x < \infty \\ 0 < t_0 \leqq t}} |u_1(x, t) - u_2(x, t)| \leqq \text{const} \|f_1 - f_2\|_p.$$

The solution $u(x, t)$ tends asymptotically to zero as $t \to \infty$ at the rate $(t^{1/2})^{1/p}$, uniformly in $-\infty < x < \infty$.

The moral of the above results is that for all data, no matter how bad (with the exception of data possessing very singular heat sources), the heat equation immediately smooths them out and begins spreading the energy as widely as possible.

Proof (outline)

Recall (see Section 1.6) that

$$\|w\|_p = \left(\int_{-\infty}^{\infty} |w(y)|^p \, dy \right)^{1/p}$$

for any function $w \in L^p(-\infty, \infty)$, $1 \leq p < \infty$, $\|w\|_\infty = \operatorname*{ess.\ sup}_{-\infty < y < \infty} |w(y)|$, and $p^{-1} + q^{-1} = 1$ defines the conjugate q to any given p. The $p = \infty$ case, and sometimes the $p = 1$ case, are often treated separately after the other p.

(1) For $p = \infty$ one has $\|G(x, y, t)\|_\infty = 1$ and for $1 \leq p < \infty$ one has

$$\|G(x, y, t)\|_p = \frac{c(p)}{(t^{1/2})^{1/q}}, \qquad c(p) = \frac{1}{(4\pi)^{1/2q} \cdot p^{1/2p}}.$$

This follows easily by the change of integration variable $s = \sqrt{(p/4t)}(y - x)$, recalling that $\int_{-\infty}^{\infty} e^{-s^2} \, ds = \pi^{1/2}$.

(2) For all $n = 1, 2, 3, \ldots$ one sees that $(\partial^n/\partial t^n)G(x, y, t)$ is $e^{-(x-y)^2/4t}$ times a polynomial in $(x - y)^2$ with coefficients that are functions of t and bounded above and below for t in any bounded positive interval $0 < \alpha \leq t \leq \beta < \infty$. Since $e^{-(x-y)^2/4\beta} \cdot (x - y)^{2k}$ can be seen to be in $L^p(-\infty, \infty)$ for all x, p, and k, so is $(\partial^n/\partial t^n)G(x, y, t)$. This calculation reduces to showing $\int_0^\infty e^{-s^2} s^{2k} \, ds < \infty$. Likewise for $(\partial^n/\partial x^n)G(x, y, t)$ for x in a bounded interval $-\infty < a \leq x \leq b < \infty$.

(3) One can now differentiate under the integral, that is, for all $n = 1, 2, 3, \ldots$,

$$\frac{\partial^n u(x, t)}{\partial t^n} = \int_{-\infty}^{\infty} \frac{\partial^n}{\partial t^n} G(x, y, t) f(y) \, dy,$$

$$\frac{\partial^n u(x, t)}{\partial x^n} = \int_{-\infty}^{\infty} \frac{\partial^n}{\partial x^n} G(x, y, t) f(y) \, dy.$$

This may be verified by induction and by consulting the Lebesgue differentiation theorems. It works as follows. Bound the difference quotient, using (2) above,

$$\left| \frac{\Delta G(x, y, t)}{\Delta t} \right| \equiv \left| \frac{G(x, y, t) - G(x, y, t_0)}{t - t_0} \right| \leq g_{t_0}(x)$$

where $g_{t_0}(x)$ is in L^q ($-\infty < x < \infty$). Thus by Hölder's inequality (see Section 1.6.2) $\Delta Gf/\Delta t$ is in L^1 since $\Delta G/\Delta t$ is in L^q and f is in L^p. Apply the Lebesgue dominated convergence theorem (see Problem 1.9.6(3)) as follows:

$$\frac{\Delta u}{\Delta t}(x, t_0) = \int_{-\infty}^{\infty} \frac{\Delta G(x, y, t_0)}{\Delta t} f(s) \, ds,$$

converges to

$$\frac{\partial u}{\partial t}(x, t_0) = \int_{-\infty}^{\infty} \frac{\partial G(x, y, t_0)}{\partial t} f(s) \, ds.$$

(4) It follows that $u(x, t)$ satisfies the heat equation and moreover is $C^{\infty}(-\infty < x < \infty, t > 0)$. The latter may be argued with ε's and δ's similarly to the considerations of (2) and (3) or alternately by recourse to the Lebesgue theory.

(5) By Hölder's inequality one has the stability statement

$$|u_1(x, t) - u_2(x, t)| \le \|f_1 - f_2\|_p \|G(x, \cdot, t)\|_q.$$

(What about other stability statements, such as with $\|u_1 - u_2\|_{p \text{ or } q}$?) Taking $u_2 = f = 0$ and by use of (1), one obtains the asymptotic convergence of $u(x, t)$ to zero as $t \to \infty$.

(6) A very interesting set for any given locally integrable function $f(x)$, and thus for any f in any L^p, is the Lebesgue set $\{x\}_f$ of $-\infty < x < \infty$ such that

$$\lim_{h \to 0} \frac{1}{|h|} \int_x^{x+h} |f(x) - f(s)| \, ds = 0.$$

The Lebesgue set $\{x\}_f$ contains almost all $-\infty < x < \infty$, excluding only a set of measure zero. One can show the solution $u(x, t)$ converges to $f(x)$ as $t \to 0$ for all x in the Lebesgue set $\{x\}_f$. In particular, this set contains all points x of continuity of $f(x)$. If desired, the convergence of $u(x, t)$ to $f(x)$ as $t \to 0$ for x in the latter smaller set of continuity points of f may be shown more directly.

Problem. (a) Verify some of the straightforward calculations in the above. (b) Those who are already familiar with elements of the Lebesgue integration theory should verify all details. (c) Interpret the Green's function for the Heat equation as the density function for the normal probability distribution of elementary statistics.

Problem 1.9.8   Nonlinear Oscillations and the Van der Pol Equation

Let us first make three points concerning perspectives.

(1) In Section 1.8 we gave a brief introduction to bifurcation theory. The point of view taken there, a currently prevalent one, was that of nonlinear boundary value problems. Historically, however, the viewpoints taken were more varied. In particular, especially for the case of ordinary differential equations, the origins go back to Poincaré and his work in celestial mechanics.* For ordinary differential equations one may write a second-order equation as a system of first-order equations and then

---

* H. Poincaré, *Les méthodes nouvelles de la mécanique céleste* (Gauthier–Villars, Paris, 1892).

use the extremely well-developed methods (see below) of phase plane analysis of trajectories of that system.

(2) In most of the examples in bifurcation theory considered thus far, for reasons of simplicity, we have considered only the ordinary differential equation versions. The original partial differential equations for the various applications carry greater physical accuracy and increased mathematical richness and are of great interest, but their systematic treatment is beyond the level of this book and in many cases has not yet been carried out.

(3) The study of nonlinear oscillations is a large one that goes far beyond bifurcation theory. Roughly speaking a bifurcation theory approach to a problem in nonlinear oscillations is justified when addressing questions in which it seems that according to physical or numerical evidence a radical change in the qualitative behavior of solutions takes place at certain critical values of a parameter.

In this final section our principal objective is to mention and draw attention to the fact that the phase plane methods for studying nonlinear equations should not be overlooked when the equations involved are, or have under additional assumptions been simplified to, autonomous ordinary differential equations.

To illustrate this method let us consider the Van der Pol equation*

$$u''(t) + \lambda(u^2(t) - 1)u'(t) + u(t) = 0, \qquad \lambda > 0,$$

which has application to self-excited oscillations in electron tube circuits as well as elsewhere.† The nonlinear term here is $\lambda(1 - u^2(t))u'(t)$. We note that it includes the parameter λ as well as the nonlinearity, and represents a damping term that will either supply or withdraw energy from the system, depending on the sign of the term. So-called stationary states (also called limit cycles, see below) are expected to prevail when the averaged increases and decreases in energy during one cycle add to zero.

In the phase plane approach to second-order ordinary differential equations, one lets

$$x_1(t) = u(t),$$
$$x_2(t) = u'(t),$$

and plots x_2 against x_1. The motivation for this approach is the basic sinusoidal oscillation $u(t) = \sin t$, which clearly has for its phase portrait a circle (Fig. 1.9f). Other oscillations for nonlinear equations depart from this phase portrait in various ways.

The unperturbed Van der Pol equation is just

$$u''(t) + u(t) = 0$$

---

* B. Van der Pol, "On relaxation oscillations," *Phil. Mag.* **2** (1926).

† Although vacuum tubes are being replaced in numerous applications by solid-state devices, the Van der Pol equation remains a classic prototype of the types of equations successfully investigated by phase plane analysis. For some nonlinear equations currently encountered in electronics theory, see Problem 2.9.8.

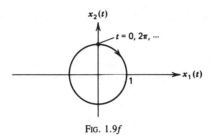

FIG. 1.9*f*

which has solutions

$$u(t) = c_1 \sin t + c_2 \cos t$$
$$= a \sin(t + \alpha).$$

where c_1, c_2, a, and α represent the usual arbitrary constants. The unperturbed problem thus has a phase portrait like that drawn in Figure 1.9*f* but with radius a. The bifurcation viewpoint would then indicate that, for small λ at least, the above unperturbed phase portrait could be an approximate "limit-cycle" for the nonlinear equation. By this reasoning one would expect that solutions of the nonlinear equation would spiral or be otherwise attracted to or from some slight distortion (depending on λ) of the above circular trajectory. The latter distorted closed loop or "limit-cycle" would represent a "periodic orbit" of the nonlinear equation. This indeed can be shown* (Fig. 1.9*g*), even though the exact solutions of the Van der Pol equation cannot be found.

Further consideration of these matters would take us too far afield. For some beautiful and complicated phase portraits of the forced Van der Pol equation

$$u''(t) + \lambda(u^2(t) - 1)u'(t) + \omega_0^2 u(t) = \omega_0^2 A \cos \omega t$$

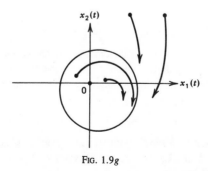

FIG. 1.9*g*

* See, for example, S. Lefschetz, *Differential Equations: Geometric Theory*, 2nd ed. (Wiley, New York, 1963).

we refer the reader to the paper of Sideriades.† For a little exercise we suggest the following.

Problem. (a) Substitute $u(t) = a(t) \sin[t + b(t)]$ as an initial approximate solution into the Van der Pol equation and see how much you can conclude. (b) Show by a little trigonometry that there are no circular solutions when $\lambda \neq 0$. (c) Establish interesting energy facts for the periodic solution, such as $\int u^2 = D(u) = \frac{1}{8}D(u^2)$, integrals taken over the period.

Problem 1.9.9   Confirmation Exercises

The reader should work the following three exercises before going on to the next chapter. See also the ninth exercise for further practice.

1. (a) On what property of limits does the linearity of d/dx and $\partial/\partial x$ depend?
   (b) Determine which of the following operators are nonlinear:

$$Lu = \frac{\partial u}{\partial t} + x^2\frac{\partial^2 u}{\partial x^2}, \qquad Lu = u\frac{\partial^2 u}{\partial x^2}, \qquad Lu = e^{x^2 t}\frac{\partial^2 u}{\partial x^2},$$

$$Lu = \Delta_2 u + \int_0^\pi \sin x \cos y \, u(x, y) \, dy, \qquad Lu = \int_a^b u(y) \, dy,$$

$$Lu = \Delta u - u\int_\Omega u^2(y) \, dy, \qquad L\begin{pmatrix} x_1 \\ x_2 \end{pmatrix} = \begin{bmatrix} a_{11} & a_{12} \\ a_{21} & a_{22} \end{bmatrix}\begin{pmatrix} x_1 \\ x_2 \end{pmatrix}.$$

   (c) Classify (as far as possible) the operators in part (b) as elliptic, parabolic, or hyperbolic.

2. (a) Show that

$$u(x, t) = \frac{1}{\sqrt{4\pi|t|}} \int_{-\infty}^{\infty} e^{-(x-y)^2/4t} f(y) \, dy$$

   satisfies the heat equation

$$u_t - u_{xx} = 0 \quad \text{for } t \neq 0, \qquad -\infty < x < \infty,$$

   assuming that one may differentiate under the integral. Then show that the following initial value problem (*backward heat equation*) is not well-posed due to lack of stability with respect to initial data:

$$\begin{cases} u_t - u_{xx} = 0, & -\infty < x < \infty, \quad t < 0, \\ u(x, 0) = f(x), & -\infty < x < \infty. \end{cases}$$

   (b) Show that the following initial value problem (Cauchy problem of Hadamard for the Laplace equation) is not well-posed due to lack of initial data stability:

---

† L. Sideriades, in *l'Onde Electrique* 463 (Editions CHIRON, Paris, 1965).

$$\begin{cases} u_{xx} + u_{tt} = 0, & 0 < x < \pi, \quad t > 0, \\ u(0, t) = u(\pi, t) = 0, & t > 0, \\ u_t(x, 0) = 0, & 0 < x < \pi, \\ u(x, 0) = f(x), & 0 < x < \pi. \end{cases}$$

(c) Why is the problem

$$\begin{cases} -u'' - 25u = \sin 5x, & 0 < x < \pi, \\ u(0) = u(\pi) = 0, \end{cases}$$

not well-posed?

3. (a) Solve formally by separation of variables the heat conduction problem (Fig. 1.9h):

$$\begin{cases} u_t - 3u_{xx} = 0, & 0 < x < \pi, \quad t > 0, \\ u(0, t) = u(\pi, t) = 0, & t > 0, \\ u(x, 0) = x, & 0 < x < \pi. \end{cases}$$

(b) Solve formally by separation of variables the Dirichlet problem (Fig. 1.9i):

$$\begin{cases} u_{rr} + r^{-1}u_r + r^{-2}u_{\theta\theta} = 0, & r < 1, \\ u(1, \theta) = \begin{cases} 1, & 0 < \theta < \pi, \\ 0, & -\pi < \theta < 0. \end{cases} \end{cases}$$

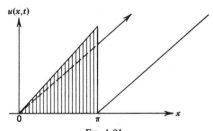

FIG. 1.9h

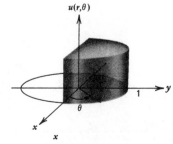

FIG. 1.9i

(c) Solve formally by separation of variables the wave equation problem (Fig. 1.9*j*):

$$
\begin{cases}
u_{tt} - u_{xx} = 0, & 0 < x < \pi, \quad t > 0, \\
u(0, t) = u(\pi, t) = 0, & t > 0, \\
u_t(x, 0) = 0, & 0 < x < \pi, \\
u(x, 0) = \begin{cases} x/\pi, & 0 < x < \pi/2, \\ 1 - x/\pi, & \pi/2 < x < \pi. \end{cases}
\end{cases}
$$

(d) Solve formally by separation of variables the Dirichlet problem of Section 1.2, Problem 1(b).

(e) From the general theory of the Dirichlet problem, what do you know about the solution found in part (d) above?

The following problems contain supplementing information and are varied in content and difficulty.

4. (a) Liouville's theorem states that a harmonic function on the whole plane must be a constant. That is, $u \in C^2(\Omega)$, $\Delta u = 0$, Ω the plane, implies that $u(x, y) = c$. Show this.

 (b) Bernstein's theorem states similarly that a C^2 solution u of the minimal surface equation $(1 + (u_y)^2)u_{xx} - 2u_x u_y u_{xy} + (1 + (u_x)^2)u_{yy} = 0$ on the plane must be linear, that is, $u = ax + by + c$. Show this.

 (c) Jörgen's theorem states in like fashion that a C^2 solution of the slicing equation (gravitational theory) $u_{xx}u_{yy} - u_{xy}^2 = 1$ is quadratic, that is, $u = ax^2 + cy^2 + bxy + dx + ey + f$. Show this.

 (d) Show Jörgen's theorem implies Bernstein's theorem and that Bernstein's theorem implies Liouville's theorem.

5. As stated in the introduction to this chapter, the minimal surface equation

$$(1 + (u_y)^2)u_{xx} - 2u_x u_y u_{xy} + (1 + (u_x)^2)u_{yy} = 0 \tag{i}$$

may be transformed into the linear equation

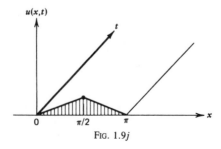

FIG. 1.9*j*

$$(1 + \xi^2)w_{\xi\xi} + 2\xi\eta w_{\xi\eta} + (1 + \eta^2)w_{\eta\eta} = 0 \tag{ii}$$

and the latter may be transformed into the linear Laplace equation

$$v_{rr} + v_{ss} + v_{tt} = 0 \tag{iii}$$

by suitable change of variable. This is done by means of the theory of the Legendre transformation, in which the solution surface u of a partial differential equation is regarded as the envelope of its tangent planes. Recalling that a surface $u(x, y)$ has a tangent plane $u_0 - u = (x_0 - x)u_x(x_0, y_0) + (y_0 - y)u_y(x_0, y_0)$ at a point $u_0(x_0, y_0)$, we see that the coordinates in the tangent plane are u_x (for x_0), u_y (for y_0), and $xu_x + yu_y - u$ (for u_0). Labeling these as new variables, namely, $\xi = u_x$, $\eta = u_y$, and $\omega = xu_x + yu_y - u$, they describe the surface by means of ω as a function of ξ and η. That is, if we know the surface $u(x, y)$, then we know ξ and η and hence ω as a function of x and y. Moreover, we may regain x and y from $x = \omega_\xi$ and $y = \omega_\eta$. The so-called Legendre transformation here is thus

$$\begin{cases} u(x, y) + \omega(\xi, \eta) = x\xi + y\eta, \\ \xi = u_x, \quad \eta = u_y; \quad x = \omega_\xi, \quad y = \omega_\eta. \end{cases}$$

The general Legendre transformation is accordingly given by

$$\begin{cases} u(x_1, \ldots, x_n) + \omega(\xi_1, \ldots, \xi_n) = \displaystyle\sum_{i=1}^n x_i\xi_i, \\ \xi_i = u_{x_i}, \quad x_i = \omega_{\xi_i}, \quad i = 1, \ldots, n. \end{cases}$$

(a) Verify the above conversions (i) to (ii) to (iii) by means of Legendre transformation theory.

(b) Read further about the related general Hamilton–Jacobi theory and about the hodograph method for problems in fluid flow.

(c) Think further about the Dirichlet problem and the minimal surface problem for parametric and nonparametric surfaces.*

6. (a) As an exercise, write down a rigorous proof of the fact that a harmonic function u on a region Ω in the plane can have no local flats, that is, u cannot be constant in any subregion.

(b) Do this using not that $\Delta u = 0$ and $u \in C^2(\Omega)$, but rather only that u satisfies the mean value property that $u(P) = (2\pi r)^{-1} \oint_{|P-Q|=r} u(Q)\, dS$ for all points P and Q on the region, given in Section 1.6.1.

---

* And for additional accounts, see the articles by J. Serrin, "The solvability of boundary value problems" (Hilbert's problem 19) and E. Bombieri, "Variational problems and elliptic equations" (Hilbert's problem 20) in *Mathematical Developments Arising from Hilbert's Problems* (American Mathematical Society, F. Browder (ed.), Providence, Rhode Island, 1976).

7. (a) Show that the only solution to the following stationary problem for the nonlinear heat equation

$$\begin{cases} (-k(u)u_{,i})_{,i} = 0 \text{ in } \Omega, \\ u = 0 \text{ on } \partial\Omega, \end{cases}$$

where $k(\cdot)$ is positive and smooth enough, is $u = 0$.

(b) Does this yield a uniqueness result for the problem

$$\begin{cases} (-k(u)u_{,i})_{,i} = F \text{ in } \Omega, \\ u = f \text{ on } \partial\Omega \text{ ?} \end{cases}$$

(c) Try your hand at uniqueness for the following four problems.

(i) $$\begin{cases} \Delta u - \int_\Omega u(y) \, dy = 0 \text{ in } \Omega, \\ u = f \text{ on } \partial\Omega. \end{cases}$$

(ii) $$\begin{cases} \Delta u - u \int_\Omega u^2(y) \, dy = 0 \text{ in } \Omega^*, \\ u = f \text{ on } \partial\Omega. \end{cases}$$

(iii) $$\begin{cases} \Delta u - u^3 = 0 \text{ in } \Omega, \\ u = f \text{ on } \partial\Omega. \end{cases}$$

(iv) $$\begin{cases} \Delta u - u^2 = 0 \text{ in } \Omega, \\ u = f \text{ on } \partial\Omega. \end{cases}$$

8. (a) Investigate further the following version of the problem $\Delta u = \lambda e^u$ in $|x| < 1$, $u = 0$ at $|x| = 1$, discussed in Section 1.8; namely, the problem

$$\begin{cases} -(r^{n-1}u')' = \lambda r^{n-1}e^u, \qquad 0 < r < 1, \\ u(1) = u'(0) = 0, \end{cases}$$

for $n = 1, 2, 3, \ldots$.

(b) How does this result from the original partial differential equation problem on a unit sphere? What additional assumptions have been made?

(c) Study it by means of the phase portrait methods of Section 1.9.8.

9. The following exercises are for additional practice as one proceeds through the first chapter.

(1) Classify

---

* Nonlinear problems of this type (although somewhat more complicated) occur in the numerical approximation theory of quantum mechanics. See for example E. Lieb and B. Simon, "The Hartree–Fock Theory for Coulomb Systems," *Comm. Math. Phys.* **53**, (1977), and K. Gustafson, "Recent progress on the nonlinear Hartree–Fock, concentration-diffusion, and Navier–Stokes equations," Proceedings of the Bielefeld Conference on Bifurcation Theory: *Applications of Nonlinear Analysis in the Physical Sciences* (Pitman, London, 1981).

(a) $u_{xx} + 5u_{yy} - 4u_{xy}$
(b) $u_{xx} + 4u_{yy} - 4u_{xy}$
(c) $u_{tt} + u_{xt} + u_{xx}$
(d) $u_{tt} - 4u_{xt} + u_{xx}$
(e) $u_{xx} + 2u_{yy} - 2u_{xy}$
(f) $u_y - u_{xx} - u_{yy} + 2u_{xy}$
(g) $u_{xx} + tu_{tt}$
(h) $u_{xx} + (u_t)^3 u_{tt} + u^3$.

(2) Find transformations C converting some of them, e.g., parts (e), (f), to canonical form.

(3) Calculate the Fourier trigonometric series representations of the following.
   (a) $f(x) = x$ on $(0, \pi)$.
   (b) $f(x) = -\pi - x$ on $(-\pi, 0)$, $f(0) = 0$, $f(x) = \pi - x$ on $(0, \pi]$.
   (c) $f(x) = x^2(\pi - x)$ on $(0, \pi)$.
   (d) $f(x) = -x$ on $(-\pi, 0)$, $f(x) = x$ on $(0, \pi)$.

(4) For S_N the first N terms of the Fourier series for $f(x) = x$ on $(0, \pi)$, find the smallest N such that $|f(\pi/2) - S_N(\pi/2)| < \varepsilon$ for (a) $\varepsilon = 0.5$, (b) $\varepsilon = 0.15$.

(5) Let $f(x) = -\pi/4$ on $[-\pi, 0]$, $f(x) = \pi/4$ on $(0, \pi)$.
   (a) Compute its Fourier series.
   (b) Investigate its convergence.
   (c) Consider $x = \pi/2$ to show $1 - 1/3 + 1/5 - 1/7 + \cdots = \pi/4$.
   (d) Find a way to show $\sum_{n=1}^{\infty} \dfrac{1}{n^2} = \pi^2/6$.

(6) Prove the *Linearity Lemma*: for a linear operator L, the following two statements are equivalent: (i) $Lu = 0 \Rightarrow u = 0$. (ii) The equation $Lu = f$ has at most one solution.

(7) Show uniqueness of the solution for the following problems

(a) $\begin{cases} \Delta u = 3xy^3 & \text{in } x^2 + y^2 < 1, \\ \dfrac{\partial u}{\partial n} + 7u = x^8 & \text{on } x^2 + y^2 = 1. \end{cases}$

(b) $\begin{cases} \Delta u = F(x, y) & \text{in } \Omega, \\ \dfrac{\partial u}{\partial n} = g(x, y) & \text{on part of } \partial\Omega, \ u = f(x, y) \text{ on the rest.} \end{cases}$

(c) $\begin{cases} \Delta u = F \text{ in } \Omega, \\ \dfrac{\partial u}{\partial n} = g \text{ on } \partial\Omega \text{ and (i) } u(P) = 3 \text{ at a point } P \text{ on } \partial\Omega, \\ \qquad\qquad\qquad \text{or (ii) } \oint_{\partial\Omega} u = 9. \end{cases}$

(d) $\begin{cases} (e^x u_x)_x + (e^y u_y)_y = 8x^3 y \sin xy \text{ in } \Omega \subset R^2, \\ u = x^2 \text{ on } \partial\Omega. \end{cases}$

(8) By mean value reasoning determine the value of $u(0, 0)$ in each of the following problems

(a) $\begin{cases} \Delta u = 0 & \text{in } r < 1, \\ u(1, \theta) = 1 & \text{for } 0 < \theta < \pi, \\ = 0 & \text{for } \pi < \theta < 2\pi. \end{cases}$

(b) $\begin{cases} \Delta u = 0 & \text{in } x^2 + y^2 + z^2 < 1, \\ u = 1 & \text{on } x^2 + y^2 + z^2 = 1. \end{cases}$

(c) $\begin{cases} \Delta u = 0 & \text{in } x^2 + y^2 < 1, \\ u(x, y) = y^2 & \text{on } x^2 + y^2 = 1. \end{cases}$

(d) $\begin{cases} \Delta u = 0 & \text{in } \Omega \text{ the square } -1 < x < 1, \ -1 < y < 1, \\ u = 1 & \text{on } x = 1, \ -1 \leq y \leq 1, \\ = 0 & \text{on the other three sides.} \end{cases}$

(9) Consider existence and uniqueness of solutions of the problem

$$\begin{cases} \Delta u = x^2 y + c \text{ in } \Omega \\ \dfrac{\partial u}{\partial n} = g(x, y) \text{ on } \partial\Omega, \end{cases}$$

where Ω is the rectangle $0 < x < 4$, $0 < y < 2$, and where g has values $g(x, y) = 0$ on the bottom and left side, $g(x, y) = x - 2$ on the top, and $g(x, y) = y - 1$ on the right side. Determine the appropriate value of the constant c, and construct the solution by separation of variables.

CHAPTER 2

FOURIER SERIES AND HILBERT SPACE

. . . one and the same . . .

Roughly speaking, given a partial differential equation and auxiliary conditions one may always try "separation of variables to ODEs" provided that:

(i) The domain Ω in question can be described in some coordinate system so that $\partial\Omega$ consists of straight line segments in that coordinate system.

(ii) The operators L of the PDE actually "separate" into functions of the separate variables.

(iii) The domain Ω is bounded.

(iv) The resulting ODEs are solvable.

Thus approaching a partial differential equation from the separation of variables point of view one usually (i) writes the problem in the most natural coordinate system for the geometry of the domain Ω on which the problem is posed, then (ii) plugs in a function $u(x_1, \ldots, x_m) = X_1(x_1) \cdots X_m(x_m)$ as a trial solution, but (iii) ends up trying a Fourier Series $u(x_1, \ldots, x_m) = \sum_n c_n u_n(x_1, \ldots, x_m)$ of such functions, wherein (iv) the $X_k(x_k)$ have been found from the resulting ODEs plus the auxiliary conditions. Let us discuss the four procedural points (i)–(iv) a bit more.

Usually the domain must be rectangular, spherical, cylindrical, or a combination thereof. In most problems L has simple coefficients and if of second order the resulting ODEs are solvable.* Although in some cases one can solve an infinite

---

* The theory of such solvable second-order ODEs is usually called "special functions" and is a branch of mathematics in itself, including as eigenfunctions and related functions the well-known function classes of Bessel functions, Hankel functions, Legendre functions, Airy functions, Hypergeometric functions, Elliptic functions, Theta functions, Gamma functions, Zeta functions, Lamé functions, Mathieu functions, Abel functions, Jacobi functions, Tchebycheff polynomials, Neumann functions, Gegenbauer functions, Hermite polynomials, Laguerre polynomials, among others. A classic treatment is Whittaker and Watson, *Modern Analysis* (Cambridge Univ. Press, Cambridge, 1902). See also Section 2.4 herein for the associated Sturm–Liouville theory.

domain problem by separation of variables,[†] in general in that case there are either not enough (or too many, depending on which viewpoint one takes) eigenfunctions $X_k(x_k)$ to adequately represent solutions to the problem. Physically this often means that, in addition to a discrete set of eigenvalues λ_n, one has also present effects caused by a continuous spectrum. For these infinite domain problems one resorts to other methods, among them the Fourier transform and spectral methods, topics we will take up briefly later.[‡]

One might ask, why solve all of these special classes of ODEs resulting from separation of variables, why not just expand everything in terms of the trigonometric functions and hope for the best? One can indeed adopt the latter viewpoint in many cases, accepting the resulting "approximate" solutions given by the partial sums of the Fourier trigonometric expansion. But those solutions are not as good as those resulting from the natural expansion functions, the error may be harder to determine, and the "physically correct" fit has been lost.

To further emphasize this point consider for example a linear problem

$$Lu = F$$

and suppose one tried a set of functions $\{\psi_n\}$ other than the eigenfunctions $\{\varphi_n\}$ of L. Writing blindly

$$u = \sum_{n=1}^{\infty} c_n \psi_n, \qquad F = \sum_{n=1}^{\infty} d_n \psi_n.$$

one would have upon formal substitution into the equation the relation

$$\sum_{n=1}^{\infty} c_n L\psi_n = \sum_{n=1}^{\infty} d_n \psi_n.$$

In the case that the $\{\psi_n\}$ are exactly the eigenfunctions $\{\varphi_n\}$ of L one has $L\varphi_n = \lambda_n \varphi_n$, and one may then equate coefficients to obtain

$$c_n = d_n/\lambda_n$$

and thus the formal solution. In the case that the $\{\psi_n\}$ are some set of functions other than the naturally occurring eigenfunctions $\{\varphi_n\}$, this equating of coefficients cannot be done. A good example of the validity of this preference for an expansion in terms of the true eigenfunctions of an operator is that of the main theorem (Jordan form) of linear algebra, in which one much prefers a basis $\{\varphi_n\}_{n=1}^{N}$ in which a given

---

† Especially if the data is periodic, in which case the problem may be essentially solved on only semi-infinite or finite domains. For a theoretical study of PDEs on infinite domains from this point of view in some sense see L. Bers, F. John, and M. Schechter, *Partial Differential Eqns.* (Wiley, New York, 1964). For a discussion of how most unbounded domains generate enough continuous spectra so that the eigenfunction expansion method is no longer by itself adequate, see I. Glazman, *Direct Methods of Qualitative Spectral Analysis of Singular Differential Operators* (Israel Sci. Transl. Ltd., Jerusalem, 1965).

‡ See Sections 2.7 and 2.8.

N by N linear transformation may be represented in diagonal, or almost diagonal, form.

This chapter has been entitled Fourier series and Hilbert space. A principal message in this chapter is that, in the abstract, Fourier series and Hilbert space are one and the same. From the more practical point of view and for solving partial differential equations, we may therefore say that each of the subjects

Separation of Variables

Fourier (Eigenfunction) Series

Hilbert Space Representations

although differing conceptually in some ways, are nonetheless concerned with the same procedures on the same underlying objects. All three of the above theories may be included in the general

Method of Eigenfunction Expansion,

which may in turn be described as a part of the

Method of Best Least Squares Fit.

Understanding the meanings and content of the above subject names is important. However, to do so, we must cut short any further discussion of semantics, and begin.

2.1 LOTS OF SEPARATION OF VARIABLES

Let us first recall the problems we have already solved (formally) by separation of variables.

Vibrating String Problem (hyperbolic, rectangular (Fig. 2.1a)):

$$\begin{cases} u_{tt} - u_{xx} = 0, & 0 < x < \pi, \quad t > 0, \\ u(0, t) = u(\pi, t) = 0, & t \geq 0, \\ u(x, 0) = f(x), & 0 < x < \pi, \\ u_t(x, 0) = 0, & 0 < x < \pi, \end{cases}$$

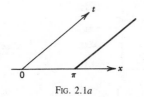

FIG. 2.1a

Solution:

$$u(x, t) = \sum_{n=1}^{\infty} c_n \sin nx \cos nt, \qquad c_n = \frac{2}{\pi} \int_0^{\pi} f(s) \sin ns \, ds.$$

Dirichlet Problem (elliptic, rectangular (Fig. 2.1*b*)):

$$\begin{cases} u_{xx} + u_{yy} = 0, & 0 < x < \pi, \quad 0 < y < \pi, \\ u(x, 0) = f_1(x), & 0 < x < \pi, \\ u(\pi, y) = f_2(y), & 0 < y < \pi, \\ u(x, \pi) = f_3(x), & 0 < x < \pi, \\ u(0, y) = f_4(y), & 0 < y < \pi, \end{cases}$$

Solution:

$$u(x, y) = u_1(x, y) + u_2(x, y) + u_3(x, y) + u_4(x, y),$$

where

$$u_1(x, y) = \sum_{n=1}^{\infty} c_n \sin nx \sinh n(\pi - y), \qquad c_n = \frac{2}{\pi \sinh n\pi} \int_0^{\pi} f_1(s) \sin ns \, ds,$$

$$u_2(x, y) = \sum_{n=1}^{\infty} c_n \sin ny \sinh nx, \qquad c_n = \frac{2}{\pi \sinh n\pi} \int_0^{\pi} f_2(s) \sin ns \, ds,$$

$$u_3(x, y) = \sum_{n=1}^{\infty} c_n \sin nx \sinh ny, \qquad c_n = \frac{2}{\pi \sinh n\pi} \int_0^{\pi} f_3(s) \sin ns \, ds,$$

$$u_4(x, y) = \sum_{n=1}^{\infty} c_n \sin ny \sinh n(\pi - x), \qquad c_n = \frac{2}{\pi \sinh n\pi} \int_0^{\pi} f_4(s) \sin ns \, ds.$$

Dirichlet Problem (elliptic, spherical (Fig. 2.1*c*)):

$$\begin{cases} u_{rr} + r^{-1}u_r + r^{-2}u_{\theta\theta} = 0, & r < 1, \\ u(1, \theta) = f(\theta), \end{cases}$$

Solution

$$u(r, \theta) = \tfrac{1}{2}a_0 + \sum_{n=1}^{\infty} r^n(a_n \cos n\theta + b_n \sin n\theta),$$

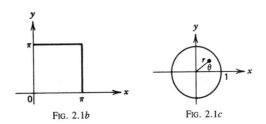

FIG. 2.1*b*   FIG. 2.1*c*

where

$$a_n = \frac{1}{\pi} \int_{-\pi}^{\pi} f(s) \cos ns \, ds, \qquad b_n = \frac{1}{\pi} \int_{-\pi}^{\pi} f(s) \sin ns \, ds.$$

Eigenvalue Problem (elliptic, rectangular (Fig. 2.1*d*)):

$$\begin{cases} -\Delta u = \lambda u \text{ in } \Omega, \\ u = 0 \text{ on } \partial\Omega, \end{cases}$$

where

$$\Delta u = u_{xx} + u_{yy},$$

Solution

eigenvalues $\lambda_{nm} = n^2 + m^2, \qquad n = 1, 2, 3, \ldots,$
$$m = 1, 2, 3, \ldots,$$

eigenfunctions $\varphi_{nm} = \sin nx \sin my, \qquad n = 1, 2, 3, \ldots,$
$$m = 1, 2, 3, \ldots.$$

In the above cases we have thus seen that the separation of variables method works for initial value problems, boundary value problems, eigenvalue problems, and combinations thereof, on rectangular and spherical geometries, for the hyperbolic and elliptic types of operators. The parabolic case (see examples below) is similar provided the domain Ω is bounded in the x variable.

There are an unlimited number of similar problems that have been, or can be, treated by this method. We list a few below, along with a few remarks about the salient properties of each.

Let us first comment that one may change the Ω lengths rather easily and we usually just take a convenient one. For example, the solution to the vibrating string problem above in the case that Ω of length π is changed to a length of l becomes

$$u_l(x, t) = \sum_{n=1}^{\infty} c_n \sin n(\pi l^{-1})x \cos n(\pi l^{-1})t, \quad c_n = \frac{2}{l} \int_0^l f(s) \sin n(\pi l^{-1})s \, ds.$$

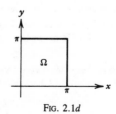

FIG. 2.1*d*

Likewise the Dirichlet problem on a sphere of radius R has solution

$$u(r, \theta) = \tfrac{1}{2}a_0 + \sum_{n=1}^{\infty} (rR^{-1})^n (a_n \cos n\theta + b_n \sin n\theta),$$

with the a_n and b_n Fourier coefficients remaining the same as before since the θ length of the domain has not changed.

In like manner the occurring of different but constant coefficients in the operators will also result in corresponding dilated variables in the Fourier series solution. For example, if in the wave equation of the vibrating string problem above one considers instead the equation

$$u_{tt} - c^2 u_{xx} = 0, \qquad 0 < x < \pi, \qquad t > 0,$$

the c makes its appearance only in the t part of the solution

$$u(x, t) = \sum_{n=1}^{\infty} c_n \sin nx \cos nct,$$

the c_n and x parts of the solution remaining the same as before since the x-domain length has not changed.

Just as the Dirichlet problem on the square above was by linearity written as four simpler problems as concerns the data, the same linearity decompositions can be applied to other problems. For example, the vibrating string problem with nonhomogeneous data present everywhere possible,

$$\begin{cases} u_{tt} - u_{xx} = F(x, t), & 0 < x < \pi, \quad t > 0, \\ u(0, t), \ u(\pi, t) = h_1(t), \ h_2(t), & \text{respectively,} \quad t > 0, \\ u(x, 0) = f(x), & 0 < x < \pi, \\ u_t(x, 0) = g(x) & 0 < x < \pi, \end{cases}$$

may be written as five separate problems with nonzero data ocurring only once in each one. Some of these separate problems are harder to solve than others, as experience will show. Once in a while one is lucky and can shift data from one less desirable place to another by a change of unknown such as $v = u + \varphi$, where φ is a known function, but for general data this is not often the case.*

Let us illustrate with a Dirichlet–Poisson problem how one formally solves a *nonhomogeneous* PDE by separation of variables in terms of eigenfunctions for the corresponding eigenvalue problem. Let Ω be as in Figure 2.1*d*.

---

* When this can be done for a general class of operators the resulting method of change of variable is often called a Stoke's or Duhamel rule or formula. Such methods are most useful when treating wave equations.

Poisson Problem (Elliptic, nonhomogeneous):

$$\begin{cases} -\Delta u = F(x, y) \text{ in } \Omega, & \text{assuming*} \quad F(x, y) = \sum_{n=1}^{\infty} \sum_{m=1}^{\infty} d_{nm} \sin nx \sin my, \\ u = 0 \quad \text{on } \partial\Omega, \end{cases}$$

Solution:

$$u(x, y) = \sum_{n=1}^{\infty} \sum_{m=1}^{\infty} c_{nm} \sin nx \sin my, \qquad c_{nm} = \frac{d_{nm}}{n^2 + m^2}.$$

The verification of the above solution formally is as follows. Remembering that by separation of variables it was found in the above eigenvalue problem

$$\begin{cases} -\Delta u = \lambda u \text{ in } \Omega, \\ u = 0 \text{ on } \partial\Omega, \end{cases}$$

that the eigenvalues were $\lambda_{nm} = n^2 + m^2, n = 1, 2, 3, \ldots, m = 1, 2, 3, \ldots,$ with the corresponding eigenfunctions $\varphi_{nm} = \sin nx \sin my$, we have (formally)

$$-\Delta u \overset{\text{(formally)}}{=} \sum_{n=1}^{\infty} \sum_{m=1}^{\infty} c_{nm}(-\Delta)\varphi_{nm}$$

$$= \sum_{n=1}^{\infty} \sum_{m=1}^{\infty} c_{nm}\lambda_{nm}\varphi_{nm} \overset{\text{(want)}}{=} \sum_{n=1}^{\infty} \sum_{m=1}^{\infty} d_{nm}\varphi_{nm} = F$$

which is solved by taking $c_{nm} = d_{nm}/\lambda_{nm}$. We have assumed that we can differentiate term by term, that is, that we can interchange the two limiting operations

$$\frac{\partial^2}{\partial x^2} + \frac{\partial^2}{\partial y^2} \quad \text{and} \quad \sum_{n=1}^{\infty} \sum_{m=1}^{\infty}$$

and this requires justification† as has been discussed previously.

---

* Since $F(x, y)$ is a function of two rather than one variable, this is called a double Fourier series for F. The solution is of the same type.

† Unless of course the eigenfunction expansion for $F(x, y)$ terminates after a finite number of terms. Then linearity alone suffices. For example, if $F(x, y) = \sin 2x \sin 307y + 5 \sin 3x \sin 7y$, then the solution is

$$u(x, y) = \frac{1}{94253} \sin 2x \sin 307y + \frac{5}{58} \sin 3x \sin 7y.$$

The third type of equation, the parabolic case, is solved similarly under separation of variables. Note that the t-components of the solution now decay exponentially, as the body cools in the heat conduction interpretation.* That is, we have

Heat Conduction Problem (domain as in Fig. 2.1a):

$$\begin{cases} u_t - u_{xx} = 0, & 0 < x < \pi, \quad t > 0, \\ u(x, 0) = f(x), & 0 < x < \pi, \\ u(0, t) = 0, & t > 0, \\ u(\pi, t) = 0, & t > 0, \end{cases}$$

Solution:

$$u(x, t) = \sum_{n=1}^{\infty} c_n e^{-n^2 t} \sin nx, \qquad c_n = \frac{2}{\pi} \int_0^{\pi} f(s) \sin ns \, ds.$$

The verification is straightforward. One tries as usual $u(x, t) = X(x)T(t)$, which upon substitution into the differential equation yields

$$\frac{X''(x)}{X(x)} = \frac{T'(t)}{T(t)} = -\lambda$$

with λ some arbitrary constant. The eigenvalue problem

$$\begin{cases} -X''(x) = \lambda X(x), & 0 < x < \pi, \\ X(0) = X(\pi) = 0, \end{cases}$$

yields as always the infinite number of solutions $X_n(x) = $ (const) $\sin nx$ and the eigenvalues $\lambda_n = n^2, n = 1, 2, 3, \ldots$, which yields the corresponding first-order linear ordinary differential equations

$$T'(t) + n^2 T(t) = 0, \qquad t > 0,$$

with solutions $T_n(t) = $ (const) $\cdot e^{-n^2 t}$. Upon then trying $u(x, 0) = \sum_{n=1}^{\infty} c_n X_n(x) T_n(t)$, one sees as before that c_n is determined by the matching of the Fourier series expansions for $u(x, 0)$ and $f(x)$†:

$$u(x, 0) = \sum_{n=1}^{\infty} c_n \sin nx \overset{\text{want}}{=} \sum_{n=1}^{\infty} d_n \sin nx = f(x).$$

---

* Corresponding physically to a linearized model for a slab of homogeneous material in the appropriate units. The initial temperature $f(x)$ is given and assumed to be the same at all horizontal levels of the slab, and the boundary (left and right sides) faces of the slab are kept cooled to zero. The slab is sufficiently tall so that the top and bottom effects may be ignored.

† We have given these arguments before but want to emphasize the comparison with the matching procedure for coefficients given in the Poisson (nonhomogeneous) equation above. Here, in the initial value problem set-up, the matching of $c_n = d_n$ is trivial because the relevant operator is just the identity operator I operating on u at $t = 0$.

Finally, as in the elliptic and hyperbolic problems discussed above, one easily generalizes the problem and solution to other lengths and constants. For example, consider the problem with thermal conductivity constant k and slab width l:

$$\begin{cases} u_t - ku_{xx} = 0, & 0 < x < l, \quad t > 0, \\ u(x, 0) = f(x), & 0 < x < l, \\ u(0, t) = 0, & t > 0, \\ u(l, t) = 0, & t > 0, \end{cases}$$

Solution:

$$u(x, t) = \sum c_n e^{-n^2(\pi l^{-1})^2 kt} \sin n(\pi l^{-1})x,$$

$$c_n = \frac{2}{l} \int_0^l f(s) \sin n(\pi l^{-1})s \, ds.$$

We now list with minimal explanation, since most of the explanation is the same as the above except for the study of the special eigenfunctions themselves,* an assortment of separation of variables solutions to PDE problems.

Vibrating Membrane Problem (hyperbolic, rectangular (Fig. 2.1e)):

$$\begin{cases} u_{tt} - \Delta u = 0 \text{ in } \Omega, & t > 0, \\ u = 0 \text{ on } \partial\Omega, & t > 0, \\ u(x, y, 0) = f(x, y) \text{ in } \Omega, \\ u_t(x, y, 0) = 0, \end{cases}$$

Solution:

$$u(x, y, t) = \sum_{n=1}^{\infty} \sum_{m=1}^{\infty} c_{nm} \sin nx \sin my \cos(m^2 + n^2)^{1/2}t,$$

$$c_{nm} = \frac{4}{\pi^2} \int_0^\pi \int_0^\pi f(s, \tau) \sin ns \sin m\tau \, ds \, d\tau.$$

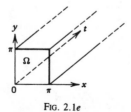

FIG. 2.1e

* Such a study is usually carried out in a course on special functions or on Sturm–Liouville theory. We will here note only certain important properties of these special classes of eigenfunctions as we need them. See however the footnote at the beginning of this chapter, and Section 2.4, where a part of the general theory is presented.

Vibrating Membrane Problem (hyperbolic, spherical (Fig. 2.1*f*)):

$$\begin{cases} u_{tt} - \Delta u = 0 \text{ in } \Omega, \qquad t > 0, \\ u = 0 \text{ on } \partial\Omega, \qquad t > 0, \\ u(r, \theta, 0) = f(r, \theta) \text{ in } \Omega, \\ u_t(r, \theta, 0) = 0, \end{cases}$$

Solution:

$$u(r, \theta, t) = \tfrac{1}{2} \sum_{n=1}^{\infty} a_{n0} J_0[(\lambda_n^0)^{1/2} r] \cos[(\lambda_n^0)^{1/2} t] + \sum_{m=1}^{\infty} \sum_{n=1}^{\infty} J_m[(\lambda_n^m)^{1/2} r]$$

$$\times [a_{nm} \cos m\theta + b_{nm} \sin m\theta] \cos[(\lambda_n^m)^{1/2} t],$$

where

$$a_{nm} = \frac{1}{d_{nm}} \int_0^1 \int_{-\pi}^{\pi} f(s, \psi) J_m[(\lambda_n^m)^{1/2} s] \cos m\psi \, d\psi \, s \, ds,$$

$$m = 0, 1, 2, 3, \ldots,$$
$$n = 1, 2, 3, \ldots,$$

and

$$b_{nm} = \frac{1}{d_{nm}} \int_0^1 \int_{-\pi}^{\pi} f(s, \psi) J_m[(\lambda_n^m)^{1/2} s] \sin m\psi \, d\psi \, s \, ds,$$

$$m = 0, 1, 2, 3, \ldots,$$
$$n = 1, 2, 3, \ldots,$$

where the normalizing factors d_{nm} are (evaluating the trigonometric part)

$$d_{nm} = \pi \int_0^1 J_m^2[(\lambda_n^m)^{1/2} s] s \, ds.$$

Here J_m is the so-called Bessel function of the first kind of order m and the λ_n^m are the squares of the zeros of J_m. Thus $J_m[(\lambda_n^m)^{1/2} r] = 0$ on the boundary when $r = 1$. To obtain this solution, one tries $u(r, \theta, t) = R(r)\Theta(\theta)T(t)$, and recalling that in polar coordinates $\Delta u = u_{rr} + r^{-1} u_r + r^{-2} u_{\theta\theta}$, one arrives at

$$\frac{T''}{T} = \frac{R''}{R} + r^{-1}\frac{R'}{R} + r^{-2}\frac{\Theta''}{\Theta} = -\lambda$$

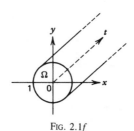

FIG. 2.1*f*

and

$$r^2\frac{R''}{R} + r\frac{R'}{R} + \lambda r^2 = \frac{-\Theta''}{\Theta} = \nu.$$

Solving first the eigenvalue problem, as we did before for the Dirichlet problem,

$$\begin{cases} -\Theta'' = \nu\Theta, & -\pi < \theta < \pi, \\ \Theta(\pi) = \Theta(-\pi), \\ \Theta'(\pi) = \Theta'(-\pi), \end{cases}$$

one obtains the eigenvalues $\nu = m^2$, $m = 1, 2, 3, \ldots$, and the corresponding pairs of eigenfunctions $\Theta_m = \cos m\theta$ and $\Theta_m = \sin m\theta$. Next the Bessel equation eigenvalue problem for λ with $\nu = m^2$ now known is considered for each fixed m,

$$\begin{cases} -R'' - r^{-1}R' + r^{-2}m^2R = \lambda R, & 0 < r < 1, \\ R(1) = 0, \\ R(0) \text{ and } R'(0) \text{ bounded}, \end{cases}$$

yielding* as solutions $R_{mn}(r) = J_m[(\lambda_n^m)^{1/2}r]$ with an infinite number of positive eigenvalues λ_n^m (the squares of the zeros of J_m), $n = 1, 2, 3, \ldots$. With the λ_n^m now determined, one solves

$$\begin{cases} T'' + \lambda_n^m T = 0, & t > 0, \\ T'(0) = 0, \end{cases}$$

obtaining, due to the positivity of λ_n^m, $T_n^m = (\text{const}) \cos[(\lambda_n^m)^{1/2}t]$.

Dirichlet Problem (elliptic, cube (Fig. 2.1h)):

$$\begin{cases} \Delta u = 0 \text{ in } \Omega, \\ u = f \text{ on } \partial\Omega, \end{cases}$$

where

$$\Delta u = u_{xx} + u_{yy} + u_{zz}$$

---

* We recall that the Bessel equations are of nonconstant coefficient but are of the type usually described as those with a regular singular point at $r = 0$. These equations are treated in ODE by seeking a solution in the form of a power series $\Sigma_{k=0}^{\infty} c_k r^k$. For example, one finds for $m = 0$ the solution $J_0(r) = \Sigma_{k=0}^{\infty} c_{2k}r^{2k}$, where the coefficient $c_{2k} = (-1)^k/(2^k k!)^2$. $J_0(r)$ looks like Figure 2.1g and is not unlike a damped cosine function. Approximate values of the first three eigenvalues λ_n^0 are $\lambda_1^0 \cong 5.7831860$, $\lambda_2^0 \cong 30.471262$, $\lambda_3^0 \cong 74.887007$.

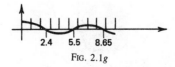

2.4   5.5   8.65

Fig. 2.1g

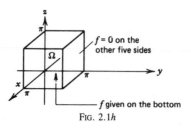

FIG. 2.1h

and

$$f(x, y) = \sum_{n=1}^{\infty} \sum_{m=1}^{\infty} d_{nm} \sin nx \sin my.$$

Solution:

$$u(x, y, z) = \sum_{n=1}^{\infty} \sum_{m=1}^{\infty} c_{nm} \sin nx \sin my \sinh[(m^2 + n^2)^{1/2}(\pi - z)].$$

Dirichlet Problem (elliptic, sphere (Fig. 2.1i)):

$$\begin{cases} \Delta u = 0 \text{ in } \Omega, \\ u = f \text{ on } \partial\Omega, \end{cases}$$

where

$$\Delta u = u_{rr} + 2r^{-1}u_r + r^{-2}(\sin \theta)^{-1}(\sin \theta u_\theta)_\theta + r^{-2}(\sin \theta)^{-2}u_{\varphi\varphi}$$

and

$$f(\theta, \varphi) = \tfrac{1}{2} \sum_{n=0}^{\infty} a_{n0}P_n(\cos \theta) + \sum_{n=0}^{\infty} \sum_{m=1}^{n} P_n^m(\cos \theta)$$
$$\times [a_{nm} \cos m\varphi + b_{nm} \sin m\varphi], \qquad 0 < \theta < 2\pi, \qquad 0 < \varphi < \pi.$$

Solution:

$$u(r, \theta, \varphi) = \tfrac{1}{2} \sum_{n=0}^{\infty} a_{n0}r^nP_n(\cos \theta) + \sum_{n=0}^{\infty} \sum_{m=1}^{n} r^nP_n^m(\cos \theta)$$
$$\times [a_{nm} \cos m\varphi + b_{nm} \sin m\varphi].$$

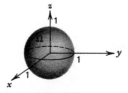

FIG. 2.1i

Here P_n^m are the so-called associated Legendre functions and the composite functions $r^n P_n^m(\cos \theta) \cdot (\cos m\varphi$ or $\sin m\varphi)$ are called spherical harmonics. The associated Legendre functions are obtained in terms of derivatives of the Legendre polynomials P_n.*

Heat Equation (parabolic, any Ω (Fig. 2.1*j*)):

$$\begin{cases} u_t - \Delta u = 0 \text{ in } \Omega, & t > 0, \\ u = 0 \text{ on } \partial\Omega, & t > 0, \\ u(x, 0) = f(x) \text{ in } \Omega, \end{cases}$$

where

$$\Delta u = u_{x_1 x_1} + \cdots + u_{x_n x_n}, \qquad x = \mathbf{x} = (x_1, \ldots, x_n).$$

Solution:

$$u(x, t) = \sum_n c_n e^{-\lambda_n t} \varphi_n(x), \qquad c_n = \int_\Omega f\varphi_n \, dx \Big/ \int_\Omega \varphi_n^2 \, dx,$$

where $\varphi_n(x)$ are the eigenfunctions to the eigenvalue problem

$$\begin{cases} -\Delta\varphi = \lambda\varphi \text{ in } \Omega, \\ \varphi = 0 \text{ on } \partial\Omega. \end{cases}$$

We have assumed that Ω is such that the $\varphi_n(x)$ are at most countable,† and we have assumed several other things, of course, such as the usual formal operations of the separation of variables method, but additionally that the φ_n, $n = 1, 2, 3, \ldots$, have a property of completeness,‡ which we discuss in the next sections.

FIG. 2.1*j*

---

* $P_n(x)$ is the nth eigenfunction of the Legendre equation

$$\begin{cases} -[(1 - x^2)u_x]_x = \lambda u, & -1 < x < 1, \\ (1 - x^2)^{1/2}u_x \text{ bounded at the end points,} \end{cases}$$

corresponding to the eigenvalue $\lambda_n = n(n + 1)$, $n = 0, 1, 2, \ldots$, and is found by the power series method similar to the Bessel equation. It turns out that $P_n(x)$ is a polynomial of degree n.

† That is, can be indexed by the integers $n = 1, 2, 3, \ldots$. For unbounded domains it is usually not the case.

‡ This property will guarantee a Fourier expansion of any reasonable data $f(x)$ in terms of the eigenfunctions $\varphi_n(x_1, \ldots, x_n)$.

In the case that Ω is rectangular, one finds the $\varphi_n(x)$ to be sine functions, that is, products of them, in accordance with the above examples. The spherical cases yield Bessel functions, Legendre functions, and the like, depending on the boundary conditions (one does not need to take a Dirichlet boundary condition as was done above). Cylinders and certain other geometries can be handled similarly to get explicitly the expansion functions φ_n. The general theory of PDEs guarantees their existence for all nice bounded domains Ω.

We close this section by noting a few of the special eigenfunctions that solve the special ODEs that come out in various PDE problems.*

Bessel's Equation:

$(xu')' - x^{-1}m^2u + \lambda xu = 0, \qquad u(0) < \infty, \qquad u(1) = 0.$
Eigenvalues: $\lambda_n = $ (zeros)^2 of J_m (see above).
Eigenfunctions: $J_m(\lambda_n^{1/2}x)$ Bessel functions.

Legendre's Equation:

$((1 - x^2)u')' - (1 - x^2)^{-1}m^2u + \lambda u = 0, \qquad u(-1) < \infty, \qquad u(1) < \infty.$
Eigenvalues: $\lambda_n = n(n + 1), \qquad n = 0, 1, 2, \ldots$.
Eigenfunctions: $P_n^m(x)$ associated Legendre functions,

$$P_0^0 = 1, \qquad P_1^0 = x, \qquad P_2^0 = \tfrac{1}{2}(3x^2 - 1), \qquad P_3^0 = \tfrac{5}{2}x^3 - \tfrac{3}{2}x,$$

$$P_4^0 = \frac{7 \cdot 5}{4 \cdot 2}x^4 - 2\frac{5 \cdot 3}{4 \cdot 2}x^2 + \frac{3 \cdot 1}{4 \cdot 2}, \qquad P_5^0 = \frac{9 \cdot 7}{4 \cdot 2}x^5 - 2\frac{7 \cdot 5}{4 \cdot 2}x^3 + \frac{5 \cdot 3}{4 \cdot 2}x,$$

$$P_n^0(x) = \frac{1}{2^n n!}\frac{d^n((x^2 - 1)^n)}{dx^n}, \qquad P_n^m(x) = \frac{1}{2^n n!}(1 - x^2)^{m/2}\frac{d^{m+n}((x^2 - 1)^n)}{dx^{m+n}}.$$

Jacobi Equation:

$((1 - x)^{p-q+1}x^qu')' + \lambda(1 - x)^{p-q}x^{q-1}u = 0, \qquad u(0) < \infty, \qquad u(1) < \infty.$
Eigenvalues: $\lambda_n = n(n + p), \qquad n = 0, 1, 2, \ldots$.
Eigenfunctions: $G_n(x)$ Jacobi functions (hypergeometric polynomials),

$$G_n(p, q, x) = F(-n, p+n; q; x) = x^{1-q}(1 - x)^{q-p}\frac{\Gamma(q+n)}{\Gamma(q)}\frac{d^n}{dx^n}(x^{q+n-1}(1 - x)^{p+n-q})$$

---

* The interval on which the ODE acts is clear from the boundary conditions. We leave it to the reader as an exercise to find the originating PDE. We also leave it to the interested student as an ODE exercise to solve the given equations for the given solutions.

Tchebycheff Equation:

$$((1 - x^2)^{1/2}u')' + (1 - x^2)^{-1/2}\lambda u = 0, \qquad u(-1) < \infty, \qquad u(1) < \infty.$$

Eigenvalues: $\lambda_n = n^2$, $\quad n = 0, 1, 2, \ldots$.

Eigenfunctions: $T_n(x)$ Tchebycheff polynomials,

$$T_0(x) = 1, \qquad T_n(x) = 2^{-(n-1)}\cos(n \cos^{-1} x).$$

Hermite Equation:

$$(e^{-x^2}u')' + \lambda e^{-x^2}u = 0, \qquad u(\pm\infty) = 0(x^{\text{some } N}).$$

Eigenvalues: $\lambda_n = n$, $\quad n = 0, 1, 2, \ldots$.

Eigenfunctions: $H_n(x)$ Hermite polynomials,

$$H_0 = 1, \qquad H_1 = 2x, \qquad H_2 = 4x^2 - 2, \qquad H_3 = 8x^3 - 12x, \qquad H_n(x) =$$

$$(-1)^n e^{x^2} d^n (e^{-x^2})/dx^n \qquad \text{(eqn. alternately: } u'' - x u' + \lambda u = 0).$$

Harmonic Oscillator Equation:

$$u'' - \frac{x^2}{4}u + \lambda u = 0, \qquad \int_{-\infty}^{\infty} u^2(x)\, dx < \infty^*.$$

Eigenvalues: $\lambda_n = n + \frac{1}{2}$, $\quad n = 0, 1, 2, \ldots$.

Eigenfunctions: $\varphi_n(x) = e^{-x^2/4} H_n (x/\sqrt{2})$ Hermite functions (see above for $H_n(x)$).

Laguerre Equation:

$$(xu')' + (\alpha - x)u' + \lambda u = 0, \qquad \int_0^{\infty} u^2(x)x^\alpha e^{-x}\, dx < \infty.$$

Eigenvalues: $\lambda_n = n$, $\quad n = 0, 1, 2, \ldots$.

Eigenfunctions: $L_n^\alpha(x) = \dfrac{e^x x^{-\alpha}}{n!} \dfrac{d^n(e^{-x}x^{n+\alpha})}{dx^n}$ Laguerre polynomials

$$L_0 = 1, \qquad L_1 = 1 - x, \qquad L_2 = (x^2 - 4x + 2)/2$$

$$L_3 = (-x^3 + 9x^2 - 18x + 6)/6 \qquad \text{(eqn. alternately: } (xe^{-x}u')' + \lambda e^{-x}u = 0).$$

---

* One may impose boundary conditions in either of two ways, by requiring a certain growth rate or finiteness at the boundary as in the examples above, or by requiring the eigenfunctions to lie in a L^2 space as was done in this example. Note that the condition that $\int_{-\infty}^{\infty} \varphi_n^2(x)\, dx < \infty$ just corresponds to $\int_{-\infty}^{\infty} H_n^2(x)e^{-x^2/2}\, dx < \infty$. The Hermite equation as first written is in the so-called Sturm–Liouville form; as alternately written, in what we shall call the ODE form. There is a third standard form called the normal form; see Section 2.4, Problem 1.

Whittaker's Equation ($\alpha > -1$):

$$u'' - \left(\frac{1}{4} + \frac{\alpha^2 - 1}{4x^2}\right)u + \lambda x^{-1}u = 0, \qquad \int_0^\infty u^2(x)x^{-1}\,dx < \infty.$$

Eigenvalues: $\lambda_n = n + \dfrac{\alpha + 1}{2}, \qquad n = 0, 1, 2, \ldots$.

Eigenfunctions: $\varphi_n(x) = \left(\dfrac{n!}{(n + \alpha)!}\right)^{1/2} x^{(\alpha + 1)/2} e^{-x/2} L_n^\alpha(x)$ Whittaker functions

(see above for $L_n^\alpha(x)$).

Balmer's Equation:

$$\tfrac{1}{2}u'' + x^{-1}u' + \left(Zx^{-1} - \frac{l(l + 1)}{2}x^{-2}\right)u + \lambda u = 0,$$

$$\int_0^\infty u^2(x)x^2\,dx < \infty, \qquad l = 0, 1, 2, \ldots.$$

Eigenvalues: $\lambda_n = -\dfrac{Z^2}{2n^2}, \qquad n = 1, 2, 3, \ldots$.

Eigenfunctions: $R_{nl}(x)$ radial eigenfunctions of hydrogen ($Z = 1$),

$R_{10} = 2e^{-x}, \qquad R_{20} = 2^{-1/2}e^{-(1/2)x} \cdot (1 - \tfrac{1}{2}x),$

$R_{30} = \dfrac{2}{3\sqrt{3}}e^{-(1/3)x} \cdot (1 - \tfrac{2}{3}x + \tfrac{2}{27}x^2),$

$R_{40} = \tfrac{1}{4}e^{-(1/4)x} \cdot (1 - \tfrac{3}{4}x + \tfrac{1}{8}x^2 - \tfrac{1}{192}x^3),$

$R_{nl} = e^{-(Z/n)x} \cdot x^l L_{n-l-1}^{2l+1}\left(\dfrac{2Z}{n}x\right)$

The following three problems serve to refresh and facilitate the reader's review of the application of the separation of variables method to particular problems. In Problem 2.9.1, the reader will find a more abstract interpretation of separation of variables as a tensor product.

Problem 1. (a) Solve formally by separation of variables the vibrating string problem (Fig. 2.1*k*)

$$\begin{cases} u_{tt} - u_{xx} = 0, & 0 < x < \pi, \quad t > 0, \\ u(0, t) = u(\pi, t) = 0, & t \geq 0, \\ u(x, 0) = 0, & 0 < x < \pi, \\ u_t(x, 0) = g(x), & 0 < x < \pi. \end{cases}$$

(b) Try to sketch the solution for the case in which $g(x) = 1$ for $\pi/3 \leq x \leq 2\pi/3$, $g(x) = 0$ elsewhere on the interval $[0, \pi]$

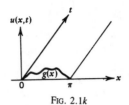

FIG. 2.1k

Problem 2. (a) Solve formally by separation of variables the Dirichlet problem (Fig. 2.1*l*)

$$\begin{cases} u_{xx} + u_{yy} = 0, & 0 < x < \pi, \quad 0 < y < \pi, \\ u(x, 0) = x(\pi - x), & 0 < x < \pi, \\ u(\pi, y) = 0, & 0 < y < \pi, \\ u(x, \pi) = 0, & 0 < x < \pi \\ u(0, y) = y, & 0 < y < \pi. \end{cases}$$

(b) Try to sketch the solution.

Problem 3. (a) Solve formally by separation of variables the heat problem (Fig. 2.1*m*)

$$\begin{cases} u_t - u_{xx} = 0, & 0 < x < \pi, \quad t > 0, \\ u(x, 0) = 0, & 0 < x \le \pi/2, \\ u(x, 0) = 1, & \pi/2 < x \le \pi, \\ u(0, t) = 0, & t > 0, \\ u(\pi, t) = 0, & t > 0. \end{cases}$$

(b) How does the Fourier series solution u behave along and near the lines $x = 0, x = \pi/2, x = \pi$? (c) Try to sketch the solution.

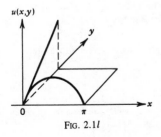

FIG. 2.1*l*

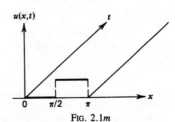

FIG. 2.1*m*

2.2 MATHEMATICAL JUSTIFICATIONS OF THE METHOD

Before proceeding to Fourier series from the Hilbert space viewpoint (Section 2.3), we return in this section to the three questions raised in Section 1.5 concerning the rigorous justification of the separation of variables method. That is, we accept for the moment the separation of variables method and the resulting Fourier series (formal) solution, and concern ourselves only with finding conditions under which its substitution into the partial differential equation may be validated. By restricting attention to sufficiently continuous functions, we need employ only methods of calculus and advanced calculus. Roughly speaking, this amounts to working in the maximum norms rather than in the more efficient Hilbert space norms.

1. Can we differentiate term by term in the Fourier solution?
2. Does the Fourier series solution provide pointwise equality with the given data on the domain and boundary?
3. Is the Fourier series solution the only "physical" solution?

As mentioned previously, Question 3 can be resolved by (i) uniqueness proofs showing at most one solution within the class of functions having enough regularity from the physical point of view, plus (ii) demonstrating that degree of regularity for the Fourier series solution. Question 2 comes down to the Dini tests (see 1.6.3 and 1.9.6.2) and will be discussed later in this section. As concerns Question 1, from calculus or advanced calculus we know that sufficient conditions for differentiating a series term by term are available by showing enough uniform convergence, and it is to this first question that we turn now.

We first make two heuristic comments to indicate how we shall proceed in this section.

As is clear from the uniform convergence theorems stated in Section 1.6.3, the concept of uniform convergence is that of convergence in the maximum norm. Thus it is not unreasonable that maximum principles could (and will) play a role in demonstrating a uniform convergence. Maximum principles are strongest for elliptic partial differential equations and weakest for hyperbolic partial differential equations. For this reason we will illustrate uniform convergence proofs of the separation of variables solution first for an elliptic problem.

A second and related comment is that, as we have seen, the solutions to elliptic (by the mean value property, for example) and parabolic (by the Poisson representation formula, for example) problems tend to immediately smooth out as one leaves the boundary, even for bad data. This makes it easier to prove uniform convergence. Solutions to hyperbolic equations do not smooth out, and indeed we would not want it to be the case, for a wave should remain a wave and a shock should remain a shock.*

---

* Our heuristic comments here were on the basis of linear intuition. For nonlinear equations, shocks and other solution discontinuities can develop regardless of type. For a recent survey of such problems, see the article by P. Lax, *Am. Math. Monthly* **79** (1972).

For these two reasons, then, the separation of variable proofs are easier in the elliptic and parabolic cases, and require less regularity of the data.

Let us now, after the above comments, give proofs of the separation of variables Fourier solutions to problems of the three types.

Dirichlet Problem (Fig. 2.2a)

$$\begin{cases} \Delta u = 0 \text{ in } \Omega, \\ u = f \text{ on } 0 \leq x \leq \pi, \quad y = 0, \\ u = 0 \text{ other three sides.} \end{cases}$$

The formal solution via separation of variables was

$$u(x, y) = \sum_{n=1}^{\infty} c_n \sin nx \sinh n(\pi - y), \qquad c_n = \frac{2}{\pi \sinh n\pi} \int_0^{\pi} f(s) \sin ns \, ds.$$

Let us for convenience write the solution also in the form

$$u(x, y) = \sum_{n=1}^{\infty} u_n(x, y) = \lim_{N \to \infty} \sum_{n=1}^{N} u_n(x, y) = \lim_{N \to \infty} s_N(x, y).$$

To apply the Weierstrass M-test of calculus we need $\Sigma \, u_n$ bounded by a $\Sigma \, M_n$ in the sense $|u_n(x, y)| \leq M_n$. To differentiate term by term* we need similarly $\Sigma \, u_n'$ bounded by a $\Sigma \, M_n'$ and $\Sigma \, u_n''$ bounded by a $\Sigma \, M_n''$, where the prime here has been used to symbolize any partial derivative.

Let us begin with the question of whether we may perform rigorously the operation of one differentiation with respect to x of the series, that is,

$$\frac{\partial}{\partial x} \sum_{n=1}^{\infty} u_n(x, y) = \sum_{n=1}^{\infty} \frac{\partial}{\partial x} u_n(x, y).$$

One has immediately the bound

$$|u_n(x, y)| \leq |c_n| \, |\sinh n(\pi - y)|$$

$$\leq \frac{2}{\pi} \int_0^{\pi} |f(s)| \, ds \cdot \left| \frac{\sinh n(\pi - y)}{\sinh n\pi} \right| \leq (\text{const}) e^{-ny},$$

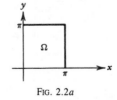

FIG. 2.2a

* See Section 1.6.3.

where we have used* that for all $n = 1, 2, 3, \ldots,$

$$\frac{\sinh n(\pi - y)}{\sinh n\pi} = \frac{e^{n\pi}e^{-ny} - e^{-n\pi}e^{ny}}{e^{n\pi} - e^{-n\pi}}$$

$$= e^{-ny}\frac{(1 - e^{-2n(\pi - y)})}{(1 - e^{-2n\pi})} \leq \frac{1}{(1 - e^{-2\pi})}e^{-ny},$$

and where we have also made the assumption that $f \in L^1(0, \pi)$, that is,

$$\int_0^\pi |f(s)|\, ds = M < \infty.$$

Let us fix a value Y of y arbitrarily small but positive and consider uniform convergence possibilities on the closed set $\overline{\Omega}_Y$ as indicated in Figure 2.2b. Letting

$$M_n = (\text{const})e^{-nY} = \frac{2\int_0^\pi |f(s)|\, ds}{\pi(1 - e^{-2\pi})}e^{-nY}$$

we have by the above bound

$$|u_n(x, y)| \leq M_n \qquad \text{for all } (x, y) \text{ in } \overline{\Omega}_Y$$

so that† by the Weierstrass M-test $\Sigma\, u_n(x, y)$ converges at each point in $\overline{\Omega}_Y$, in fact, uniformly. For term by term differentiation we also need $|\partial u_n(x, y)/\partial x|$ bounded by some M_n' from a convergent series of constants $\Sigma\, M_n'$. But since

$$\frac{\partial u_n(x, y)}{\partial x} = nc_n \cos nx \sinh n(\pi - y)$$

and

$$\frac{\partial^2 u_n(x, y)}{\partial x^2} = -n^2 c_n \sin nx \sinh n(\pi - y)$$

we must just take M_n' and M_n'' to be nM_n and n^2M_n, which are from convergent

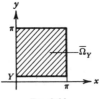

Fig. 2.2b

---

* Recall that $\sinh u = (e^u - e^{-u})/2$.

† The point here is that $\Sigma\, M_n$, that is, $\Sigma\, e^{-nY}$ apart from the common term constant, converges for $Y > 0$ by the ratio test of calculus, since one has $M_{n+1}/M_n = e^{-Y} < 1$.

series in the same way as M_n by the ratio test. Thus* we may apply the operator $\partial/\partial x$ to the series by applying it term by term, throughout $\overline{\Omega}_Y$; and, in turn, regarding $\Sigma \, \partial u_n(x, y)/\partial x$ as the convergent series to be differentiated, $\partial/\partial x$ may be applied again term by term to obtain

$$\frac{\partial^2}{\partial x^2} \sum_{n=1}^{\infty} u_n(x, y) = \sum_{n=1}^{\infty} \frac{\partial^2}{\partial x^2} u_n(x, y).$$

We proceed in the same fashion for differentiation with respect to y. Since

$$\frac{\partial u_n(x, y)}{\partial y} = -nc_n \sin nx \cosh n(\pi - y)$$

and

$$\frac{\partial^2}{\partial y^2} u_n(x, y) = -n^2 c_n \sin nx \sinh n(\pi - y)$$

one may take $M_n = M_n' = M_n'' = 2(\text{const})n^2 e^{-nY}$, where the constant is the same as above, the n^2 accommodates all three series $\Sigma \, u_n$, $\Sigma \, u_n'$, and $\Sigma \, u_n''$, and the extra 2 comes from

$$\frac{\cosh n(\pi - y)}{\sinh n\pi} = e^{-ny} \frac{(1 + e^{-2n(\pi - y)})}{1 - e^{-2n\pi}} \leq \frac{2}{(1 - e^{-2\pi})} e^{-ny}$$

in the first derivative series. Therefore, again† by the Weierstrass M-test we arrive at the validity of the term by differentiation

$$\frac{\partial^2}{\partial y^2} \sum_{n=1}^{\infty} u_n(x, y) = \sum_{n=1}^{\infty} \frac{\partial^2}{\partial y^2} u_n(x, y).$$

Thus

$$\Delta u = \Delta \sum u_n = \sum \Delta u_n = 0$$

provided that the data $f \in L^1(0, \pi)$, that is, that $\int_0^\pi |f(s)| \, ds < \infty$.

We also note that since each of the $u_n(x, y)$ is zero on the other three sides, the same is true of the series $u = \Sigma \, u_n$. Also (as noted in Section 1.6.3), since the limit of a sequence of uniformly continuous functions is continuous, by the above arguments, we have $u(x, y) \in C^2(\Omega)$ (since $u \in C^2(\overline{\Omega}_Y)$ for arbitrary $Y > 0$) and also that $u(x, y) \to 0$ as $(x, y) \to$ any boundary point on the other three sides of Ω. In fact, the above differentiation arguments may be repeated indefinitely,

---

* In the viewpoint of a more advanced course in functional analysis, the meaning of this is that the *derivative operator* $\partial/\partial x$ is a *closed operator* in the sense that $s_N \xrightarrow{\text{unif}} u$ and $\partial/\partial x \, s_N \xrightarrow{\text{unif}}$ "something" imply that the "something' is of necessity $\partial/\partial x \, u$. The concept of the closed operator is weaker than that of the *continuous operator*, the latter being best exemplified by *integral operators* and the fact that $s_N \xrightarrow{\text{unif}} u$ implies that $\int s_N \to \int u$. (See Problems 2.9.2 and 2.9.5.)

† The $\Sigma \, M_n$ again converges by the ratio test since $M_{n+1}/M_n = (1 + 2/n + 1/n^2)e^{-Y} < 1$ for sufficiently large n.

showing that $u(x, y) \in C^\infty(\Omega)$. This of course we also know from the fact that u is harmonic, hence analytic. To complete the answering of the three questions raised at the beginning of this section, it therefore remains only to investigate the *regularity* of u on the "data" line segment $y = 0, 0 \leq x \leq \pi$.

For the latter we call upon the Dini tests* and assume that f is continuously differentiable (i.e., $f \in C^1[0, \pi]$) and that $f(0) = f(\pi) = 0$. Then, since the separation of variables construction made the Fourier series for $f(x)$ and $u(x, 0)$ equal, namely,

$$u(x, 0) = \sum u_n(x, 0) \equiv \sum d_n \sin nx, \quad d_n = \frac{2}{\pi \sinh n\pi} \int_0^\pi f(s) \sin ns \, ds,$$

we have $u(x, 0) = f(x)$ everywhere on $0 \leq x \leq \pi$. Moreover, and as just mentioned, by the additional assumed regularity on f we know its Fourier series $\sum d_n \sin nx$ converges uniformly to $f(x)$ on $[0, \pi]$, which in terms of the Cauchy criterion for convergence means that for arbitrary $\varepsilon > 0$ one can guarantee that

$$|s_M(x, 0) - s_N(x, 0)| < \varepsilon$$

for all sufficiently large M and N. Since for any given M and N, $M > N$,

$$\Delta(s_M(x, y) - s_N(x, y)) = \Delta\left(\sum_{N+1}^M u_n(x, y)\right) = \sum_{N+1}^M \Delta u_n(x, y) = 0,$$

by the maximum principle applied to the function $v(x, y) = s_M(x, y) - s_N(x, y)$ on the domain Ω we have $\max_{\bar{\Omega}}|v(x,y)| \leq \max_{\partial\Omega}|v(x, y)|$, that is,

$$|s_M(x, y) - s_N(x, y)| < \varepsilon, \qquad (x, y) \in \bar{\Omega}$$

which by the Cauchy convergence criterion means that $s_N(x, y)$ converges uniformly to $u(x, y)$ on $\bar{\Omega}$. In particular, then, $u(x, y)$ is the limit of a uniformly convergent sequence of sums $s_N(x, y)$ of continuous functions and is hence itself continuous on $\bar{\Omega}$; thus as the point $(x, y) \to (x_0, 0)$ along any path (Fig. 2.2c), $u(x, y) \to u(x_0, 0) = f(x_0)$.

Let us summarize what has thus been shown in the above for the separation of variables solution $u(x, y) = \sum u_n$ to the Dirichlet problem with data present on all four sides, as shown in Figure 2.2d.

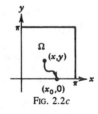

FIG. 2.2c

* See Section 1.6.3 and Problem 1.9.6(2).

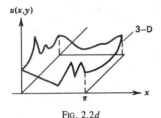

FIG. 2.2d

Problem (Dirichlet):

$$\begin{cases} \Delta u = 0 \text{ in } \Omega, \\ u = f \text{ on } \partial\Omega, \\ f \in C^0(\partial\Omega), \qquad f \in PC^1(\partial\Omega), \\ f(0, 0) = f(\pi, 0) = f(\pi, \pi) = f(0, \pi).* \end{cases}$$

Answers (to the six questions):

(1) *Existence* (by the verification above that the separation of variables "solution" $u = \Sigma\, u_n$ *is* a solution)

(2) *Uniqueness* (by the maximum principle, among functions $C^2(\Omega) \cap C^0(\overline{\Omega})$)

(3) *Stability* (in the maximum norms)

(1') *Construction* ($u = \Sigma\, u_n$)

(2') *Regularity* ($u \in C^\infty(\Omega) \cap C^0(\overline{\Omega})$)

(3') *Approximation* ($u = \lim s_N$ partial Fourier series, both pointwise (uniformly) and in L^2 convergence)

Proofs in the second (i.e., parabolic) case are similar, although different majorizing series $\Sigma\, M_n$, $\Sigma\, M_n'$, and $\Sigma\, M_n''$, appear of course. The heat initial boundary value problem is left as an exercise (see Problem 1 following). Again, the equation immediately smooths the data into the solution and uniform convergences are easily shown.

The third type of equation, the hyperbolic one, does not smooth out the data. Indeed it is desired that a sharp wave impulse produce a sharp wave, so that smoothing would be unexpected. Data discontinuities are thus preserved in the solutions, and uniform convergence is not so easily verified. Moreover, there is no maximum principle to transfer uniform convergence from the Fourier series for the data to the Fourier series for the solution.

Let us proceed however to prove in the same manner as above the validity of the separation of variables Fourier solution to a problem, the wave initial boundary

---

* This somewhat artificial restriction at the corners can be removed by subtracting from the solution a function $a_0 + a_1 x + a_2 y + a_3 xy$ that fits any given corner values. In general, one expects a Fourier series representation of a discontinuous function to converge to the average value at a jump.

value problem, of hyperbolic type. To do so, we shall assume too much regularity on the data.*

Vibrating String Problem (Fig. 2.2e)

$$\begin{cases} u_{tt} - u_{xx} = 0, & 0 < x < \pi, \quad t > 0, \\ u(0, t) = u(\pi, t) = 0, & t \geqq 0, \\ u(x, 0) = f(x), & 0 \leqq x \leqq \pi, \\ u_t(x, 0) = 0, & 0 \leqq x \leqq \pi, \\ \text{where} \\ f \in C^4[0, \pi] \text{ and} \\ f(0) = f(\pi) = f''(0) = f''(\pi) = 0. \end{cases}$$

The formal solution via separation of variables was found to be

$$u(x, t) = \sum_{n=1}^{\infty} \underbrace{c_n \sin nx \cos nt}_{u_n(x, t)} \qquad c_n = \frac{2}{\pi} \int_0^{\pi} f(s) \sin ns \, ds.$$

For finding majorizing series we first note that

$$|u_n(x, t)| \leqq |c_n|, \qquad \left| \frac{\partial}{\partial x} u_n(x, t) \right| \leqq n|c_n|, \qquad \left| \frac{\partial^2}{\partial x^2} u_n(x, t) \right| \leqq n^2|c_n|,$$

$$\left| \frac{\partial}{\partial t} u_n(x, t) \right| \leqq n|c_n|, \qquad \left| \frac{\partial^2}{\partial t^2} u_n(x, t) \right| \leqq n^2|c_n|.$$

In order to force $M_n \sim c_n$, $M_n' \sim nc_n$, and $M_n'' \sim n^2c_n$ to come from convergent series, we integrate by parts† to see if some regularity of the data f will be reflected

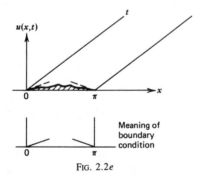

Fig. 2.2e

---

* By slightly more involved arguments it is sufficient that $f \in C^2[0, \pi]$ and that $f(0) = f(\pi) = f''(0) = f''(\pi) = 0$. This is most easily seen from a "Green's function" approach to the problem as discussed in Section 2.5.

† Once again we see in treating partial differential equations that integration by parts is often the key. The meaning of the calculation can be stated in another way, namely, that the Fourier cosine coefficients c_n' for the differentiated function $f'(x)$ are nc_n in terms of the Fourier (sine) coefficients c_n for the given function $f(x)$. Likewise the Fourier (sine) coefficients c_n'' for the function $f''(x)$ are $-n^2c_n$. (Footnote continued on page 165)

in sufficiently rapid decrease of the Fourier coefficients c_n for the data f. Recalling that we assumed $f \in C^4[0, \pi]$, $f(0) = f(\pi) = 0$, we have

$$\int_0^\pi f(s) \sin ns \, ds = \frac{1}{n} \int_0^\pi f'(s) \cos ns \, ds$$

$$= -\frac{1}{n^2} \int_0^\pi f''(s) \sin ns \, ds$$

$$= \frac{f''(\pi)(-1)^n - f''(0)}{n^3} - \frac{1}{n^3} \int_0^\pi f'''(s) \cos ns \, ds$$

$$= \frac{f''(\pi)(-1)^n - f''(0)}{n^3} + \frac{1}{n^4} \int_0^\pi f''''(s) \sin ns \, ds.$$

Recalling that we also assumed $f''(0) = f''(\pi) = 0$,* we thus have the bounds

$$|u_n(x, t)| \leq \frac{2}{\pi} \int_0^\pi |f''(s)| \, ds \cdot \frac{1}{n^2}, \qquad \left|\frac{\partial u_n}{\partial x}(x, t)\right| \leq \frac{2}{\pi} \int_0^\pi |f'''(s)| \, ds \cdot \frac{1}{n^2},$$

$$\left|\frac{\partial^2}{\partial x^2} u_n(x, t)\right| \leq \frac{2}{\pi} \int_0^\pi |f''''(s)| \, ds \cdot \frac{1}{n^2},$$

$$\left|\frac{\partial}{\partial t} u_n(x, t)\right| \leq \frac{2}{\pi} \int_0^\pi |f'''(s)| \, ds \cdot \frac{1}{n^2},$$

$$\left|\frac{\partial^2}{\partial t^2} u_n(x, t)\right| \leq \frac{2}{\pi} \int_0^\pi |f''''(s)| \, ds \cdot \frac{1}{n^2},$$

so that taking the M_n, M_n', and M_n'' to be the right-hand sides of the above inequalities, and remembering that $\Sigma(1/n^2)$ converges,† we have by the Weierstrass M-test that

$$\left(\frac{\partial^2}{\partial t^2} - \frac{\partial^2}{\partial x^2}\right) \sum u_n(x, t) = \sum \left(\frac{\partial^2}{\partial t^2} - \frac{\partial^2}{\partial x^2}\right) u_n(x, t) = 0.$$

Thus the wave equation is satisfied. The boundary conditions are satisfied by the separation of variables construction. The convergences above being uniform everywhere for $0 \leq x \leq \pi$ and $t \geq 0$ guarantees that the solution $u(x, t) \in C^2[0, \pi] \times [0, \infty)$, that is, for all $0 \leq x \leq \pi$ and all $t \geq 0$. We have thus answered affirmatively the questions of (1) *Existence*, (1') *Construction*, (2') *Regularity*, and (3') *Approximation*. *Uniqueness* (2) and a result on *Stability* (3) may be shown by the energy method in the following Problem 3.

Problem 1. Justify mathematically the separation of variables solution to the heat conduction problem (Fig. 2.2*f*)

(Footnote continued from page 164)

The latter integration by parts provides also a clue as to the use of Fourier transforms (see the later Section 2.7.3) under which the operation $d^2u/dx^2 \rightarrow -k^2u$, that is, $F(d^2u/dx^2) = -k^2F(u)$, where F denotes Fourier transform and k is a real variable playing a role analogous to n in the above.

* This assumption was not used in the calculation above, for the sake of the following Problem 2.

† By the integral test of calculus, for example.

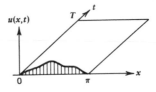

FIG. 2.2f

$$\begin{cases} u_t - u_{xx} = 0, & 0 \leq x < \pi, \quad t > 0, \\ u(x, 0) = f(x), & 0 \leq x \leq \pi, \\ u(0, t) = 0, & t \geq 0, \\ u(\pi, t) = 0, & t \geq 0, \end{cases}$$

in the following two ways:

(a) Use a maximum principle and the Dini test as was done for the Dirichlet problem above. Assume $f \in C^1[0, \pi]$ and $f(0) = f(\pi) = 0$. First recall (see Problem 1.9.3) the heat equation maximum principle: A smooth solution to the heat equation attains its maximum and minimum values (e.g., temperatures) either initially or at the boundary. More precisely: $u \in C^2([0, \pi] \times [0, T])$, that is, u twice continuously differentiable on the domain of the figure, $u_t - u_{xx} = 0$ in the interior of that domain, implies that

$$\max_{\substack{0 \leq x \leq \pi \\ 0 \leq t \leq T}} |u(x, t)| \leq \max\{ \max_{0 \leq x \leq \pi} |u(x, 0)|, \ \max_{0 \leq t \leq T} u(0, t), \ \max_{0 \leq t \leq T} u(\pi, t)\}$$

for arbitrary $T > 0$.

(b) Use an excess of regularity on f as was done for the vibrating string problem above. Assume $f \in C^4[0, \pi]$ and $f(0) = f(\pi) = f''(0) = f''(\pi) = 0$.

Problem 2. Examine the term $[f''(\pi)(-1)^n - f''(0)]/n^3$ in the integration by parts above. (a) What is wrong with the argument that one needs only $f''(0) = 0$ because the alternating series with terms $(-1)^n/n$ is a convergent one? (b) For what values $f''(\pi)$ and $f''(0)$ is this term zero. (c) A question related to another solution method (see Section 2.5) in which f needs to be extended to all $-\infty < x < \infty$ is the following: Given $f(x) \in C^2[0, \pi]$ and $f(0) = f(\pi) = 0$, and then if possible extended oddly and periodically about 0 and π to $f \in C^2(-\infty, \infty)$, does it follow that necessarily $f''(0) = f''(\pi) = 0$?

Problem 3. Using the total energy functional

$$E(t) \equiv E(u(t)) = \frac{1}{2} \int_0^\pi \left[\left(\frac{\partial u(x, t)}{\partial x} \right)^2 + \left(\frac{\partial u(x, t)}{\partial t} \right)^2 \right] dx$$

show *uniqueness* and a *stability* result for the vibrating string problem

$$\begin{cases} u_{tt} - u_{xx} = 0, & 0 < x < \pi, \quad t > 0, \\ u(0, t) = u(\pi, t) = 0, & t \geq 0, \\ u(x, 0) = f(x), & 0 \leq x \leq \pi, \\ u_t(x, 0) = g(x), & 0 \leq x \leq \pi. \end{cases}$$

Assume $u \in C^2$ on the closed T domain shown in Figure 2.2f and make use of differentiation under the integral, the divergence theorem, Schwarz's inequality, or whatever.

2.3 FOURIER SERIES AND HILBERT SPACE

In this section we first give an axiomatic description of Hilbert space, and then explain what it means in terms of Fourier series. The main result is given in Theorem H. In Section 2.4 we couple the present section concerning Hilbert space to the method of separation of variables and to a variety of resulting important Hilbert spaces, many of which correspond to examples given previously (Section 2.1).

Although Section 2.2 may have seemed to depend only upon methods of calculus and advanced calculus, it should be noted that Hilbert space methods also entered there, in the integration by parts calculations and in the orthogonality arguments implicit in proving a Dini test. Thus it may be said that whenever one writes down a Fourier series, one is, like it or not, working in a Hilbert space.

One may view the Hilbert space structure as being built up in four stages

algebraic
topological
analytical
geometrical

which we describe as follows in terms of the notions of

linear space
normed linear space
complete normed linear space
complete normed inner product linear space

and for which an instructive sequence of corresponding examples is

$C[0, \pi]$, the continuous functions* on $0 \leq x \leq \pi$,
$C[0, \pi]$, with RMS† norm $\|u\| = [\int_0^\pi |u(x)|^2 dx]^{1/2}$,

---

* We will usually state function space concepts for the case of real valued functions. The case of complex valued functions $u(x) + iv(x)$ is similar, but one needs a complex conjugate inserted at the right places, for example, $(u, v) = \int_0^\pi u(x)\overline{v(x)} \, dx$ (mathematicians) or $(u, v) = \int_0^\pi \overline{u(x)}v(x) \, dx$ (physicists). Most functional analysis books either explicitly or tacitly treat primarily the complex case because the resulting theorems are stronger, and one must be careful in applying those results to the real case.

† Root mean square, and alternately, L^2 norm. See Section 1.6.3.

$C[0, \pi]$, with maximum norm $\|u\| = \max|u(x)|$, $0 \leqq x \leqq \pi$,

$L^2[0, \pi]$, with inner product $(u, v) = \int_0^\pi u(x)v(x)\,dx$.

As another instructive example sequence in only the RMS norm one may consider

$R^2[0, \pi]$, the Riemann (square) integrable functions on $0 \leqq x \leqq \pi$,

$R^2[0, \pi]$, with the above RMS norm $\|u\| = [\int_0^\pi |u(x)|^2\,dx]^{1/2}$,

$L^2[0, \pi]$, the Lebesgue square integrable functions on $0 \leqq x \leqq \pi$,

$L^2[0, \pi]$, with the inner product $(u, v) = \int_0^\pi u(x)v(x)\,dx$.

Historically the continuous functions, being Riemann integrable, were extended to all Riemann integrable functions, which, lacking the completeness property, were in turn extended to the Lebesgue integrable functions.

The linear or vector space structures mentioned above are the following.

Linear Space†

Vectors such that:

$$\left.\begin{array}{l} u + v = v + u \\ u + (v + w) = (u + v) + w \\ \text{unique } 0 \text{ such that } u + 0 = u \\ \text{unique negative such that } -u + u = 0 \end{array}\right\} \begin{array}{l} \text{that is, a commutative} \\ \text{additive group} \end{array}$$

and *scalars* such that

$$\left.\begin{array}{l} a(u + v) = au + av \\ (a + b)u = au + bu \\ (ab)u = a(bu) \\ 1 \cdot u = u \end{array}\right\} \begin{array}{l} \text{scalars taken from a} \\ \text{"field" that is usually} \\ \text{the real or complex numbers.} \end{array}$$

Normed Linear Space

Linear space equipped with a norm $\|u\|$, norm properties are:

$$\|u\| \geqq 0, \qquad \|u\| = 0 \Leftrightarrow u = 0 \qquad \text{positive definite,}$$

$$\|u + v\| \leqq \|u\| + \|v\| \qquad \text{triangle inequality,}$$

$$\|\alpha u\| = |\alpha|\,\|u\| \qquad \text{homogeneous.}$$

---

* Also called a *vector space*. One need not take infinite dimensional examples such as $C[0, \pi]$ or $R^2[0, \pi]$; n-space R^n will suffice. However, the point is that almost any *function space* satisfies the linear space axioms; for example, $C^0(\Omega)$, $C^m(\Omega)$, $C_0^\infty(\Omega)$, $L^p(\Omega)$, and many others. An exception to this occurs however if one attempts to include a specific nonhomogeneous boundary condition; for example, $f(1) = g(1) = 1 \Longrightarrow\!\!\!\!\!\Longrightarrow$ the same for $f + g$.

Complete Normed Linear Space

Normed linear space that is Cauchy complete*: $\|u_n - u_m\| \to 0$ as $n \to \infty$, $m \to \infty$ implies the existence of a limit vector u *in the space* such that $\|u_n - u\| \to 0$ as $n \to \infty$.

Complete Normed Inner Product Linear Space (= Hilbert Space)

Complete normed linear space in which the norm is given by an inner product $\|u\|^2 = (u, u)$, where the inner product† has the following properties:

$$\left.\begin{array}{ll} (au + bv, w) = a(u, w) + b(v, w) & \\ (u, w) = \overline{(w, u)} & \text{symmetric} \\ (u, u) = \|u\|^2 & \text{norm compatible.} \end{array}\right\} \text{bilinear,}$$

A few remarks about the completeness property of a Hilbert space.

The notion of completeness is an idealization in which one defines a limit element even if it is not very constructible or realizable. Thus the axioms of the real number system define the irrational numbers as "limits" of Cauchy sequences $|a_n - a_m| \to 0$ of rational numbers, it then being shown (in the traditional treatments‡) that after one has the rationals and irrationals, there are no more. That is, if one takes a Cauchy sequence of irrationals or of mixed irrationals and rationals one does not generate any further idealized limits. This fact is fundamental in the same way to the completeness of 2-space, 3-space, n-space, l_2 (the infinite dimensional coordinate Hilbert space), L^2, and practically all complete function spaces used in analysis and differential equations.

Likewise the notion of Lebesgue integrability is an idealization created in order to complete the Riemann integrable functions.§ In the above examples of normed linear spaces one may thus regard $L^2[0, \pi]$ as the completion of $R^2[0, \pi]$ in the RMS norm. It turns out that the smaller function space $C[0, \pi]$ also has as RMS completion $L^2[0, \pi]$. This is even the case for $C^1[0, \pi]$, $C^n(0, \pi]$, and even for $C_0^\infty[0, \pi]$, the infinitely differentiable functions of compact support strictly within $[0, \pi]$.¶ Any such subspace of $L^2[0, \pi]$ with completion $L^2[0, \pi]$ is called a *dense subspace* of $L^2[0, \pi]$ and is an example of what are sometimes called *pre-Hilbert spaces*: a linear space, with norm given by an inner product, but not complete. It should be stressed that in both the theory and application of Hilbert space methods one usually ends up going back to some dense pre-Hilbert space of nice functions

---

* The fundamental example from which all others follow is the one-dimensional vector space of real numbers with norm $\|a\| = |a|$, the absolute value of a. The completeness follows from the construction of the irrationals from the (incomplete) rationals.

† Scaler (real, or complex) valued.

‡ In "nonstandard" analysis one creates an enlargement of the traditional real number system.

§ Consult books on integration theory for examples of Riemann integrable functions converging in various senses to nonRiemann integrable functions.

¶ See Problem 2.9.3, where a proof of this important fact is given.

to do all calculations, arguing then for the general case by going to the limit to obtain the desired result for the whole Hilbert space.

The maximum norm $\|u\| = \max|u(x)|$, $0 \leq x \leq \pi$, is stronger than the L^2 norm.* In fact, it is strong enough to make $C[0, \pi]$ complete, since the uniform pointwise limit of a sequence of continuous functions is continuous. The price one pays for this completeness is the lack of an inner product consistent with the norm.†

The existence of an inner product in a Hilbert space allows generalization of the usual geometric interpretations that one uses in discussing analytical vector dynamics in 2-space or 3-space to the infinite dimensional function spaces. This is the content of the following basic theorem, namely, that any function may be regarded as the (infinite) sum of its projected components on a basis of perpendicular vectors, just as in the Pythagorean theorem for 2- and 3-space.

Definition. If $(u, v) = 0$, u and v are called *orthogonal* (i.e., perpendicular to each other). If also $\|u\| = \|v\| = 1$, u and v are called *orthonormal*. A set of vectors $\{\varphi_i\}$ is called an *orthonormal set* if they are all mutually orthonormal, that is, if any one of them is orthonormal to all the rest. A set of vectors $\{\varphi_i\}$ is called a *maximal‡ orthonormal set* (m.o.s.) in a given Hilbert space H if it is impossible to find another vector in H that is orthonormal to all the $\{\varphi_i\}$.

Examples of orthogonal vectors are $(1, 0)$ and $(0, 3)$ in 2-space and $\sin nx$, $n = 1, 2, 3, \ldots$ in $L^2(0, \pi)$. They become orthonormal when normalized, that is, when divided by their length or norm, for example, $(1, 0)$ and $(0, 1)$ in 2-space and $(2\pi^{-1})^{1/2} \sin nx$, $n = 1, 2, 3, \ldots$ in $L^2(0, \pi)$. The most traditional maximal orthonormal set in Fourier analysis is the so-called trigonometric basis $\{\varphi_i\} = \{1/(2\pi)^{1/2}, \quad \pi^{-1/2} \cos nx, \quad n = 1, 2, 3, \ldots, \quad \pi^{-1/2} \sin nx, \quad n = 1, 2, 3, \ldots\}$. The fact that it is maximal in $L^2(-\pi, \pi)$ is of course nontrivial and was not proved until the beginning of this century.

Theorem H

Let $\{\varphi_i\}$ be an orthonormal set in a Hilbert space H. Then equivalently:

(1) $\{\varphi_i\}$ is a maximal orthonormal set (m.o.s.);
(2) Any u in H orthogonal to all φ_i is necessarily 0;
(3) Every u in H has (Fourier) representation $u = \Sigma \, (u, \varphi_i)\varphi_i$;

---

* See Section 1.6.3, Problem 1, where it is shown that uniform convergence implies L^2 convergence.

† Complete normed linear spaces are called Banach spaces. $C[0, \pi]$ with the maximum norm is a Banach space but is not a Hilbert space.

‡ Many books use the term *complete orthonormal set* (c.o.s.). The property is however one of *maximality*, although both adjectives, maximal and complete, may be used loosely to mean the same thing, namely, that there are no "missing" basis elements that have been left out of the basis $\{\varphi_i\}$. The adjective "complete" has already been used above to describe a convergence (analytical) property of Hilbert space so we prefer the adjective "maximal" for the above (geometrical) property.

(4) Every u in H has norm given by the (Parseval's) equation $\|u\|^2 = \sum |(u, \varphi_i)|^2$.

Before proving the theorem, we note the following three facts stated as lemmas and to be proved below.

Lemma 1

The inner product is jointly continuous, that is, when u_n converges in H to u and v_n converges in H to v, then the inner products

$$(u_n, v_n) \rightarrow (u, v).$$

Lemma 2

Whether or not the orthonormal set $\{\varphi_i\}$ is maximal in H, one has the *Bessel's Inequality*

$$\|u\|^2 \geqq \sum |(u, \varphi_i)|^2. \tag{4'}$$

Lemma 3*

For any Hilbert space H the sums (3) and (4) make sense and involve only a countable (i.e., indexed by $i = 1, 2, 3, \ldots$) number of the $\{\varphi_i\}$ for each u.

Proof of the Theorem†

We may show $(1) \Rightarrow (2) \Rightarrow (3) \Rightarrow (4) \Rightarrow (1)$.

$(1) \Rightarrow (2)$. Suppose not. Then there exists some $u \perp \{\varphi_i\}$, $u \neq 0$. We let $\varphi = u/\|u\|$, note that $(\varphi, \varphi_i) = 0$ for all φ_i, so that $\{\varphi, \varphi_i\}$ is an enlarged orthonormal set, which is a contradiction.

$(2) \Rightarrow (3)$. Consider $u - \sum_{i=1}^{\infty} (u, \varphi_i)\varphi_i$. This is a vector in the Hilbert space because by Lemma 3 there are only a countable number of the φ_i for which $(u, \varphi_i) \neq 0$ and because the infinite sum $\sum_{i=1}^{\infty} (u, \varphi_i)\varphi_i$ is the limit of the partial sums $s_N = \sum_{i=1}^{N} (u, \varphi_i)\varphi_i$ that converge by Lemma 2 (see the

---

* We were purposely vague in the sums in (3), (4), (4') and in the indexing $\{\varphi_i\}$ by i. Some Hilbert spaces need an *uncountable* number of $\{\varphi_i\}$ but as Lemma 3 shows, in all cases each u has a countably indexed representation. Most Hilbert spaces used here will have a single countable $\{\varphi_i\}$ like the sine functions $\varphi_n(x) = 2^{1/2}\pi^{-1/2} \sin nx$, $n = 1, 2, 3, \ldots$, for $H = L^2[0, \pi]$.

† To avoid complex conjugation we just consider real Hilbert spaces, although Theorem H and its proof remain the same for the complex and any other scalar field.

discussion below). "Dotting" this difference $u - \sum_{i=1}^{\infty} (u, \varphi_i)\varphi_i$ by any φ_j yields

$$
\begin{aligned}
\left(\varphi_j, u - \sum_{i=1}^{\infty} (u, \varphi_i)\varphi_i \right) &= (\varphi_j, u) - \left(\varphi_j, \sum_{i=1}^{\infty} (u, \varphi_i)\varphi_i \right) \\
&= (\varphi_j, u) - \lim_{N \to \infty} \left(\varphi_j, \sum_{i=1}^{N} (u, \varphi_i)\varphi_i \right) \\
&= (\varphi_j, u) - (u, \varphi_j) \\
&= 0
\end{aligned}
$$

which by (2), implies (3). In the second step above we used the separate continuity of the inner product, which is implied by the joint continuity of Lemma 1.

(3) $\Rightarrow$ (4). This follows from the joint continuity of the inner product by

$$
\|u\|^2 \equiv \lim_{N \to \infty} \left(\sum_{i=1}^{N} (u, \varphi_i)\varphi_i, \sum_{i=1}^{N} (u, \varphi_i)\varphi_i \right)
$$

$$
= \lim_{N \to \infty} \sum_{i=1}^{N} |(u, \varphi_i)|^2 = \sum_{i=1}^{\infty} |(u, \varphi_i)|^2.
$$

(4) $\Rightarrow$ (1). Suppose the $\{\varphi_i\}$ are not maximal. For an enlarged set $\{\varphi, \varphi_i\}$ applying (4) to φ yields the contradiction

$$
1 = \|\varphi\|^2 = \sum_{i=1}^{\infty} |(\varphi, \varphi_i)|^2 = 0.
$$

Proof of Lemma 1

Let $\|u_n - u\| \to 0$ and $\|v_n - v\| \to 0$. Then, employing the Schwarz inequality, we have

$$
\begin{aligned}
|(u_n, v_n) - (u, v)| &= |(u_n - u, v_n) + (u, v_n - v)| \\
&\leq |(u_n - u, v_n)| + |(u, v_n - v)| \\
&\leq \|u_n - u\| \|v_n\| + \|u\| \|v_n - v\| \\
&\leq \|u_n - u\| \|v_n - v\| + \|u_n - u\| \|v\| + \|u\| \|v_n - v\| \\
&\to 0.
\end{aligned}
$$

Proof of Lemma 2

By Lemma 3 we know that for a given u we need consider at most a countable number of the φ_i, those for which the components $(u, \varphi_i) \neq 0$. We consider any finite number n of these φ_i and obtain Bessel's inequality as follows:

$$0 \le \left\| u - \sum_{i=1}^{n} (u, \varphi_i)\varphi_i \right\|^2$$

$$= \|u\|^2 - 2(u, \sum_{i=1}^{n} (u, \varphi_i)\varphi_i) + \sum_{i=1}^{n} |(u, \varphi_i)|^2$$

$$= \|u\|^2 - \sum_{i=1}^{n} |(u, \varphi_i)|^2.$$

Thus for any finite number n of the φ_i we have $\sum_{i=1}^{\infty} |(u, \varphi_i)|^2 \le \|u\|^2$. This means that the series $\sum_{i=1}^{\infty} |(u, \varphi_i)|^2$ converges absolutely, and to some limit no greater than $\|u\|^2$. It also means that the partial sums $s_n = \sum_{i=1}^{n} (u, \varphi_i) \varphi_i$ converge (a fact used above in the proof of (2) $\Rightarrow$ (3))since they form a Cauchy sequence by

$$\|s_n - s_m\|^2 = \sum_{i=n+1}^{m} |(u, \varphi_i)|^2 \le \sum_{i=n+1}^{\infty} |(u,\varphi_i)|^2 \to 0.*$$

Proof of Lemma 3

Let us begin with any maximal orthonormal set $\{\varphi_\alpha\}$, where we use the index α to denote either the countable or a possibly uncountable index set. ¶ For any given u in H, and let us take $\|u\| = 1$, the set

$$\Phi(u) = \{ \text{all } \varphi_\alpha \text{ such that } (u, \varphi_\alpha) \ne 0\}$$

is clearly given by $\Phi(u) = \bigcup_{n=1}^{\infty} \Phi_n(u)$, where

$$\Phi_n(u) = \{\text{all } \varphi_\alpha \text{ such that } |(u, \varphi_\alpha)|^2 \ge 1/n\}.$$

By the finite Bessel's inequality $\sum_{i=1}^{n} |(u, \phi_i)|^2 \le \|u\|^2$ (which stands independently of these Lemmas) there can be at most n vectors φ_α in $\Phi_n(u)$. Since $\Phi(u)$ is the countably indexed union of finite numbers of vectors, it is itself countable (see Problem 2 for an exercise on this type of reasoning).

*Moreover, since $\sum_{i=1}^{\infty} |(u, \varphi_i)|^2$ converges absolutely and since by calculus absolutely convergent series have the same limit independent of rearrangement, it follows that $s_n \to u$ regardless of the order in which the φ_i are put in.

¶ An interesting example of a nonseparable Hilbert space is the Hilbert space of almost periodic functions on $-\infty < x < \infty$. One begins with all continuous functions $f(x)$ such that for every $\varepsilon > 0$ there exists an interval length l such that with every l-interval there exists at least one number τ such that $|f(x + \tau) - f(x)| < \varepsilon$ for all $-\infty < x < \infty$. It can be shown that an inner product is given by

$$(f, g) = \lim_{T \to \infty} \frac{1}{2T} \int_{-T}^{T} f(x + t)g(x + t) \, dt,$$

which exists, independent of x. The completion is called the almost periodic function Hilbert space and was introduced by Harold Bohr, mathematician and brother of the physicist Niels Bohr. An orthonormal basis is $\varphi_\alpha = e^{i\alpha x}$, $-\infty < \alpha < \infty$. Lemma 3 asserts that for any given almost periodic function $f(x)$, only a countable number of the φ_α are required to Fourier represent $f(x)$.

Problem 1. (a) Show from the inverse triangle inequality that when $u_n \to u$ in a Hilbert space H (or more generally in any normed linear space) one has the norms converging also, that is, $\|u_n\| \to \|u\|$. (b) Use part (a) to shorten the proof of Lemma 1 by one line.

Problem 2. In the proof of Lemma 3 we used a set theory fact that a countable union of finite sets is countable. Prove that in fact a countable union of countable sets is countable.

Problem 3. One uses a set-theoretic fact-axiom known as Zorn's lemma to prove that any Hilbert space contains at least one maximal orthonormal set. One can also use it to guarantee the existence of such things as a maximal dissipative extension for every dissipative partial differential operator,* and so on. In general one can assert that it forms an interesting bridge between analysis and logic. We have accordingly included this problem to provide some practice on this bridge for those who are so inclined.

The argument establishing the existence of a maximal orthonormal set is as follows. Let Φ be the class of all orthonormal sets in H. H nonempty implies Φ nonempty. One partially orders Φ by saying $\{\varphi_\alpha\} \subset \{\varphi_\beta\}$ if all the φ_α's are already φ_β's. By the following two sentence ordering argument Zorn's lemma now asserts the existence of a maximal orthonormal set. For any chain of orthonormal sets the union Ψ of all elements thereof is an upper bound for the chain, because any $\{\varphi_\alpha\}$ in the chain is upperbounded by Ψ in the set inclusion sense. Moreover, $\Psi \subset \Phi$ because any φ_a and φ_b in Ψ are elements of orthonormal sets $\{\varphi_\alpha\}$ and $\{\varphi_\beta\}$ in the chain such that either $\{\varphi_\alpha\} \subset \{\varphi_\beta\}$ or $\{\varphi_\beta\} \subset \{\varphi_\alpha\}$, and in either case $(\varphi_a, \varphi_b) = 0$.

Consult the literature while doing the following exercises. (a) Complete the checking of the details for the above proof. (b) It was long conjectured that a similar basis result held in Banach spaces, but recently a counterexample was given.† (i) State the properties of such a Banach space basis (Schauder basis) and write down one concrete example. (ii) Look up the counterexample and determine some of its properties. (c) Every normed linear space contains an algebraic (linear) basis. (i) Determine the properties and dimension of such a basis (Hamel basis). (ii) Does Zorn's lemma enter into its existence? (d) Think about why Zorn's lemma cannot be used to guarantee a Schauder basis for Banach spaces.

2.4   FOURIER SERIES AND STURM–LIOUVILLE EQUATIONS

Coupled with Theorem H of the previous section, which asserts that we may Fourier-expand all reasonable functions in terms of any maximal orthonormal set $\{\varphi_n\}$, is

---

* For the notion of dissipative differential operators see R. Phillips *Trans. Am. Math. Soc.* **90** (1959).

† For this counterexample, see P. Enflo, *Acta Mathematica* **130** (1973).

the Sturm–Liouville theorem, which asserts that a large class of ordinary differential equations have the property that their eigenfunctions are indeed a maximal orthonormal set $\{\varphi_n\}$. As nature (or perhaps just mathematical consistency) would have it, this class of ordinary differential equations contains most of those that we are led to when solving a partial differential equation on a given domain Ω by separation of variables. Moreover, the $\{\varphi_n\}$ thus obtained are usually the most natural and the most efficient basis for Fourier-expanding solutions to the given partial differential equation.

Let $L^2(a, b, r)$ be those functions $u(x)$ defined on any given interval (a, b)* and square integrable there with respect to the "weight-function" $r(x)$, that is,

$$\int_a^b |u(x)|^2 r(x)\, dx < \infty.$$

$L^2(a, b, r)$ is always a separable Hilbert space (see Problem 2.9.3). Consider any ordinary differential equation of the form

$$-(pu')' + qu = \lambda ru, \qquad a < x < b \qquad (1)$$

where u is in $L^2(a, b, r)$ and moreover is sufficiently nice so that the operations in (1) can be performed while staying within $L^2(a, b, r)$, and such that the following conditions† hold:

  (i) p is a real measurable function defined on (a, b), $p(x) \neq 0$ there, and p and p^{-1} are locally integrable in (a, b).‡
  (ii) q is a real measurable function defined on (a, b) and locally integrable there.
  (iii) r is a real measurable function defined on (a, b), $r(x) > 0$ there, and r is locally integrable in (a, b).

Definition.   Under the above conditions the Equation (1) is called a Sturm–Liouville equation. If additionally (a, b) is finite and p^{-1}, q, and r are integrable on all of (a, b), then (1) is called *regular*. Otherwise (1) is called *singular*.

---

* Finite, a half line, or the whole real line. Actually these functions must be complex valued functions $u(x) + iv(x)$, u and v real valued, but except for technical considerations one may think of real valued functions and real valued solutions of Equation (1).

† The conditions, in order to be sufficiently comprehensive, are stated here in the technical language of Lebesgue integration theory. However, one may for example just think in terms of reasonable conditions such as all of p, p', p^{-1}, q, and r being positive continuous functions on (a, b).

The standard example is our familiar (Rayleigh's) equation

$$-u''(x) = \lambda u(x), \qquad 0 < x < \pi.$$

For other examples see Section 2.1.

‡ That is, integrable on any bounded interval $[c, d]$ contained in (a, b).

For simplicity in stating the following theorem we restrict attention to the regular case* and so-called separated boundary conditions of the form

$$u(a) \cos \alpha - u'(a)p(a) \sin \alpha = 0, \qquad (2)$$
$$u(b) \cos \beta - u'(b)p(b) \sin \beta = 0,$$

where α and β are any numbers $\alpha \in [0, \pi)$, $\beta \in [0, \pi)$.

Theorem (Sturm–Liouville)

The eigenfunctions of the Equation (1) subject to the boundary conditions (2) form a maximal orthonormal set (m.o.s.) $\{\varphi_n\}_{n=1}^{\infty}$ for the Hilbert space $H = L^2(a, b, r)$.

In other words, the Equation (1) with boundary conditions (2) behaves just like the "canonical" Rayleigh equation with Dirichlet boundary conditions in permitting a Fourier representation

$$u(x) = \sum_{n=1}^{\infty} c_n \varphi_n(x)$$

of an arbitrary square integrable function u. This justifies then in principle the separation of variables procedure of trying a solution candidate of the form

$$u(x_1, x_2, \ldots) = \sum_{n_1, n_2, \ldots} c_{n_1 n_2 \ldots} u_{n_1 n_2 \ldots}(x_1, x_2, \ldots)$$

where the $u_n(x_1, x_2, \ldots) \equiv \varphi_{n_1}(x_1) \varphi_{n_2}(x_2) \cdots$ is the product of special Sturm–Liouville eigenfunctions and related initial value problem solutions depending on the original partial differential equation and the given boundary and initial conditions. Of course for a mathematically complete treatment, one must go further in each case and actually solve for (or estimate) the solutions in order to establish enough properties for the associated eigenfunctions so as to be able to prove (e.g., by verifying differentiation term by term) that the formal solution constructed in this way is in fact the bona fide regular solution of the partial differential equation satisfying the existence, uniqueness and stability statements.

Let us give here briefly the essentials of a proof of the above Sturm-Liouville theorem. The details are left to the student (Problem 3).

Proof

One begins with the Wronskian

---

* In the singular case, the eigenfunctions sometimes are and sometimes are not a maximal orthonormal set. For example, if one has the so-called limit circle case at both ends of the interval, then under appropriate boundary conditions one obtains a maximal orthonormal set of eigenfunctions as in the regular case. In the limit point case, it may go either way. See Problem 2.9.4.

$$W(u, \bar{v})(x) = p(x)[u(x)\overline{v'(x)} - u'(x)\overline{v(x)}]$$

which comes from the integration by parts

$$\int_a^b Lu\bar{v}\, dx - \int_a^b u\overline{Lv}\, dx = \int_a^b \frac{d}{dx} W(u, \bar{v})(x) = W(u, \bar{v})|_a^b.$$

Here

$$Lu = -(pu')' + qu$$

is the Sturm–Liouville operator given in (1) above. In order that L be a symmetric operator in $L^2(a, b),$* it is both necessary and sufficient that

$$W(u, \bar{v})|_a^b = 0.$$

For this, the separated boundary conditions (2) above, among others, suffice. The Sturm-Liouville operator (1), (2) is thus seen to be symmetric on any reasonable subspace of sufficiently regular functions. Moreover, it may be said to be formally self-adjoint (meaning that its closure on the $C_0^\infty(a, b)$ functions will be self-adjoint), and by extending its domain $D(L)$ to distributional derivatives (see Problem 1.9.7(2)), it will be a self-adjoint operator. Any self-adjoint operator has only real eigenvalues, as may be seen for the Sturm–Liouville operator (1), (2) by substituting $Lu = \lambda u$ into the integration by parts above.

From the theory of ordinary differential equations under the suitable continuity or weaker assumptions on the coefficients we know that there exists for each λ a two-parameter family $\varphi(x) = c_1\varphi_1(x) + c_2\varphi_2(x)$ of solutions to the homogeneous equation

$$Lu - \lambda ru = 0.$$

Let a particular nontrivial solution $\varphi(x, \lambda)$ be determined for all λ by the left end condition

$$\varphi(a, \lambda) = \sin \alpha, \qquad \varphi'(a, \lambda) = p^{-1}(a) \cos \alpha.$$

Note then that $\varphi(x, \lambda)$ satisfies the left boundary condition of (2). Thus $\varphi(x, \lambda)$ will be an eigenfunction for the operator (1), (2) and λ the corresponding eigenvalue if and only if $\varphi(x, \lambda)$ satisfies the right boundary condition of (2), that is, if and only if

---

* See Problem 1.9.7. L symmetric in $L^2(a, b)$ is easily seen to be equivalent to $(1/r)L$ being symmetric in $L^2(a, b, r)$.

$$f(\lambda) \equiv \varphi(b, \lambda) \cos \beta - \varphi'(b, \lambda)p(b) \sin \beta = 0.$$

There can be no other linearly independent eigenfunction ψ for (1), (2) at λ. For if there were, by the Wronskian test for linear independence of solutions, one must have $W(\varphi, \psi) \neq 0$ on the interval (a, b), but to the contrary we would have

$$p(a)[\psi(a)\varphi'(a, \lambda) - \psi'(a)\varphi(a, \lambda)] = 0.$$

It will be advantageous now to construct in the same way a particular nontrivial solution $\psi(x, \lambda)$ determined by the right end condition

$$\psi(b, \lambda) = \sin \beta, \qquad \psi'(b, \lambda) = p^{-1}(b) \cos \beta.$$

This solution $\psi(x, \lambda)$ satisfies the right boundary condition of (2) and will be an eigenfunction for the operator (1), (2) if and only if it satisfies the left boundary condition of (2), in which case it will be a multiple of $\varphi(x, \lambda)$. Otherwise φ and ψ are two fundamental solutions for the homogeneous equation.

Substituting the assigned right-end boundary values for $\psi(x, \lambda)$ into the above $f(\lambda)$ yields

$$f(\lambda) = W_\lambda(\varphi, \psi)(b).$$

By the integration by parts above and the symmetry of the operator (1), (2), for λ complex and not real $f(\lambda)$ may be seen to be nonzero. Moreover, $f(\lambda)$ has the important property of being analytic in λ, and belongs to the class of entire functions of order less than or equal to one-half. As seen above the zeros of $f(\lambda)$ are exactly the eigenvalues of the operator (1), (2). It may be deduced from theorems in analytic function theory that the number of zeros of $f(\lambda)$ in any bounded interval function is finite and moreover can be estimated. This, coupled with the observation that the operator is self-adjoint but unbounded, guarantees that the totality of eigenvalues λ_n is an unbounded but countable set of real numbers. For each λ_n we have as constructed above the eigenfunction $\varphi(x, \lambda_n)$.

To show the maximality of the eigenfunctions $\{\varphi_n\}$ one may now proceed in a number of ways. One approach is to consider the Green's function

$$-G_\lambda(x, y) = \begin{cases} \dfrac{\varphi(x, \lambda)\psi(y, \lambda)}{W_\lambda(\varphi, \psi)(b)}, & a \leqq x \leqq y \leqq b, \\[2ex] \dfrac{\psi(x, \lambda)\varphi(y, \lambda)}{W_\lambda(\varphi, \psi)(b)}, & a \leqq y \leqq x \leqq b, \end{cases}$$

as is done in the variation of parameters solution to the inhomogeneous problem. One may then show that for any λ the corresponding integral operator (see Problem 2.9.5)

$$(r^{-1}L - \lambda)^{-1}w(x) = \int_a^b G_\lambda(x, y)w(y)\, dy$$

is "compact" on $L^2(a, b, r)$ and indeed as indicated by notation does invert the operator (1), (2) at any noneigenvalue λ. From elementary spectral theory one knows that the eigenfunctions of a compact self-adjoint operator are maximal, and from the relation

$$L_{\varphi_n} = \lambda_n r \varphi_n \text{ iff } (r^{-1}L)^{-1}\, \varphi_n = \lambda_n^{-1}\, \varphi_n$$

one sees that the eigenfunctions of L and those of $(r^{-1}L)^{-1}$ are the same.*

One can sometimes prove the maximality of a given orthonormal set directly without recourse to the Sturm–Liouville theorem. In so doing one avoids solving a Sturm–Liouville equation. On the other hand, one usually needs some correspondingly strong result. We illustrate such direct proofs of the maximality of a given orthonormal set by two examples.

First consider our familiar sine function basis $\{\varphi_n = (2/\pi)^{1/2} \sin nx\}$ in $L^2(0, \pi)$. By the Sturm–Liouville theorem above applied to the Rayleigh equation with Dirichlet boundary conditions, $\{\varphi_n\}$ is a maximal orthonormal set for $L^2(0, \pi)$.

To prove this fact directly, we use the strong Dini test (see Problem 1.9.6) and the fundamental fact (see Problem 2.9.3) that the C_0^∞ functions are always dense in $L^2(a, b, r)$. By Theorem H of the previous section, it suffices to show that for any f in $L^2(0, \pi)$ one can make

$$\|f - s_N^f\|_{L^2(0,\pi)}$$

arbitrarily small for N sufficiently large, where we recall

$$s_N^f(x) = \sum_{n=1}^N c_n \varphi_n(x) \text{ and } c_n = (f, \varphi_n) = \int_0^\pi f(s)\, \varphi_n(s)\, ds.$$

For any given $\varepsilon > 0$ one may find by the above-mentioned denseness of the C_0^∞ functions an approximating function h in $C_0^\infty(0, \pi)$ such that

$$\|f - h\| < \varepsilon/2.$$

As in the proof of Lemma 2 (Bessel's inequality) of the previous section, one has†

---

* Variations on this type of proof occur. For example, to avoid showing the inverse of L to be compact self-adjoint with discrete spectra, one may find a recourse to the variational characterization of the eigenvalues λ_n. But the rigorous equivalence of the spectral properties of the differential equation and those of its variational characterization is roughly equivalent to the demonstrating of a compact inverse. For further information about spectral theory see Section 2.7.2.

† This completing of the squares shows also that the best least squares approximation to f by a weighted sum of the φ_n is obtained by using the Fourier coefficients for the weights.

$$\|f - s_N^h\|^2 = \|f\|^2 - 2\left(f, \sum_{n=1}^{N} (h, \varphi_n)\varphi_n\right) + \sum_{n=1}^{N} (h, \varphi_n)^2$$

$$= \|f - s_N^f\|^2 + \sum_{n=1}^{N} [(h, \varphi_n) - (f, \varphi_n)]^2$$

and thus

$$\|f - s_N^f\| \leq \|f - s_N^h\|.$$

By the strong Dini test the Fourier series $s_N^h(x)$ for $h(x)$ converges uniformly point-wise to h on the interval $[0, \pi]$, and hence (see Problem 1 of Section 1.6.3) also in the mean, so that

$$\|h - s_N^h\| < \varepsilon/2$$

for N sufficiently large. Combining these facts by the triangle inequality we therefore have

$$\|f - s_N^f\| \leq \|f - s_N^h\|$$

$$\leq \|f - h\| + \|h - s_N^h\|$$

$$< \varepsilon/2 + \varepsilon/2 = \varepsilon$$

for N sufficiently large.

For a second example of a direct proof of the maximality of a given sequence of functions $\{\phi_n\}$, consider the powers $\varphi_n(x) = 1, x, x^2, x^3, \ldots, x^n, \ldots, n = 0, 1, 2, \ldots$, suitably normalized for $L^2(a, b)$ for any finite interval (a, b). For simplicity let us consider the case $L^2(-1, 1)$. We use the Weierstrass approximation theorem, which asserts that for any h in $C^0[-1, 1]$ there exists a polynomial p such that

$$\max_{-1 \leq x \leq 1} |h(x) - p(x)| < \varepsilon/2\sqrt{2}$$

for any given $\varepsilon > 0$. This clearly implies that

$$\|h - p\|_{L^2(-1,1)} < \varepsilon/2.$$

By employing also the denseness of the C_0^∞ functions in $L^2(-1, 1)$, we therefore have for any f in $L^2(-1, 1)$ a suitable approximating h in C_0^∞ and a corresponding polynomial p such that

$$\|f - p\| \leq \|f - h\| + \|h - p\|$$

$$< \varepsilon/2 + \varepsilon/2 = \varepsilon.$$

This means that we can approximate f arbitrarily closely by a linear combination

$$\sum_{n=0}^{N} c_n \varphi_n$$

of the $\varphi_n = x^n$, $n = 0, 1, 2, \ldots$. These $\{\varphi_n\}$ are thus maximal in the sense (3) of Theorem H even though they cannot be mutually orthonormal. Properly combined (see Problem 2 below) they become the maximal orthonormal set of Legendre polynomials given in Section 2.1.

In summary then we have seen three ways to show $\{\varphi_n\}$ to be a maximal orthonormal set.

1. By use of:
    (i) the Fundamental Approximation Theorem (Problem 2.9.3) that the C_0^∞ test functions are dense, plus
    (ii) some other already established fundamental auxiliary fact such as a Dini test or a variational characterization, plus
    (iii) some basic orthogonality argument such as Bessel's inequality.
2. By use of:
    (i) the Fundamental Approximation Theorem for C_0^∞ functions, plus
    (ii) another powerful theorem such as the Weierstrass theorem, plus
    (iii) orthogonalizing in some way such as Gram–Schmidt (see Problem 2.9.3).
3. By use of:
    (i) the Sturm–Liouville theorem of this section, plus
    (ii) the establishing that the operator L involved is self-adjoint, plus
    (iii) the determining of its regularity or acceptable type of singularness (Problem 2.9.4).

Problem 1. (a) One can eliminate the coefficient p and weight function r by transforming the equation from the Sturm–Liouville form (1) to the so-called normal form

$$-v''(y) + c(y)v(y) = \lambda v(y), \qquad a' < y < b', \qquad (1)'$$

by the change of independent variable

$$y = \int_{x_0}^{x} r(s)^{1/2} p(s)^{-1/2} \, ds$$

and the change of dependent variable

$$v(y) = u(x)r(x)^{1/4} p(x)^{1/4},$$

the new interval (a', b') depending on the choice of x_0 in (a, b). Do this, finding the new coefficient $c(y)$.

(b) Convert the Hermite and Legendre equations (see Section 2.1) in and out of normal form $(1)'$ and Sturm–Liouville form (1).

(c) For additional practice convert both equations in and out of the third or ordinary differential equation form

$$a_0(x)w''(x) + a_1(x)w'(x) + a_2(x)w(x) = 0. \tag{1}''$$

Problem 2. (a) Check that the Rayleigh equation

$$-u''(x) = \lambda u(x), \qquad 0 < x < \pi,$$

is a regular Sturm–Liouville equation. (b) Verify that the Hermite and Legendre equations (see Section 2.1) are singular Sturm–Liouville equations. (c) By Gram–Schmidt (see Problem 2.9.3) orthonormalize the $\varphi_n = x^n$ given above.

Problem 3. (a) Identify α and β in the general separated boundary condition (2) to obtain (i) Dirichlet, (ii) Neumann, and (iii) Robin boundary conditions for the Rayleigh equation and the interval $(0, \pi)$. (b) From the separated boundary condition (2), what additional spectral information does one obtain, beyond the self-adjointness of the operator and the existence of a countable number of eigenvalues λ_n? (c) Fill in the details in the proof of the Sturm–Liouville theorem given above.

2.5 FOURIER SERIES AND GREEN'S FUNCTIONS

Quite clearly there must exist certain connections between the (1) Fourier series (via separation of variables) representations of solutions of partial differential equations and (2) Green's function representations of these solutions. Although we do not go too far into those connections, we will illustrate what one may expect by looking at three problems: a hyperbolic one, and both a homogeneous and a nonhomogeneous elliptic one. The parabolic case can be approached in similar ways (see Problem 3).

Let us remark that, since we have the three basic methods of solutions for partial differential equations, namely (1) separation of variables, (2) Green's functions, and (3) variational methods, one could discuss going back and forth between them in all six directions, and for all three classes of equations, for the homogeneous and nonhomogeneous cases, for different boundary conditions, and so on. Thus our discussion here is of necessity limited. In Section 2.6, however, one will find some connections with the third basic solution method, for an elliptic problem.

Let us consider first the wave initial boundary value problem (Fig. 2.5a)

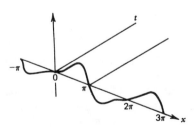

FIG. 2.5a

$$\begin{cases} u_{tt} - u_{xx} = 0, & 0 < x < \pi, \quad t > 0, \\ u_t(x, 0) = 0, & 0 < x < \pi, \\ u(x, 0) = f(x), & 0 < x < \pi, \\ u(0, t) = u(\pi, t) = 0, \end{cases}$$

where $f(x)$ has Fourier sine expansion

$$f(x) = \sum_{n=1}^{\infty} c_n \sin nx, \qquad c_n = \frac{2}{\pi} \int_0^{\pi} f(s) \sin ns \, ds.$$

By separation of variables (see Sections 1.5 and 2.1) we have found the solution

$$u(x, t) = \sum_{n=1}^{\infty} c_n \sin nx \cos nt.$$

If we extend $f(x)$ oddly about $x = 0$ and $x = \pi$ as shown in the figure, it is easily seen that the Fourier expansion above for $f(x)$ remains valid for all $-\infty < x < \infty$, since it just repeats itself periodically. Using the fact

$$\sin nx \cos nt = \tfrac{1}{2} \sin n(x - t) + \tfrac{1}{2} \sin n(x + t)$$

we have, using also the fact that the limit of a sum is the sum of the limits when all exist,

$$u(x, t) = \frac{1}{2} \sum_{n=1}^{\infty} c_n \sin n(x - t) + \frac{1}{2} \sum_{n=1}^{\infty} c_n \sin n(x + t)$$

$$= \frac{f(x - t) + f(x + t)}{2},$$

the so-called *d'Alembert Formula* for the solution to the wave initial value problem. This formula may be, in accordance with the point of view being taken in this section, regarded as a Green's function representation for the solution

$$u(x, t) = \int_{-\infty}^{\infty} G(x, y, t) f(y) \, dy$$

to the given problem, where

$$G(x, y, t) = \frac{\delta(x - t, y) + \delta(x + t, y)}{2},$$

δ denoting the "delta function" measure that gives the point evaluation of an integrand at the point at which the two arguments of the delta function coincide.

The general wave initial value problem

$$\begin{cases} u_{tt} - u_{xx} = F(x, t), & -\infty < x < \infty, \quad t > 0, \\ u(x, 0) = f(x), & -\infty < x < \infty, \\ u_t(x, 0) = g(x), & -\infty < x < \infty, \end{cases}$$

may be integrated directly (see Problem 1) to obtain the more general d'Alembert formula

$$u(x, t) = \frac{1}{2}\left[f(x - t) + f(x + t) + \int_{x-t}^{x+t} g(s) \, ds \right.$$

$$\left. + \int_0^t \int_{x-(t-s)}^{x+(t-s)} F(y, s) \, dy \, ds \right].$$

This expression may also be regarded, from the point of view being taken in this section, as the Green's function representation formula for the solution to the given problem. It would be, however, more complicated to arrive at this expression via the separation of variables solution as was done in the instance above, and in fact, it should be regarded as a more general solution valid for data that need not be periodic.

For our second illustration of the connections between Fourier series and Green's functions, we next show how we may deduce the Poisson formula for the solution of the Dirichlet problem on the two-dimensional unit sphere by summing the Fourier series solution obtained previously by separation of variables. We recall that the Dirichlet problem (Fig. 2.5b)

$$\begin{cases} \Delta u = 0, & r < 1, \\ u(1, \theta) = f(\theta), \end{cases}$$

has separation of variables solution

$$u(r, \theta) = \tfrac{1}{2}a_0 + \sum_{n=1}^{\infty} r^n(a_n \cos n\theta + b_n \sin n\theta)$$

with coefficients

$$a_n = \pi^{-1} \int_{-\pi}^{\pi} f(\varphi)\cos n\varphi \, d\varphi, \qquad b_n = \pi^{-1} \int_{-\pi}^{\pi} f(\varphi) \sin n\varphi \, d\varphi.$$

Let us assume that f is C^1 and of period 2π, and recall (see Problem 1.9.6) that when deriving the Dini test the first step was the insertion of a_n and b_n so that the partial Fourier sums were in the form

$$s_N^f(x) = \frac{1}{\pi} \int_{-\pi}^{\pi} f(s)\left(\frac{1}{2} + \sum_{n=1}^{N} \cos n(x - s) \right) ds.$$

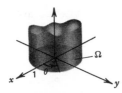

Fig. 2.5b

Using the same approach here, we have for $s_N^u(r, \theta)$, after inserting the values of the Fourier coefficients a_n and b_n into the partial sum for $u(r, \theta)$, the expression

$$s_N^u(r, \theta) = \frac{1}{2\pi} \int_{-\pi}^{\pi} f(\varphi) \, d\varphi$$

$$+ \sum_{n=1}^{N} r^n \left(\int_{-\pi}^{\pi} f(\varphi)(\cos n\theta \cos n\varphi + \sin n\theta \sin n\varphi) d\varphi \right)$$

$$= \frac{1}{\pi} \int_{-\pi}^{\pi} f(\varphi) \left(\frac{1}{2} + \sum_{n=1}^{N} r^n \cos n(\theta - \varphi) \right) d\varphi.$$

By the assumption that $f \in C^1$ we know from the strong Dini test that the Fourier series for f on the boundary converges uniformly pointwise to f, that is, $s_N(1, \theta) \to f(\theta)$ uniformly for $-\pi \leqq \theta \leqq \pi$. By use of the maximum principle it then follows as in Section 2.2 that the partial sums $s_N(r, \theta) \to u(r, \theta)$ uniformly on the closed unit sphere $\overline{\Omega}$. Thus pointwise we have for $r \leqq 1$

$$u(r, \theta) = \lim_{N \to \infty} s_N(r, \theta) = \lim_{N \to \infty} \frac{1}{\pi} \int_{-\pi}^{\pi} f(\varphi) \left(\frac{1}{2} + \sum_{n=1}^{N} r^n \cos n(\theta - \varphi) \right) d\varphi.$$

The integration has as usual strengthened convergence; the geometric type series under the integral need not converge at $r = 1$. Let us continue, however, and consider this series, for $r < 1$, as analogous to the consideration for the Dirichlet kernel we encountered in the proof of the Dini test, namely,

$$\frac{1}{2} + \sum_{n=1}^{N} \cos n\theta = \frac{\sin(N + \frac{1}{2})\theta}{2 \sin \frac{1}{2}\theta},$$

and show that the series in the expression above may be summed into the so-called Poisson kernel,

$$\frac{1}{2} + \sum_{n=1}^{\infty} r^n \cos n(\theta - \varphi) = \frac{1 - r^2}{2[r^2 + 1 - 2r \cos(\theta - \varphi)]}. \tag{*}$$

If so, we will have arrived at the *Poisson Formula* of the Green's function representation of the solution to the Dirichlet problem,

$$u(r, \theta) = \int_{-\pi}^{\pi} \frac{1}{2\pi} \left[\frac{1 - r^2}{r^2 + 1 - 2r \cos(\theta - \varphi)} \right] f(\varphi) \, d\varphi, \qquad r < 1.$$

The summing of the series is obtained by the method of "telescoping," which was also employed in the case of the Dirichlet kernel and the Dini tests. The idea is to cross-multiply the left-hand side of the claimed identity (*) by the denominator of the right-hand side and see what drops out. The first term in this cross multiplication is just

$$[r^2 + 1 - 2r \cos(\theta - \varphi)].$$

Let us look then at the series term. We have, upon multiplying the series term of the left-hand side by r^2, 1, and $- 2r \cos(\theta - \varphi)$ in that order,

$$r^2 \cdot \sum_{n=1}^{\infty} r^n \cos n(\theta - \varphi) = r^3 \cos(\theta - \varphi) + r^4 \cos 2(\theta - \varphi)$$
$$+ r^5 \cos 3(\theta - \varphi) + \cdots,$$

$$1 \cdot \sum_{n=1}^{\infty} r^n \cos n(\theta - \varphi) = r \cos(\theta - \varphi) + r^2 \cos 2(\theta - \varphi)$$
$$+ r^3 \cos 3(\theta - \phi) + r^4 \cos 4(\theta - \phi) + \ldots,$$

$$-2r \cos(\theta - \varphi) \sum_{n=1}^{\infty} r^n \cos n(\theta - \varphi) = -2r^2 \cos(\theta - \varphi) \cos(\theta - \varphi)$$
$$- 2r^3 \cos(\theta - \varphi) \cos 2(\theta - \varphi)$$
$$- 2r^4 \cos(\theta - \varphi) \cos 3(\theta - \varphi)$$
$$- 2r^5 \cos(\theta - \varphi) \cos 4(\theta - \varphi) + \cdots,$$

all series converging absolutely and uniformly for $r < 1$ by the ratio test. Recalling that

$$2 \cos(\theta - \varphi) \cos n(\theta - \varphi) = \cos(n + 1)(\theta - \varphi) + \cos(n - 1)(\theta - \varphi),$$

we see that the term $- 2r^2 \cos(\theta - \varphi) \cos(\theta - \varphi)$ cancels the term $r^2 \cos 2(\theta - \varphi)$ leaving the term $-r^2$, the term $-2r^3 \cos(\theta - \varphi) \cos 2(\theta - \varphi)$ cancels the terms $r^3 \cos(\theta - \varphi)$ and $r^3 \cos 3(\theta - \varphi)$, the term $- 2r^4 \cos(\theta - \varphi) \cos 3(\theta - \varphi)$ cancels the terms $r^4 \cos 2(\theta - \varphi)$ and $r^4 \cos 4(\theta - \varphi)$, and so on, so that upon adding the three series above we have

$$2[r^2 + 1 - 2r \cos(\theta - \varphi)] \sum_{n=1}^{\infty} r^n \cos n(\theta - \varphi) = 2r \cos(\theta - \varphi) - 2r^2.$$

Upon adding to this the first term already cross-multiplied above, we thus have the desired identity (*) and the stated Poisson kernel formula.

Let us relate the latter formula more explicitly to a Green's function. According to Section 1.6.1 (see also Remarks 1, 2, and 3 there), from Green's third identity for the case of two dimensions we should have the solution of the Dirichlet problem

$$\begin{cases} \Delta u = 0 \text{ in } \Omega \\ u = f \text{ on } \partial\Omega \end{cases}$$

represented in terms of the Green's function $G(P, Q)$ for the problem as (Fig. 2.5c, where $P = (r, \theta)$ in general, $Q = (1, \varphi)$ on $\partial\Omega$, and $Q = (\rho, \varphi)$ in general)

$$u(P) = -\int \frac{\partial G}{\partial n_Q} (P, Q) f(Q) \, ds_Q,$$

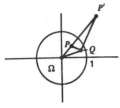

Fig. 2.5c

the other terms having vanished in Green's third identity due to the conditions $\Delta_Q u = 0$ in Ω and $G(P, Q) = 0$ for $Q \in \partial\Omega$. We also know that the Green's function will have general form

$$G(P, Q) = \frac{1}{2\pi} \ln\frac{1}{r_{PQ}} + g(P, Q),$$

where the term $g(P, Q)$ must be harmonic and selected to adjust the fundamental singularity $(1/2\pi)\ln(1/r_{PQ})$ so that the sum $G(P, Q)$ vanishes for Q on the boundary. As in Section 1.6.1, Remark 3, we see that we may take

$$G(P, Q) = \frac{1}{2\pi} \ln \frac{1}{r_{PQ}} - \frac{1}{2\pi} \ln \frac{1}{r_{OP}r_{QP'}},$$

where by the law of cosines and the definition of the image point $P' = r_{OP}^{-2}P$ one has

$$r_{PQ}^2 = r^2 + \rho^2 - 2r\rho \cos(\theta - \varphi),$$

$$r_{QP'}^2 = \frac{1}{r^2} + \rho^2 - \frac{2\rho}{r} \cos(\theta - \varphi),$$

$$r_{OP}^2 r_{QP'}^2 = 1 + r^2\rho^2 - 2r\rho \cos(\theta - \varphi),$$

and thus

$$G(P, Q) = -\frac{1}{4\pi} \ln[r^2 + \rho^2 - 2r\rho \cos(\theta - \varphi)]$$

$$+ \frac{1}{4\pi} \ln[1 + r^2\rho^2 - 2r\rho \cos(\theta - \varphi)].$$

On the boundary $\partial\Omega$ of the unit sphere we have

$$\frac{\partial G}{\partial n_Q} = \frac{\partial G}{\partial \rho}\bigg|_{\rho=1};$$

therefore,

$$\frac{\partial G(P, Q)}{\partial n_Q}\bigg|_{Q \in \partial\Omega} = -\frac{1}{4\pi} \cdot \frac{2\rho - 2r \cos(\theta - \varphi)}{r^2 + \rho^2 - 2r\rho \cos(\theta - \varphi)}\bigg|_{\rho = 1}$$

$$+ \frac{1}{4\pi} \cdot \frac{2r^2\rho - 2r \cos(\theta - \varphi)}{1 + r^2\rho^2 - 2r\rho \cos(\theta - \varphi)}\bigg|_{\rho = 1}$$

$$= \frac{-1}{2\pi}\left[\frac{1 - r^2}{r^2 + 1 - 2r \cos(\theta - \varphi)}\right],$$

thus verifying that the Poisson kernel arrived at above by separation of variables is in fact in agreement with the direct Green's function representation of the solution of the Dirichlet problem.

For a third illustration of the connections between Fourier series solutions and Green's function solutions consider the Dirichlet–Poisson problem (Fig. 2.5d)

$$\begin{cases} -\Delta u = F(x, y) \text{ in } \Omega, \\ u = 0 \text{ on } \partial\Omega. \end{cases}$$

As seen in Section 2.1, if we Fourier expand the data

$$F(x, y) = \sum_{n=1}^{\infty} \sum_{m=1}^{\infty} d_{nm} \sin nx \sin my,$$

we arrive at the solution

$$u(x, y) = \sum_{n=1}^{\infty} \sum_{m=1}^{\infty} \frac{d_{nm}}{n^2 + m^2} \sin nx \sin my.$$

However, we may also expect in general from Green's third identity that the solution may be given in terms of a Green's function representation

$$u(P) = \int_{\Omega} G(P, Q)F(Q) \, dV_Q$$

$$= \int_0^\pi \int_0^\pi G(x, y, \xi, \eta)F(\xi, \eta)d\xi \, d\eta,$$

where $P = (x, y)$ and $Q = (\xi, \eta)$. Substituting the values

$$d_{nm} = \frac{4}{\pi^2}\int_0^\pi \int_0^\pi F(\xi, \eta) \sin n\xi \sin m\eta \, d\xi \, d\eta$$

FIG. 2.5d

into the separation of variables solution $u(x, y)$, we arrive* at the Green's function representation

$$u(x, y) = \int_0^\pi \int_0^\pi \underbrace{\sum_{n=1}^\infty \sum_{m=1}^\infty \frac{(\sin n\xi \sin m\eta \sin nx \sin my)}{(n^2 + m^2)(\pi^2/4)}}_{G(P, Q) \equiv G(x, y, \xi, \eta)} F(\xi, \eta) \, d\xi \, d\eta.$$

Stated another way, if we let φ_{nm} denote the normalized orthonormal basis vectors $(2/\pi) \sin nx \sin my$, we have the eigenfunction expansion

$$G(x, y, \xi, \eta) = \sum_{n,m} \frac{1}{\lambda_{nm}} \varphi_{nm}(x, y)\varphi_{nm}(\xi, \eta)$$

representing the Green's function in terms of the eigenfunctions and eigenvectors of the problem

$$\begin{cases} -\Delta\varphi = \lambda\varphi \text{ in } \Omega, \\ \varphi = 0 \text{ on } \partial\Omega. \end{cases}$$

This is typically the case for such problems on bounded domains for which the eigenfunctions form a maximal orthonormal set.

Problem 1. (a) Derive (Fig. 2.5*e* should help) the d'Alembert formula by making the change of variable

$$\begin{pmatrix} \xi \\ \eta \end{pmatrix} = \begin{bmatrix} 1 & 1 \\ 1 & -1 \end{bmatrix} \begin{bmatrix} x \\ t \end{bmatrix}$$

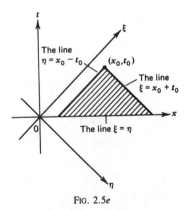

FIG. 2.5*e*

---

* Formally, by interchanging the sums and integrals. However, this may be made rigorous by reference to Parseval's equation and Theorem H of Section 2.3. See the next Pause.

to change canonical forms (see Problem 1.9.1)

$$u_{tt} - u_{xx} = -4u_{\xi\eta}.$$

Note that the Jacobian of the transformation above changing variables is

$$J = \begin{vmatrix} 1 & 1 \\ 1 & -1 \end{vmatrix} = -2,$$

which explains the factor $\frac{1}{2}$ in the d'Alembert formula.

    (b) Verify the general d'Alembert formula given for the wave pure initial value problem by differentiating and substitution into the problem.

    (c) Justify that the solution $u(x_0, t_0)$ depends only on the data present in the shaded region of Fig. 2.5e. This region is called the domain of dependence.

Problem 2. (a) Without worrying about convergence show that

$$\int_0^\pi \int_0^\pi \int_0^\pi \int_0^\pi G^2(x, y, \xi, \eta) \, d\xi \, d\eta \, dx \, dy = \sum_{n,m} \frac{1}{\lambda_{nm}^2}$$

in the case of the Dirichlet–Poisson problem considered above. (b) Verify that this quantity exists, that is, that it is indeed finite. (c) Why for this problem may we therefore write

$$u(x, y) = \int_0^\pi \int_0^\pi G(P, Q)F(Q) \, dV_Q = L^{-1} F(x, y)?$$

Problem 3. For the following wave initial value problems, find the specific value $u(x_0, t_0)$:

    (a) $u_{tt} - c^2 u_{xx} = 0$, $-\infty < x < \infty$, $t > 0$,
    $u(x, 0) = x^2$, $u_t(x, 0) = 0$,   find $u(100, 100)$;

    (b) $u_{yy} - 4u_{xx} = e^x$, $-\infty < x < \infty$, $y > 0$,
    $u(x, 0) = 2$, $u_y(x, 0) = 1$,   find $u(1/2, 3/2)$;

    (c) $u_{tt} - u_{xx} = 0$, $0 < x < 1$, $t > 0$,
    $u(x, 0) = 1$, $u_t(x, 0) = \sin \pi x$,
    $u_x(0, t) = u_x(1, t) = 0$,   find $u(1/2, 2)$.

THIRD PAUSE: EXAMPLES, EXPLANATIONS, EXERCISES

In Sections 2.3 and 2.4 the presentation deviated toward the abstract.* This was done in order to group together within two general theorems most of the special function expansions that occur in applications. As we mentioned in Section 2.1, and will illustrate again in Section 2.6, all of this theory can be thought of in terms of a single classical *metatheorem*: try to obtain a *best least squares fit.*†

Example

Parseval's Equation of Theorem H of Section 2.3 states that

$$\sum_{n=1}^{\infty} c_n^2 = \int_{\Omega} u^2 r\, dx = \|u\|_{L^2(\Omega, r)}^2$$

when $u(x)$ is given a Fourier expansion $u = \sum_1^{\infty} c_n \phi_n$ in terms of a maximal orthonormal set $\{\phi_n\}_{n=1}^{\infty}$ in a Hilbert space $L^2(\Omega, r)$. The Sturm–Liouville Theorem of Section 2.4 indicates the wide range of $L^2(a, b, r)$ spaces available in the case when Ω is an interval (a, b). If possible, the best choice for applications will be to take the $\{\phi_n\}$ to be the exact eigenfunctions of the partial differential equation under treatment, if you know them.

As indicated above, Parseval's Equation should be regarded as an instance of the Principle of Best Least Squares Fit. Show this, given that the $\{\phi_n\}_{n=1}^{\infty}$ has already been chosen.

Solution. This fact is already implicit in each of

   (i) the answer to Exercise 7 in the Second Pause,

   (ii) the proof of Lemma 2 in Section 2.3,

   (iii) the direct proof of maximality of $\{\phi_n = (2/\pi)^{1/2} \sin nx\}$ in Section 2.4.

The general proof is therefore evident. Consider an approximant

$$u_N(x) = \sum_{n=1}^{N} c_n \phi_n(x)$$

and the corresponding mean square error

---

* Abstract, elegant, general mathematics possesses an advantage of precision of statement, with a corresponding disadvantage of loss of specific information. Between the two extremes of abstraction and detail, where one should choose to work will depend on the task at hand.

†This point of view can hardly be overemphasized in mathematical science.

$$\int_\Omega (u - u_N)^2\, dx = \int_\Omega u^2\, dx - 2 \sum_{n=1}^N c_n \int_\Omega u\phi_n dx + \sum_{n=1}^N c_n^2$$

$$= \int_\Omega u^2\, dx + \sum_{n=1}^N \left(\int_\Omega u\phi_n dx - c_n \right)^2$$

$$- \sum_{n=1}^N \left(\int_\Omega u\phi_n dx \right)^2,$$

where we have completed the square of the last two terms of the error. The only adjustable term is the second, which gives the best least squares fit when the c_n are chosen to be the Fourier coefficents

$$c_n = \int_\Omega u(x)\phi_n(x)dx.$$

Exercises

1. Prove Parseval's Equation with the $\{\phi_n\}$ orthogonal but not normalized and for the case when the weight r is not 1.

2. Prove that $\displaystyle\int_0^\pi (lnx)\,(\cos nx)dx \to 0$ as $n \to \infty$.

3. Prove that $\displaystyle\int_{-\pi}^\pi \frac{\cos nx}{\sqrt{|x|}}\, dx \to 0$ as $n \to \infty$.

4. Justify that for the nonhomogeneous Dirichlet Poisson problem $-\Delta u = F$ in Ω, $u = 0$ on $\partial\Omega$ of Section 2.5, one can perform the step

$$u = \sum\sum \frac{d_{nm}^F}{n^2 + m^2} \sin nx \sin my = \iint GF.$$

5. Prove the inner product form of Parseval's Equality:

$$\langle u, v \rangle = \sum_{n=1}^\infty c_n d_n,$$

where $u = \Sigma\, c_n\phi_n$ and $v = \Sigma\, d_n\phi_n$, $\{\phi_n\}$ a maximal orthonormal sequence.

There are far too many special function Fourier eigenfunction expansions (as evidenced by the examples given in Section 2.1 and Section 2.4) coming out of mathematical physics from a separation of variables of a partial differential equation, to consider here. However, let us inspect more closely one example, namely, that requested in Problem 1(b) of Section 2.4, the Hermite functions as eigenfunctions coming out of a quantum mechanical harmonic oscillator.

Example

The behavior of a quantum mechanical particle in a three-dimensional isotropic harmonic oscillator potential $V = m\omega^2 r^2/2$ is governed by the Schrodinger equation

$$\frac{\hbar^2}{2m} \Delta\psi(x, y, z) + \frac{1}{2} m\omega^2 r^2 \psi(x, y, z) = E\psi(x, y, z)$$

where $\hbar = h/2\pi \approx 1.054 \times 10^{-34}$ joule second, h being Planck's constant, m the mass of the particle, E its energy, ω its angular frequency, r the radial space coordinate $\sqrt{x^2 + y^2 + z^2}$. Reduce this partial differential equation to a system of ordinary differential equations by separation of variables.

Solution. From $\psi(x, y, x) = X(x)Y(y)Z(z)$ we obtain

$$\frac{-h^2}{2m} \left(\frac{X''}{X} + \frac{Y''}{Y} + \frac{Z''}{Z} \right) + \frac{1}{2} m\omega^2(x^2 + y^2 + z^2) = E$$

from which we conclude that

$$\frac{-h^2}{2m} X'' + \frac{1}{2} m\omega^2 x^2 X = E_x X$$

$$\frac{-h^2}{2m} Y'' + \frac{1}{2} m\omega^2 y^2 Y = E_x Y$$

$$\frac{-h^2}{2m} Z'' + \frac{1}{2} m\omega^2 z^2 Z = E_z Z$$

where E_x, E_y, E_z denote as yet arbitrary constants which must sum to the energy E.

Exercises

6. Convert the one-dimensional harmonic oscillator equation

$$\frac{-h}{2m} u''(x) + \frac{m}{2} \omega^2 x^2 u = Eu$$

   to dimensionless normal form.
7. The postulates of quantum mechanics require that such equations possess a complete set of eigenfunctions. Show this for the harmonic oscillator equation.
8. Show that the resulting eigenfunctions are the Hermite polynomials.

Although we have emphasized that the most useful maximal orthonormal sets $\{\phi_n\}$ arise from such differential equations, one should not conclude that all do.

Example

The Haar system

$$\phi_1(t) = \chi_{[0,1]}(t)$$
$$\phi_{2^n+j}(t) = 2^{n/2}[\chi_{[0,1]}(2^{n+1}t - 2j + 2)$$
$$- \chi_{[0,1]}(2^{n+1}t - 2j + 1)]$$

where $j = 1, \ldots, 2^n$, $n = 0, 1, 2, \ldots$, forms a basis for all spaces $L^p_{[0,1]}$, $1 \leq p < \infty$. Haar introduced this system to show that maximal orthonormal sets need *not* come from a differential equation. Write out a few of these functions.

Solution. These functions just repeatedly cut the preceding signed support intervals in half so as to remain orthogonal. We have

$$\phi_1(t) = 1, \quad 0 \leq t \leq 1,$$

$$\phi_2(t) = \begin{cases} 1, & 0 \leq t < \frac{1}{2} \\ -1, & \frac{1}{2} < t \leq 1, \end{cases}$$

$$\phi_3(t) = \begin{cases} \sqrt{2} & 0 \leq t < \frac{1}{4} \\ -\sqrt{2} & \frac{1}{4} < t < \frac{1}{2} \\ 0 & \frac{1}{2} < t < \frac{3}{4} \\ 0 & \frac{3}{4} < t \leq 1 \end{cases}$$

$$\phi_4(t) = \begin{cases} 0, & 0 \leq t < \frac{1}{4} \\ 0, & \frac{1}{4} < t < \frac{1}{2} \\ \sqrt{2}, & \frac{1}{2} < t < \frac{3}{4} \\ -\sqrt{2} & \frac{3}{4} < t \leq 1 \end{cases}$$

and so on.

Exercises

9. (a) Sketch a few of these functions and prove that they are orthonormal.
   (b) How can you be sure that they do not come as eigenfunctions of some (strange) differential equation? (c) What are two of Haar's other contributions to mathematics?

2.6 FOURIER SERIES AND VARIATIONAL AND NUMERICAL METHODS

The idea of numerical approximation of a solution of a partial differential equation can be illustrated by the following simple boundary value problem:

$$\begin{cases} \Delta u = 0 \text{ in } \Omega, \\ u = f \text{ on } \partial\Omega. \end{cases}$$

Here for example $u(x, y)$ may be regarded as the temperature at the point (x, y) of a square slab of material whose boundary temperature f is known at the 12 boundary points shown in Figure 2.6a, it being desired to approximate or estimate the temperature u at the four interior lattice points. For example, let f be zero on three sides, that is, $f(0, 0) = f(0, 1) = f(0, 2) = f(0, 3) = f(1, 3) = f(2, 3) = f(3, 3) = f(3, 2) = f(3, 1) = f(3, 0) = 0$, and let $f(1, 0) = 1$ and $f(2, 0) = 2$. It is now desired to estimate the temperature $u(1, 1)$, $u(1, 2)$, $u(2, 1)$, and $u(2, 2)$.

In order to determine an efficient way in which to proceed with this temperature estimation, we accumulate evidence from three different viewpoints, all leading to the same conclusion. For numerical methods, it is always nice to have corroborating evidence.

First, from the mean value theorem for harmonic functions (see Section 1.6.1) a reasonable approximation to $u(x, y)$ at interior lattice points is the average of the values of u at the four adjacent lattice points.* Second, it is quite reasonable from the physical point of view to ask that a steady state temperature obey this isotropic self-averaging law. There is, moreover, a third point of view that puts a little more precision into these reasonability statements and illustrates the so-called *finite differences method* of approximating solutions of differential equations.† For if one replaces the partial derivatives by their approximating difference quotients as follows:

$$\frac{\partial u(x, y)}{\partial x} \cong \frac{u(x + h, y) - u(x, y)}{h},$$

$$\frac{\partial^2 u(x, y)}{\partial x^2} \cong \frac{\dfrac{u(x + h, y) - u(x, y)}{h} - \dfrac{u(x, y) - u(x - h, y)}{h}}{h},$$

that is, if one takes first a single forward difference, and then, for the second

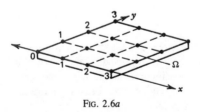

FIG. 2.6a

---

* This amounts to an approximation of the integral $\oint u$, taken over the unit circle centered at the lattice point (x, y), by the Riemann sum for the integral

$$\sum_{i=1}^{4} u(x_i, y_i) \, \Delta_i \theta \text{ with } \Delta_i \theta = \pi/2.$$

† This method clearly works in a similar way for the parabolic and hyperbolic cases.

derivative, if one takes a single backward difference, and likewise for the other independent variable, one has

$$\frac{\partial^2 u(x, y)}{\partial y^2} \cong \frac{u(x, y + k) - 2u(x, y) + u(x, y - k)}{k^2}.$$

Then upon solving (taking $h = k$ for simplicity) the equation

$$0 = \Delta u \cong u(x + h, y) + u(x - h, y) + u(x, y + k)$$
$$+ u(x, y - k) - 4u(x, y)$$

one arrives at the same approximation discussed above by taking $h = k = 1$.

In this way then the boundary value problem reduces to a problem in matrix algebra as follows:

$$u(1, 1) = \tfrac{1}{4}(u(0, 1) + u(1, 2) + u(2, 1) + u(1, 0)),$$

$$u(1, 2) = \tfrac{1}{4}(u(0, 2) + u(3, 1) + u(2, 2) + u(1, 1)),$$

$$u(2, 1) = \tfrac{1}{4}(u(1, 1) + u(2, 2) + u(3, 1) + u(2, 0)),$$

$$u(2, 2) = \tfrac{1}{4}(u(1, 2) + u(2, 3) + u(3, 2) + u(2, 1)),$$

which in matrix form becomes

$$\begin{bmatrix} 4 & -1 & -1 & 0 \\ -1 & 4 & 0 & -1 \\ -1 & 0 & 4 & -1 \\ 0 & -1 & -1 & 4 \end{bmatrix} \begin{bmatrix} u(1, 1) \\ u(1, 2) \\ u(2, 1) \\ u(2, 2) \end{bmatrix} = \begin{bmatrix} 1 \\ 0 \\ 2 \\ 0 \end{bmatrix},$$

which has (see Problem 1) the solution

$$u(1, 1) = \tfrac{11}{24}, \qquad u(1, 2) = \tfrac{4}{24}, \qquad u(2, 1) = \tfrac{16}{24}, \qquad \text{and } u(2, 2) = \tfrac{5}{24}.$$

The *finite difference method* has been a mainstay in the numerical treatment of differential equations. By taking smaller mesh size h and k and by using more refined differencing approximations the accuracy is greatly increased.* In the remainder of this section we will consider briefly a more recent development in the numerical solution of differential equations, namely, the *finite element method*, which may be regarded in some cases as a variational form of the finite difference method.† We will also consider in a similar context other variational methods, all of which depend on a best least squares approximation in a suitable energy norm.

---

* See the following Pause.

† A further introduction to numerical methods, more from the implementation point of view, is given in Appendix B of Chapter 3. For more extensive treatments let us refer to books such as Forsythe and Wasow, *Finite-Difference Methods for Partial Differential Equations* (Wiley, New York, 1960)

Let us return for a moment to the above example, and let us now suppose for simplicity that we know the boundary value f at all points on the boundary; and to be specific, let $f(x)$ be given as the piecewise linear function (Fig. 2.6*b*)

$$f(x) = \begin{cases} x, & 0 \leq x \leq 2, \\ 2(3 - x), & 2 \leq x \leq 3. \end{cases}$$

Or, if you wish, we may just assign this linear interpolation to fit the original data given at the original boundary lattice points. As mentioned above, the finite difference method then decomposes the square Ω into finer and finer grids of mesh size h and k and approximates $u(x, y)$ at the lattice grid points by the solutions to the corresponding larger and larger algebraic linear systems. The data as just interpolated will now be available during this process at the finer and finer boundary lattice points.

Now we wish to similarly interpolate to establish values for our approximation to $u(x, y)$ at all of the interior nonlattice points (x, y). By the reasoning that led to the finite difference method, we would approximate $u(x, y)$ over each of the nine squares by a step function with value equal to the average of the finite difference values found at the corners. For the above example we would then arrive at the approximate solutions as shown in Figure 2.6*c*, (where heights $= \frac{27}{96}, \frac{91}{96}, \frac{64}{96}$, and so on).

A better interpolation, and one consistent with the above interpolation of the data f, is obtained by using piecewise linear, or more correctly, piecewise planar, interpolations, as shown in Figure 2.6*d*. This better accommodates both the boundary data and the finite difference values found at the interior corners, which are

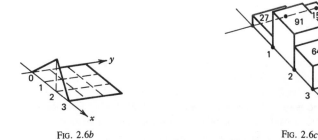

FIG. 2.6*b*　　　　　　　FIG. 2.6*c*

and R. Varga, *Matrix Iterative Analysis* (Prentice-Hall, Englewood Cliffs, New Jersey, 1962), as standard texts for further reading on the finite difference methods and other numerical methods used in treating partial differential equations. For more recent developments, see O. Zienkiewicz, *The Finite Element Method in Structural and Continuum Mechanics* (McGraw-Hill, New York, 1971) and B. Finlayson, *The Method of Weighted Residuals and Variational Principles* (Academic Press, New York, 1972). For a comprehensive treatment of the subject, see Ames, *Numerical Methods for Partial Differential Equations*, 2nd Ed. (Academic Press, New York, 1977).

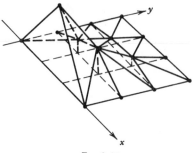

FIG. 2.6*d*

both fit exactly in this example. There is an ambiguity in the choice of diagonals used, but that does not offend us; one can try any of the possible configurations on the computer. The fact that three points determine a plane thereby inducing a triangular grid is characteristic (although far from the full story*) of the *finite element method*, of which the scheme just discussed is an example.

There now appears the question of convergence of these approximate solutions to the actual solution $u(x, y)$. One may investigate this question in a variety of convergence norms, but we will restrict attention to an energy norm related to a variational approach.

The key to the next step is to go back to the Dirichlet (variational) principle (see 1.5.3 and 1.9.5(3)), which asserted that the solution to the Dirichlet problem (Fig. 2.6*e*)

$$\begin{cases} \Delta u = 0 \text{ in } \Omega, \\ u = f \text{ on } \partial\Omega, \end{cases}$$

FIG. 2.6*e*

is obtained by the function u that minimizes the energy integral $D(v) = \int_\Omega |\text{grad } v|^2 \, d\Omega$ over all functions v which have value f on $\partial\Omega$.

Indeed, as shown in Section 1.5.3 for the case of one dimension, and as follows in exactly the same way in higher dimensions (see Problem 2), by appropriately integrating by parts one shows immediately that the solution u to the Dirichlet problem on any nice domain Ω has the minimizing property§:

---

* In more complicated or higher order problems one may wish to use rectangular or other grids, depending on the boundary, and higher order piecewise polynomial interpolations, leading to the theory of "splines."

§ The harder part is to go in the other direction, showing that the variational solution u determined by

$$D(u) \le D(v), \qquad v = f \text{ on } \partial\Omega,$$

v smooth enough for the integration by parts.

It is typical of variational methods that one minimizes successfully by taking the largest possible class $\{v\}$ that could reasonably be considered. For the case of $D(v) = \int_\Omega |\text{grad } v|^2 \, dx$, this means the class of functions v in $L^2(\Omega)$ with first partial derivatives square-integrable. This class includes, for example, the $C^1(\overline{\Omega})$ functions and further, many other functions that possess square integrable distributional first derivatives; see the discussion in Problem 1.9.7 concerning weak or distributional derivatives. The space of such functions $\{v\}$ is called the Sobolev space

$$H^1(\Omega) = \left\{ v \in L^2(\Omega) \text{ such that } \frac{\partial v}{\partial x_i} \in L^2(\Omega), \qquad i = 1, \ldots, n \right\}*$$

with inner product

$$(u, w) = \int_\Omega uw + \int_\Omega \text{grad } u \cdot \text{grad } w,$$

and norm (squared)†

$$\|u\|_1^2 = \int_\Omega u^2 + \int_\Omega |\text{grad } u|^2 \equiv \|u\|^2 + \|\text{grad } u\|^2,$$

where $\|u\|$ denotes the L^2 or root-mean-square norm $[\int_\Omega u^2]^{1/2}$. It is a property of functions v in $H^1(\Omega)$, for Ω a one-dimensional open set, that a suitable representative of v will be continuous (see Third Pause). For $n = 2$ and 3 dimensions one must go to $H^2(\Omega)$ to assure the continuity of the functions therein. Similar regularity results hold for $H^m(\Omega)$ in higher dimensions.‡

$$D(u) = \min_{v=f \text{ on } \partial\Omega} D(v) = \text{infimum}_{v=f \text{ on } \partial\Omega} D(v)$$

actually exists and moreover that u is C^2 and $\Delta u = 0$ (see Problem 1.9.5(3)). To show this one needs to be more careful and one must in particular specify exactly over which class the infimum is obtained.

* Other notations are $H^{1,2}(\Omega)$, $W^1(\Omega)$, $W^{1,2}(\Omega)$. In the same way one has

$$H^2(\Omega) = \left\{ v \in L^2(\Omega), \frac{\partial v}{\partial x_i} \in L^2(\Omega), \frac{\partial^2 v}{\partial x_i \partial x_j} \in L^2(\Omega) \right\},$$

the "second" Sobolev space, and so on for the "higher" Sobolev spaces $H^m(\Omega)$.

† For convenience we will take the liberty of using the term "norm" hereafter in cases involving a bonafide-norm, or a norm squared, or $D(u)$, or $D(u)^{1/2}$, and so on, when the entity involved is clear.

‡ In fact the Sobolev lemma asserts that v in $H^{[n/2]+1+k}(\Omega)$ will be $C^k(\Omega)$, where n is the dimension and where $[n/2]$ means "rounded down to an integer." Thus as dimension increases one must take higher order Sobolev spaces to guarantee the continuity of functions therein.

The Dirichlet principle now states (for reasonable data f, see below) that

$$D(u) = \min_{\substack{v=f \text{ on } \partial\Omega \\ v\in H^1(\Omega)}} D(v) = \inf_{\substack{v=f \text{ on } \partial\Omega \\ v\in H^1(\Omega)}} D(v),$$

Looked at in another way, upon subtracting off the data f (let us assume for the moment that f as given is a function (e.g., a polynomial) already defined throughout Ω), the Dirichlet   principle asserts that

$$D(u) = \inf_{v-f\in H_0^1(\Omega)} D(v).$$

Here

$$H_0^1(\Omega) = \{h \in L^2(\Omega) \text{ such that grad } h \in L^2(\Omega) \text{ and } h = 0 \text{ on } \partial\Omega\},$$

with the same norm and inner product as $H^1(\Omega)$. The principal part of this norm (squared) is the energy term

$$D(v) = \int_\Omega |\text{grad } v|^2$$

and we will sometimes purposely neglect the other part,* both for simplicity and in order to emphasize the principal role of the Dirichlet energy integral in these considerations.

The principal links between the Dirichlet principle, the methods of orthogonal projection (which include least squares, Galerkin, Rayleigh–Ritz, and others), and finite element approximation, may now be seen via Green's identity as follows.

We have for any two C^2 functions h and w (and by taking limits, for any two $H^1(\Omega)$ functions h and w) by Green's first identity that

$$D(h, w) = \oint_{\partial\Omega} h \frac{\partial w}{\partial n} - \int_\Omega h \, \Delta w.$$

If $h = 0$ on $\partial\Omega$ and w is harmonic so that $\Delta w = 0$ in Ω, then

$$D(h, w) = 0.$$

Since $D(h, w)$ is an inner product,† this means that h and w are orthogonal to each other.

---

* The two norms (squared, and hence the norms themselves, as in Section 1.6.2, Problem 3) $D(v)$ and $\|v\|_1^2 = \|v\|^2 + D(v)$ are equivalent for v in $H_0^1(\Omega)$. That is, there are constants c_1, c_2 such that $c_1 D(v) \leq \|v\|_1^2 \leq c_2 D(v)$. One may take $c_1 = 1$, and c_2 follows from the Poincaré inequality, namely, that $\|v\|^2 \leq c_3 D(v)$, where the constant c_3 turns out to be (see Section 1.6.2, Problem 2) λ_1^{-1}. We recall that the first eigenvalue λ_1 for the problem is strictly positive. For $H^1(\Omega)$ (no boundary condition) one has $D(v) = 0$ for v nonzero and constant, but this causes no essential difficulty and moreover seldom occurs for the functions encountered in practice. In the theory it may also always be taken care of (e.g., see the next footnote) without difficulty.

† Again for simplicity (see the previous footnote) we just do everything modulo the constant functions. If $v = c$ a constant one has $D(v) = 0$, which (as a technicality) interferes with the idea that for a norm $D(v) = 0$ should imply $v = 0$. To get around this point many authors consider instead of

$$h \perp w$$

in the sense of this inner product. One may write this fact as

$$H^1(\Omega) = H_0^1(\Omega) \overset{\perp}{\oplus} U(\Omega)$$

where $H_0^1(\Omega)$ is the subspace of $H^1(\Omega)$ functions vanishing on $\partial\Omega$ and where U is the subspace of harmonic functions.*

Now we go back to the Dirichlet problem and *we assume that the boundary data f can be extended (see Problem 2) into the domain Ω as an $H^1(\Omega)$ function* f_e. By the above orthogonal decomposition the extended function f_e has unique decomposition into components†

$$f_e(x) = h(x) \overset{\perp}{\oplus} w(x),$$

where $h(x) = 0$ on $\partial\Omega$ and where $\Delta w(x) = 0$ in Ω. On $\partial\Omega$ we have $w(x) = f_e(x) = f(x)$ and thus w is the solution u to the Dirichlet problem.

In geometric language one may say that the solution u is obtained as the projection of f_e in H^1 onto the subspace U, that is,

$$u = P_U^{H^1} f_e.$$

For a useful picture of the Dirichlet principle, see Figure 2.6f, which shows that "the shortest way up, is straight up," that is, that indeed

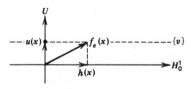

FIG. 2.6f

---

the operator Δv the operator $\Delta v + v$, which "agrees" exactly with the $H^1(\Omega)$ norm; but this remains nonetheless an artifice. In terms of the decomposition given on this page, constant functions find themselves in U and functions w in U may differ by constants without affecting the norm $D(w)$.

* Actually the fact that functions h vanishing on the boundary and harmonic functions w are orthogonal in the energy inner product is not enough to write $H^1 = H_0^1 \oplus^\perp U$, for it still must be shown that their orthogonal sum fills out the whole space. Thus one *defines* U as the orthogonal complement in H^1 of H_0^1 so that all of H^1 is guaranteed in the sum. Then one shows as follows that functions w in U are indeed harmonic. Let h be an arbitrary $C_0^\infty(\Omega)$ "test function" and let w be in U defined as the orthogonal complement of H_0^1. Thus $D(h, w) = 0$ and hence by Green's first identity $\int_\Omega w \, \Delta h = 0$ for all such h. Hence $\Delta w = 0$ in the weak or distributional sense (see Problem 1.9.7). This is the sense in which an $H^1(\Omega)$ function is harmonic.

† We may assume the boundary data f is not constant, the Dirichlet problem otherwise being trivial. Then w is not constant and the arbitrary constant referred to in the previous footnote may be taken equal to zero.

$$D(u) \leqq D(v)$$

for v in $\{v\} = f_e + \{H_0^1\}$.*

Let us summarize our considerations up to this point. We started the section with the question of how to numerically approximate the solution to a Dirichlet problem. The method of finite differences was first considered, and this then led in turn rather naturally to a finite element approximation method. This was followed by the question of convergence, or put another way, the question of the amount of error in any approximation. To measure the error one needed a norm, and going back to the Dirichlet principle provided us with such a norm. Namely, if u is the solution to the Dirichlet problem and v_N is the Nth approximant to the solution, the error in "energy" is given by

$$D(u - v_N) = \int_\Omega |\text{grad}(u - v_N)|^2 \, dx.$$

One could for example take the v_N to be the successively refined finite difference or finite element approximations to the solution. Or one could take the v_N (more variationally) to be a minimizing sequence according to the Dirichlet principle.† Each procedure for choosing v_N leads to a convergence question, for which the energy norm may be used as one measure of error.

More generally the Dirichlet problem (Fig. 2.6g)

$$\begin{cases} \Delta u = 0 \text{ in } \Omega, \\ u = f \text{ on } \partial\Omega \end{cases}$$

---

* The critically intelligent student might at this point object that the extension f_e of f into the domain Ω is not at all unique, so that the picture is not complete in this respect. But any other extension f'_e will be represented by a vector ending somewhere on the "dotted line" of $\{v\}$ as represented in Figure 2.6f, since any such extension will have the same harmonic component in U. Let $f_e = h \oplus w$ and $f'_e = h' \oplus w'$. Then

$$\begin{aligned} D(w - w') &= \oint_{\partial\Omega} (w - w') \frac{\partial}{\partial n} (w - w') \, ds \\ &= \oint_{\partial\Omega} (f_e - f'_e) \frac{\partial}{\partial n} (w - w') \, dx \\ &= \oint_{\partial\Omega} (f - f) \frac{\partial}{\partial n} (w - w') \, ds = 0, \end{aligned}$$

because h and h' vanish on $\partial\Omega$. Thus $w = w' = u$ in the above picture.

† For the reasons evident from Figure 2.6f, the *Dirichlet principle* is sometimes called the *method of orthogonal projection*. But the latter is more general and need not be variational, and is itself a special case of a still more general Hilbert space approximation method called the *method of best least squares fit*. This method applied in diverse contexts such as statistics and elsewhere, and is itself just one (although fundamental) aspect of *Fourier series*. When applied to partial differential equations the latter method is often called the *Galerkin* method, especially when the projections and operators involved are not necessarily self-adjoint. The self-adjoint case when approximating eigenvalues is called the *Rayleigh–Ritz* method.

FIG. 2.6*g*

for an arbitrary bounded domain in any number of dimensions can be similarly approached numerically as we have discussed here (in order of progress) via finite differences, finite elements, and then convergence considerations. Finite differences provide approximate solutions at a grid of interior points. Finite elements provide interpolation by approximate solutions satisfying

$$\Delta u_{\text{approx.}} = 0$$

everywhere except on the grid lines. The Dirichlet principle provides a convergence proof by taking approximating solutions in a variational prescription as shown in Figure 2.6*h*.

To close this section we wish to relate the Dirichlet principle and variational methods to the general method of best least square's approximation, of which they are special cases.

The general idea for any *method of least squares approximation* to an operator equation*

$$Lu = F$$

is to choose a maximal orthonormal set $\{\varphi_n\}$ in the sense of Theorem H of Section 2.3 in an appropriate† Hilbert space, and approximate the solution u by a partial sum

$$u_N = \sum_{n=1}^{N} c_n \varphi_n$$

with the coefficients c_N determined by the requirement

$$\langle Lu_N - F, \varphi_m \rangle = 0, \qquad m = 1, 2, 3, \ldots, N.$$

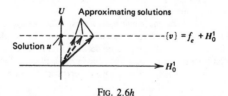

FIG. 2.6*h*

---

* Here it is perhaps useful to think first in terms of the Poisson problem $-\Delta u = F$ in Ω, $u = 0$ on $\partial\Omega$, although we will later come back to the Dirichlet problem again. See also Problem 2.9.5 for some of the concepts used here.

† Finding the "appropriate" Hilbert space is often the trick.

If P_N is the projection onto the span of the $\varphi_1, \ldots, \varphi_N$, this amounts to requiring that

$$P_N(Lu_N - F) = 0,$$

that is, that the operator equation* be solved approximately when "projected" onto the approximating subspace. Since the $\{\varphi_n\}$ were maximal one has $P_N \to I$ as $N \to \infty$ and, depending on how well the P_N and the operator L "commute," one can reasonably expect the u_N to often converge to the solution. We give here a little argument† to show why this is so, assuming for simplicity that the operator L is continuous,‡ that is, assuming that there exists a number M such that

$$\|Lw\| \leq M\|w\| \text{ for all } w.$$

The infimum of all such M is called the norm $\|L\|$ of the operator L. We also suppose that the projections P_N are symmetric and that L is strongly positive (also called coercive): There exists a number $m > 0$ such that

$$(Lw, w) \geq m\|w\|^2 \text{ for all } w.$$

The convergence argument then goes as follows.

We first note that if the $\{\varphi_n\}$ are maximal, one has as noted above the convergence

$$\|u - P_N u\| \to 0$$

for any u and in particular for the solution u. The projected equation above may be written as

$$P_N L P_N u_N = P_N F$$

from which the least squares approximant u_N is obtained as

$$u_N = (P_N L P_N)^{-1} P_N F,$$

the inverse therein being guaranteed§ by the assumed strong positiveness of L, the orthogonality (i.e., the symmetry) of P_N and Schwarz's inequality:

---

* When the operator equation is a partial differential equation, this is often called Galerkin's method, after the Russian mathematician Galerkin.

† This argument is indeed a basic one in numerical analysis. The fact that $\{\varphi_n\}$ is maximal is called *consistency*; the fact that the projected operators are uniformly invertible is called *stability*; and then the method converges. Partial differential operators are not continuous in a Hilbert space (although they can be made continuous in other spaces by going to the theory of distributions, which is done, however, at a certain cost in geometry). Integral operators and matrix operators are often continuous in a Hilbert space norm.

‡ For a discussion of continuous, that is, bounded, operators, see Problem 2.9.5. Note that in the present discussion we are in fact assuming that both L and L^{-1} (the latter by the strong positivity of L; see part (b) of 2.9.5) are continuous operators.

§ This is a rather standard technique in handling operators. Strongly positive operators (such as, for example, most elliptic differential operators with appropriate boundary conditions) always have bounded inverses by this argument. Let us also note that $\|P_N\| = 1$ if P_N is an orthogonal projection,

$$\|w\| \, \|P_N L P_N w\| \geqq \langle P_N L P_N w, \, w \rangle$$

$$= \langle L P_N w, \, P_N w \rangle$$

$$\geqq m \| P_N w \|^2.$$

The last inequality states in particular that for w which lie within the range of P_N, that is, w in the span of the ϕ_n, $n = 1, \ldots, N$, then if $P_N L P_N w = v$, one has $w = (P_N L P_N)^{-1} v$, and moreover

$$\|w\| \leqq \frac{1}{m} \|(P_N L P_N) w\|.$$

Thus the projected operator $P_N L P_N$ is invertible on the range of P_N as claimed, with inverse bounded (uniformly in N) by

$$\|(P_N L P_N)^{-1}\| \leqq \frac{1}{m}.$$

Since we may write $P_N u = (P_N L P_N)^{-1} (P_N L P_N) u$ and $u_N = (P_N L P_N)^{-1} P_N F$, we therefore have

$$\|u - u_N\| \leqq \frac{1}{m} \|P_N L P_N u - P_N F\| + \|(I - P_N) u\|$$

$$\leqq \frac{1}{m} \|L P_N u - L u\| + \|(I - P_N) u\|$$

$$\leqq \left(\frac{\|L\|}{m} + 1 \right) \|P_N u - u\| \to 0,$$

the desired convergence of the method.

Let us illustrate the above general method by looking first at the Poisson problem with Dirichlet boundary conditions

$$\begin{cases} -\Delta u = F \text{ in } \Omega, \\ u = 0 \text{ on } \partial\Omega. \end{cases}$$

We assume that Ω is bounded and nice, for which there are a countable number of eigenfunctions $\{\varphi_n\}$ for the given operator L, which in this case means $-\Delta\varphi_n = \lambda_n \varphi_n$ in Ω, $\varphi_n = 0$ on $\partial\Omega$. Let P_N be the projection of $H = L^2(\Omega)$ onto the span $sp(\varphi_1, \ldots, \varphi_N)$ of the first N eigenfunctions. The set $\{\varphi_n\}$ is a maximal orthonormal set in the sense of Theorem H of Section 2.3, as may be shown from

---

and also that we have not worried at all about the domain $D(L)$ when taking compositions of operators in these arguments. As noted in the previous footnote, L is usually not bounded if it is a differential operator. Nonetheless the above argument, suitably fixed up, is a common one in numerical convergence considerations for differential operators. $L = \Delta$ for example becomes a bounded operator if one uses the H^2 norm $\|u\|_2^2 = \|u\|^2 + \|\text{grad } u\|^2 + \|\Delta u\|^2$, but then one loses other properties desired in the simplified argument given above, such as positivity and ease of projection.

more advanced considerations in higher dimensions similar to the Sturm–Liouville theory given in Section 2.4. The operator L is also strongly positive, as seen from

$$\langle Lu, u \rangle = -\int_\Omega \Delta u(x)u(x)\, dx \geqq \lambda_1 \int_\Omega u^2(x)\, dx$$

and the positiveness of the first eigenvalue λ_1 for the corresponding eigenvalue problem. Letting $u_N = \Sigma_{n=1}^N c_n \varphi_n$, the equation

$$P_N L P_N u_N = P_N F$$

becomes

$$\sum_{n=1}^N c_n \lambda_n \varphi_n = \sum_{n=1}^N d_n \varphi_n,$$

where the data F has been Fourier expanded $F = \Sigma_{n=1}^\infty d_n \varphi_n$ in terms of the eigenfunctions φ_n. Thus we may take $c_n = d_n/\lambda_n$. By the strong positivity of L we have a continuous inverse L^{-1} and thus can show convergence directly:

$$u = L^{-1}F = L^{-1} \sum_{n=1}^\infty d_n \varphi_n = L^{-1} \lim_{N\to\infty} \sum_{n=1}^N d_n \varphi_n$$

$$= \lim_{N\to\infty} L^{-1} \sum_{n=1}^N d_n \varphi_n$$

$$= \lim_{N\to\infty} \sum_{n=1}^N \frac{d_n}{\lambda_n} \varphi_n = \lim_{N\to\infty} u_N.$$

This may be done for all data F in the range of L, and the latter is in fact (see Problem 2.9.5(d)) all of $L^2(\Omega)$.

From this example we would like to make three remarks:

(i) In being able to take $c_n = d_n/\lambda_n$, we should be led to realize that in this case the Galerkin method is exactly the Fourier method.

(ii) It should also be remarked that for many domains Ω one will not be able to find the φ_n explicitly. In that case normally one will be forced to use some approximations to the φ_n or some other base ψ_n for the approximation procedure.

(iii) Finally, we note that the convergence proof for the above Poisson problem worked because $L^{-1}P_N = P_N L^{-1}$. In the previously mentioned general situation wherein some inexact or convenient φ_n other than the true eigenfunctions φ_n must be used, one will not have this exact commuting of the operators L and L^{-1} with the projections P_N. Note also that the commuting relation $LP_N = P_N L$ gives convergence, as in the above Poisson problem, whether L is continuous or not.

As a second illustration of the general least squares approximation method, let us mention any eigenvalue problem

$$Lu = \lambda u.$$

For this problem and any chosen $\{\varphi_n\}$ the Galerkin equation becomes, with $u_N = \sum_{n=1}^{N} c_n \varphi_n$, the $N \times N$ system

$$\langle (L - \lambda)u_N, \varphi_m \rangle = 0, \qquad m = 1, 2, 3, \ldots, N,$$

which yields the coefficients c_n. When L is self-adjoint, this method usually goes by the name Rayleigh-Ritz method. It has been extremely valuable and very much used in applications.*

For a third illustration† of the general method of least squares approximation let us return to the Dirichlet problem

$$\begin{cases} \Delta u = 0 \text{ in } \Omega, \\ u = f \text{ on } \partial\Omega. \end{cases}$$

The data f are extended to a function f_e defined on $\overline{\Omega}$ and in $H^1(\Omega)$. Let $\{\varphi_n\}$ be a maximal orthonormal set in $H_0^1(\Omega)$: This can always be arranged by smoothing step functions into C_0^∞ functions and then orthonormalizing as in Problems 2.9.3 and 2.9.4, but using of course the norm $D(w)$. Let P_N be the projection of $H^1(\Omega)$ onto the span of $\varphi_1, \ldots, \varphi_N$. What is the operator L here? It is comprised of two parts, the Laplacian Δ applied to functions over Ω and the identity I on $\partial\Omega$, with data F being thought of as also comprised of two parts, namely, 0 in Ω and f on $\partial\Omega$. If we take (admittedly apparently ad hoc at this point)

$$u_N = f_e - \sum_{n=1}^{N} c_n \varphi_n$$

and require that the c_n be determined by

$$\langle \Delta u_N, \varphi_m \rangle = 0, \qquad m = 1, 2, \ldots, N,$$

we are in spirit at least following the above general method, in that the u_N "approximates the identity on the boundary" and the orthogonality condition just given follows the general Galerkin coefficient-determining procedure as applied to this case of zero data F on Ω.

That is, in operator language, we have required that $P_N \to I$ in H_0^1 and that $P_N \Delta u_N = 0$, which is the Galerkin prescription for the case of zero domain data.

From this it is easily seen that $c_m = D(f_e, \phi_m)/D(\phi_m, \phi_m)$. Moreover, we have

$$0 \leq D(u_N - u) = D(u_N) - D(u),$$

the picture (Fig. 2.6i), and (upon further verification) the fact that the $\{u_N\}$ form what is called a *Rayleigh–Ritz sequence*, which indeed converges to the solution u from *above*:

---

* See for example H. Weinberger, *Variational Methods for Eigenvalue Approximation* (SIAM, Philadelphia, 1974).

† The reader should fill in some of the details here, see Problem 3(c) at the end of this section.

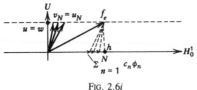

FIG. 2.6*i*

$$D(u) \leqq \cdots D(u_{N+1}) \leqq D(u_N).$$

In general, any such approximating sequences from above are called Ritz sequences.*

In general, and in order to complete the verification of the above, we note that one may thus obtain a Ritz sequence as follows. Since $H_0^1(\Omega)$ is separable, there are many maximal orthonormal sets $\{\varphi_n\}$ in $H_0^1(\Omega)$ in the $D(\varphi)$ norm. Let P_N be the projection onto the span $sp\{\varphi_1, \ldots, \varphi_N\}$ and note that $P_N \to$ identity in $H_0^1(\Omega)$, so that $P_N \to I - P$ in $H^1(\Omega)$, where $P \equiv P_u^{H^1}$ is the previously noted Dirichlet projection of $H^1(\Omega)$ onto U. Letting

$$u_N = (I - P_N)f_e,$$

one has therefore that

$$u_N \to Pf_e \equiv u$$

in $H^1(\Omega)$, so that

$$D(u_N - u) \to 0.$$

By the Pythagorean theorem one has then the desired convergence from above,

$$D(u_N) = D((I - P_N)f_e) \geqq D((I - P_{N+1})f_e) \geqq D(Pf_e) = D(u).$$

One may obtain an approximating sequence from below in the same way. Such lower-bounding sequences are called, among others, Trefftz sequences.† Let $\{\varphi_n\}$ be a maximal orthonormal set in U in the $D(\varphi)$ norm, and let P_N be the projection onto the span $sp\{\varphi_1, \ldots, \varphi_N\}$ so that $P_N \to$ identity in U and thus $P_N \to P$ in $H^1(\Omega)$. Letting

$$u_N = P_N f_e,$$

one has therefore that

$$u_N \to Pf_e \equiv u$$

so that

* Or Rayleigh–Ritz sequences, among others.

† Lower bounding sequences, that is, sequences approximating from below are sometimes also called Thompson sequences (for the Thompson principle). A general method for getting lower bounds is the Weinstein–Aronszajn method (see A. Weinstein and W. Stenger, *Methods of Intermediate Problems for Eigenvalues* (Academic, New York, 1972)).

$$D(u_N - u) = \oint_{\partial\Omega} (P_N f_e - u) \frac{\partial}{\partial n} (P_N f_e - u) \to 0.$$

By the Pythagorean theorem

$$D(u_N) = D(P_N f_e) \leq D(P_{N+1} f_e) \leq D(P f_e) = D(u),$$

anu thus the convergence from below.

Other approximating sequences, such as those of H. A. Schwarz and H. Poincaré, may converge in a manner alternating from above and below $D(u)$.

Problem 1. (a) Verify and solve the 4×4 linear system yielding the finite difference approximation in the temperature estimation problem at the beginning of this section. (b) Approximating the data by the piecewise linear function $f(x)$, solve the problem by separation of variables. (c) Change (several times) the value of $f(1, 0)$ and $f(2, 0)$ and repeat part (a) of this problem, noting the ease with which the finite difference values may be cranked out.

Problem 2. (a) Prove the converse to the Dirichlet principle for an arbitrary Dirichlet problem on any reasonable (i.e., Dirichlet) domain Ω, as was shown in Section 1.5.3 for the case of one dimension, namely, that the solution of

$$\begin{cases} \Delta u = 0 \text{ in } \Omega, \\ u = f \text{ on } \partial\Omega \end{cases}$$

FIG. 2.6j

has the property that

$$D(u) \leq D(v)$$

for "any" v such that $v = f$ on $\partial\Omega$. (b) Concern yourself with the "any". (c) Prove for $\partial\Omega \in C^m$ and $f \in C^m(\partial\Omega)$ that there exists an extension f_e of f such that $f_e \in C^m(R^n)$. (d) Investigate other sufficient conditions for the extension of f on $\partial\Omega$ to $f_e \in H^1(\Omega)$.

Problem 3. Show Thompson's Principle: among all harmonic functions,

$$R(v) = 2 \oint_{\partial\Omega} f \, \partial v / \partial n - \int_\Omega |\text{grad } v|^2$$

is maximized by the solution of the Dirichlet problem.

FOURTH PAUSE: EXAMPLES, EXPLANATIONS, EXERCISES

The variational setting taken in Section 2.6 allowed us to cast all at once the three basic numerical approaches of finite difference, finite element, and least squares (i.e., spectral) approximations within a single theoretical frame. The finite difference approximations are the classical "tried and true" method. To get more accuracy, in practice one usually just takes a finer grid.

Example

Double the mesh on the sample problem of Section 2.6 and inspect the accuracy improvement.

Solution. We refine the previous in which $h = k = 1$ to the finer mesh $h = k = 0.5$. This 6×6 grid creates 25 interior mesh points, which we have numbered as shown in Fig. 4Pa. Using the same (centered difference) approximation, we arrive at the 25×25 coefficient matrix below, with the data shown to the right. This is a 6-band matrix due to the ordering used above. The matrix equation is solved by the usual Gauss–Jordan elimination which yields the solution vector given below..

Below are the mentioned 25×25 matrix problem, right-hand-side data, solution, and, with no claims of coding expertise, the codes* that produced these results.

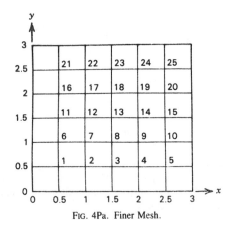

FIG. 4Pa. Finer Mesh.

---

* The few codes given in this book are now somewhat archaic but are retained for illustration, their original purpose in the book.

Coefficient Matrix																									Right-hand Vector
4.	-1.	0	0	0	-1.	0	0	0	0	0	0	0	0	0	0	0	0	0	0	0	0	0	0	0	.500
-1.	4.	-1.	0	0	0	-1.	0	0	0	0	0	0	0	0	0	0	0	0	0	0	0	0	0	0	1.000
0	-1.	4.	-1.	0	0	0	-1.	0	0	0	0	0	0	0	0	0	0	0	0	0	0	0	0	0	1.500
0	0	-1.	4.	-1.	0	0	0	-1.	0	0	0	0	0	0	0	0	0	0	0	0	0	0	0	0	2.000
0	0	0	-1.	4.	0	0	0	0	-1.	0	0	0	0	0	0	0	0	0	0	0	0	0	0	0	1.000
-1.	0	0	0	0	4.	-1.	0	0	0	-1.	0	0	0	0	0	0	0	0	0	0	0	0	0	0	0
0	-1.	0	0	0	-1.	4.	-1.	0	0	0	-1.	0	0	0	0	0	0	0	0	0	0	0	0	0	0
0	0	-1.	0	0	0	-1.	4.	-1.	0	0	0	-1.	0	0	0	0	0	0	0	0	0	0	0	0	0
0	0	0	-1.	0	0	0	-1.	4.	-1.	0	0	0	-1.	0	0	0	0	0	0	0	0	0	0	0	0
0	0	0	0	-1.	0	0	0	-1.	4.	0	0	0	0	-1.	0	0	0	0	0	0	0	0	0	0	0
0	0	0	0	0	-1.	0	0	0	0	4.	-1.	0	0	0	-1.	0	0	0	0	0	0	0	0	0	0
0	0	0	0	0	0	-1.	0	0	0	-1.	4.	-1.	0	0	0	-1.	0	0	0	0	0	0	0	0	0
0	0	0	0	0	0	0	-1.	0	0	0	-1.	4.	-1.	0	0	0	-1.	0	0	0	0	0	0	0	0
0	0	0	0	0	0	0	0	-1.	0	0	0	-1.	4.	-1.	0	0	0	-1.	0	0	0	0	0	0	0
0	0	0	0	0	0	0	0	0	-1.	0	0	0	-1.	4.	0	0	0	0	-1.	0	0	0	0	0	0
0	0	0	0	0	0	0	0	0	0	-1.	0	0	0	0	4.	-1.	0	0	0	-1.	0	0	0	0	0
0	0	0	0	0	0	0	0	0	0	0	-1.	0	0	0	-1.	4.	-1.	0	0	0	-1.	0	0	0	0
0	0	0	0	0	0	0	0	0	0	0	0	-1.	0	0	0	-1.	4.	-1.	0	0	0	-1.	0	0	0
0	0	0	0	0	0	0	0	0	0	0	0	0	-1.	0	0	0	-1.	4.	-1.	0	0	0	-1.	0	0
0	0	0	0	0	0	0	0	0	0	0	0	0	0	-1.	0	0	0	-1.	4.	0	0	0	0	-1.	0
0	0	0	0	0	0	0	0	0	0	0	0	0	0	0	-1.	0	0	0	0	4.	-1.	0	0	0	0
0	0	0	0	0	0	0	0	0	0	0	0	0	0	0	0	-1.	0	0	0	-1.	4.	-1.	0	0	0
0	0	0	0	0	0	0	0	0	0	0	0	0	0	0	0	0	-1.	0	0	0	-1.	4.	-1.	0	0
0	0	0	0	0	0	0	0	0	0	0	0	0	0	0	0	0	0	-1.	0	0	0	-1.	4.	-1.	0
0	0	0	0	0	0	0	0	0	0	0	0	0	0	0	0	0	0	0	-1.	0	0	0	-1.	4.	0

Solution Vector

U(1) = .354779
U(2) = .684266
U(3) = .943837
U(4) = 1.022902
U(5) = .589059
U(6) = .234848
U(7) = .438447
U(8) = .568182
U(9) = .558712
U(10) = .333333
U(11) = .146168
U(12) = .266492
U(13) = .331731
U(14) = .310431
U(15) = .185562
U(16) = .083333
U(17) = .149621
U(18) = .181818
U(19) = .165720
U(20) = .098485
U(21) = .037544
U(22) = .066841
U(23) = .080201
U(24) = .072145
U(25) = .042657

```
UNIVERSITY OF MINNESOTA FORTRAN COMPILER (VERSION 5.4 - 79/03/01) ON THE 6400 UNDER
KRONOS 2.1.0 ON 81/09/24 AT 19.52|UNIVERSITY COMPUTING CENTER - UNIVERSITY OF
COLORADO |
                    MNF(I=GAUSS,PL,BL)

              C   ******************************************************
              C   * THIS PROGRAM CALCULATES THE SOLUTION OF THE DIRICHLET *
              C   * PROBLEM IN A (3X3) SPACE DOMAIN BY MEANS OF A FINITE  *
              C   * DIFFERENCE PROCEDURE. THE NUMBER OF INTERIOR MESH     *
              C   * POINTS IS 25.                                         *
              C   ******************************************************
  1.   000000B      DIMENSION C(25,25),D(25),U(25)
              C
              C       ...INPUT NUMBER OF INTERIOR MESH POINTS
  2.   002131B      READ (5,108) N
              C
              C       ...SET COEFFICIENT MATRIX (C)
  3.   003400B      DO 1 I=1,25
  4.   003401B      DO 1 J=1,25
  5.   003403B        C(I,J)=0.0
  6.   003403B    1 CONTINUE
  7.   003413B      DO 2 I=1,24
  8.   003415B      J=I+1
  9.   003415B      L=I+5
 10.   003417B      C(I,I)=4.0
 11.   003421B      IF (I.LE.20) C(I,L)=-1.0
 12.   003431B      IF (I.LE.20) C(L,I)=-1.0
 13.   003436B      R=FLOAT(I)/5.
 14.   003437B      IF (R.EQ.IFIX(R)) GO TO 2
 15.   003442B      C(I,J)=-1.0
 16.   003446B      C(J,I)=-1.0
 17.   003450B    2 CONTINUE
 18.   003453B      C(25,25)=4.0
              C
              C       ...SET RIGHT-HAND VECTOR (D)
 19.   003454B      DO 4 I=1,25
 20.   003456B      IF (I.LE.4) D(I)=0.5*FLOAT(I)
 21.   003463B      IF (I.EQ.5) D(I)=1.0
 22.   003466B      IF (I.GT.5) D(I)=0.0
 23.   003471B    4 CONTINUE
              C
              C       ...PRINT COEFFICIENT MATRIX AND RIGHT-HAND VECTOR
 24.   003473B      WRITE (6,101)
 25.   003477B      DO 3 I=1,25
 26.   003500B      WRITE (6,100) (C(I,J),J=1,25),D(I)
 27.   003516B    3 CONTINUE
              C
              C       ...CALCULATE SOLUTION VECTOR (U)
 28.   003520B      CALL GAUSS(C,D,U,N)
              C
              C       ...PRINT SOLUTION VECTOR
 29.   003522B      WRITE (6,102)
 30.   003525B      DO 5 I=1,25
 31.   003526B      WRITE (6,103) I,U(I)
 32.   003535B    5 CONTINUE
 33.   003537B      STOP
              C
              C
              C       ***************** FORMAT STATEMENTS *****************
 34.   003540B  100 FORMAT (/25(1X,F3.0),15X,F6.3)
 35.   003540B  101 FORMAT (////37X,18HCOEFFICIENT MATRIX,58X,
                 1            17HRIGHT-HAND VECTOR//)
 36.   003540B  102 FORMAT (1H1////60X,15HSOLUTION VECTOR//)
 37.   003540B  103 FORMAT (/59X,2HU(,I2,4H) = ,F10.6)
 38.   003540B  108 FORMAT (I5)
 39.   003540B      END
```

The Gauss Elimination linear solver called is the following code. For very large systems (i.e., very fine meshes) one would prefer other linear solvers.

```
1.    000000B          SUBROUTINE GAUSS(A,B,X,N)
                 C     *************************************************************
                 C     * SOLVE UP TO 25 SIMULTANEOUS LINEAR ALGEBRAIC EQUATIONS *
                 C     * BY GAUSS-JORDAN ELIMINATION METHOD                     *
                 C     *************************************************************
                 C
2.    000000B          DIMENSION A(25,25),B(25),X(25)
                 C
                 C     ...PERFORM MANIPULATIONS
3.    000000B          DO 4 I=1,N
4.    000002B            DO 2 K=1,N
5.    000004B              IF (K.EQ.I) GO TO 2
6.    000006B              CONST=-A(K,I)/A(I,I)
7.    000014B              DO 1 J=1,N
8.    000017B                A(K,J)=A(K,J)+CONST*A(I,J)
9.    000026B                IF (J.EQ.I) A(K,J)=0.0
10.   000033B          1     CONTINUE
11.   000035B              B(K)=B(K)+CONST*B(I)
12.   000041B          2   CONTINUE
13.   000044B            CONST=A(I,I)
14.   000047B            DO 3 J=1,N
15.   000051B              A(I,J)=A(I,J)/CONST
16.   000051B          3   CONTINUE
17.   000057B            A(I,I)=1.0
18.   000062B            B(I)=B(I)/CONST
19.   000064B          4 CONTINUE
                 C
                 C     ...OBTAIN SOLUTION VECTOR
20.   000067B          DO 5 I=1,N
21.   000070B            X(I)=B(I)
22.   000070B          5 CONTINUE
23.   000074B          RETURN
24.   000076B          END
```

Let the separation of variables solution (see Problem 1 of Section 2.6)

$$u(x, y) = \sum_{n=1}^{\infty} \frac{18 sin(2n\pi/3)}{(n^2 \pi^2 \, sinh \, n\pi)} \, sin \, \frac{n\pi x}{3} \, sinh \left(\frac{(3 - y)n\pi}{3}\right)$$

summed to 50 terms be called the "analytic solution." Then we have the following comparison

	Analytical Solution	FD Solution $h = k = 1$	Error (%)	FD Solution $h = k = 0.5$	Error (%)
$u(1, 1)$	0.43051	0.45833	6.5	0.43845	1.8
$u(1, 2)$	0.14285	0.16667	16.7	0.14962	4.7
$u(2, 1)$	0.51732	0.66667	28.9	0.55871	8.0
$u(2, 2)$	0.15310	0.20833	36.1	0.16572	8.2

The percentage error is reduced by a factor of about 4 as the discretization length h is halved. This bears out the fact that this method is $0(h^2)$, i.e., halving h quarters the error.

Exercises

1. Do the above problem with $h = k = 0.75$.
2. Do Problem 2 of Section 2.1, first using $h = k = \pi/3$.
3. Repeat (2) with $h = k = \pi/4$.
4. Inspect the error in (3) by computing a partial sum of the analytic solution.
5. Prepare a sketch or a graph of the solutions of Problem 2.

We return to the other main theme of Section 2.6, the analytic variational method. As seen there, a proper theoretical understanding of the Dirichlet Principle requires the notion of weak or distributional derivatives and spaces, such as $H^1(\Omega)$. Let us prove one instance of the Sobolev Lemma mentioned in Section 2.6.

Example

Show $H^1[0, 1] \subset C^0[0, 1]$, that is, show that functions u (that are square integrable along with their derivative) are in fact continuous.

Solution. We use the fact (see e.g., Section 2.9.3, for the arguments needed to establish this) that $H^1[0, 1]$ is the completion of $C^1[0, 1]$ in the $H^1[0, 1]$ norm. Accepting this, we then obtain the bound

$$\|u\|_{C[0,1]} \equiv \max_{0 \leq x \leq 1} |u(x)| = |u(t_0)| \qquad \text{(for some } t_0 \in [0, 1])$$

$$= \left| u(s_0) + \int_{s_0}^{t_0} u'(\tau) \right| \qquad (u(s_0) = \min u, \text{ for some } s_0 \in [0, 1])$$

$$\leq |u(s_0)| + \int_0^1 |u'(\tau)| \leq \int_0^1 |u(\tau)| + \int_0^1 |u'(\tau)|$$

$$\leq \left(\int_0^1 |u(\tau)|^2 \right)^{1/2} + \left(\int_0^1 |u'(\tau)|^2 \right)^{1/2} = \|u\|_{H^1[0,1]} .$$

For any u in H^1 there exists a sequence u_m in C^1 such that

$$u_m \xrightarrow{L^2} u \quad \text{and} \quad u'_m \xrightarrow{L^2} \text{(something, say } w \text{ or } u').$$

By the bound just established,

$$\|u_n - u_m\|_{C[0,1]} \leq \|u_n - u_m\|_{H^1[0,1]} \to 0.$$

Hence (u_m) is uniformly convergent to some continuous function v. However, uniform convergence implies (see Section 1.6.3) L^2 convergence

$$\int_0^1 |u_m - v|^2 \, dx \leq \max_{0 \leq x \leq 1} |u_m - v| \to 0$$

so by the uniqueness of L^2 limits $u = v$ (almost everywhere). Thus, we may select the continuous v to represent u within H^1.

Exercises

6. Give an alternate proof, using the notion of weak derivative, to show that the H^1 functions are in fact absolutely continuous.
7. Construct a counterexample to show that H^1 functions are not continuously differentiable.
8. For w in $L^2[0, \pi]$ and $\phi_n = (2/\pi)^{1/2} \sin nx$, $n = 1, 2, 3, \ldots$, let

$$u(x) = \sum_{n=1}^{\infty} \frac{1}{n} (w, \phi_n) \phi_n(x).$$

Show that $u(x)$ is continuous on $(0, \pi)$.

Weak solutions of partial differential equations were defined in Section 1.9.5(3) and mentioned in Sections 1.9.7 and 2.6. Generally, the finite element and other numerical solutions (no matter how fine the mesh or how high the order of the elements or how large the series partial sum) are not weak solutions.

Example

Show that "tent functions" such as those of Figure 2.6d are not weak solutions.

Solution. We recall that a locally integrable function on Ω is said to be a weak solution of $\Delta u = 0$ if $\int_\Omega u \Delta \phi = 0$ for all test functions ϕ in $C_0^\infty(\Omega)$. Project the planar pieces of the function u vertically downward onto Ω. This creates a triangulation of $\overline{\Omega}$, the particular form of it not being important.

Within one of the cut rectangles, place a ball B of radius r centered at the center x_0 of the diagonal γ. Construct an approximate characteristic function $\phi \in C_0^\infty(B)$ such that its support is in the shaded region shown in Fig. 4Pb. This can be done by molification, see Section 2.9.3.

Let $u_i = u_{T_i}$ where T_1 and T_2 denote the two triangular parts of the cut rectangle. With the T_1 and T_2 outer normals as shown, for each of the two subdomains $i = 1, 2$, we have

$$\int_{T_i} \Delta \phi u = \int_{T_i} \phi \Delta u_i - \oint_{\partial T_i} \phi \frac{\partial u_i}{\partial n_i} + \oint_{\partial T_i} u_i \frac{\partial \phi}{\partial n_i}$$

$$= \oint_{\partial T_i} \left(u_i \frac{\partial \phi}{\partial n_i} - \phi \frac{\partial u_i}{\partial n_i} \right).$$

Consequently

$$0 = \int_\Omega \Delta \phi u = -\oint_{\partial T_1} \phi \frac{\partial u_1}{\partial n_1} - \oint_{\partial T_2} \phi \frac{\partial u_2}{\partial n_2}$$

$$= \int_\gamma \phi \operatorname{grad} (u_2 - u_1) \cdot n_1$$

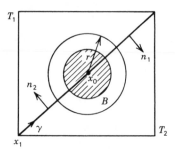

FIG. 4Pb. Projected Triangulation.

because

$$\oint_{\partial T_1} u_1 \frac{\partial \phi}{\partial n_1} + \oint_{\partial T_2} u_2 \frac{\partial \phi}{\partial n_2} = \int_\gamma u_1 \frac{\partial \phi}{\partial n_1} - \int_\gamma u_2 \frac{\partial \phi}{\partial n_1} = 0,$$

using $u_1 = u_2$ and $n_1 = -n_2$ along the diagonal γ. Because grad $(u_2 - u_1) \cdot n_1$ is a constant for the piecewise-linear tent functions, and $\int_\gamma \phi \neq 0$ for ϕ chosen as a molified characteristic function, grad $(u_2 - u_1) \cdot n_1 = 0$. As $u_1 = ax + by$ and $u_2 = \alpha x + \beta y$ this means that $(\alpha - a, \beta - b) \cdot \vec{n}_1 = 0$. Because $\vec{x}_0$ and $\vec{x}_1$ lie under the intersection of the u_1 and u_2 planes, $(\alpha - a, \beta - b) \cdot (\vec{x}_0 - \vec{x}_1) = 0$. But $\vec{n}_1$ and $\vec{x}_0 - \vec{x}_1$ are orthogonal so necessarily $(\alpha - a, \beta - b) = 0$, that is, $\alpha = a, \beta = b$. This says that the only piecewise planar continuous weak solution of $\Delta u = 0$ is a plane.*

Exercises

9. (a) The above idea is that the condition $\int_\Omega u \Delta\phi = 0$ as $\Delta\phi$ runs through all $\phi \in C_0^\infty(\Omega)$ is a very strong one, as the latter are dense in $L^2(\Omega)$. One cannot quite conclude that the $\Delta\phi$ form a dense set, but the residual subspace allowed for u is a small one indeed. Verify this in one dimension.
   (b) Show that $e^{-x^{-2}}$ is C^∞ but not analytic.
   (c) Show that the following function is C_0^∞ with support $[-1, 1]$.

$$j(x) = \begin{cases} e^{-\frac{1}{1-x^2}} & -1 < x < 1, \\ 0 & \text{otherwise.} \end{cases}$$

* For the theory of spaces of weakly differentiable functions see, e.g., R. Adams, *Sobolev Spaces* (Academic, New York, 1975).

In Section 2.6 and this Fourth Pause we have discussed numerical solution methods for elliptic partial differential equations. In Problem 2.9.6 we give a marching (i.e., ODE) method for numerically solving parabolic equations. We also present there an important differential equation in the secondary recovery of oil, the Buckley-Leverett equation. The solution of this and related equations requires combinations of elliptic, parabolic, and hyperbolic numerical methods, and continues to spur their development.

A more complete introduction to numerical methods from the computational point of view has been deferred, to alow instructor and reader choice, to Chapter 3, Appendix B. Included there one will find computational solutions of the following problems already studied analytically in this book:

Section 1.2 Problem 1; Section 1.9.6(2)c (Gibb's Effect computed to 1,000,000 terms); Section 2.1 Problems 1, 2, and 3; Section 2.9.8(a) (Traffic flow problem).

Additionally, some of the most recent results in computational fluid dynamics are illustrated in Chapter 3. The Sixth Pause presents the essentials of shock capturing methods. Appendix C considers briefly the theory and computation of the full (viscous) Navier-Stokes equations.

2.7 SOME UNBOUNDED DOMAIN CONSIDERATIONS (AND CONTINUOUS SPECTRA)

Partial differential equations, when encountered on an unbounded domain, require additional considerations. In this section we wish to consider some of these new features. In Section 2.8 we look at a particularly important problem occurring on unbounded domains: that of scattering theory.

In particular, in this present section we wish to emphasize the following natural progression out of which the Fourier transform may be seen to come:

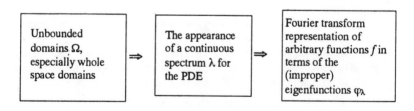

To begin let us recall our three basic solution methods of separation of variables, Green's functions, and variational methods.

The separation of variables method often breaks down for unbounded domains because one no longer has a countable number of square integrable eigenfunctions.* However, the Fourier series $\sum_{n=1}^{\infty} c_n \varphi_n$ may often be replaced by a Fourier integral $\int c_\lambda \varphi_\lambda$ to accommodate the "uncountable number of eigenfunctions." Although it is not quite that simple, the occurrence and use of Fourier transforms may indeed be viewed conceptually as the extension of the separation of variables method to unbounded domains Ω that are the whole space ($\Omega = n$-space E^n for some n). In like manner one may similarly employ the Laplace transform, especially for initial value problems and for problems in which Ω is a half-space. For arbitrary unbounded domains both the Fourier and Laplace integral methods have severe limitations. The latter fact should not surprise us since the same limitations occurred for the Fourier series (how to find the exact eigen functions for a domain with irregular geometry).

For Ω an unbounded domain the Green's function for a problem often exists and can sometimes be found.† In those cases this method remains a good one, although one must, as we shall see in some examples, exercise more care in its use. The finding of a Green's function sometimes requires use of the Fourier transform.

For unbounded domains one may still sometimes employ variational methods. In practice even their formal application often leads to physically important results. However, their validity and convergence is generally harder to prove due to loss of "compactness." The latter can sometimes be replaced by arguments of convexity and lower semicontinuity of certain functionals.

One can get around some of these difficulties of unbounded domains by working locally, that is, by working on bounded subdomains of the given domain Ω. Much of the modern theory of partial differential equations does just that. Such treatments yield general existence proofs but do not in general provide the construction of actual solutions.

We will proceed via three subsections. In (1) *Recapitulation and Initial Observations* we recount what we already know or may expect. In (2) *Continuous Spectrum* we discuss some aspects of elementary spectral theory from the points of view of eigenvalues λ_n as the so-called point spectra of an operator and continuous spectra λ as other λ for which the partial differential equation

$$(L - \lambda)u = f$$

is not well posed. In (3) *Fourier Transform* we take a very brief look at how the

---

* This depends on whether one is in the "limit-circle" or "limit-point" case, for example, when the separated ordinary differential equation is of singular Sturm–Liouville type on an infinite interval. (See 2.4 and 2.9.4.)

† Indeed we treated an unbounded domain in Problem 2 of Section 1.2 at the beginning of the book. But how did we find the Green's function?

Fourier transform plays the same role for "integrating" the equation over the continuous spectrum as did the Fourier series for the case of point spectra.

2.7.1 Recapitulation and Initial Observations

In order to consolidate our thinking, and to develop a little preliminary intuition for problems on unbounded domains, let us at this point pause to collect certain observations already available to us, and to consider (at least on an initial basis) some exterior domain problems.

To begin, let us consider the first problem considered in the book—the Dirichlet problem (see Section 1.2)—and for simplicity let us consider the one-dimensional case. For Ω the interval $(-1, 1)$ and boundary condition $u(-1) = a$ and $u(1) = b$ we had the solution as shown in Figure 2.7a. The exterior problem, that is, for Ω the exterior of the interval $[-1, 1]$, with the same boundary conditions $u(-1) = a$ and $u(1) = b$, is different in that we no longer get unique solutions (see Figure 2.7b). Thus we are reminded that $\partial\Omega$ in the exterior domain problem includes also the points at infinity, and to get a well-posed problem we need appropriate boundary conditions at $x = \pm\infty$. Notice that in the one-dimensional problem that we have just considered one needs a Neumann boundary condition at $x = \pm\infty$ rather than a Dirichlet condition there, and in no case (other than $a = b = 0$) is the solution square integrable on Ω.*

The second problem considered in this book, that of the heat equation initial value problem for $-\infty < x < \infty$ and $t > 0$, was as mentioned above already an unbounded domain problem. Although it is true that unboundedness in t is indeed present, as we saw in the heat equation initial boundary value problems for x in a

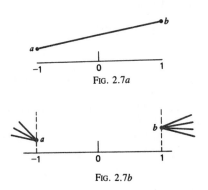

FIG. 2.7a

FIG. 2.7b

---

* The $n = 2$ and higher-dimensional Dirichlet problems for Ω the exterior of the unit ball are actually a little better due to a combination of facts such as the connectedness of Ω in those cases and the very much larger number of harmonic functions that occur in higher dimensions. However, the square integrability is lost. See Problem 1 at the end of this subsection, and Problem 1 at the end of the next subsection.

finite interval $[a, b]$ as solved by separation of variables, and as we saw in showing uniqueness via the maximum principle (see Section 2.2), the t-unboundedness (which is always present in initial value problems regardless of type) is not the distinguishing factor. Rather it is whether the problem has bounded or unbounded domain Ω in the x variable. The discussion of uniqueness and stability, for example, for the unbounded domain $-\infty < x < \infty$ as in Problem 2 of Section 1.3 was made more difficult due to the unboundedness of the domain, even when given the existence and explicit representation of a solution by means of the Poisson kernel Green's function.

Let us turn now to exterior domain considerations for the third problem considered in the book, the Rayleigh eigenvalue problem. To begin let us consider the problem on an interior domain of length $2l$:

$$\begin{cases} -u''(x) = \lambda u(x), & -l < x < l, \\ u(-l) = u(l) = 0. \end{cases}$$

The eigenfunctions, except for the normalizing constant, were

$$u_n(x) = \cos \frac{(2n - 1)}{2l} \pi x,$$

with eigenvalues

$$\lambda_n = \left(\frac{(2n - 1)\pi}{2l} \right)^2.$$

Suppose $l \to \infty$. Note (as shown in Fig. 2.7c) that the eigenvalues slide to the left toward zero and at the same time begin filling up the enlarging interval $(0, l)$. Moreover, the norm (squared) of the eigenfunctions

$$\|u_n(x)\|^2 = \int_{-l}^{l} \cos^2 \frac{(2n - 1)\pi x}{2l} \, dx = l$$

blows up as $l \to \infty$.

Let us consider this last example a bit further. Also, rather than getting involved in the exterior (disconnected) domain considerations that would prevail if we considered Ω the exterior of the interval $[-l, l]$, such as those that we already discussed

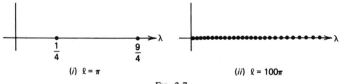

(i) $l = \pi$ (ii) $l = 100\pi$

Fig. 2.7c

relative to the exterior Dirichlet problem above, we will look only at the whole space problem

$$\begin{cases} -u''(x) = \lambda u(x), & -\infty < x < \infty, \\ u(\pm\infty) = 0. \end{cases}$$

The Dirichlet boundary condition at $\pm\infty$, $u(\pm\infty) = 0$, can be specified more precisely in terms of the functions actually in the domain of the operator, via Sobolev spaces as in Section 2.6; but let us forego that for now and ask only that u be good enough so that integration by parts asserts as usual* that

$$\lambda = \int_{-\infty}^{\infty} (u')^2 \, dx \Big/ \int_{-\infty}^{\infty} u^2 \, dx \geq 0$$

for any nontrivial solution to the problem. Then from elementary ordinary differential equations we know the solution to be

$$u_\lambda(x) = c_1 \sin \lambda^{1/2} x + c_2 \cos \lambda^{1/2} x, \qquad \lambda > 0,$$

$$u_0(x) = c_1 + c_2 x, \qquad \lambda = 0,$$

and we see immediately that the problem has no nontrivial solutions that go to zero at $\pm\infty$. Moreover, none of the solutions is square integrable on $\Omega = (-\infty, \infty)$, whether or not they go to zero at $\pm\infty$.

Yet, continuing with this last example, the $u_\lambda(x)$ are clearly eigenfunctions satisfying the Rayleigh differential equation. Such eigenfunctions, satisfying the differential eigenvalue problem in question but not lying within the L^2 space of square integrable functions in which the problem is set, are called improper eigenfunctions. As we shall see later there is nothing improper about them physically except that they describe scattering states rather than bound states, radiation rather than vibration, continuous spectrum rather than point spectrum, to use the terminology of differing important contexts in which they are found.

Some further recapitulation and review will be found in the observation of the next subsection as we pursue the related occurrences of unbounded domains and continuous spectrum.

Problem 1. Let Ω = the exterior of the unit ball in R^3, that is, $\Omega = \{x = (x_1, x_2, x_3) | x_1^2 + x_2^2 + x_3^2 > 1\}$. (a) Consider the uniqueness question for Ω for Dirichlet, Neumann, and Robin boundary conditions, given that $\Delta u = 0$ in Ω. Assume that you can integrate by parts, as in the uniqueness questions for the corresponding interior problem. (b) Write down your "boundary conditions at ∞" as you have employed them in part (a). What regularity have you assumed on u? (c) Can you use maximum principles or any other "interior problem" uniqueness techniques?

---

* See Problem 2.

Problem 2. (a) Try to state exactly the condition on u needed at $x = \pm \infty$ to ensure that $\lambda \geqq 0$ for the problem considered above in the text, namely, $-u''(x) = \lambda u(x)$ in $\Omega = (-\infty, \infty)$. (b) Do you believe the eigenfunctions to be symmetric, that is, even functions? (c) Consider the same problem as in part (a) on $\Omega = (0, \infty)$ subject to the left boundary condition $u'(0) = 0$.

Problem 3. The problem of symmetry of solutions is an interesting one in general and extends beyond the problems mentioned above, for both bounded and unbounded symmetric domains and for both linear and nonlinear problems.

(a) For the Problem 2(a) above, show how the question of symmetry of solutions is related to the complex valued solutions to the ordinary differential equations, namely, $e^{\pm i\lambda^{1/2}x}$.

(b) Show the nonlinear eigenvalue problem $-u'' = \lambda f(u)$ on $-1 < x < 1$, $u(-1) = u(1) = 0$, has only symmetric solutions. Assume f to be positive and continuous.

(c) Repeat (b) if possible on the unbounded domain $-\infty < x < \infty$, with appropriate boundary conditions at $\pm \infty$.

(d) Investigate whether (b) and (c) remain true for the "nonautonomous" equation $-u''(x) = \lambda \rho(x) f(u)$, where $\rho(x)$ is a nice positive continuous weight function.

2.7.2 Continuous Spectrum

Problems on unbounded domains bring directly before us examples of what is called the continuous spectrum. In our previous use of Fourier series and separation of variables we were working essentially with the easier case of point spectra (e.g., the eigenvalues λ_n).

In this section we want to make these terms precise, for they permeate much of differential equations and the more abstract operator theory.[†] We will illustrate them in terms of operators with which we are already familiar.

We consider an equation

$$Lu - \lambda u = f, \tag{*}$$

[†] Historically, the differential equations came first, and therefore with considerable justification they should be presented first. The more abstract theory (called spectral theory) is clarifying but often not needed for the actual solution of the partial differential equations. It is our intent here to present some of the essential elements of the spectral theory, without a full development. For a more complete discussion, although the literature is very large, we can recommend for the reader's convenience the classic F. Riesz and B. Sz. Nagy, *Functional Analysis* (Ungar, New York, 1955), and the encyclopedic Dunford and Schwartz, *Linear Operators* I, II, III (Wiley-Interscience, New York, 1958, 1963, 1971), among others.

where λ is a real or complex number. If (*) is *well-posed*, which we remember from Section 1.3 means that (*) has all the good qualities of (1) existence, (2) uniqueness, and (3) stability of solutions, then we say that λ is in the *resolvent set* $\rho(L)$ of L. That is, we can "resolve" the problem completely. It can be easily shown that the resolvent set for any closed operator (see Problem 2.9.2) is an open subset of the complex plane. The complement in the complex plane of the resolvent set $\rho(x)$ is the *spectrum* $\sigma(L)$ of L. As the complement of an open set we know that the spectrum $\sigma(L)$ is always a closed set.

All self-adjoint operators $L = L^*$ must be closed operators, since it can be shown that all adjoint operators are necessarily closed. The closedness of an operator, as important also to the convergence of separation of variables solutions, was indicated in Problem 2.9.2. For spectral theory one must use closed operators and almost all operators encountered in practice in differential equations are either closed or closeable.

For the Rayleigh problem $-u'' = \lambda u$ on $0 < x < \pi$, $u(0) = u(\pi) = 0$, the spectrum $\sigma(L)$ was exactly the eigenvalues $\lambda_n = n^2 = 1, 4, 9, \ldots$. Similarly all of the Sturm-Liouville problems of Section 2.4 possessed a similar spectrum $\sigma(L)$ consisting only of the eigenvalues of the operator. This is typical of bounded domain problems (although there are exceptions). The *point spectrum* $\sigma_p(L)$ of an operator consists of those λ such that there exists a nontrivial solution of

$$(L - \lambda)u = 0$$

and consists, as we have just seen, of the eigenvalues of the operator. Thus $\sigma_p(L)$ is those λ for which the *uniqueness* property (2) of well-posed problems *fails*.

It can occur that the uniqueness property (2) holds but the existence property (1) fails. An example of this is given in Problem 1 at the end of this subsection, and indeed as mentioned above the occurrence of this situation is important to a number of physical phenomena. Such λ, that is, λ such that the equation (*)$(L - \lambda)u = f$ is solvable for most (a properly dense set) but not quite all data f and yet *uniquely* solvable in each such case, are called the *continuous spectrum* $\sigma_c(L)$ for the operator L.† The term continuous derives from the fact that in most cases such λ form a continuum in the plane, that is, the set $[0, \infty)$ mentioned in the above subsection in the example describing "improper eigenfunctions." We will see further examples of this type later in this section (the Helmholtz equation in Problem 3 below) and in the next section. There is also a precise "continuous" meaning of $\sigma_c(L)$ for self-adjoint operators L in terms of the associated spectral measures but

---

† We must insist here in fact that (*) is solvable for a *properly dense* set of data f in the space in question. It can happen that (*) possesses the uniqueness property but can be solved only for a certain nondense subspace of data f. That third spectrum $\sigma_r(L)$ is called the residual spectrum, which we shall ignore here. For self-adjoint L there is no residual spectrum.

we do not discuss this meaning here.† Finally, there are examples in which σ_c occurs as isolated points; good examples of the latter are $\sigma_c(L^{-1})$ for L a Sturm-Liouville operator.

Consider, for example, the Rayleigh operator $Lu = -u''$ with boundary conditions $u(0) = u(\pi) = 0$, which had for its spectrum $\sigma(L) \equiv \sigma_p(L) = 1, 4, 9,$ $\ldots, n^2, \ldots$, as shown in Fig. 2.7d. A general theorem (the spectral mapping theorem) asserts that $\sigma(L^{-1}) = \sigma(L)^{-1}$ so that the inverse of the Rayleigh operator

$$L^{-1}f = \int_0^\pi G(x, s)f(s) \, ds$$

has spectrum $\lambda_n = n^{-2}$. Because the spectrum is closed, the value $\lambda = 0$ is also in $\sigma(L^{-1})$. In Problem 1 it will be demonstrated directly that L^{-1} cannot be defined for all $f \in L^2(0, \pi)$. Hence $\sigma_p(L^{-1}) = n^{-2}$ and $\sigma_c(L^{-1}) = 0$, as shown in Fig. 2.7e.

One does not need the spectral mapping theorem for the above example. Clearly for $\lambda \neq 0$ and L invertible one has

$$(L - \lambda)u = 0 \Leftrightarrow \left(L^{-1} - \frac{1}{\lambda}\right)u = 0.$$

More generally one easily verifies that for a closed operator L one has $\lambda \in \rho(L)$ iff $\lambda^{-1} \in \rho(L^{-1})$. The spectral mapping theorem is more comprehensive and very useful when considering "functions of an operator." The "function" considered here was the inverse, a relatively simple one.

It was shown in Section 2.5, Problem 2 that, similarly, the partial differential operator $Lu = -\Delta u$ on $\Omega = (0, \pi) \times (0, \pi)$ with $u = 0$ on $\partial\Omega$ had a compact

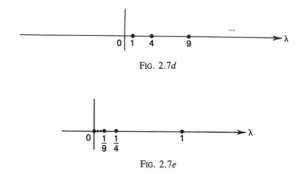

Fig. 2.7d

Fig. 2.7e

† A self-adjoint operator L may be written as a Lebesgue-Stieltjes integral in the sense that the inner product $(Lu, u) = \int \lambda \, d(E_\lambda u, u)$ for all u in the domain $D(L)$. The E_λ are called the spectral family for L, the $d(E_\lambda u, u)$ the associated spectral measures, and for λ in the continuous spectrum $\sigma_c(L)$ the $(E_\lambda u, u)$ are continuous for all u.

inverse $L^{-1} = \int_\Omega G(P, Q)$, since the expressibility of $\|G\|^2$ as the sum of the eigenvalues squared is a known (Hilbert–Schmidt) condition for L^{-1} to be compact. *Compactness* of an operator is a stronger condition than just *boundedness**; that L^{-1} in this case is bounded is easier to establish (see the discussions in Problem 2 of Section 2.5 and in Problem 2.9.5). The eigenvalues and hence $\sigma_p(L)$ were already found (see Section 2.1) and were $n^2 + m^2$, $n = 1, 2, 3, \ldots$, $m = 1, 2, 3,$ $\ldots$. Thus $\sigma_p(L^{-1}) = (n^2 + m^2)^{-1}$ and $\sigma_c(L^{-1}) = 0$ and the qualitative picture of $\sigma_c(L^{-1})$ is the same as for the Rayleigh operator inverse. It can be shown that this qualitative picture is characteristic of the spectrum of any (nonsingular) self-adjoint compact operator.

As argued in the previous subsection, the operator $Lu = -u''$ on the infinite domain $\Omega = (-\infty, \infty)$ with appropriate conditions on u at $\pm\infty$ is expected to have continuous spectrum $[0, \infty)$ as shown in Figure 2.7f. This will be demonstrated in Problem 2.

An important operator with the same spectrum (see Problem 2) is the "free Hamiltonian" $Lu = -\Delta u$ in $L^2(R^3)$ already encountered in Section 1.6.3. Upon introduction of the Coulomb potential $1/r$, the hydrogen operator $Lu = -\frac{1}{2}\Delta u - (1/r)u$ retains the same continuous spectrum but also possesses the eigenvalues $\lambda_n = -1/2n^2$, $n = 1, 2, 3, \ldots$ (Fig. 2.7g). See the discussion already given in Section 1.6.3.

The Helmholtz operator, namely, the Laplacian $Lu = -\Delta u$ on an unbounded domain occurs in a wide variety of physical situations, and is analyzed in one of them in Problem 3. As will be seen in Section 2.8 the eigenvalue problems for the Helmholtz operators

$$-\Delta u = \lambda u, \qquad \Omega = R^n,$$
$$(-\Delta + V(x))u = \lambda u, \qquad \Omega \text{ an exterior domain,}$$
$$(-\Delta + V(x) - \lambda)u = f, \qquad \Omega \text{ an exterior domain,}$$

Fig. 2.7f

Fig. 2.7g

---

* The precise meaning of compactness of an operator A is that the set $\{A(x){:}\|x\| = 1\}$ has a compact closure. From this it follows indeed that $\{A(x){:}\|x\| = 1\}$ is bounded, that is, that A is a bounded operator.

are important and arise from separation of variables of the corresponding wave equation

$$u_{tt} - \Delta u + V(x)u = f(x, t).$$

Problem 1. (a) Show directly that $0 \in \sigma_c(L^{-1})$ for the Rayleigh problem as discussed above. (b) Explain why then $0 = \sigma_c(L^{-1})$, that is, why there are no other points in the continuous spectrum of that L^{-1}. (c) For λ not in $\sigma_p(L)$ for the Rayleigh operator, write out explicitly the solution to

$$(L - \lambda)u = f$$

for all f in $L^2(0, \pi)$. You may just assume $f \in C_0^\infty(0, \pi)$ in order not to worry about the Lebesgue integration classes, if you wish.

Problem 2. (a) Consider the operator $Lu = -u''$ on $\Omega = (-\infty, \infty)$ and show that each $\lambda \geq 0$ fulfills the conditions needed for λ to be in the continuous spectrum $\sigma_c(L)$. (b) Consider the Hamiltonian $Lu = -\Delta u$ in $L^2(R^3)$, that is, $\Omega = $ the whole space R^3 and we consider only data and solutions in $L^2(R^3)$. Show $\lambda = 0$ is not an eigenvalue but that on the other hand $-\Delta u = F$ cannot be solved for all data F in $L^2(\Omega)$. (c) If interested, show the same for all $\lambda > 0$, and that for all other complex λ the equation $(L - \lambda)u = F$ is well posed.

Problem 3. (a) Solve the Helmholtz equation

$$-\Delta u = \lambda u$$

for $\Omega = R^1, R^2, R^3$, and R^n in general, by formal separation of variables in polar coordinates. You may use complex valued solutions if you wish. (b) Comment on the L^2-ness or non-L^2-ness of the solutions. (c) Solve formally the hydrogen equation (for $\Omega = R^3$)

$$-\frac{1}{2}\Delta u - \frac{1}{r}u = \lambda u$$

by separation of variables. (d) Distinguish the L^2 solutions form the non-L^2 solutions.

2.7.3 Fourier Transform

Let us define and utilize the Fourier transform to illustrate its use on whole space problems.

Let $f(x)$ be integrable and square integrable, that is, $f \in L^1(-\infty, \infty) \cap L^2(-\infty, \infty)$. The Fourier transform $F(f) \equiv \hat{f}$ is defined as:

$$F(f)(k) = \hat{f}(k) = \frac{1}{\sqrt{2\pi}} \int_{-\infty}^{\infty} e^{ikx}f(x)\, dx, \qquad -\infty < k < \infty.$$

It may be shown* that F maps $L^2(-\infty, \infty)$ one to one onto $L^2(-\infty, \infty)$, preserving norms and inner products:

$$\|\hat{f}\| = \|f\|, \qquad \langle \hat{f}, \hat{g} \rangle = \langle f, g \rangle.$$

Such a transformation is called a unitary transformation and may be thought of in a number of useful ways, such as a change of basis, a rotation, and so on.

Thus every f in $L^2(-\infty, \infty)$ has an equivalent representation given by its Fourier transform $\int e^{ikx} f(x) \, dx$. This is analogous to what was done by means of Fourier series in Theorem H of Section 2.3. Recalling that

$$e^{ikx} = \cos kx + i \sin kx$$

and letting $k = \lambda$, the Fourier integral in the transform may thus be regarded as the "continuous" Fourier coefficients of f,

$$c_\lambda = (2\pi)^{-1/2} \langle f, \varphi_\lambda \rangle$$

where ϕ_λ are $L^\infty(-\infty, \infty)$ "basis" functions $\varphi_\lambda(x) = e^{i\lambda x}$, $-\infty < \lambda < \infty$. To recover f from its Fourier coefficients $f(\lambda)$ and to see more explicitly the analogy with Theorem H, we apply the inverse Fourier transform:

$$f = F^{-1}(\hat{f}) = (2\pi)^{-1/2} \int_{-\infty}^{\infty} e^{-ix\lambda} \hat{f}(\lambda) \, d\lambda = (2\pi)^{-1/2} \int \overline{\varphi}_\lambda c_\lambda.$$

Compare Theorem H (3) of Section 2.3.

One defines similarly the Fourier transform $F : L^2(R^n) \to L^2(R^n)$ by

$$\hat{f}(k) = (2\pi)^{-n/2} \iiint_{R^n} e^{ik \cdot x} f(x) \, dx.$$

Except for convergence questions it is quite easy to show (Problem 1) the important property of the Fourier transform that

$$\widehat{\frac{\partial u}{\partial x_i}} = -i \, k_i \hat{u}$$

From this one has the resulting transformation of the Lapalacian operator

$$-\widehat{\Delta u} = \left(\sum_{i=1}^{n} k_i^2 \right) \hat{u}.$$

---

* This is also a Parseval's theorem; proofs may be found in many books. Theorem H of Section 2.3 may be regarded as a special case of a general Parseval's Theorem. However we shall adhere to the other point of view that the latter is a generalization of the former, as is the inner product form of Parseval's equation in the Third Pause.

Thus the Fourier transform converts a differential operator into a mutiplication operator. In particular, $-\Delta$ goes to multiplication by $|k|^2$.*

Let us illustrate then the use of this transform on the unbounded domain problem

$$-u'' = \lambda u, \qquad \Omega = (-\infty, \infty).$$

By Fourier transform (which is a linear operator) we have

$$k^2 \hat{u} = \lambda \hat{u},$$

which shows immediately that the eigenfunction u must be such that $\hat{u}(k) = 0$ for all k except $k = \pm\lambda^{1/2}$. From this the solutions $u = \sin \lambda^{1/2} x$ and $\cos \lambda^{1/2} x$ may be deduced.

In like manner a Poisson problem

$$-\Delta u = f, \qquad \Omega = R^n$$

has solution by Fourier transform

$$u = F^{-1}\left(\frac{1}{|k|^2}(Ff)(k)\right) = \overset{\frown}{\frac{1}{|k|^2}}^{-1} \hat{f}$$

where we have used two common notations to illustrate them. The singularity at zero of the inverse L^{-1} of the Laplacian is shown well here in terms of the kernel $1/|k|^2$, and lends force to the fact that we cannot invert this problem for all data f.

By means of the Fourier transform one can establish the Malgrange–Ehrenpreis theorem in the abstract theory of partial differential operators. This result asserts that for whole space problems on $\Omega = R^n$ every linear partial differential equation with constant coefficients

$$Lu = f$$

possesses a Green's function $G(P, Q)$ in the sense that the solution is given by

$$u(P) = \int_{R^n} G(P - Q)f(Q)\, dv_Q$$

for all test functions $f \in C_0^\infty(R^n)$. Recall (see Section 1.5.2) that we gave three interpretations to Green's functions $G(P, Q)$:

(1) $\int G(P, Q) = L^{-1}$,
(2) $G(P, Q)$ is a fundamental singularity,
(3) $L_Q G(P, Q) = \delta(P, Q)$ the delta function.

The sense (1) here means that the integral of the Green's function provides a formal right inverse for L; for smooth functions f one has $L(\int G(P, Q)f(Q)\, dv_Q) = f(P)$.

---

* To be more precise here, especially when transforming differential equations, one should designate exactly which u are being treated. For $\partial u/\partial x_i$ it is sufficient that u possess square integrable distributional derivatives, and for Δu, that u possess square integrable second order distributional derivatives. See Problem 1.9.7(2) and Section 2.6.

The sense (2) is borne out, for example, by the fact that for the Laplacian $L = -\Delta$ on R^3 one has $G(P, Q) = 1/4\pi|P - Q|$, as the student will be asked to show in Section 2.8, Problem 1. The sense (3) states that $G(P, Q)$ is the fundamental solution for the problem, in that G satisfies the differential equation except at $P = Q$ where it then combines the senses (2) and (1) by possessing just the right amount of singularity to integrate the problem in sense (1).

Note that for whole space problems one finds generally that the notions of Green's function, fundamental singularity, and fundamental solution all coincide. That is, the adjusting function $g(P, Q)$ (see Section 1.6.1, Remark 2) is not needed.

As a final remark, for the Poisson problem above for R^3 we found (using $P = x$ and $Q = y$ now, and putting in the variables)

$$u(x) = \int_k \int_y \frac{e^{-ik\cdot x} e^{ik\cdot y}}{|k|^2} f(y) \, dy \, dk, \tag{1}$$

which we know by the Malgrange-Ehrenpreis theorem must also be given by

$$u(x) = \int_y G(x - y) f(y) \, dy, \tag{2}$$

which must also, as noted above, be given by

$$u(x) = \frac{1}{4\pi} \int_y \frac{1}{|x - y|} f(y) \, dy. \tag{3}$$

From this it should be apparent to the reader that the uses (1) of the Fourier transform, (2) of the theory, and (3) of explicit solution representation, while in principle equivalent, may appear in different forms and may be of different efficacity.

Problem 1. (a) Show the properties of the Fourier transform

$$\widehat{\frac{du}{dx}} = -ik\hat{u} \text{ and } \widehat{\frac{d^2u}{dx^2}} = -k^2\hat{u}$$

by means of one and two integration by parts respectively. (b) Do the same for $\partial u/\partial x_i$ and $-\Delta u$ in several dimensions. (c) Think about the fact that the use of Fourier transform seems to require that Ω be a whole space.

Problem 2. (a) Prove the so-called convolution theorem:

$$\widehat{f * g} = (2\pi)^{-n/2} \hat{f}\hat{g},$$

where $f * g$ is the so-called convolution product

$$(f * g)(x) = \int_{-\infty}^{\infty} f(x - y) g(y) \, dy.$$

(b) Justify to some extent the following four steps to show how this may be used to generate the Poisson kernel solution

$$u(x, t) = (4\pi t)^{-1/2} \int_{-\infty}^{\infty} e^{-(x-y)^2/4t} f(y) \, dy$$

for the heat equation initial value problem

$$\begin{cases} u_t - u_{xx} = 0, & -\infty < x < \infty, \qquad t > 0 \\ u(x, 0) = f. \end{cases}$$

1. Taking transforms $\Rightarrow \begin{cases} \hat{u}_t = k^2 \hat{u}, \\ \hat{u}(0) = \hat{f}. \end{cases}$

2. Solution of latter is by ODE $\hat{u}(k) = \hat{f}(k) e^{-k^2 t}$.

3. $e^{-k^2 t} = \overparen{(4\pi t)^{-1/2} e^{-x^2/4t}}$.

4. Convolution theorem $\Rightarrow$ the result.

(c) Justify from the discussion at the end of the section above that

$$\overparen{|k|^{-2}}^{-1} = \frac{1}{4\pi |x|}.$$

Problem 3. (a) Consult the literature for the Laplace transform. (b) Use it as in Problem 2(b) above to resolve in a similar way the heat initial value problem there.

2.8 ELEMENTS OF SCATTERING THEORY

Scattering theory is concerned with solutions to the wave equation

$$u_{tt} - \Delta u = f(x, t, u), \qquad x \in \Omega,$$

in exterior domains, and as such is a proper subject for investigation as a mathematical question involving hyperbolic partial differential equations. Its roots are however so deep within the context of physical experiments that one cannot ignore them when treating the subject. Therefore we shall begin this section by looking at three physical settings from which scattering problems arise: (a) classical scattering; (b) quantum scattering; and (c) inverse scattering. We will then close the section with a brief look at the original work of Rayleigh on classical scattering. This work explains for example why the sky is blue, and, although done approximately 100 years ago, remains historically valuable for a basic introductory understanding of the subject of scattering.

As in Section 1.8 (Elements of Bifurcation Theory), we can do no more here than to scratch the surface of this important subject. Scattering methods have developed over the last century as a fundamental tool in physics and engineering. This can be expected to continue due to the very basic occurrences of propagation of energy by scattering, whether that energy comes out of an elementary particle experiment, off a power transmission line, or from the sunlight scattered out of the rainbow.*

(a) Classical Scattering

There are many types of classical scattering. One that is rather easy to visualize is that of the scattering of acoustic waves off an obstacle. In Figure 2.8a we have indicated a schematic illustration of this situation, in which incoming (plane) waves are reflected off a nonabsorbing rigid obstacle as outward going spherical waves. In the optical (ray) approximation, also drawn in Figure 2.8a, Snell's reflection law governs the basic direction of reflection, but in the actual wave mechanics the situation is much richer and much more complicated.

To continue with this description, if we consider the case in which the obstacle is submerged in an infinite homogeneous fluid, if the incoming wave packet consists of sound waves, and if we accept the linearized model from fluid dynamics with constant speed of sound c in Ω, we are led to the wave equation

$$\begin{cases} u_{tt} - c^2 \Delta u = 0 \text{ in } \Omega, & -\infty < t < \infty, \\ \dfrac{\partial u}{\partial n} = 0 \text{ on } \partial\Omega. \end{cases}$$

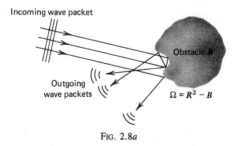

Incoming wave packet

Obstacle B

Outgoing wave packets

$\Omega = R^3 - B$

Fig. 2.8a

*We mention for the reader a few references, with apologies for all others not listed. For the general theory of scattering from the point of view of mathematics and physics, there is the excellent treatise of R. Newton, *Scattering Theory of Waves and Particles* (McGraw-Hill, New York, 1966). For classical scattering of obstacles see the monograph by P. Lax and R. Phillips, *Scattering Theory* (Academic, New York, 1967). For quantum scattering and inverse scattering one is fortunate to have the very recent books by W. Amrein, J. Jauch, and K. Sinha, *Scattering Theory in Quantum Mechanics* (W. A. Benjamin, Inc., Reading, Massachusetts, 1977) and K. Chadan and P. Sabatier, *Inverse Problems in Quantum Scattering Theory* (Springer, Berlin, 1977). For general surveys of recent results see also *Scattering Theory in Mathematical Physics*, J. La Vita and J. Marchand, eds. (Reidel, Doortrecht, 1974).

This is an exterior domain problem of the type discussed in Section 2.7. Here u $= u(x, y, z, t)$ is the velocity potential, and the Neumann boundary condition states that although the velocity need not be zero on $\partial\Omega$ when being reflected, there is no movement of the obstacle B. Thus the velocity components of the sound wave are $v_i(x, t) = c\partial u(x, t)/\partial x_i$, $i = 1, 2, 3$, the quantity $\partial u(x, t)/\partial t$ represents the excess pressure caused by the sound wave, $\partial^2 u(x, t)/\partial t^2$ measures the rate of change of said pressure, and the wave equation states the conservation law that the time rate of change in pressure is equal to the spatial rate of change (c times the divergence) of the velocity.

One may think of the problem of submarine detection by sonar, for example, as a concrete setting for this type of scattering problem.

Separation of variables for this problem yields the Helmholtz equation

$$\Delta u + k^2 u = 0 \text{ in } \Omega.$$

We will return briefly to this aspect of the problem after discussing the two other physical situations (b) and (c) below.

(b) Quantum Scattering

Here one scatters a particle off another. Two situations are easily envisioned (Fig. 2.8b), in which there is (i) an attraction and (ii) a repulsion. In accelerator experiments one usually must direct a whole cloud of particles against a cloud of the other type. However for the theory we may consider single particles, and we may also assume the larger is at rest by taking into account only their relative motion. Moreover we may assume for the description here that the attractive or repulsive force may be given by a potential function $V(r)$ dependent only on the relative distance r between the particles.

One of the simplest examples is the hydrogen atom equation we encountered in Section 1.6, namely,

$$-\tfrac{1}{2}\Delta u - \frac{1}{r} u = \lambda u.$$

The time dependence and certain units have been factored out of the original Schrödinger equation

$$i\frac{\partial u(t)}{\partial t} = \frac{-\hbar^2}{2m} \Delta u - \frac{Ze^2}{r} u$$

(i) (ii)

FIG. 2.8b

by a separation of variables and the use of so-called atomic (Hartree) units. In the original equation e is the charge of the electron, Ze is the charge on the nucleus, m is the mass of the electron, the nucleus has a presumed infinite rest mass, and $\hbar$ is Planck's constant. The above reduced stationary Schrödinger equation for u is an eigenvalue problem that plays the same role as did the Helmholtz equation for the classical scattering described in (a) above. Its spectrum was given in Section 1.6.

In Problem (3) of Section 2.7.2 one was asked to formally solve the Helmholtz equation and the stationary Schrödinger equation by separation of variables.

(c) Inverse Scattering

This is one of the most important types of scattering from the experimental point of view. In the accelerators various types of prepared particles are sent in and detectors measure what comes out (Fig. 2.8c). Sometimes different types of particles come out, sometimes the same as went in come out but with perhaps a phase change. The inverse scattering problem is: What happened in the "scattering center," that is, in the target region?

In particular one would like to recover from all of the data collected in such experiments, run as repetitively as necessary, the potentials $V(r)$ between the various particles as mentioned in the previous section. One of the important questions in physics, the answer in many cases elusive to this day, is to determine the "interaction laws" $V(r)$ from the data. Most of the large physical accelerator experiments may be thought of as being concerned with this problem.

Looked at mathematically, and in a simple example, given the boundary value problem

$$\begin{cases} -u''(x) + V(x) = \lambda u(x), & 0 < x < \pi, \\ u(0) = u(\pi) = 0, \end{cases}$$

how much does one need to know in order to determine completely $V(x)$? In some cases this problem can be solved. It would take us too far afield to discuss further this interesting problem. Physically, its solution depends on knowing enough "scattering data." Mathematically, one needs to be able to deduce enough about the

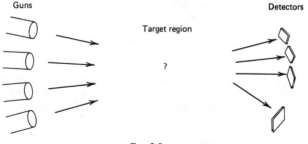

FIG. 2.8c

spectrum and eigenfunctions for the operator. Intuitively and classically, the question can be thought of in such ways as: (i) What information (sounds) are required to "hear" the shape of a drum,* or in terms of the acoustic scattering described in (a) above, (ii) how much information do we need scattered back to distinguish a submarine from a whale?

Rayleigh Scattering Off a Small Sphere

Rayleigh observed and proved† that when energy is propagated against a small hard sphere, the higher frequency waves scatter more. His argument proceeded as follows (see Fig. 2.8d).

Let an incident plane wave-train be directed downward along the z axis toward a small sphere centered at the origin; let the resulting scattered wave-train be spherical with amplitude depending on the angle θ that the outgoing direction makes with the z axis; and let the full wave $u(t)$ satisfy the wave equation

$$u_{tt} - c^2 \Delta u = 0 \text{ in } \Omega,$$

Ω the exterior of the small sphere. As in part (a) of this section, here $u(t)$ is a velocity potential (c a propagation speed in the exterior medium) of a wave propagating with a given fixed wave length μ, and correspondingly $k = \omega/c = 2\pi/\mu$ is the given fixed single frequency of the propagation. Thus the full wave $u(t)$ consists of a mixture of the incoming wave-train and the resulting scattered wave, the latter being assumed (as is often justified) to be spherically scattered and retaining the same propagation frequency. We may therefore write

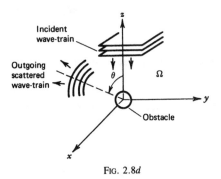

FIG. 2.8d

---

* See for example M. Kac, "Can one hear the shape of a drum," *Am. Math. Monthly* **73** (1966), and R. Prosser, addendum to "Can one see the shape of a surface," *Am. Math. Monthly* **84** (1977). The answer turns out to be no. See C. Gordon, D. Webb, S. Wolpert, *Invent. Math.* **110** (1992). But the counterexamples would make strange music, at best.

† A good accounting of this is given in Rayleigh's book *The Theory of Sound* (Macmillan, London, 1877; Dover, New York, 1945), Vol. II, pp. 272–277.

$$u(t) = u_{\text{incident}} + u_{\text{scattered}}$$

$$= e^{i(kz-\omega t)} + f(\theta)\frac{e^{i(kr-\omega t)}}{r}$$

$$= e^{-i\omega t}u,$$

where

$$u = e^{ikz} + f(\theta)\frac{e^{ikr}}{r}.$$

Here we have factored $u(t)$ into its time-dependent part and its stationary part, which amounts to a preseparation of variables. The scattered amplitude $f(\theta)$ depends only on the outgoing angle θ. An elementary dimensional analysis shows that for such spherical scattering the radial amplitude attenuation by r^{-1} is appropriate.

Substitution of $u(t)$ into the wave equation yields

$$c^2 e^{-i\omega t}\,\Delta u = c^2\Delta u(t) = u_{tt} = -\omega^2 e^{-i\omega t}u,$$

so that u satisfies the Helmholtz equation

$$\Delta u + k^2 u = 0 \text{ in } \Omega.$$

Thus separation of variables, with the t-part predetermined, has been performed on the original wave equation, leading immediately to the Helmholtz eigenvalue problem

$$-\Delta u = \lambda u \text{ in } \Omega,$$

where $\lambda = k^2$ is now the eigenvalue parameter. Boundary conditions of Neumann, Dirichlet, or other type may now be imposed, depending on the application.

Rayleigh was interested for example in the application of electromagnetic radiation impinging on a small water droplet in the earth's outer atmosphere. From the above model, by an expansion of e^{ikz} in terms of spherical harmonics (which are essentially the associated Legendre functions mentioned in Section 2.1), and by a similar expansion of $f(\theta)e^{ikr}/r$ in terms of spherical harmonics and Bessel functions, it was then concluded by Rayleigh that for a sphere sufficiently small relative to the frequency band under consideration one has indeed a scattered amplitude

$$f(\theta) \sim k^2(1 - \tfrac{3}{2}\cos\theta).$$

Hence higher frequency waves do scatter more.

This type of analysis would indicate for example in the sonar application mentioned in (a) that the scattering amplitude is very frequency dependent. In sonar one sends out a band of frequencies but gets back (essentially) only one frequency from the target. The dissipation of the other frequencies is caused by damping and other more complicated effects in the water. The same frequency selection is observed in listening to a bell struck under water.

In the application of Rayleigh to electromagnetic radiation on the atmosphere the above analysis shows why the sky is blue. Recall (Fig. 2.8*e*) that blue is near the upper end of the color spectrum, the lower colors being too attenuated to make it down.

A few aspects of a nonlinear scattering theory will be given in Problem 2.9.8, together with a number of interesting examples.

In Section 2.7 the separation of variables and Fourier transform methods were used for exterior problems. The first problem below illustrates the Green's function method for such problems. Variational methods can also be employed in some cases but we do not discuss them here.

The second and third problems below introduce the student to the wave operators $W_\pm$ and to the scattering operator S.

Problem 1. (a) Show for $-\Delta u = f$ in $L^2(R^3)$ the solution is given by

$$u(P) = \iiint\limits_{R^3} G_0(P, Q)f(Q)\, dV_Q,$$

where $G_0(P, Q)$ is the Green's function

$$G_0(P, Q) = \frac{1}{4\pi|P - Q|}.$$

(b) Show that for any complex λ such that λ is not in the real spectrum $\sigma(-\Delta) = [0, \infty)$, the solution u of the inhomogeneous Helmholtz equation

$$(-\Delta - \lambda)u = f$$

is given by the Green's function representation for the resolvent operator $(-\Delta - \lambda)^{-1}$ as

$$u(P) = (-\Delta - \lambda)^{-1}f = \iiint\limits_{R^3} G_\lambda(P, Q)f(Q)\, dV_Q,$$

where

$$G_\lambda(P, Q) = \frac{e^{i\sqrt{\lambda}|P-Q|}}{4\pi|P - Q|}.$$

Red   Yellow   Blue Violet   Increasing
     Orange Green   frequency

Fig. 2.8*e*

(c) Investigate the validity of the limit of the above as $\lambda = x + iy$ tends to the nonnegative real axis, for example, as $x \geq 0$ is fixed and as $0 < y \to 0$.

Problem 2. Two fundamental entities in scattering theory are the wave operators

$$W_{\pm}\varphi = \lim_{t \to \pm\infty} e^{it(H_0 + V)} e^{-itH_0} \varphi$$

and the scattering operator

$$S = W_+^* W_-.$$

These operators provide a base for the mathematical study of waves propagating according to an "interaction Hamiltonian" $H_0 + V$ as compared to waves propagating "freely" under a "bare Hamiltonian" H_0. We have here used the formalism of quantum mechanics in which an interaction is governed by the potential V, but similar formulations of $W_{\pm}$ and S hold for acoustic scattering, by replacing V with the effect of an obstacle, and in other scattering theories. For the present situation the reader may think in terms of Figure 2.8b, imagining a "free trajectory" $e^{itH_0}\varphi$ as a straight line and an "interaction trajectory" as a straight line coming into the target region, there being bent by the interaction, then coming out eventually as a straight line in a changed direction.

There are a number of interesting and profound mathematical considerations involved in defining and investigating the above quantities. However, working formally,† one may see here some of the basic ideas.

Let us suppose that the interaction V commutes with H_0, so that

$$W_+\varphi = \lim_{t \to \infty} e^{itV}\varphi.$$

Let us consider a standing plane state $\varphi = e^{ikx}$, k a fixed frequency. Then formally

$$W_+\varphi = \lim_{t \to \infty} e^{i(kx + Vt)}.$$

Thus V sets up a time-dependent oscillation. If $V = V(x) \to 0$ for large x (far away from the target region), then the phase $kx + Vt$ eventually settles down to the original time-independent phase kx and $W_+\varphi$ reapproaches φ asymptotically. If $V = k_0 x/t$, the basic frequency of φ is changed so that $W_+\varphi = e^{i(k + k_0)x}$. (a) Find formally $S\varphi$ for the $\varphi = e^{ikx}$ and $V(x) \to 0$ discussed above. (b) Repeat (a) for $V(x, t) = k_0 x/t$.

Problem 3. To see how cavalier we have been in the above, go back to linear algebra and let $H_0 = A$ and $H_0 + V = B$, where A and B are two $n \times n$

---

† That is, treating all objects as scalars.

matrices on a finite-dimensional Hilbert space. (a) Define e^A. (b) Investigate $W = e^{A+B}e^{-A}$. (c) Consider also $S = (e^{-A})*(e^{A+B})*(e^{-(A+B)})(e^A)$.

2.9 SUPPLEMENTARY DISCUSSIONS AND PROBLEMS

As in Section 1.9, this section contains problem and reading material supplementing each of the eight previous sections 2.1 through 2.8, and then there is a final Problem 2.9.9 containing three "confirmation" exercises and further exercises and problems.

In Problem 2.9.1 a brief exposition is given to hint at the connections between separation of variables, tensor products, and group representations. Problem 2.9.2 relates the convergence of our solutions found as limits of Fourier partial sums to the notion of a closed operator. The related and fundamental facts that the Hilbert spaces $L^2(a, b, r)$ are all separable and that the test functions C_0^∞ are always dense are proved in Problem 2.9.3. The Sturm–Liouville theory of Section 2.4 is augmented in Problem 2.9.4 by the inclusion of a short discussion of the limit-point and limit-circle cases for use on singular problems. The notion of Green's functions as kernels of bounded inverses of the given differential operators is elaborated in Problem 2.9.5. Problem 2.9.6 contains additional numerical considerations to supplement Section 2.6, including (1) a connection between finite element convergence and Fourier series, (2) a typical numerical solution of an initial value problem (Section 2.6 did only boundary value problems and eigenvalue problems), and (3) a short treatment of the Buckley–Leverett equation, which occurs importantly and recently in moving interface problems such as those encountered in secondary oil recovery and in other significant applications. The use of the Riemann mapping theorem as a clasical method for solving the Dirichlet problem on both bounded and unbounded domains in two dimensions is explained in Problem 2.9.7(1). The aspects of spectral theory discussed in Section 2.7 are augmented in Problem 2.9.7(2) by a brief look at the resolvent equation. Problem 2.9.7(3) contains Kirchoff's formula and Huygen's principle for the wave equation. Some elements of the recently developing theories of nonlinear waves and solutions are presented in Problem 2.9.8.

Problem 2.9.1   Separation of Variables and Tensor Products

From the strictly practical point of view, one may regard separation of variables as a way of solving a partial differential equation by reduction to ordinary differential equations. The solutions to the latter are then pieced together in a way so as to fit the initial or boundary data. In a historical perspective, this corresponds to the physical context and the notion, as used for example by Bernoulli and DesCartes, that the fundamental modes in the data reproduce themselves in the solutions.

To more fully understand a concept it is often both amusing and enlightening to try to cast it in more than one context. We have seen in.this chapter for example that the separation of variables scheme may be placed in the contexts of Fourier series and Hilbert space, and related to Green's functions, among others. We wish now to make a further conceptual connection not often made in the literature, namely, to try to view the separation of variables solution in the context of tensor products.*

This asserted connection between separation of variables and tensor products may seem a bit esoteric at first though and indeed for most applications it would be an indulgence rather than a necessity. On the other hand it provides a fundamental conceptual link between the method of separation of variables and the more advanced theory of group representations (useful in theoretical physics for example) and may be taken as a motivation of the latter by the former, or of the former by the latter, depending on your disposition.† We wish here only to indicate this connection, leaving its further processing to the interested student.

We have already become familiar (e.g., Sections 1.6.2 and 2.3) with the notion of the inner product (u, v) of two vectors or functions u and v. For example, if $u = (a_1, a_2, a_3)$ and $v = (b_1, b_2, b_3)$ are each vectors in 3-space E^3, then one may find the inner product calculation symbolically displayed in matrix notation as

$$(u, v) = \begin{pmatrix} a_1 \\ a_2 \\ a_3 \end{pmatrix} (b_1, b_2, b_3) = a_1b_1 + a_2b_2 + a_3b_3.$$

The tensor product ‡ of u and v consists of all multiples a_ib_j and may symbolically be displayed in matrix notation as

$$u \otimes v = \begin{pmatrix} a_1b_1 & a_1b_2 & a_1b_3 \\ a_2b_1 & a_2b_2 & a_2b_3 \\ a_3b_1 & a_3b_2 & a_3b_3 \end{pmatrix}$$

The tensor product should not be confused with the outer product§ $u \times v$, which is vector valued. Recall that the inner product is scalar valued and the tensor product as above is matrix valued.

Actually the tensoring of two vectors is more complicated and one must be a bit more careful. First one forms the tensor product $U \otimes V$ of the two vector spaces in question (in the case above, $U = V = E^3$) by taking all pairs of elements of U and V as an algebraic basis, forming their linear combinations, then factoring that vector space into equivalence classes so that the tensor product operation $\otimes$ possesses the bilinearity property

$$\sum_{i=1}^{n} c_iu_i \otimes \sum_{j=1}^{m} d_jv_j = \sum_{i=1}^{n} \sum_{j=1}^{m} c_id_ju_i \otimes v_j.$$

---

* One may wish to have a book on tensor ¡products at hand in working through this problem.

† See K. Gustafson, review of the book *Unitary group representations in physics, probability, and number theory* by G. Mackey, *Bull. Amer. Math. Soc.* 2 (1980).

‡ Another name used: direct product.

§ Other names for the outer product are: cross product, vector product.

There is also another (less direct) prescription for $U \otimes V$, using linear functionals.

Without going further into the algebraic details let us continue, and in particular let us note two further examples of tensor products.

If U is the vector space of polynomials $X(x)$ of degree n or less and V is the vector space of polynomials $Y(y)$ of degree m or less, then $U \otimes V$ may be identified* with a vector space of polynomials $Z(x, y)$ in the two variables x and y so that the "simple tensor" $X(x) \otimes Y(y)$ of two vectors $X(x)$ and $Y(y)$ may be identified with the polynomial product $X(x)(Y(y)$.

If $U = L^2(a, b)$ and $V = L^2(c, d)$, then $U \otimes V$ may be identified with $L^2((a, b) \times (c, d))$.

Let us now return to partial differential equations and in particular to the Dirichlet problem on a square Ω,

$$\begin{cases} \Delta u = 0 \text{ in } \Omega, \\ u = f \text{ on } \partial\Omega, \end{cases}$$

where $f = 0$ on the three sides other than the base $y = 0, 0 < x < \pi$. The separation of variables solution was

$$u(x, y) = \sum_{n=1}^{\infty} c_n X_n(x) Y_n(y)$$

in terms of known functions $X_n(x)$ and $Y_n(y)$ that solve associated ordinary differential equations. We may thus regard the solution as a "simple tensor" in $L^2((0, \pi) \times (0, \pi))$.

> **Problem.** Do one of the following, if it interests you. (a) Do some reading and then elaborate further on the above connection. (b) Look at the way one may tensor product two operators on given vector spaces. Then demonstrate for example how $\partial^2/\partial x\, \partial y$ may be regarded as the tensor product of the operators $\partial/\partial x$ and $\partial/\partial y$. (c) Experienced mathematicians often believe in the "meta-rule" that "when things don't commute the right way," just "tensor the heck out of them."† The corollaries of this are that the resulting tensor products can be hard to work with and that it may be hard to put things back together again. Think about this, especially as concerns separation of variables, and comment upon it.

Problem 2.9.2   Fourier Series Convergence and Closed Operators

In showing that a Fourier series partial sum converges to the solution of a differential equation, one is in fact working with the concept of a closed operator. Most

---

* In mathematical precision, "may be identified" means "is isometrically isomorphic"

† We have already accepted this "meta-rule" for the case of partial differential equations, as evidenced by our growing faith in the separation of variables method. This amounts to a procedure of, when in doubt, just "eigenfunction-expand the heck out of them".

differential operators L are closed operators when placed in any correct setting. Very often (see 2.9.5) their inverses L^{-1} are continuous operators.

The concept of closed operator is weaker than that of continuous operator, so let us recall the latter first. A linear operator A from one normed linear space X (see Section 2.3) to another normed linear space Y is called *continuous* if $u_n \to u$ implies $Au_n \to Au$. This is no different than the original concept of a continuous function as encountered in a first calculus course. All matrix operators on finite dimensional spaces have this property, as well as many integral operators.

Differentiation tends to weaken convergence, and for that reason differential operators usually fail to be continuous in their natural settings.* A linear operator A from a domain $D(A)$ in one normed linear space X to another normed linear space Y is called *closed* if $u_n \in D(A)$, $u_n \to u$, and $Au_n \to y$ imply that $u \in D(A)$ and $Au_n \to Au$.

Problem. (a) Show that $L = d/dx$ is a closed operator from $X = Y = C^0[0, 1]$, the space of continuous functions on the interval $0 \leq x \leq 1$, if one takes $D(L) = C^1[0, 1]$ and uses convergence in the maximum norm $\|u\| = \max_{0 \leq x \leq 1}|u(x)|$, the so-called uniform convergence norm.

The above problem is solved by reference to the fact (see any advanced calculus book) that uniform convergence of $u'_n(x)$ to anything and uniform convergence of $u_n(x)$ to $u(x)$ imply that $\lim_{n\to\infty}u'_n(x) = u'(x)$. We used this property extensively in Section 2.2 in the proofs of the validity of the separation of variables solutions.

In the same way $\partial/\partial x$ and $\partial^2/\partial x^2$ turn out to be closed operators in the right settings. In the $L^2(0, 1)$ norm to get d/dx and d^2/dx^2 closed one needs to use domains $H^1(0, 1)$ and $H^2(0, 1)$, the Sobolev spaces of weak derivatives (see Section 2.6 for more information on these spaces). Even though the continuous function space $C^n(\Omega)$ norms as in the problem above are easier to work with initially, the $L^2(\Omega)$ and Sobolev space $H^n(\Omega)$ norms are often more appropriate physically.

Problem. (b) Given that the Laplacian Δ is a closed operator in $X = Y = L^2(R^3)$, as in Section 1.7, Problem (3), show that the full hydrogen operator given there is also a closed operator.

In Section 2.2, in the proofs that $\Delta(\lim s_N) = 0$ the argument was really an aspect of the fact that Δ was a closed operator. That is, let $s_N \to u$ and note that $\Delta s_N = 0$ for each N. Then if Δ is known to be closed, one has immediately that $\Delta u = 0$.

Problem. (c) Elaborate on the paragraph immediately above, in terms of both the maximum norm and the L^2 norm.

---

* Differential operators can be forced to be continuous operators. The doing of it underlies much of the modern theory of distributions and topological vector spaces. A certain naturalness is however lost.

Problem 2.9.3 Separability and Test Functions

In this problem we establish the related and fundamental facts that:

(i) $L^2(a, b, r)$ is always a *separable* Hilbert space;

(ii) the *test functions* $C_0^\infty(a, b)$ are always *dense* therein.

For (i) we also clarify, in a preliminary lemma, the equivalence of the notions of:

(iii) (1) *separability*;
      (2) countable *denseness*;
      (3) countable *completeness*;
      (4) countable *maximalness*.

The latter brings in, in a natural way, the

(iv) *Gram–Schmidt* orthogonalization procedure.

The proof of (ii) introduces the reader to the key technique of the

(v) *mollification*.

of a given function into a C_0^∞ function

This section can thus be seen to contain the principal key ingredients for a further study of the abstract theory of ordinary and partial differential operators in function spaces.

A Hilbert space H is said to be separable if there exists a dense countable set $\{\psi_n\}$. By this we mean that for every h in H and for every $\varepsilon > 0$ there is a ψ_k such that

$$\|h - \psi_k\| < \varepsilon.$$

For a full understanding of this concept let us establish the following lemma.

Lemma (Separability)

The following are equivalent for a Hilbert space H:

(1) H is separable, that is, there exists a dense countable $\{\psi_n\}$ as stated above.

(2) There exists a "complete" countable sequence $\{\eta_n\}$: $\overline{sp\{\eta_n\}} = H$.*

(3) There exists a "maximal" countable orthonormal sequence $\{\varphi_n\}$ in the sense of Theorem H of Section 2.3.

---

* Recall $\overline{sp\{\eta_n\}}$ means the closure of the span of the $\{\eta_n\}$. The span of the $\{\eta_n\}$ consists of all *finite* linear combinations $\Sigma_{i=1}^n a_i \eta_i$ of the $\{\eta_n\}$, said span then required to be dense in H under condition (2) of the lemma.

Proof

(1) $\Rightarrow$ (2). Let H be separable. Then $\{\psi_n\}$ is a "complete" sequence in the sense (2), even without taking the linear combinations. That (2) $\Rightarrow$ (3) may be seen with the aid of the Gram–Schmidt orthogonalization procedure as follows. Let $\{\eta_n\}$ be independent and complete. Let

$$\varphi_n = \sum_{i=1}^{n} a_{ni}\eta_i, \qquad n = 1, \ldots,$$

where by prescription

$$\langle \varphi_n, \eta_i \rangle = 0 \qquad \text{for } i = 1, \ldots, n-1,$$

$$\|\varphi_n\| = 1, \qquad a_{nn} > 0.$$

Solving this system for η_i yields

$$\eta_n = \sum_{i=1}^{n} b_{ni}\varphi_i, \qquad n = 1, \ldots,$$

and thus $sp\{\varphi_n\} = sp\{\eta_n\}$. The implication (3) $\Rightarrow$ (1) follows by the countability and denseness of the span $sp_{\text{rat.}}\{\varphi_n\}$ of linear combinations $\Sigma_{i=1}^{n} a_i\varphi_i$ of the $\{\varphi_n\}$ taken over rational coefficients $\{a_i\}$ only.

Problem. (a) Write out the above proof in more detail.

In Section 2.4 the Hilbert spaces $L^2(a, b, r)$ play a fundamental role. $L^2(a, b, r)$ consists of those functions defined on any interval (a, b) and square integrable there with respect to the weight function $r(x)$, that is, $\int_a^b |u(x)|^2 r(x)\,dx < \infty$. The weight function is to be strictly positive and locally integrable in (a, b).

Proposition. (**Separability of** $L^2(a, b, r)$). Let $a < c < b$, and let (Fig. 2.9a)

$$\eta_t(x) = \begin{cases} \chi_{[c, t]}(x), & c \leq t, \\ -\chi_{[t, c]}(x), & t < c. \end{cases}$$

Let $\{t_i\}$ be an ordering of all rationals in (a, b). Then $\{\eta_{t_i}\}$ is "complete" in $L^2(a, b, r)$ in the sense (2) of the lemma above.

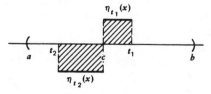

FIG. 2.9a

Proof

Each η_{t_i} is in $L^2(a, b, r)$ because the weight function r is positive and locally integrable, since then

$$\int_a^b (\eta_{t_i})^2 r(t)\, dt \leqq \int_c^t r(t)\, dt < \infty.$$

It is sufficient to show that $(f, \eta_{t_i}) = 0$ for all i implies in turn that $f = 0$, as in the proof of Theorem H of Section 2.3. Proceeding, we have:

$(f, \eta_{t_i}) = 0$ for all $i \Rightarrow f \perp sp\{\eta_{t_i}\}$

$\Rightarrow f \perp sp\{\varphi_{t_i}\}$ via Gram–Schmidt (see above)

$\Rightarrow \{f\|f\|^{-1}, \varphi_{t_i}\}$ is an enlargement of $\{\varphi_{t_i}\}$

if $f \neq 0$. Proceeding backwards in the Gram–Schmidt relations from $\{f\|f\|^{-1}, \varphi_{t_i}\}$, we reach a contradiction of $(f, \eta_{t_i}) = 0$ for all η_{t_i}.

Problem. (b) Write out in more detail the argument just given. (c) Validate the details in the following alternate proof. With the situation as above, let $F(t) = (f, \eta_t) = \int_c^t f(s) r(s)\, ds$. F is a continuous function of t in (a, b), $F(t_i) = 0$ on a dense set of $\{t_i\}$ in (a, b), so F is identically zero on (a, b). Therefore the integrand $f(s)r(s) = 0$ almost everywhere by differentiation, and hence $f = 0$ almost everywhere.

The separability of $L^2(a, b, r)$ established above depended on constructing a dense set of step-functions. For those who know the Lebesgue theory of integration, that is an entirely natural way to proceed. For dense sets of functions for use in differential equations one prefers instead very smooth functions that can be differentiated without worry.

The fact, to be proved below, that $C_0^\infty(\Omega)$, the infinitely differentiable functions of compact support, are dense in any $L^p(\Omega)$ space, $1 \leqq p < \infty$, is the basic connection between the classical analysis and the modern functional analysis treatments of differential equations. Roughly stated, it means that with additional work and care the usual formal classical arguments, which just assume as much smoothness as needed, go over at least in a so-called weak form to all functions involved in the problem. That is, one may work with smooth functions first, then try by limiting arguments (using the denseness) to complete the discussion needed for all data involved and for all functions (e.g., only the weakly differentiable ones, or distributions) in the domain of the operator under study.

The proofs that $C_0^\infty(\Omega)$ is dense in $L^p(\Omega)$ and in particular in $L^2(\Omega)$ are usually similar to the proof given below for $L^2(a, b, r)$ in one dimension.*

---

* For a general proof in the general situations, see for example N. Dunford and J. Schwartz, *Linear Operators II* (Wiley, New York, 1963).

Fundamental Approximation Theorem. (Test functions)

$C_0^\infty(\Omega)$ is dense in $L^p(\Omega)$ for any reasonable Ω and any $1 \leqq p < \infty$. In particular, $C_0^\infty(a, b)$ is a dense subspace of $L^2(a, b, r)$.

Proof

As mentioned, we will prove only the latter, for simplicity. The idea here is a key one: that of *mollifying* the η_t of the above proposition into C_0^∞ functions while at the same time retaining their denseness. Mollification is a fundamental tool in the theory of partial differential equations.*

Let (Fig. 2.9b)

$$\delta_\varepsilon(x) = \begin{cases} c_\varepsilon e^{-1/(\varepsilon^2 - x^2)}, & |x| < \varepsilon, \\ 0, & |x| > \varepsilon, \end{cases}$$

where

$$c_\varepsilon = \frac{1}{\int_{-\varepsilon}^{\varepsilon} e^{-1/(\varepsilon^2 - x^2)}\, dx}.$$

Then $\delta_\varepsilon(x)$ is a C_0^∞ function, $\delta_\varepsilon(x) > 0$, and $\int_{-\infty}^\infty \delta_\varepsilon(x)\, dx = 1$. For any y in (a, b), $\delta_\varepsilon(x - y)$ is in $C_0^\infty(a, b)$ if ε is small, that is, if $\varepsilon < d(y, \partial(a, b))$. Recall that for c in (a, b), t in (a, b), we had η_t in $L^2(a, b, r)$ in the proposition above. Let

$$\eta_{t,\varepsilon}(x) = \int_a^b \eta_t(y)\delta_\varepsilon(x - y)\, dy.$$

This is the *mollified* (*mollification of*, if you wish) η_t. It may be verified that $\eta_{t,\varepsilon}(x)$, as a C_0^∞ smoothing of the step function η_t, is in $C_0^\infty(a, b)$ for ε small, that is, for $\varepsilon < \min[b - c, c - a, b - t, t - a]$. Since $\int_{-\infty}^\infty \delta_\varepsilon(x)\, dx = 1$,

$$\eta_{t,\varepsilon}(x) - \eta_t(x) = \int_a^b (\eta_t(y) - \eta_t(x))\delta_\varepsilon(x - y)\, dy$$

and

$$|\eta_{t,\varepsilon}(x) - \eta_t(x)| \leqq \max_{y \in [x-\varepsilon,\, x+\varepsilon]} |\eta_t(y) - \eta_t(x)|$$

$$\leqq \begin{cases} 2, & \text{for all } x \text{ in } (a, b), \\ 0, & \text{for all } x \text{ such that } |x - c| > \varepsilon \\ & \text{and } |x - t| > \varepsilon. \end{cases}$$

Thus $|\eta_{t,\varepsilon}(x) - \eta_t(x)| \to 0$ for almost all x as $\varepsilon \to 0$.

This fact, that the mollification $\eta_{t,\varepsilon}(x)$ converges pointwise almost everywhere to the original $\eta_t(x)$, is interesting in itself.

---

* This is an additional reason for exposing it here.

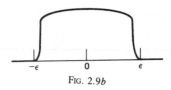

FIG. 2.9b

To complete the proof of the theorem, we now need to show in addition that $\eta_{t,\varepsilon}(x)$ converges to $\eta_t(x)$ in the $L^2(a, b, r)$ norm. This follows immediately now by recourse to the Lebesgue dominated convergence theorem (see Section 1.6.3), which asserts that for the case that $|\eta_{t,\varepsilon_n}(x) - \eta_t(x)|^2$ converges boundedly to zero almost everywhere as $\varepsilon_n \to 0$, then also

$$\|\eta_{t,\varepsilon_n} - \eta_t\|^2_{L^2(a,b,r)} = \int_a^b |\eta_{t,\varepsilon}(x) - \eta_t(x)|^2 r(x)\, dx \to 0.$$

In summary: Any function f in $L^2(a, b, r)$ can be approximated arbitrarily closely by a finite linear combination of the η_{t_i}, which can in turn be approximated arbitrarily closely by their C_0^∞ mollifications.

Problem. (d) Prove that $\delta_\varepsilon(x)$ is indeed C_0^∞. (e) For more exercise, prove that $\eta_{t,\varepsilon}(x)$ is also C_0^∞.

Problem 2.9.4   Limit-Point and Limit-Circle

No discussion of Sturm–Liouville theory would be complete without at least brief mention of the limit-point and limit-circle cases for singular problems.*

Recall that the Sturm–Liouville theorem given in Section 2.4 was stated for the regular case: The eigenfunctions $\{\varphi_n\}$ of

$$-(pu')' + qu = \lambda r u, \qquad a < x < b, \tag{1}$$

with a self-adjoint boundary condition provide a Fourier expansion for all functions f in $L^2(a, b, r)$. On the other hand, we saw in Problem 2 of Section 2.4 that familiar equations such as the Hermite and Legendre equations are not regular, and yet they do have maximal eigenfunction sets.

Herman Weyl† analyzed this situation in the complex domain, that is, for λ in the complex plane, and found that there are two cases, which are in fact independent of λ. Let λ be nonreal, and let c be chosen in (a, b) so that we may look at the singular behavior first near the endpoint b, (then similarly near a).

---

* For more information see for example E. Titchmarsh, *Eigenfunction Expansions Associated With Second-Order Differential Equations* I, 2nd ed. (Oxford Press, Oxford, 1962), and E. Coddington and N. Levinson, *Theory of Ordinary Differential Equations* (McGraw-Hill, New York, 1955).

† H. Weyl, *Math. Annalen* **68** (1910).

Case 1. *Limit-circle case.* Every solution of (1) is square integrable, that is, is in $L^2(c, b, r)$.

Case 2. *Limit-point case.* There is one fundamental solution of (1) that is square integrable, that is, which is in $L^2(c, b, r)$, and there is another fundamental solution that is not square integrable.

Recall that under very general conditions one knows what an ordinary differential equation such as (1) possesses two linearly independent (thereafter called fundamental) solutions (not at all unique) and that all solutions are then linear combinations of any two fundamental solutions. The two cases in Weyl's result are exhaustive since it can be shown that there always is at least one square-integrable solution. The terminology "limit-circle" and "limit-point" come from the proof, so let us look at that now. However, let us first state somewhat more precisely the facts described above for the equation (1), regular or singular, $a < c < b$. Thinking in terms of pairs of fundamental solutions, by "the other" solution we mean any other linearly independent solution, similarly by "both" we mean all solutions.

Theorem (Weyl)

(i) For any λ nonreal there always exists one solution in $L^2(c, b, r)$.

(ii) If for any λ_0, real or nonreal, both solutions are in $L^2(c, b, r)$, then both solutions are in $L^2(c, b, r)$ for all λ. In this case the Sturm–Liouville equation (1) is said to be in the limit-circle case at b.

(iii) If for some λ_0, real or nonreal, one solution is not in $L^2(c, b, r)$, the Sturm–Liouville equation (1) is said to be in the limit-point case at b. Subcases are: (a) λ_0 nonreal; then by (i) the other solution is in $L^2(c, b, r)$. (b) λ_0 real; then the other solution may or may not be in $L^2(c, b, r)$.

Although we do not give here a proof of the theorem in its entirety (its proof may be found in most books on the theory of ordinary differential equations), let us see how the limit-circle and limit-point terminology comes out of the proof. In so doing we will in fact see all the essentials of the proof. We begin as usual with a key fact using the Wronskian.

1. For λ nonreal a solution u of (1) is in $L^2(c, b, r)$ iff $\lim_{x \to b^-} W(u, \bar{u})(x)$ is finite. This follows immediately from the integration by parts

$$(\lambda - \bar{\lambda}) \int_c^x |u|^2 r \, ds = W(u, \bar{u})(x) - W(u, \bar{u})(c).$$

2. Define two fundamental solutions $\varphi_1(x)$ and $\varphi_2(x)$ of (1) by: $\varphi_1(c) = 1$, $p(c)\varphi_1'(c) = 0$, and $\varphi_2(c) = 0$, $p(c)\varphi_2'(c) = 1$. Note that from 1 above, for any solution u the quantity $(\lambda - \bar{\lambda})^{-1} W(u, \bar{u})(x)$ is an increasing function of x. In particular, consider any $u = z\varphi_1 + \varphi_2$, z an arbitrary complex number. Then a straightforward calculation shows

$$(\lambda - \bar{\lambda})^{-1} W(u, \bar{u})(x) = (\lambda - \bar{\lambda})^{-1} W(\varphi_1, \overline{\varphi}_1)(x)[|z - z_0(x)|^2 - r_0^2(x)],$$

where

$$z_0(x) = -\frac{W(\varphi_2, \overline{\varphi}_1)(x)}{W(\varphi_1, \overline{\varphi}_1)(x)}$$

and where

$$r_0^2(x) = \frac{|W(\varphi_2, \overline{\varphi}_1)(x)|^2 + W(\varphi_2, \overline{\varphi}_2)W(\varphi_1, \overline{\varphi}_1)}{-W(\varphi_1, \overline{\varphi}_1)^2} = \frac{1}{|W(\varphi_1, \overline{\varphi}_1)(x)|^2}.$$

3. Let $z_0(x)$ be the center and $r_0(x)$ the radius of a circle $C(x)$ in the complex plane (Fig. 2.9c). From 2 above, the increasing function of x is seen to be negative for z inside the circle, from which we may conclude that $C(x_2)$ is contained inside $C(x_1)$ for $x_2 > x_1$. Thus these circles nest inward as x increases.

4. Now as $x \to b$, either (a) the circles contract to a circle $C(b)$ with center $z_0(b)$ and radius $r_0(b)$, or (b) the contraction goes all the way down to the point $z_0(b)$. In case (a) one is in the "limit-circle" case and from 1 and 2 above, all quantities, and in particular, $(\lambda - \bar{\lambda}) \int_c^b |u|^2 r\, ds$ for z inside the limit-circle $C(b)$, are finite. In case (b) one is in the "limit-point" case, and from 1 and 2 above one has

$$|\lambda - \bar{\lambda}| \int_c^x |\varphi_1|^2 r\, ds = |W(\varphi_1, \overline{\varphi}_1)(x)| = r_0^{-1}(x) \to \infty.$$

Problem. (a) Complete the details in the proof sketched above. (b) Show that the Legendre equation is in the limit-circle case and the Hermite equation is in the limit-point case. (c) Prove that the regular case of Section 2.4 is limit-circle.

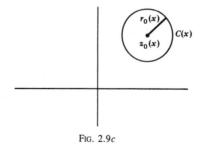

Fig. 2.9c

Problem 2.9.5   Green's Functions and Continuous Operators

In Problem 2.9.2 we discussed briefly the notion of closed operators. Recall an operator L is said to be closed if $u_n \to u$ and $Lu_n \to f$ imply that u is in the domain $D(L)$ and that $Lu = f$. For the Dirichlet problem this concept was illustrated as follows. The partial sums s_N of the separation of variables solution-candidate u converging to u, $s_N \to u$, and the fact that $\Delta s_N = 0$ for all s_N, yield that u is in the domain $D(\Delta)$ and that $\Delta u = 0$, provided that it has been shown that Δ acts as a closed operator in that situation. The latter can be shown explicitly, although we do not do so here.*

Many differential operators L have continuous inverses L^{-1}, and these inverses are often represented as integral operators with Green's functions as kernels. We have already seen (Section 1.6.1, Remark 2) how this can come about by deducing Green's III identity for the Dirichlet–Poisson problem

$$\begin{cases} -\Delta u = F \ \text{ in } \Omega, \\ u = 0 \ \text{ on } \partial\Omega, \end{cases}$$

namely, that the solution is given by

$$u(P) = \int_\Omega G(P, Q)F(Q) \, dV_Q.$$

As shown there, letting L denote the negative Laplacian $-\Delta$ acting on functions that are zero on $\partial\Omega$, L^{-1} was given by the integral operator $\int_\Omega G(P, Q)(\cdot)\,dV_Q$.

Let us show now that this L^{-1} is indeed a continuous operator. Recall† that an operator A is *continuous* if $v_n \to v$ implies $Av_n \to Av$. For the case at hand this would mean that as Poisson data $F_n \xrightarrow{L^2(\Omega)} F$,‡ we would have the resulting solutions u_n of the Dirichlet–Poisson problem also converging, that is,

$$u_n = L^{-1}F_n \xrightarrow{L^2(\Omega)} L^{-1}F = u.$$

This can be shown for example by working directly with the Green's formula and showing that as F_N converges in the $L^2(\Omega)$ norm to F, then $\int G(P, Q)F_N(Q)\,dV_Q$ converges in the $L^2(\Omega)$ norm to $\int G(P, Q)F(Q)\,dV_Q$. The argument goes as follows and depends essentially on the fact that $G(P, Q)$ is jointly square-integrable. By Schwartz's inequality, we have

---

* For a concise proof using the Lax–Milgram theorem see Yosida, *Functional Analysis* (Springer, Berlin, 1968). An important early paper was that of Garding, *Math. Scand.* **1** (1953). For an extensive general treatment see F. Browder, *Math. Ann.* **142** (1961). A recent survey and L^p theory is Simader, "On Dirichlet's Boundary Value Problem," *Springer Lec. Notes* **268** (1972).

† See Problem 2.9.2. See also Section 2.6 for related considerations.

‡ To speak of continuity one needs to have chosen a metric. Although such a choice is usually not unique, it is natural to use the Hilbert space norm $L^2(\Omega)$. In this norm L^{-1} is, as shown, continuous.

$$\|L^{-1}F_n - L^{-1}F\|^2_{L^2(\Omega)} = \int_\Omega \left(\int_\Omega G(P, Q)F_N(Q)\,dV_Q \right.$$

$$\left. - \int_\Omega G(P, Q)F(Q)\,dV_Q \right)^2 dV_P$$

$$\leq \int_\Omega \left(\int_\Omega G^2(P, Q)\,dV_Q \right) \left(\int_\Omega |F_N(Q) - F(Q)|^2\,dV_Q \right) dV_P$$

$$= \|G\|^2_{L^2(\Omega \times \Omega)} \cdot \|F_N - F\|^2_{L^2(\Omega)} \to 0.$$

Problem. (a) In Hilbert space or Banach space[†] an operator A is continuous if and only if it is *bounded*: There exists a number M such that

$$\|Au\| \leqq M\|u\|$$

for all u in the space. The smallest such number M is called the norm $\|A\|$ of the operator A. Show for A on a Hilbert space H that A is continuous if and only if A is bounded.

(b) An operator A on a Hilbert space is said to be strongly positive[‡] if there exists a positive number m such that

$$(Au, u) \geqq m\|u\|^2$$

for all u in the space. Show that then A^{-1} is continuous and that $\|A^{-1}\| \leqq m^{-1}$.

(c) In the above discussion we showed directly that L^{-1} for the Dirichlet–Poisson problem was continuous. To do so we assumed without proof that the Green's function $G(P, Q)$ was jointly square integrable (see Problem 2, Section 2.5). Show L^{-1} continuous for that problem by the method of part (b). (*Hint*: Recall the variational characterization of the eigenvalues of L.)

(d) A strongly positive self-adjoint operator maps onto the whole space. Try to give a rough proof of this important fact.

(e) A more general theorem of Banach asserts that $(L^*)^{-1}$ bounded implies that L is onto, for all closed operators L. Relate this fact[§] to the comments about a priori estimates made in the footnote of Section 1.6(2)1″.

Problem 2.9.6   Additional Numerical Considerations

The field of numerical solution of partial differential equations is a relatively new and rapidly growing one, and far from complete. Here we supplement Section 2.6

[†] In more general topological vector spaces the boundedness of A and the continuity of A need not be equivalent concepts.

[‡] Such A are called in some contexts "coercive." Perhaps a better terminology would be "uniformly positive."

[§] More information may be found in K. Gustafson, *J. Math. Mech.* (now *Indiana Univ. Math. J.*) **18** (1968), and K. Gustafson, "Operator Spectral States," *Computers Math. Applic.* **34** (1997).

with (1) an illustration of finite element convergence, (2) an example of the iterative marching methods for initial value problems, and (3) a treatment of the Buckley–Leveritt equation as it occurs in oil recovery problems. In so doing, we have barely scratched the surface of these subjects: see Chapter 3 Appendix B for further numerical considerations.

(1) Finite Element Convergence. One can sometimes establish convergence rates for finite element approximations by Fourier expansions. We give here an illustrative example of this type of connection between variational methods and Fourier methods.

Consider the Poisson problem

$$\begin{cases} -u''(x) = f(x), & 0 < x < \pi, \\ u(0) = u(\pi) = 0, \end{cases}$$

to be numericaly approximated (Fig. 2.9d) by the finite element method using piecewise linear continuous functions as the approximating functions (usually called the elements). A basis (mutually orthogonal except for adjacent ones) are the so-called roof functions (Fig. 2.9e) so that the approximating piecewise linear function u_N over a grid of N subintervals can always be written

$$u_N(x) = \sum_{n=1}^{N} u(nh)\varphi_n(x).$$

Note that at each "node" we have $u_N(nh) = u(nh)$, thus tacitly assuming that we "know" the unknown solution u; but close estimates can be obtained by finite differences and in fact what we wish to show is the convergence of the finite element approximations to any given function u, whether u originates from a differential equation or not.

Problem. (a) Derive the Poincaré inequality

$$\int_0^h v^2(x)\, dx \leqq h^2 D_h(v), \qquad D_h(v) = \int_0^h (v'(x))^2\, dx$$

for any function (C^1) on an interval of length h and vanishing at the end-points.

(b) Letting $v(x) = u(x) - u_N(x)$ on any grid interval $(n-1)h \leqq x \leqq nh$, show by Fourier expansion,

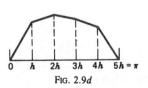

FIG. 2.9d

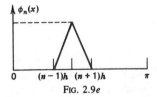

FIG. 2.9e

$$v = \sum_{m=1}^{\infty} c_m \sin \frac{m\pi}{h} x, \qquad v' = \sum_{m=1}^{\infty} c_m \frac{m\pi}{h} \cos \frac{m\pi}{h} x,$$

$$v'' = -\sum_{m=1}^{\infty} c_m \frac{m^2 \pi^2}{h^2} \sin \frac{m\pi}{h} x,$$

that for u a C^2 function one has the estimates

$$\int_{(n-1)h}^{nh} (v'(x))^2 \, dx = \frac{h}{2} \sum_{m=1}^{\infty} c_m^2 \left(\frac{m\pi}{h}\right)^2,$$

$$\int_{(n-1)h}^{nh} (v''(x))^2 \, dx = \frac{h}{2} \sum_{m=1}^{\infty} c_m^2 \left(\frac{m\pi}{h}\right)^4$$

and hence

$$D_h^n(v) \equiv \int_{(n-1)h}^{nh} (v'(x))^2 \, dx \le \frac{h^2}{\pi^2} \int_{(n-1)h}^{nh} (v''(x))^2 \, dx.$$

Upon adding and noting that $v''(x) = u''(x)$ in the intervals one thus has

$$D(u - u_N) \le \frac{h^2}{\pi^2} \int_0^{\pi} (f''(x))^2 \, dx.$$

Thus the finite element approximations converge in the Dirichlet energy norm at the rate h^2 in terms of mesh size.

(2) Marching Methods for Initial Value Problems. As an additional exercise on numerical methods, consider the initial boundary value problem

$$\begin{cases} u_t - u_{xx} = 0, & 0 \le x \le 1, \quad t \ge 0, \\ u(x, 0) = f(x), & 0 \le x \le 1, \\ u(0, t) = u(1, t) = 0. \end{cases}$$

Let x_i, $i = 0, \ldots, N$ be a regular partition, increment length h, of the interval $[0, 1]$, and approximate by finite differences as in Section 2.6, that is,

$$\begin{cases} \dfrac{d}{dt} u(x_i, t) = \dfrac{u(x_{i+1}, t) - 2u(x_i, t) + u(x_{i-1}, t)}{h^2}, & 1 \le i \le N - 1, \\ u(x_i, 0) = f(x_i), & 1 \le i \le N - 1, \\ u(x_0, t) = u(x_N, t) = 0, & t \ge 0, \end{cases}$$

Problem. (a) Solve this system by the methods of ordinary differential equations. (b) Discretize also in time by letting t_j, $j = 0, \ldots, M$, be a regular partition, increment length k, of $[0, T]$, arriving at the system

$$\begin{cases} u(x_i, t_{j+1}) = u(x_i, t_j) + \dfrac{k}{h^2}(u(x_{i+1}, t_j) - 2u(x_i, t_j) + u(x_{i-1}, t)), \\ \qquad\qquad\qquad\qquad\qquad\qquad\qquad\qquad 1 \leqq i \leqq N - 1, \\ u(x_i, 0) = f(x_i), \qquad 1 \leqq i \leqq N - 1, \\ u(x_0, t_j) = u(x_N, t_j) = 0, \qquad 0 \leqq j \leqq M. \end{cases}$$

(c) Assigning the initial value $f(x) = x(1 - x)$ and mesh sizes $N = M = 2$, 5, or 10, begin at $t = 0$ and iterate (march) upward one step at a time in t.

(d) Read about ADI methods for such problems in higher space dimensions.*

(3) The Buckley–Leverett Equation in Secondary Oil Recovery. An important but lesser-known equation occurs in secondary oil recovery problems, in certain military applications in which a light liquid is forced into a heavier liquid, and as a general equation in the theories of the mixing of immiscible fluids.† Numerical methods have been rather important for the study of equations of this type.

Let us briefly describe an interesting situation in which this equation occurs, that of secondary oil recovery.‡ We may imagine a situation as shown in Figure 2.9f, in which (i) depicts a horizontal flow section in a porous media in which no flow takes place in the y direction out of Ω (due to imagined solid rock walls there), water being forced into Ω on the left (from a hole far above), oil being forced by the water to the right (to a recovery hole somewhere). We have ignored gravity and any vertical flow aspects of the problem. The water–oil interface appears as a rightward-moving shock in the water saturation function $s(x, y)$ as shown in (ii).

For plane horizontal flow with an internal interface as indicated in Figure 2.9f, and with the assumptions of no gravity, capillary, or density variations, one arrives at the partial differential equations

$$\begin{cases} (Au_x)_x + (Au_y)_y = 0 \quad \text{in } \Omega, \\ ms_t - Au_x F_x - Au_y F_y = 0 \quad \text{in } \Omega, \end{cases}$$

with boundary conditions

---

* In Section 2.6 and the present 2.9.6 we have only briefly introduced numerical methods. See Appendix B of Chapter 3 for more about numerical techniques for partial differential equations. A good basic reference is for example R. Richtmyer and K. Morton, *Difference Methods for Initial-Value Problems*, 2nd ed. (Wiley, New York, 1967).

† See for example the initial papers and photographs by Taylor and Lewis, *Proc. Royal Soc. London, Ser. A* **201, 202** (1950), respectively, and the recent paper by Pimbley, *J. Math. Anal. Appl.* **55** (1976). The Buckley–Leverett equation itself was formulated earlier in Buckley and Leverett, *Amer. Inst. Mining Eng.* **146** (1942). For an account of such oil recovery problems see the book of Peaceman, *Fundamentals of Reservoir Engineering* (Elsevier, N.Y., 1977). The relations between the Rayleigh–Taylor interface instability and the oil displacement in sand problems seem to be not generally known.

‡ In the common terminology, primary oil recovery is that resulting from natural oil pressures, secondary recovery comes by the forcing of water into the oil field, tertiary recovery by adding chemical sulfactants to the secondary recovery procedure to break down surface tensions and other impediments to the flow.

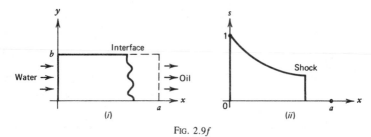

Fig. 2.9f

$$\begin{cases} u_y = 0, & y = 0 \text{ or } y = b, \quad 0 \le x \le a, \\ u = f(y, t), & x = a, \quad\quad 0 \le y \le b, \\ u_x = g(y, t)/A, & x = 0, \quad\quad 0 \le y \le b, \\ s = h(y, t), & x = 0, \quad\quad 0 \le y \le b, \end{cases}$$

and with initial condition

$$\begin{cases} s = s_0(x, y), & t \ge 0, \quad (x, y) \in \Omega, \\ \text{where } \Omega = \{(x, y)| 0 \le x \le a, 0 \le y \le b\}. \end{cases}$$

The first of the coupled partial differential equations is second-order elliptic and the second is first-order quasilinear. These equations are derived from Darcy's law for porous media, which states that filtration rates are proportional to pressure gradients, and from conservation of mass (continuity) equations, in a manner similar to that of Section 1.7.

The principal unknowns are

$$u(x, y, t) = \text{the pressure at } (x, y) \text{ at time } t$$

and

$$s(x, y\ t) = \text{the saturation at } (x, y) \text{ at time } t,$$

where s is taken to be the percent saturation by water. The corresponding oil saturation is thus given by $1 - s$. The other two principal quantities are the permeability coefficient $A(s, x, y)$, and the Buckley–Leverett function $F(s)$, which is a measure of the percent water flow. More specifically,

$$A = k\left(\frac{k_w}{\mu_w} + \frac{k_0}{\mu_0}\right) \text{ and } F = \frac{\mu k_w}{k_0 + \mu k_w},$$

where $k_w(s)$ = water permeability, $k_0(x)$ = oil permeability, μ_w = water viscosity, μ_0 = oil viscosity, $\mu = \mu_0/\mu_w$, all presumed known experimentally, and where $k(s, x, y)$ is the absolute permeability coefficient for the material. The function $m(x, y)$ represents an average (known) porosity, and f, g, h, and s_0 known data.

The boundary conditions correspond to no pressure variation in the (nonmoving) lateral walls, to a known (e.g., atmospheric) pressure on the outlet side, and to controllable pressure rates and saturations on the input side. The initial condition presumes a known (e.g., all oil) relative saturation at the beginning.

Problem. (a) Take A and m constant (and for simplicity equal to unity) and investigate the uniqueness of solutions for this problem. Make some simplifications if you like and consider the uniqueness question only for the first equation (Laplacian) for the portion of the domain Ω to the left of the interface at a fixed time, along with the boundary conditions.

A large amount of numerical work has been done on such equations and on more complicated versions thereof. As a simple illustration, let us ignore the y direction and write down the first-order difference scheme for the second equation,

$$m\frac{s_i^{n+1} - s_i^n}{k} - u_x\frac{F(s_i^n) - F(s_{i-1}^n)}{h} = 0,$$

where k denotes the time step discretization length and h the space step length. The schemes become more complicated when the pressure and second-order equation are also discretized. More efficient (e.g., Lax–Wendroff) schemes have also been employed.*

Problem. (b) Try to write down a discretization of the first (elliptic) equation.

Ignoring the y direction in the first equation, we have $(Au_x)_x = 0$, from which $Au_x = q(t)$. Substituting this back into the second (Buckley–Leverett) equation yields the first-order equation

$$m(x)s_t(x, t) - q(t)F_x(s(x, t)) = 0.$$

Problem. (c) Investigate the analytical solution of such equations by the method of characteristics (see Appendix A).

Problem 2.9.7 Additional Analytical Considerations

(1) Two-Dimensional Domains and the Riemann Mapping Theorem. Historically, the first step from ordinary differential equations to partial differential equations came in going from dimension $n = 1$ and ordinary differential equations to dimension $n = 2$ and partial differential equations. Although many physical problems occur only on domains Ω in dimension $n = 3$, by additional physical as-

---

* For a finite element approach see J. Douglas, T. Dupont, and H. Rachford, *J. Canadian Pet. Tech.* **8** (1969). For a further reference see V. Entov and V. Taranchuk, "Numerical simulation of process of unstable displacement of oil by water," *Iz. Akad. Nauk SSSR, Mech. Zhid. i Gaza*, No. 3, 1979 (translated as *Fl. Dyn.* **14**, 3 (1979)).

sumptions they often can be reduced to two-dimensional or even one-dimensional problems.

For treatment of partial differential equations on two-dimensional domains Ω, there was developed in the latter half of the nineteenth century the powerful analytic function theory (i.e., complex variables). There are too many facets of the interplay between analytic function theory and partial differential equations to treat here. We therefore focus attention on an important and illustrative example, the Riemann mapping theorem and its use in attacking the Dirichlet problem on an arbitrary domain Ω in two dimensions.

Consider such a Dirichlet problem and suppose by a mapping $w = g(z)$, that is, $w(\xi, \eta) = g(x, y)$, one could transform it to the Dirichlet problem for the unit sphere (Fig. 2.9g). In trying such a change of coordinates, in the same way as was done in the classification procedures of Problem 1.9.1 one lets $\xi = \xi(x, y)$, $\eta = \eta(x, y)$ for the mapping g, and for the inverse mapping $f = g^{-1}$, $x = x(\xi, \eta)$, $y = y(\xi, \eta)$. One then plugs these into the Laplacian operator and uses the chain rule of calculus, from which (see the problem at the end of this subsection)

$$\Delta_{\xi,\eta} u(\xi, \eta) = u_{xx}[x_\xi^2 + x_\eta^2] + 2u_{xy}[x_\xi y_\xi + x_\eta y_\eta] + u_{yy}[y_\xi^2 + y_\eta^2]$$
$$+ u_x[x_{\xi\xi} + x_{\eta\eta}] + u_y(y_{\xi\xi} + y_{\eta\eta}].$$

Thus, in the same fashion as in Problem 1.9.1, we may try to find such a 1–1 transformation by setting

(i) $x_\xi^2 + x_\eta^2 = y_\xi^2 + y_\eta^2$,

(ii) $x_\xi y_\xi + x_\eta y_\eta = 0$,

(iii) $x_{\xi\xi} + x_{\eta\eta} = y_{\xi\xi} + y_{\eta\eta} = 0$.

If we can somehow solve the system (i), (ii), (iii) to obtain a 1–1 change of variables, then $\Delta_{\xi,\eta} u(\xi, \eta) = 0$. Of course we do not yet have any idea where Ω will go under such a transformation.

The system (i), (ii), (iii) involves only derivatives of $x(\xi, \eta)$ and $y(\xi, \eta)$. Solving (i) and (ii) we arrive at $x_\xi = \pm y_\eta$ and $x_\eta = \mp y_\xi$, which are then seen to

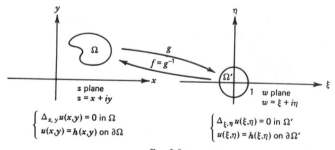

$$\begin{cases} \Delta_{x,y} u(x,y) = 0 \text{ in } \Omega \\ u(x,y) = h(x,y) \text{ on } \partial\Omega \end{cases} \qquad \begin{cases} \Delta_{\xi,\eta} u(\xi,\eta) = 0 \text{ in } \Omega' \\ u(\xi,\eta) = h(\xi,\eta) \text{ on } \partial\Omega' \end{cases}$$

Fig. 2.9g

satisfy (iii). Although either sign choice will work in the following, we choose the first sign and thus a transformation $x = x(\xi, \eta), y = y(\xi, \eta)$ satisfying the equations

$$\frac{\partial x}{\partial \xi} = \frac{\partial y}{\partial \eta}, \qquad \frac{\partial x}{\partial \eta} = -\frac{\partial y}{\partial \xi}.$$

These equations are called the Cauchy–Riemann equations for the transformation $z = f(w)$, that is,

$$x(\xi, \eta) + iy(\xi, \eta) = f(\xi + i\eta).$$

Whenever they are satisfied it can be shown that f is an *analytic function*.

An analytic function possesses a power series representation (as in Problem 1.9.4). However, here we have an analytic function of a single complex variable, and as such it may be shown that f possesses the additional property that its derivative df/dw exists and is independent of direction.* In our application we want f to represent a 1–1 transformation. From the theory in calculus of mappings from two variables to two variables we know that a sufficient condition for a 1–1 mapping is the nonvanishing of the Jacobian of the mapping, which in our case comes to

$$0 \neq J\begin{bmatrix} x & y \\ \xi & \eta \end{bmatrix} = \begin{vmatrix} x_\xi & x_\eta \\ y_\xi & y_\eta \end{vmatrix} = x_\xi^2 + y_\xi^2 = |x_\xi^2 + iy_\xi|^2 = |f'(w)|^2.$$

Thus we wish the additional property beyond $f(w)$ analytic that $f'(w) \neq 0$ on the domain in question, which is in our case the unit sphere Ω_c. Such a mapping $z = f(w)$ is said to be a *conformal mapping*, the name deriving from the additional property that such mappings preserve angles between arcs.

The Riemann mapping theorem† now guarantees such maps. Let Ω be simply connected (no holes) and not the whole complex plane, and let $z_0 = x_0 + iy_0$ be any chosen point in Ω. Then there exists a unique conformal mapping $g(z)$ taking z_0 to the origin 0 in the w plane such that $|g'(z_0)| > 0$ and such that $g(z)$ is 1–1 from Ω to the unit sphere $|w| < 1$. Under additional assumptions on $\partial\Omega$ it can be shown that $g(z)$ has a continuous 1–1 extension mapping $\partial\Omega$ to $|w| = 1$.

Thus for application to the Dirichlet problem on any such (bounded or unbounded) domain Ω we may map $g: \Omega$ in z plane $\to$ unit sphere in w plane and we may take the f derived above to be $f \equiv g^{-1}$: unit sphere in w plane $\to \Omega$ in the z plane. This gives us the desired change of variables $x = x(\xi, \eta)$ and $y = y(\xi, \eta)$. Since we know the Green's function for the Dirichlet problem on the unit sphere, namely,

$$G(p, q) = -\frac{1}{2\pi}\ln|p - q| + \frac{1}{2\pi}\ln|1 - p\bar{q}|$$

---

* Such a derivative in the more general setting of Banach spaces is called a Frechet derivative. One reason why analytic function theory is so powerful is because one is working with Frechet differentiable functions from a one-dimensional (complex) Hilbert space to itself.

† For a short proof see for example L. Ahlfors, *Complex Analysis* (McGraw-Hill, New York, 1953).

where p and q are two points in $|w| \leq 1$, we then have the Green's function for the Dirichlet problem on Ω, namely,

$$G(P, Q) = -\frac{1}{2\pi} \ln|g(P) - g(Q)| + \frac{1}{2\pi} \ln|1 - g(P)\overline{g(Q)}|$$

where P and Q are the two points in Ω that are mapped by the conformal mapping g to p and q in $|w| \leq 1$, respectively. Upon getting the Green's function for a problem, as we know, the problem is, in principle, completely solved. Such then is the case for the Dirichlet problem on Ω treated above, although it remains to check the details, e.g., to verify that $G(P, Q)$ is indeed the Green's function, to check the Poisson representation for the solution to the boundary value problem given in terms of $\partial G/\partial n_Q$, and so on. It also remains to calculate actual solutions for given data.

> **Problem.** (a) Verify the chain rule calculation, and if you like, other calculations in the above. (b) Conformal maps can be easy or hard to compute. Try the following ones: (i) Ω the z upper half plane $\rightarrow$ unit sphere $|w| < 1$. Find the map g. (ii) Unit sphere $|w| < 1 \rightarrow \Omega$ in z plane given by $z = f(w) = (w + 2)^2$. Find the domain Ω. (c) For (i) and (ii) of part (b) solve the Dirichlet problem for Ω by finding the Green's function and writing the solution $u(P)$ at a point P in Ω by means of the Poisson integral formula
>
> $$u(P) = -\oint_{\partial\Omega} \frac{\partial G}{\partial n_Q}(P, Q)h(Q) \, ds_Q.$$

(2) Resolvent Operator and Resolvent Equation. In Section 2.7.2 the spectrum $\sigma(L)$ of an operator L was defined to be those complex λ such that $(L - \lambda)u = f$ was not well-posed. The remaining λ are called the resolvent set $\rho(L)$ for L. For λ in the latter set $\rho(L)$ the equation $(L - \lambda)u = f$ is well-posed, with solution given formally by

$$u = (L - \lambda)^{-1}f.$$

In many cases by means of a Green's function one has a specific solution representation

$$u(P) = \int G_\lambda(P, Q)f(Q) \, dQ$$

the integral taken over the appropriate region (e.g., Ω or $\partial\Omega$) on which the data f are given. In any case, with or without a Green's function representation, the operator

$$R_\lambda \equiv (L - \lambda)^{-1}$$

for λ in the resolvent set $\rho(L)$ is called the *resolvent operator*.

A very important equation holds for the resolvent operator, the so-called resolvent equation

$$R_{\lambda_1} - R_{\lambda_2} = (\lambda_1 - \lambda_2)R_{\lambda_1}R_{\lambda_2}.$$

From this equation it can be seen that R_λ possesses analytic function properties like those discussed in subsection (1) above.*

Resolvent equations and their variations enter in many important applications involving partial differential equations. For example, for the Helmholtz equation

$$(-\Delta + V - \lambda)u = f \text{ on } \Omega = R^3$$

mentioned in Section 2.7.2, in scattering theory one attempts to compare solutions u to those of the (easier) unperturbed problem

$$(-\Delta - \lambda)w = f \text{ on } \Omega = R^3.$$

For λ complex and not real, by the self-adjointness of $-\Delta$ and $-\Delta + V$, V taken here to be a nice multiplicative real potential $V = V(x)$, we know that both equations are well-posed in $L^2(R^3)$. For such λ, letting

$$R_\lambda^0 = (-\Delta - \lambda)^{-1}$$

and

$$R_\lambda = (-\Delta + V - \lambda)^{-1},$$

we see that we may write the modified resolvent equation

$$R_\lambda f = (R_\lambda^0 - R_\lambda^0 V R_\lambda)f.$$

Assuming that we can explicitly write down solutions $w = R_\lambda^0 f$ for the unperturbed problem, we see that the Helmholtz equation with potential V has now been "inverted" by this resolvent equation to the equation

$$u = w - R_\lambda^0 V u.$$

This often is an integral equation

$$(I - K_\lambda)u = w,$$

where $w = R_\lambda^0 f$ is the presumed known data and where $K_\lambda = R_\lambda^0 V$ is, for the example, compact by virtue of a Green's function representation for R_λ^0 and suitably nice properties of V. Integral equations of this type are called Fredholm equations, for which there exists an extensive theory.

Problem. (a) Verify formally the resolvent equation above. (b) Verify formally the modified resolvent equation for the Helmholtz operator given above. (c) The so-called Lippmann–Schwinger equation of quantum physics is the formal limit

---

* For more on resolvent operators see any of the references given in Section 2.7.2.

of the above-given modified resolvent equation, as $\lambda = x + iy$ is allowed to tend to the real axis. Why is there any problem in writing such a limit, often written in the physics notation as

$$p^+ = p - G^0(\lambda + i0)Vp^+,$$

the $\lambda + i0$ denoting the limit from the upper half plane?

(3) Kirchoff's Formula and Huygen's Principle. By use of the Fourier transformation introduced in Section 2.7.3 one may deduce some important Green's function representations for solutions of partial differential equations. Here we illustrate how this is done by deriving the Poisson integral solution to the three-dimensional wave equation

$$\begin{cases} u_{tt} - \Delta_3 u = 0 \text{ in } \Omega = R^3, & t > 0, \\ u(x, 0) = f \text{ in } \Omega, \\ u_t(x, 0) = g \text{ in } \Omega, \end{cases}$$

where $x = (x_1, x_2, x_3)$ and where $f = f(x) = f(x_1, x_2, x_3)$ and similarly g are the initial data. Before doing so, however, let us recall the d'Alembert formula for the solution to the corresponding one-dimensional problem

$$\begin{cases} u_{tt} - u_{xx} = 0, & -\infty < x < \infty, & t > 0, \\ u(x, 0) = f(x), & -\infty < x < \infty, \\ u_t(x, 0) = g(x), & -\infty < x < \infty, \end{cases}$$

obtained in Section 2.5. There the solution was given by the "Poisson Integral" formula*

$$u(x, t) = \frac{f(x + t) + f(x - t)}{2} + \frac{1}{2}\int_{x-t}^{x+t} g(x)\, ds.$$

From this we can see that the solution u at point x_0 and time t_0 depends only on the data on the interval $[x_0 - t_0, x_0 + t_0]$. This interval (Fig. 2.9h) is called the

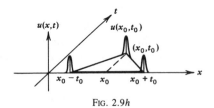

FIG. 2.9h

---

* To distinguish it from all other "Poisson Integral" formulas, it is commonly called the d'Alembert formula. In like manner the "Poisson Integral" formula to be found in the following for the three-dimensional problem is often called Kirchoff's formula.

domain of dependence of the solution. As drawn schematically in Figure 2.9h for the case in which g is taken to be zero, the physical meaning of the domain of dependence is just the fact that for the wave equation (with propagation speed c taken equal to one as we have done here), only the waves at $x_0 - t_0$ and $x_0 + t_0$ at time $t = 0$ will be felt at position x_0 at later time t_0. Had we considered the wave equation $u_{tt} - c^2 u_{xx} = 0$ with local wave velocity c, the only change in d'Alembert's formula would be to average on the interval $[x_0 - ct_0, x_0 + ct_0]$, from which

$$u(x_0, t_0) = \frac{f(x_0 + ct_0) + f(x_0 - ct_0)}{2} + \frac{1}{2c} \int_{x_0 - ct_0}^{x_0 + ct_0} g(s) \, ds.$$

The only change in Figure 2.9h would be that the characteristic lines from the point (x_0, t_0) would angle back to the initial data line $t = 0$ with slope $1/c$. For larger c, the initial waves influencing the solution u at (x_0, t_0) just arrive more quickly.

Turning now to the case of the three-dimensional wave initial value problem, we will see the same "averaging of initial data" taking place, yielding the so-called Kirchoff's "Poisson Integral" formula for the solution $u(x, t)$. Let us recall the problem, namely,

$$\begin{cases} \square_3 u = 0 \text{ in } R^3, & t > 0, \\ u(x, 0) = f, \\ u_t(x, 0) = g, \end{cases}$$

where we have here used the rather standard notation $\square_n u$ for the wave equation in n-space dimensions. From the properties of the Kirchoff formula to be found below it can be verified (see the problem at the end of this section) that the solution u is given by

$$u = u_g + \frac{\partial}{\partial t}(u_f),$$

where u_g and u_f are the solutions of

$$\begin{cases} \square u_g = 0 \text{ in } R^3, \\ u(x, 0) = 0, \\ u_t(x, 0) = g, \end{cases} \qquad \begin{cases} \square u_f = 0 \text{ in } R^3, \\ u(x, 0) = 0, \\ u_t(x, 0) = f. \end{cases}$$

Accepting this fact (which is not a priori very obvious), it therefore suffices for the general problem to solve the initial value problem with only $u_t(x, 0)$ data g present. The Kirchoff formula for the solution in that case is

$$u(x, y, z, t) = \frac{t}{4\pi} \int_0^\pi \int_0^{2\pi} g(x + t \sin \theta \cos \varphi, \ y + t \sin \theta \sin \varphi,$$
$$z + t \cos \theta) \sin \theta \, d\varphi \, d\theta.$$

This states that at any point (x_0, y_0, z_0, t_0) the solution $u(x_0, y_0, z_0, t_0)$ is the average of g on a "domain of dependence" the $S_{t_0}(x_0, y_0, z_0)$ sphere (Fig. 2.9i) and may be written as

$$u(x_0, y_0, z_0, t_0) = \frac{t_0}{4\pi} \oiint_{S_{t_0} \text{ sphere}} g(s)\, ds.$$

Note the resemblance between the Kirchoff formula for $n = 3$ and the d'Alembert formula for $n = 1$, namely, in the three-dimensional case

$$u(x_0, y_0, z_0, t_0) = \text{avg. of } g(s) \text{ on a two-sphere } t_0 \text{ units away}$$
$$\text{in space from } (x_0, y_0, z_0),$$

whereas in the one-dimensional case

$$u(x_0, t_0) = \text{avg. of } g(s) \text{ on a one-ball of radius } t_0 \text{ units in}$$
$$\text{space surrounding } x_0.$$

Note also a fundamental difference: For $n = 3$ the initial data g (and even f) arrives at (x_0, y_0, z_0) t_0 units later and *thereafter has no effect*, whereas for $n = 1$ the initial data g continues to affect the solution for all times $t \geq t_0$. This feature of solutions of the wave initial value problem is called Huygen's principle and asserts more generally that for $\square_n u = 0$ and n odd and $n > 1$, the initial data represents sharp signals with no after-effect. Without Huygen's principle radio communication would be rather messy because after the initial sounds reached you, you would be obliged to continue listening to them while trying to ungarble from them the signals transmitted later. Thus it is apparently fortunate that we live in a three-dimensional world.

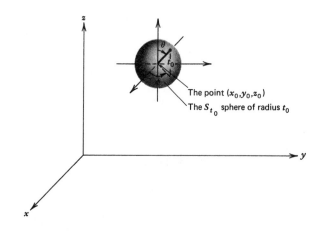

The point (x_0, y_0, z_0)
The S_{t_0} sphere of radius t_0

FIG. 2.9i

Let us close this section by formally deriving Kirchoff's formula by means of the Fourier transform (see Section 2.7.3). From the given initial value problem we have after transforming

$$
\begin{cases}
\hat{u}_{tt} + (k_1^2 + k_2^2 + k_3^2)\hat{u} = 0 \text{ in } \Omega_k = R^3, & t > 0 \\
\hat{u}(k_1, k_2, k_3, 0) = 0, \\
\hat{u}_t(k_1, k_2, k_3, 0) = \hat{g}(k_1, k_2, k_3),
\end{cases}
$$

an easily solved ordinary differential equation initial value problem with solution

$$
\hat{u}(k_1, k_2, k_3, t) = \hat{g}(k_1, k_2, k_3)\frac{\sin|k|t}{|k|}
$$

where $|k|$ denotes $|k|^2 = k_1^2 + k_2^2 + k_3^2$. By inverse Fourier transform we have the solution

$$
u(x, y, z, t) = (2\pi)^{-3/2}\iiint\limits_{R^3} \hat{g}(k_1, k_2, k_3)\frac{e^{i(k_1x_1 + k_2x_2 + k_3x_3)}\sin|k|t}{|k|} dk_1\, dk_2\, dk_3.
$$

At this intermediate point, recalling the formula $e^{i\theta} = \cos\theta + i\sin\theta$, and in the shorthand vector notation, we may note that the solution

$$
u(\mathbf{x}, t) = (2\pi)^{-3/2}\iiint\limits_{R^3} \hat{g}(\mathbf{k})\frac{e^{i[\mathbf{k}\cdot\mathbf{x} + |k|t]} - e^{i[\mathbf{k}\cdot\mathbf{x} - |k|t]}}{2i|k|} dk
$$

is the transform of the data $\hat{g}$ multiplied by both "incoming" and "outgoing" plane waves and averaged over all frequencies k, and as such is still a three-dimensional integral.

There are several ways to proceed from this point. Perhaps the shortest is to recall the convolution theorem of Section 2.7.3 in the form

$$
f * g = \widehat{\hat{f} \cdot \hat{g}}^{-1}.
$$

It is an easy exercise to show in one dimension that $(\sin kt)/k$ is the Fourier transform of a suitable constant times the characteristic function $\chi[-t, t]$ of the interval $[-t, t]$. In this way we may write the above inverse Fourier transform formula for the solution as

$$
u(x, y, z, t) = \text{const} \cdot \widehat{\hat{g} \cdot \hat{\chi}_{t\text{-ball}}}^{-1}
$$

which by the convolution theorem yields

$$
u(x, y, z, t) = \text{const} \cdot \iiint\limits_{R^3} \chi_t(\mathbf{y})g(\mathbf{x} - \mathbf{y})\, dy_1\, dy_2\, dy_3.
$$

Changing to spherical coordinates and noting that $\chi_t(\mathbf{y}) = 1$ for $|\mathbf{y}| \le t$ and 0 elsewhere then yields the Kirchoff formula.

Problem. (a) Verify carefully, by eventually differentiating under the integral, that Kirchoff's formula is indeed the solution to the wave initial value problem with initial data $u_t(x, 0) = g$. (b) Show that the formula $u = u_g + (u_f)_{,t}$ is valid for the problem in which initial data $u(x, 0) = f$ is also present. (c) Consider if you like the wave initial value problem with domain forcing data $\Box_3 u = F(x, t)$ present, and attempt to obtain a formula similar to the d'Alembert formula of Section 2.5 in that case. (d) For the student further interested in Fourier transform techniques, complete the calculation indicated above in using the convolution theorem to arrive at Kirchoff's formula.

Problem 2.9.8 Nonlinear Waves and Solitons

Because it is a rapidly growing and increasingly important theory, from both the physical and mathematical points of view, we wish in this final problem to discuss certain rudiments of the theory of nonlinear wave motion. Our discussion will be of necessity limited to some examples and basic concepts.*

Let us go back to the one-dimensional linear wave equation

$$\begin{cases} u_{tt} - c^2 u_{xx} = 0, & -\infty < x < \infty, \quad t > 0, \\ u(x, 0) = f(x), & -\infty < x < \infty, \\ u_t(x, 0) = 0, & -\infty < x < \infty, \end{cases}$$

with solution

$$u(x, t) = \frac{f(x + ct) + f(x - ct)}{2}.$$

One way to find this solution was (Section 2.5, Problem 1) by change of coordinates

$$\xi = x + ct, \qquad \eta = x - ct,$$

according to the characteristics (Problem 1.9.2) for the equation. This corresponds to factoring the wave operator as

$$\left(\frac{\partial}{\partial t} - c\frac{\partial}{\partial x}\right)\left(\frac{\partial}{\partial t} + c\frac{\partial}{\partial x}\right) u = 0.$$

If one takes this latter point of view, then the simplest linear wave equation is seen to be the first-order equation

$$u_t + cu_x = 0.$$

Following this line of reasoning, the simplest nonlinear wave equation would then be of the form

$$u_t + c(u)u_x = 0.$$

---

* An excellent reference for further study is the book of G. Whitham, *Linear and Nonlinear Waves* (Wiley-Interscience, New York, 1974). Other material concerning similarity solutions and local group invariance has been added in Appendix A, parts A.2 and A.3.

Such equations are indeed occurring more often now in the modeling of various real problems, and correspond to situations where u represents a small disturbance on some basic flow and where the local speed of propagation depends on the disturbance. We look at an example below, that of traffic congestion.

For the approach taken above, solution of either the linear or nonlinear versions or systems thereof of nonlinear wave equations would proceed by attempting to integrate along the characteristic curves of the problem, in the same manner by which one obtains the d'Alembert formula by changing the wave equation to characteristic coordinates. Because they are susceptible to treatment by such hyperbolic methods, equations of this type, or those composed of factors of this type, may be categorized as "hyperbolic."

A second and growing class of nonlinear wave equations are the so-called "dispersive" equations. These may be of "hyperbolic" type but can as well be of parabolic or mixed type. The term "dispersive" has come to mean that the phase speed is significantly nonconstant with respect to wave length, or, stated another way, modes (solutions) with different spatial oscillation lengths will propagate with different speeds. Moreover, if a nonlinear equation has a solution of the form

$$u(x, t) = A \cos(kx - \omega t),$$

then to be "nonlinearly dispersive" it is required that $\omega = \omega(k)$ vary with k to the extent that $\omega''(k) \neq 0$. We will see some concrete examples below.

Note that the linear wave equation

$$u_{tt} - c^2 u_{xx} = 0$$

with solutions (as discussed in Section 2.8)

$$u(x, t) = A e^{i(kx - \omega t)} = A \cos(kx - \omega t) + iA \sin(kx - \omega t)$$

wherein the frequency $\omega = ck = 2\pi c\lambda$ depended linearly on the wave number (or length) k (or λ, as you prefer), is not "nonlinearly dispersive" even though it is dispersive and its phase $\theta = kx - \omega t$ has phase speed $c = \omega/k$ dependent on k. On the other hand, the linear vibrating beam equation

$$u_{tt} + c^2 u_{xxxx} = 0.$$

which by dimensional analysis must enjoy the relation $\omega(k) = \pm ck^2$ between frequency and wave numbers, is "nonlinearly dispersive" with phase speed $c = \pm \omega/k^2$. In both cases, different Fourier components will travel with different speeds.

Let us note a third and interesting class of nonlinear wave equations, the "dissipative" equations. Dissipative operators usually in mathematics connote operators corresponding to physical situations in which conservation (e.g., of energy) laws are not satisfied, the dissipation (e.g., of energy) usually resulting in solution decay unless counterbalanced by continuous sources (e.g., of energy). Moreover, different Fourier components will travel with different amplitudes. An important example of this type of equation is found below.

The three categories of equations outlined above are neither mutually exclusive nor exhaustive, but serve rather as a useful guide to intuition for the various equations to be encountered. Let us close this final section of Chapter 2 by a brief look at some important examples of each of the three types of equations described above.

(i) *Traffic Flow.* Let $\rho(x, t)$ be the density of cars at point x and time t on a stretch of highway between two points A and B. For example, $\rho(x)$ may be the number of cars per kilometer. We assume a continuous density, and a conservation law (of traffic):

$$\frac{d}{dt} \int_a^b \rho(x, t) \, dx = F(a, t) - F(b, t).$$

This states that the net time rate of change (e.g., three cars per minute) of the number of cars in any subinterval $[a, b]$ of the highway is given by the net flux of cars out of and into the ends of that stretch of highway. Proceeding exactly as in the derivation of the heat equation in Section 1.7, we may use the divergence theorem to convert the right-hand side "boundary integral" into a "domain integral." Differentiating under the integral on the left-hand side at the same time, we thus have

$$\int_a^b [\rho_t(x, t) + F_x(x, t)] \, dx = 0.$$

Making the additional assumption now that at each x the flux F is a function only of the density ρ at x, we have by the chain rule that $F_x = F_\rho \rho_x$. Letting $c(\rho) = F_\rho$ and by the usual infinitesimal argument as in Section 1.7 for the above conservation law on arbitrarily small intervals $[a, b]$, we thus have the density equation

$$\rho_t + c(\rho)\rho_x = 0, \qquad A \le x \le B, \qquad t \ge 0.$$

This is of hyperbolic type.

The overall traffic flow velocity is density dependent and is given by the flux/density, that is,

$$v(\rho) = F(\rho)/\rho.$$

The propagation velocity for waves of traffic is therefore given by

$$c(\rho) = v(\rho) + \rho v'(\rho)$$

the first term representing the general flow velocity. Since the velocity $v(\rho)$ must clearly decrease with increasing density ρ, the second term $\rho v'(\rho)$ is negative. Thus the propagation velocity $c(\rho)$ is negative as waves warning the drivers of higher densities ahead are propagated back through the column of traffic.

Experiments have shown the validity of a logarithmically dependent flux law $F(\rho) = a_1 \rho \ln(a_2/\rho)$, where a_1 and a_2 are constants dependent on number of lanes and so on. From this, $c(\rho)$ is of the form $b_1 - b_2 \ln \rho$, a decreasing function of ρ. Any local increase in density propagates into a shock forming near the back of the column (Fig. 2.9j).

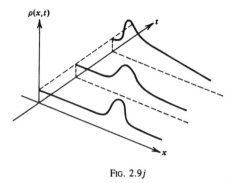

FIG. 2.9j

(ii) *The Burgers' Equation.* One may go from the traffic flow model (shown in Figure 2.9j) to a second equation,

$$u_t + u u_x - \nu u_{xx} = 0.$$

In the traffic flow model the added term νu_{xx}, where $\nu > 0$ is a suitable constant, could describe a "viscosity correction" by which drivers reduce their speed to account for an increasing density ahead. This equation is parabolic, and the effect of adding singular perturbations in this way is usually to greatly smooth the solutions. In problems of type (i), one therefore sometimes first solves the perturbed problem and then lets $\nu \to 0$ to obtain if possible the less regular solution of the original problem.*

Burgers was interested in the equation as a one-dimensional model of viscous compressible flow. It can be shown that solutions for $\nu > 0$ do tend to a solution for $\nu = 0$, which possesses a discontinuity representing a shock. The equation satisfies the conservation law (i.e., it is an exact divergence in some coordinate system):

$$(u)_t = (\nu u_x - \tfrac{1}{2} u^2)_x.$$

Hence there exists a "stream function" φ such that $\varphi_x = u$ and $\varphi_t = \nu u_x - \tfrac{1}{2} u^2$, from which

$$\varphi_t - \nu \varphi_{xx} + \tfrac{1}{2} \varphi_x^2 = 0.$$

Then by the change of variable $\ln w = -\varphi/2\nu$ one obtains the linear heat equation

---

* Just as experience shows that the addition of a little viscosity can yield smooth solutions of a nonlinear equation, experience also shows that often the addition of a little nonlinearity can stabilize a linear equation. It depends always on a close scrutiny of the application to determine whether the viscosity or the nonlinearity really deserves to be present.

$$w_t - \nu w_{xx} = 0,$$

and thus the solutions to Burgers' equation.

(iii) *The Born–Infeld Equation*.

$$(1 - (u_t)^2)u_{xx} + 2u_x u_t u_{xt} - (1 + (u_x)^2)u_{tt} = 0.$$

This equation possesses wave solutions moving with positive and negative velocities. It resembles the minimal surface equation mentioned at the beginning of the book, and in the hyperbolic region it may be transformed in a similar manner (see Problem 1.9.9, Exercise 5) by a Legendre transformation to a linear wave equation

$$w_{rs} = 0.$$

Perhaps the most well-known nonlinear equation of the second order or "dispersive" type is the following:

(i) *Korteweg–deVries Equation*.   This equation

$$u_t + uu_x + u_{xxx} = 0$$

models the lossless propagation of shallow water waves, as well as other phenomena. It satisfies the conservation law

$$(u)_t = (-u_{xx} - \tfrac{1}{2}u^2)_x,$$

as well as

$$(\tfrac{1}{2}u^2)_t = -(\tfrac{1}{3}u^3 + uu_{xx} - \tfrac{1}{2}u_x^2)_x,$$

and in fact an infinite number of such conservation laws. While we cannot go into it, this is the case for most such equations.

The Korteweg–deVries equations are best known by their property of possessing soliton solutions. A *soliton* is usually defined to be a solitary traveling wave with the additional property of persisting through an interaction with another soliton. After they pass through one another, they emerge in the same shape and velocity, having suffered no more than perhaps a phase shift. Let us see how such a solitary wave solution looks.

We presume that there is a solitary traveling wave solution with velocity c,

$$u(x, t) = u(\xi = x - ct).$$

When substituted into the partial differential equation this yields the ordinary differential equation

$$u_\xi(u - c) + u_{\xi\xi\xi} = 0.$$

One integration yields

$$u_{\xi\xi} = c_1 + cu - u^2/2,$$

and after multiplying both sides by u_ξ, a second integration yields

$$\frac{1}{2}(u_\xi)^2 = c_2 + c_1 u + \frac{c}{2}u^2 - \frac{u^3}{6} .$$

By a third integration this may be solved implicitly (in a manner similar to that of elementary ordinary differential equations for the nonlinear pendulum) in terms of an elliptic integral

$$\int_{u_0}^u \frac{du}{\sqrt{2c_2 + 2c_1 u + cu^2 - u^3/3}} = x - ct,$$

where u_0 is the initial value $u(0 = x - ct)$.

We now place a boundary condition on any such solitary wave solution u by requiring that it be solitary indeed (i.e., localized along the characteristics), that is, we require that u_ξ and $u_{\xi\xi}$ must tend to zero for large ξ. This forces the arbitrary constants of integration above to satisfy $c_1 = c_2 = 0$. Then the integral above may be evaluated explicitly, from which

$$u(x - ct) = 3c \, \text{sech}^2\left(\frac{\sqrt{c}}{2}(x - ct)\right).$$

(ii) A second important nonlinear dispersive equation that occurs for example in solid-state electronics is the *Sine–Gordon Equation*:

$$u_{tt} - u_{xx} + \sin u = 0,$$

The nonlinearity here is interesting because it contains arbitrary (albeit odd) high powers of u.

(iii) An equation historically important in this theory is the *Fermi–Pasta–Ulam Equation*:

$$u_{tt} - (1 + u_x)u_{xx} + u_{xxxx} = 0.$$

One may go (approximately) from this equation to the Korteweg–de Vries equations by means of the variable changes $w = u_x$ or u_t. The original version of the Fermi–Pasta–Ulam equation consisted of ordinary differential equations in a lattice as a model of a nonlinear vibrating string, and numerical simulation yielded not an expected ergodic behavior but rather a preservation of nonlinear mode behavior similar to that of solitons.

Time and space permit very little discussion of the third or "dissipative" type of wave equations. A rather important one in the biological study of nerve impulse propagation is the following.

(i) *The Hodgkin–Huxley Equations* for the state of a nerve axon, wherein $u(x, t)$ represents a voltage across a membrane, is of the form

$$\begin{cases} u_t - u_{xx} = N(u, w), \\ w_t - A(u)w = M(u), \end{cases}$$

where $w = (w_1, w_2, w_3)$ measures the degree of permeability of the membrane for the ions being described. Here $A(u)$ is a 3×3 matrix, $M(u) = (m_1(u), m_2(u), m_3(u))$, both are nonlinear in u, and $N(u, w)$ is nonlinear in w although linear in u. These equations are rather complicated and a simplified model is the following.

(ii) *Nagumo Equations*,

$$\begin{cases} u_t - u_{xx} = u(1 - u)(u - \alpha) + vw, \\ w_t - u = 0 \end{cases}$$

wherein $0 < \alpha < 1$, $v > 0$. Presuming solitary traveling wave solutions with the same velocity c, namely,

$$u(x, t) = u(\xi = x - ct),$$
$$w(x, t) = w(\xi = x - ct),$$

yields upon substitution into the partial differential equation the ordinary differential equations

$$\begin{cases} u' = v, \\ v' = -cv - u(1 - u)(u - a) + vw, \\ w' = -c^{-1}u. \end{cases}$$

The localizing requirement of solitary waves will specify also, as in the Korteweg–deVries equation above, that u, w, u_ξ, and w_ξ tend asymptotically to zero. This system of ordinary differential equation is thus amenable to treatment by the phase portrait approach of Problem 1.9.8. One would like to find propagation speeds c such that orbits which begin at $(0, 0, 0)$ and end near the same point exist.*

(iii) *Nonlinear Diffusion Equations.* It is not necessary that traveling wave solutions exist only along paths $\xi = x - ct$. In a number of applications, for example, of flow in porous media under continuous input and in other problems in chemistry and biology, one finds solutions along quadratic and other paths. We mention a specific example (along the lines of Section 1.7), namely, the initial boundary value problem

$$\begin{cases} u_t - (k(u)u_x)_x = 0, & 0 \leqq x < \infty, \quad t > 0, \\ u(x, 0) = f(x), & 0 \leqq x < \infty, \\ u(t, 0) = \alpha, \\ u(t, \infty) = 0, \end{cases}$$

where $k(u) \geqq 0$ is a solution-dependent diffusion coefficient, where $\alpha > 0$ is a maintained (e.g., concentration) boundary value along the left side of an infinite strip of media ($0 \leqq x < \infty$), where it is presumed that the filtration process slows down asymptotically as $x \to \infty$, and where $f(x)$ is the initial concentration.†

---

* For some results on this problem see for example S. Hastings, "Some mathematical problems from neurobiology," *Am. Math. Monthly* **82** (1975).

† Such problems may be called "concentration-diffusion" equations. A particularly important one is the porous media equation $u_t = (u^m)_{xx}$, which we will encounter later in Problems A-5 and A-6.

Problem. (a) In the traffic flow problem, put in a specific nice symmetric density profile $\rho(x, 0)$ as shown in Figure 2.9j and try the propagation velocity $c(\rho) = 76.184 - 17.2 \ln \rho$ and determine (numerically, or other) how soon and where the shock develops. (b) Go backward as indicated from solutions of the linear heat equation to solutions of Burgers' equation. (c) Check the integrations performed in obtaining the solitary wave solution of the Korteweg–deVries equation, and sketch the solution. (d) For the nonlinear diffusion problem above, assume a traveling wave solution of the form $u = u(\xi = x/\sqrt{t + 1})$ and thereby reduce the partial differential equation to the ordinary differential equation

$$\begin{cases} \frac{1}{2}\xi u_\xi + (k(u)u_\xi)_\xi = 0, & 0 < \xi < \infty, \\ u(0) = \alpha, \\ u(\infty) = 0, \end{cases}$$

and solve this problem for some specified data $f \geq 0$ and $k(u) = u$. (e) Look at the traffic flow equation of part (a) via the theory in Appendix A.1.

Problem 2.9.9 Confirmation Exercises

Any reader who has completed a significant part of this book should work the following three exercises. See also the ninth exercise for more practice.

1. (a) Solve by separation of variables the vibrating membrane problem

$$\begin{cases} u_{tt} - (u_{xx} + u_{yy}) = 0 \text{ in } \Omega = \begin{cases} 0 < x < \pi, \\ 0 < y < \pi, \end{cases} & t > 0, \\ u = 0 \text{ on } \partial\Omega, & t > 0, \\ u(x, y, 0) = f(x)f(y), & f(s) = \begin{cases} s \text{ for } 0 < s \leq \pi/2, \\ \pi - s \text{ for } \pi/2 < s < \pi, \end{cases} \\ u_t(x, y, 0) = 0. \end{cases}$$

   (b) Repeat part (a) with $f(s) = s^2$, $0 < s < \pi$. Does that initial data give you any misgivings about the solution?
   (c) Repeat part (a) with Ω now given by $0 < x < 1$, $0 < y < 1$, and $f(s) = s(1 - s)$, $0 < s < 1$.
   (d) Let Ω be a square plate $0 < x < \pi$, $0 < y < \pi$ whose four edges are kept at a constant temperature (which we normalize to be zero). Assume there are no heat sources within the plate and that the top and bottom

---

Difficulties can arise when the diffusion coefficient $k(u)$ nears zero. For another class of nonlinear equations, the reaction-diffusion equations, in which the derivative portion is linear (e.g., the first of the Nagumo equations given above), see P. Fife, *Mathematical Aspects of Reacting and Diffusing Systems* (Springer, Berlin, 1979).

faces are completely insulated. Let the initial temperature be $u(x, y, 0)$ $= f(x, y)$ and find the temperature $u(x, y, t)$ for $t \geqq 0$.

(e) Evaluate the answer for part (d) in the case that $f(x, y) = f(x)f(y)$ with $f(s) = s$ for $0 < s < \pi$.

(f) Sketch the initial temperature distribution of part (e), calculate a few values of the solution for $t = 1$, and sketch the latter.

(g) One states that the vibrating membrane of part (a) gives off a musical note if the solution is periodic in t. Show that this is the case if $f(s) =$ $\sin ks$ for any integer k, and find the solution and the period.

2. (a) Solve the eigenvalue problem

$$\begin{cases} -u'' = \lambda u, & 0 < x < \pi, \\ u(0) = u(\pi), \\ u'(0) = u'(\pi). \end{cases}$$

(b) Comment on the multiplicity of the eigenvalues for part (a). Order and graph the first few eigenfunctions so that $\varphi_n(x)$ has $n - 1$ nodes (i.e., interior zeros).

(c) Are the boundary conditions of the separated type that were considered in Section 2.4?

(d) Show whether or not the problem is self-adjoint.

(e) Consider the equation of Rayleigh type

$$-u'' = \lambda u, \qquad 0 < x < \infty,$$

and determine whether it is of limit-circle or limit-point type.

(f) Solve completely the problem of part (e) for λ real and complex, commenting on square integrability.

(g) Arrive at the Legendre equations by solving by separation of variables the Dirichlet problem on the unit sphere Ω in three dimensions,

$$\begin{cases} \Delta u = 0 \text{ in } \Omega, \\ u = f \text{ on } \partial\Omega. \end{cases}$$

3. (a) (i) Solve the one-dimensional vibrating string problem

$$\begin{cases} u_{tt} - u_{xx} = 0, & -\infty < x < \infty, \quad t > 0, \\ u(x, 0) = 0, & -\infty < x < \infty, \\ u_t(x, 0) = g(x), & -\infty < x < \infty, \end{cases}$$

by separation of variables. (ii) In particular, solve for $g(x) = \sin x$. (iii) Then, solve for $g(x) = x \sin x$.

(b) Solve by integrating, that is, by the Green's function approach.

(c) Show, with suitable decay assumptions on the solution, that the total energy in the string is conserved while the string vibrates.

(d) Solve the problem of part (a) by Fourier transform.

(e) For the problem of part (a), let the initial conditions by changed to $u_t(x, 0) = 0$ and $u(x, 0) = x(\pi - x)$ for $0 \le x \le \pi$, then extended oddly to all $-\infty < x < \infty$. Using this data, show that the wave operator lacks the smoothing properties of the potential and heat operators.

(f) For the Dirichlet problem for the square slab of Section 2.6, extend the data $f(x)$ to $f_e(x, y) = f(x) (1 - y/3)$ on Ω. Why isn't f_e the solution? Compare f_e to the numerical and to the analytical (i.e., separation of variables) solutions.

(g) Prove directly that $D(u_{N+1}) \le D(u_N)$ in the Rayleigh–Ritz approximation procedure for the Dirichlet problem given in Section 2.6.

The following problems contain supplementing information and are varied in content and difficulty.

4. (a) Show the result of Section 2.6, Problem 3, that $H_0^1(0, 1)$ functions are continuous, another way.

(b) Show that $u \in H_0^1(0, 1)$ implies that $|u| \in H_0^1(0, 1)$.

5. (a) Consider the two-dimensional Dirichlet problem

$$\begin{cases} \Delta u = 0 \text{ in } \Omega: y > 0, \\ u = f \text{ on } \partial\Omega: y = 0, \\ u \to 0 \text{ as } x^2 + y^2 \to \infty, \end{cases}$$

for Ω the upper half-plane. Derive in some way the Poisson integral formula

$$u(x, y) = \frac{1}{\pi} \int_{-\infty}^{\infty} \frac{y}{(x - s)^2 + y^2} f(s) \, ds$$

for the solution.

(b) From this write down the Green's function $G(P, Q)$ for the problem.

(c) For general application to the spectral theory of an arbitrary self-adjoint operator L with "spectral family" $E(\lambda)$, $-\infty < \lambda < \infty$, show that the "matrix element" $\langle R_z h, h \rangle$ for the resolvent operator $R_z = (L - z)^{-1}$, $z = \lambda + i\varepsilon$, possesses the very important properties:

$$\text{Im}\langle R_z h, h \rangle = \varepsilon \|R_z h\|^2 = \pi^{-1} \int_{-\infty}^{\infty} \frac{\varepsilon}{(\lambda - s)^2 + \varepsilon^2} \frac{d\langle E(\lambda)h, h \rangle}{d\lambda}.$$

6. The theory of "pseudodifferential operators" is becoming a useful tool in the study of partial differential equations. The ideas therein are related to the solution of partial differential equations by transform techniques. For example, for the Poisson problem

$$-\Delta u = f \text{ in } R^3$$

the solution by Fourier transform from Section 2.7 was

$$u = F^{-1}|k|^{-2}(F(f)),$$

which one would express in the pseudodifferential operator terminology as

$$u = T^{-1}f$$

thinking of the composite operator as a single operator. Unfortunately, the "symbol" $|k|^{-2}$ is not a good one. Converting the Poisson problem directly gives

$$f = (F^{-1}|k|^2 F)u = Tu$$

and a good "symbol" $|k|^2$. General pseudodifferential operators including

$$p(x, D)u = F^{-1}p(x, k)Fu$$

for a wide class of symbols $p(x, k)$ can be considered.
(a) Consult the recent literature for more about pseudodifferential operators.
(b) Consult the classical literature for other types of integral transforms and try to relate them to pseudodifferential operators.
(c) Fourier integral operators are also a recent development, and may be thought of as corresponding to the replacement of the Fourier transform kernel $e^{i\lambda \cdot x}$ by $e^{i\phi(\lambda)\cdot x}$. Think about this, and what it might mean in terms of general eigenfunction expansions.

7. Consider the Helmholtz equation in $L^2(\Omega)$,

$$(-\Delta - \lambda)u = f \text{ in } \Omega$$

for $\Omega = R^3$ and for $\lambda > 0$. Let $\chi_R(0)$ denote the characteristic function of the ball $B(0, R)$ with center at the origin and radius R, that is,

$$\chi_R(0) = \begin{cases} 1 \text{ for } x^2 + y^2 + z^2 \leqq R^2, \\ 0 \text{ otherwise.} \end{cases}$$

(a) Show that some characteristic functions $\chi_R(0)$ are in the range of the Helmholtz operator, and others are not.
(b) Find exactly those that are in the range.
(c) What do you then suspect concerning whether or not the test functions $C_0^\infty(R^3)$ are all in the range?*

8. There are interesting linear models of nerve-axon equations that do not involve the nonlinear features found in those models discussed in Problem 2.9.8. One assumes a conservative electrostatic field $E = -\text{grad } \varphi$ with potential φ and a model in which the neuron is visualized as a cylindrical core surrounded by a negligibly thin cylindrical membrane. Then employing

---

* For further information see K. Gustafson and G. Johnson, "A Study of the Helmholtz operator," *Proceedings of the Boulder AMS Special Session on Topics in Mathematical Physics, Quantum Mechanics in Mathematics, Chemistry and Physics* (Plenum, New York, 1981).

the classical dipole surface potential theory and Green's identities, one obtains (ignoring end effects)

$$\varphi(P) = \frac{1}{4\pi} \oint_{\partial\Omega} \frac{\partial}{\partial n} \left(\frac{1}{|P - Q|} \right) \left(\varphi_0(Q) - \frac{\sigma_1}{\sigma_0} \varphi_1(Q) \right) ds_Q,$$

where $\partial\Omega$ is the outer surface of the neuron, $\varphi(P)$ is the exterior (extra-cellular) potential, and where φ_0 is the outside surface potential, φ_1 is the inside surface potential, σ_0 is the outside conductivity, and σ_1 is the inside surface conductivity. By voltage clamp techniques the latter data is all measurable to some extent.

(a) Further study the classical dipole potential theory of partial differential equations.

(b) If interested read further about the nerve models.*

9. The following exercises are for additional practice as one proceeds through the second chapter of the book.

1. It should not be thought that the separation of variables ODE solutions are always easily compiled into the PDE solution infinite sum. Compare

$$\begin{cases} u_t - u_{xx} = 0 & -\pi < x < \pi \\ u(x, 0) = 1 + \cos x \\ u'(-\pi, t) = u'(\pi, t) = 0 & \text{(a)}, \end{cases}$$

and the same problem with the Neumann boundary condition (a) replaced by the Robin boundary condition

$$\begin{cases} -u'(-\pi, t) + 2u(-\pi, t) = 0 \\ u'(\pi, t) + 2u(\pi, t) = 0 \end{cases} \quad \text{(b)}.$$

2. Approximate the "ground state" of the one-dimensional quantum mechanical harmonic oscillator by means of a Rayleigh–Ritz trial function

$$\phi(x) = \begin{cases} 1 - x^2/l^2 & \text{for } |x| \le l, \\ 0 & \text{for } |x| > l. \end{cases}$$

3. Prove the convergence of the polar partial sums $s_N(r, \theta) \to u(r, \theta)$, alluded to in Section 2.5, along the lines of Section 2.2.

4. Show for $u \in C^2(\Omega) \cap C^1(\overline{\Omega})$ in two dimensions

$$u(P) = \frac{-1}{2\pi} \int_\Omega \int \ln \frac{1}{r_{PQ}} \Delta u + \frac{1}{2\pi} \oint_{\partial\Omega} \ln \frac{1}{r_{PQ}} \frac{\partial u}{\partial n_Q}$$

$$- \frac{1}{2\pi} \oint_{\partial\Omega} u \frac{\partial}{\partial n_Q} \ln \frac{1}{r_{PQ}},$$

harkening back to the three-dimensional proof given in Section 1.6.1.

---

* E.g., see C. Stevens, *Neurophysiology: A Primer* (Wiley, New York, 1966).

5. Let Ω_n be the open square in the plane with sides of length n parallel to the coordinate axes and with southeast corner at $(n^2, 0)$ on the x-axis. Let Ω be the union $\cup_{n=1}^{\infty} \Omega_n$. (a) Find all eigenvalues λ_i for the problem

$$\begin{cases} -\Delta u = \lambda u \text{ in } \Omega, \\ \quad u = 0 \text{ on } \partial\Omega. \end{cases}$$

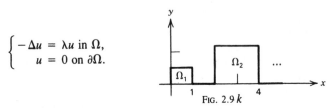

FIG. 2.9 k

(b) Accepting that the spectrum of an operator must be a closed set, what may you conclude?

6. Let $u \; \varepsilon \; C^2(\Omega)$, $\Delta u = 0$ on Ω the whole space R^n. Suppose u is not identically 0. Show $\int_\Omega u^2$ does not exist.

7. Show by appeal to the domains of dependence and influence how d'Alembert's formula may be used to accommodate boundary conditions for the wave initial value problem.

8. Going back to the equation of Problem 3 of Section 1.3, namely, $u_{xx} + 4xu = 16$, show that it is of Airy's type, which comes from partial differential equations of caustics, fluid dynamics, and other applications. Hence its solutions may be expressed in terms of the Airy functions. Write down an ODE two point boundary value problem which does not come from any application and which may not be solved.

9. It sometimes seems that a nonlinear partial differential equation is never completely solved. Let us go back to the first equation mentioned in this book, the minimal surface equation

$$(1 + (u_y)^2)u_{xx} - 2u_x u_y y_{xy} + (1 + (u_x)^2)u_{yy} = 0.$$

(a) Verify that the helicoid and catenoid are solutions.

(b) Read about the third type of complete embedded minimal surface recently obtained* with the aid of computer graphics.

(c) Try an "additive separation of variables"

$$u(x, y) = X(x) + Y(y)$$

to obtain a new local solution.

(d) Try the same trick on the Born–Infeld equation of Section 2.9.8 and on the slicing equation of Problem 1.9.9.4(c).

---

* An excellent exposition is given in *Science News* 127 (March 16, 1985). See also *Omni Magazine* 8: 7, 1986, pp. 88–90 for nice color graphics of these surfaces. Graphics are becoming a valuable tool for understanding solutions of nonlinear partial differential equations. See Appendix B.3 for graphical views of the "sag" difference between Laplacian and minimal surface solutions of Problem 1, Section 1.2, and Problem 2, Section 2.1.

CHAPTER 3

APPENDICES

... past, present, and future ...

For expositional convenience we have placed here, later in the book, introductions to three important topics:

        A. First-Order Equations
        B. Computational Methods
        C. Advanced Fluid Dynamics

These can provide a basis for a third-semester course. If one wishes, they may be taken into the second or even the first semester. To permit this flexibility we have written them to be essentially self-contained, although for continuity we often explicitly connect their development to other parts of the book.

(A) The treatment of first-order partial differential equations is intimately connected a fortiori to the notions of characteristics first encountered in the classification procedures at the beginning of the book, see Sections 1.1, 1.9.1, 1.9.2. For nonlinear equations (e.g., such as those treated in Sections 1.8, 1.9.8, 2.9.8), a very important related concept is that of a self-similar solution. Current attempts to fully understand all such solutions lead us to a theory of local transformation groups, currently developing along lines originally introduced by Lie.

(B) The explosive acceleration of high-speed computing technology in recent years has greatly enhanced the viability of using computational methods to enable practical solutions to otherwise intractible partial differential equations. The three principal computational methods—namely, finite difference methods, finite element methods, and (finite) spectral (least square) methods—were already presented in Section 2.6. Here we wish to introduce some basic elements of their implementation as well as mentioning a few of the important modern schemes, in a way that leaves their actual coding as an option not necessarily pursued here.

(C) An important area of current research in partial differential equations is that of fluid dynamics. Whole journals are now devoted, for example, just to computational fluid dynamics. In this final Appendix we present a glimpse of the current state of the art for important aspects of this topic (viz., the Navier–Stokes equations).

Each of Appendices A, B, and C has been arranged in three parts. In a sense one may perceive those parts 1, 2, 3, as reflections of the past, present, and future study of partial differential equations.

APPENDIX A: FIRST-ORDER EQUATIONS

The word *characteristic* is overused in mathematics. Thus the French *characteristic vector* is giving way to the German *eigenvector*, at least in part to decrease the confusion of meanings of the adjective *characteristic*. On the other hand, for the treatment of firrst-order partial differential equations, or systems of them, the uses of terms such as *characteristic curves, characteristic directions, characteristic manifolds, characteristic strip, characteristic surface, method of characteristics, characteristic data*, and so on, is so ingrained as to constitute integral and important parts of the subject.

The philosophy of treatment of first-order equations and systems is in many ways different from that of the more commonly encountered second-order and higher-order partial differential equations of physics, engineering, and science. As will be seen below, it is more akin to the viewpoints found in the theories of systems of nonlinear ordinary differential equations. Because it goes in directions other than those we have wanted to emphasize heretofore, and because their separate study can be quite time-consuming,* we have postponed where possible their treatment unitl now.

In the first part[§] of this first appendix we provide some essentials of the theory of firstorder partial differential equations. In the second part we then go on to an important topic, that of self-similar solutions of first - and higher - order equations. An introduction to a general theory of transformation group treatment of partial differential equations, currently under intense development and use, is given in the third part.

These three parts of the appendix are intimately related. Because whole books have been or currently are being written about just parts of each, our goal here has been to provide some accessibility to the key ideas, and no generality.

A.1 PDE to ODE

First-order partial differential equations may always be reduced to a system of ordinary differential equations. Under rather minimal smoothness conditions on the partial differential equation, its solvability via the ordinary differential equations is then guaranteed, in principle. A number of interesting geometrical considerations appear, however, and the complete integration of the equation should not be regarded as always trivial.

At this point the reader should recall his first course in ordinary differential equation. Let us help in that regard by posing the following straightforward review exercises.

> **Problem A.1**
>
> (a) Solve the first order inhomogeneous linear equation initial value problem
>
> $$\begin{cases} u'(x) + p(x)u(x) = q(x), \\ u(x_0) = u_0, \end{cases}$$
>
> by the method of integrating factor.
>
> (b) Show that the initial value problem
>
> $$\begin{cases} u'(x) = f(x, u), \\ u(0) = u_0, \end{cases}$$

---

*Indeed, this author remembers as a student his first course in partial differential equations, which in a whole semester never got beyond first-order equations and systems.

§ This part, Part A.1, is essentially the *Appendix* of the first edition of the book.

where $f(0, u) = 0, f(x \neq 0, u) = 1$, has no solution.
(c) For the initial value problem

$$\begin{cases} u'(x) = e^{xu(x)}, \\ u(0) = 0, \end{cases}$$

solve by the power series method, compute the first five coefficients, and determine the radius of convergence.

Generally speaking, first-order partial differential equations are usually encountered and treated as initial value problems. If they are twice continuously differentiable in the unknowns and first derivatives thereof, the further property of analyticity is not needed for their local integrability. Rather, what turns out to be needed for their exact integrability is that the initial data is not *characteristic*, in the sense to be described below.

Let us consider the general quasilinear partial differential equation of first order in two independent variables,

$$a(x, y, u) \frac{\partial u}{\partial x} + b(x, y, u) \frac{\partial u}{\partial y} = c(x, y, u). \tag{1}$$

It will be useful to begin with a picture (Fig. A.1). The solution $u(x, y)$ to Equation (1), as a three vector (x, y, u), determines the three vector $(\partial u/\partial x, \partial u/\partial y, -1)$, and the corresponding perpendicular three vector (a, b, c). The vector (a, b, c) is called the characteristic direction at (x, y, u). Thus the solution $u(x, y)$ as x, y ranges over a region in the plane determines the vector field (a, b, c) lying in the tangent plane to the surface $u(x, y)$ at each point (x, y). In other words, $u(x, y)$ as a surface in the (x, y, u) space is a solution to Equation (1) if and only if the direction field $(a(x, y, u), b(x, y, u), c(x, y, u))$ lies in the tangent planes at each (x, y, u). This is the necessary geometry of any solutions of (1).

Let us now consider the initial value problem

$$\begin{cases} au_x + bu_y = c, \\ u = l(x, y) \text{ a given curve.} \end{cases}$$

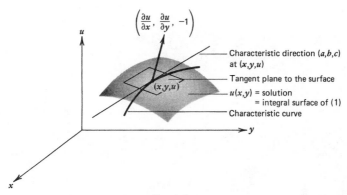

FIG. A.1

By this we mean that u is given as an initial curve parameterized by t, t in some interval $[a, b]$:

$$l \begin{cases} x = x(t) \\ y = y(t) \\ u = u(t) \end{cases} l',$$

where for future reference we let l' denote the projection of l down onto the x, y, plane. We wish to extend the initially given curve l into an integral surface representing the solution $u(x, y)$ over the x, y plane. See Figure A.2, where for simplicity we have drawn the integral surface extending from l in one direction only.

It turns out that this initial value problem is well-posed if the initial data curve l is not characteristic, that is, if the projected characteristic direction line does not fall on the tangent line of the projected initial curve l. Let us sketch the proof of this fact, which we state here as a theorem.

Theorem

For Equation (1) $au_x + bu_y = c$ with initial curve $l \in C^1$ and with coefficients a, b, $c \in C^1$, if the tangent line to $l' \neq$ the projected characteristic direction line, then there exists exactly one solution $u(x, y)$ defined in a neighborhood of l'.

The C^1 conditions have to do with a use of the inverse function theorem, as will be seen below.

Let us see now why the theorem just stated is so. Imagining that we already have the solution $u(x, y)$ as an integral surface, we may fix a point (x_0, y_0, u_0) on the initial curve l. We wish to consider the characteristic curve $(x(s), y(s), u(x(s), y(s)))$ emanating from (x_0, y_0, u_0), and we therefore parametrize it by s, as we have just done. We know that necessarily, by Equation (1), that along this characteristic curve we must satisfy the equation

$$a \frac{\partial u}{\partial x} + b \frac{\partial u}{\partial y} = c.$$

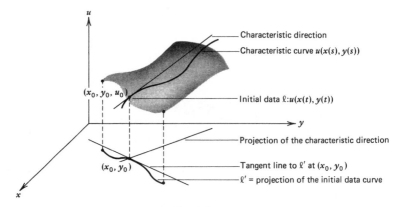

FIG. A.2

The projected characteristic direction line has direction numbers (a, b) and hence the left-hand side of the equation is just the directional derivative of u in the characteristic direction. Equation (1) thus states the requirement that

$$\left(\frac{du(x(s), y(s))}{ds}\right)_{(a,\ b)} = c(x(s), y(s), u(s)),$$

the left-hand side denoting the directional derivative of u in the characteristic direction (a, b). If the tangent line to l' coincides with the projected characteristic direction line, then we cannot integrate the above and at the same time move off l into the integral surface. That is, we are constrained to the situation $u(x(s), y(s)) = u(x(t), y(t))$ on l' and in l. This is why the theorem works, for in all other cases we may integrate the above directional derivative and in so doing move off the initial data curve l.

The unique determination of the characteristic curve $u(x, y)$ then follows from unique integrability of ordinary differential equations and the inverse function theorem. Let s be a coordinate parameter for the projection of the characteristic curve. By the nontangential condition of the theorem, s is independent of t. That is, we fix a value t_0 in $[a, b]$, the corresponding (x_0, y_0) in l', the corresponding (x_0, y_0, u_0) in l, and set $s = 0$. We may then consider the ordinary differential equation system

$$\begin{cases} \dfrac{dx(s)}{ds} = a(x(s), y(s), u(x(s), y(s))), \\[2mm] \dfrac{dy(s)}{ds} = b(x(s), y(s), u(x(s), y(s))), \\[2mm] \dfrac{du(s)}{ds} = c(x(s), y(s), u(x(s), y(s))), \\[2mm] x(0) = x_0, \\ y(0) = y_0, \\ u(0) = u_0. \end{cases}$$

This initial value problem, with C^1 right-hand side, possesses a unique C^1 solution for small s.

As t runs from a to b, the initial point $(x_0(t), y_0(t), u_0(t))$ runs through the initial data curve l, and the characteristic curves $(x(s), y(s), u(s))$ solving the above system may be written, with t dependence included, as

$$\begin{cases} x = x(s, t), \\ y = y(s, t), \\ u = u(s, t). \end{cases}$$

Since s and t are independent by the assumed nontangential condition, the Jacobian

$$\frac{\partial(x, y)}{\partial(s, t)} = \begin{vmatrix} \dfrac{\partial x(s, t)}{\partial s} & \dfrac{\partial x(s, t)}{\partial t} \\[3mm] \dfrac{\partial y(s, t)}{\partial s} & \dfrac{\partial y(s, t)}{\partial t} \end{vmatrix}$$

at $s = 0$ and $t \in [a, b]$ is nonvanishing:

$$\frac{\partial(x, y)}{\partial(s, t)}\bigg|_{\substack{s=0 \\ t \in [a,b]}} = a(x_0(t), y_0(t), u_0(t)) \frac{dy_0(t)}{dt} - b(x_0(t), y_0(t), u_0(t)) \frac{dx_0(t)}{dt} \neq 0.$$

Therefore by the inverse function theorem, in a neighborhood of $s = 0$, $t = t_0 \in [a, b]$, we may express s and t as functions of x and y, and hence we may write the solution $u(s, t)$ as an integral surface $u(x, y)$ in a neighborhood of l':

$$u(s, t) = u(s(x, y), t(x, y)) = u(x, y).$$

To recapitulate: Given the ODE system with a, b, c all C^1, we obtain a local solution in s for each $t \in [a, b]$. Considering only the first two components $x(s, t)$ and $y(s, t)$, we invert them to obtain $s(x, y)$ and $t(x, y)$. The dependence of $t(x, y)$ was already present in l' and we have done the inversion of coordinates principally to obtain $s(x, y)$ and the corresponding $u(x, y)$. The integral surface $u(x, y)$ is indeed a surface in the sense of being a C^1 function. It satisfies the initial data because $u(x_0(t), y_0(t)) = u(0, t) = u_0(t)$. It satisfies the partial differential equation (1) because

$$
\begin{aligned}
au_x + bu_y &= a(u_s s_x + u_t t_x) + b(u_s s_y + u_t t_y) \\
&= u_s(as_x + bs_y) + u_t(at_x + bt_y) \\
&= u_s(x_s s_x + y_s s_y) + u_t(x_s t_x + y_s t_y) \\
&= u_s(s_s) + u_t(t_s) \\
&= u_s(1) + u_t(0) \\
&= c
\end{aligned}
$$

Let us consider the following example:

$$
\begin{cases}
uu_x + u_y = 1, \\
l: x_0(t) = t, \quad y_0(t) = t, \quad u_0(t) = t/2, \quad 0 \leq t \leq 1.
\end{cases}
$$

First we check the nontangential condition. In this example we have $a = u$, $b = 1$, $c = 1$, and

$$
\begin{aligned}
\frac{\partial(x_0(t), y_0(t))}{\partial(0, t)} &= a(y_0(t))_t - b(x_0(t))_t \\
&= u_0(t)(y_0(t))_t - (x_0(t))_t \\
&= (t/2) - 1,
\end{aligned}
$$

which is nonvanishing for $t \in [0, 1]$. The associated ODE initial value problem

$$
\begin{cases}
x_s = u \Rightarrow x(s, t) = s^2/2 + u(0)s + x_0(t) = s^2/2 + st/2 + t, \\
y_s = 1 \Rightarrow y(s, t) = s + y(0) = s + t, \\
u_s = 1 \Rightarrow u(s, t) = s + u(0) = s + t/2, \\
x(0) = t, \\
y(0) = t, \\
u(0) = t/2,
\end{cases}
$$

is solved as just indicated (by moving in direction upward in the implications). The solution

$u(x, y)$, as a solution in terms of x and y rather than in terms of s and t, is easily seen to be

$$u(x, y) = \frac{2y - x - y^2/2}{2(1 - y/2)}.$$

Problem A.2

(a) Sketch the above example, indicating l, l' and the solution $u(x, y)$. For which (x, y) does $u(x, y)$ remain a solution to the differential equation (1)?

(b) We assumed in the above disccussion that the nontangential condition was satisfied all along the initial curve l. There are three other possibilities: (i) First, the Jacobian vanishes all along l' and the initial value problem nonetheless possesses a solution. Then the initial data curve l is itself a characteristic curve and the problem has in fact an infinite number of solutions. (ii) The Jacobian vanishes along l' but no solutions exist. (iii) The Jacobian vanishes on a proper subset of l'.

(i) Show that exceptional case (i) prevails for the example above with

$$l: x_0(t) = t^2/2, \qquad y_0(t) = t, \qquad u_0(t) = t, \qquad 0 \le t \le 1.$$

Find all solutions and sketch some of them.

(ii) Show that exceptional case (ii) prevails for the example above with

$$l: x_0(t) = t^2, \qquad y_0(t) = 2t, \qquad u_0(t) = t, \qquad 0 \le t \le 1.$$

Try to solve anyway, to see what happens.

(iii) Examine the example above for $0 \le t \le 3$.

(c) Give a uniqueness proof, for the nonexceptional case, by showing that any characteristic curve with a point in common with an integral surface of Equation (1), lies completely within the integral surface.

The general first-order partial differential equation in two independent variables

$$F(x, y, u, \partial u/\partial x, \partial u/\partial y) = 0$$

may be investigated in like manner, and similarly, that in n independent variables

$$F(x_1, \ldots, x_n, u, u_{x_1}, \ldots, u_{x_n}) = 0,$$

as well as systems of such equations in more than one dependent variable. The theory so found is extensive and will not be developed here. We confine ourselves in closing this appendix to some remarks about the case $n = 2$.

Adopting a rather common notation, we may write the general first-order equation in the form ($p = \partial u/\partial x$, $q = \partial u/\partial y$)

$$F(x, y, u, p, q) = 0. \tag{2}$$

As in the geometrical picture for the special quasilinear case of Equation (1), a tangent plane at a point (x_0, y_0, u_0) is given by (see Fig. A.1)

$$(p, q, -1) \begin{pmatrix} x - x_0 \\ y - y_0 \\ u - u_0 \end{pmatrix} = 0,$$

where $(p, q, -1)$ are direction numbers of the normal to the plane. Given the values of x_0, y_0, and u_0 in Equation (2), different values of p therein will yield different values of q and hence a one parameter family of tangent planes parametrized by p. This envelope of tangent planes is called the Monge cone for Equation (2) at (x_0, y_0, u_0). The Monge cone locus may be described by:

$$\begin{cases} p(x - x_0) + q(y - y_0) - (u - u_0) = 0, \\ (x - x_0) + \dfrac{dq}{dp}(y - y_0) = 0, \end{cases}$$

where the second equation was derived from differentiating the first with respect to p and requiring that $du/dp = 0$, recalling the definition of p and the fact that u is to lie in the tangent plane.

In the quasilinear case, Equation (1) described a vector field of characteristic direction lines. These lines were degenerate Monge cones. The general equation (2) thus describes a cone field as shown in Figure A.3. A surface in 3-space solves Equation (2) if and only if it remains tangent to the cones, as the latter vary from point to point.

Practice with specific equations shows that a constructive way to proceed is to differentiate Equation (2) with respect to p, solve for dq/dp, thereby arriving at the Monge cone. From $F(p, q) = 0$ we have

$$\frac{dF}{dp} = \frac{\partial F}{\partial p} + \frac{\partial F}{\partial q}\frac{dq}{dp} = 0,$$

which, together with the Monge cone determining equations above, yields the line-of-contact equation

$$\frac{x - x_0}{F_p} = \frac{y - y_0}{F_q} = \frac{u - u_0}{pF_p + qF_q}$$

for the intersection of the integral surface and the Monge cone.

Turning now to the initial value problem corresponding to Equation (2),

$$\begin{cases} F(x, y, u, p, q) = 0, \\ u = l(x, y) \text{ a given curve,} \end{cases}$$

proceeding exactly as in the quasilinear case we are led to consider the associated ordinary differential equation initial value problem

$$\begin{cases} \dfrac{dx}{ds} = F_p, \\[2mm] \dfrac{dy}{ds} = F_q, \\[2mm] \dfrac{du}{ds} = pF_p + qF_q, \\[2mm] x(0) = x_0, \\[2mm] y(0) = y_0, \\[2mm] u(0) = u_0, \end{cases}$$

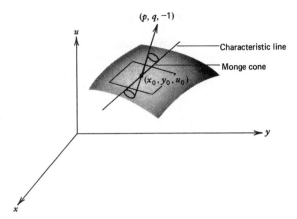

$(p, q, -1)$

Characteristic line

Monge cone

(x_0, y_0, u_0)

FIG. A.3

for each fixed t. Unfortunately, this system has three equations in five unknowns. Implicit differentiation of Equation (2) yields two more equations,

$$\begin{cases} \dfrac{dp}{ds} = -(F_x + pF_u), \\ \dfrac{dq}{ds} = -(F_y + qF_u), \end{cases}$$

and we assume that we can find in some way initial values for p and q on the initial curve l,

$$\begin{cases} p(0) = p_0, \\ q(0) = q_0. \end{cases}$$

From these five equations we may then in principle integrate (2).

Thus, in practice, one need not find the general Monge cone, line of contact equations, and so on. Rather, one may proceed as follows:

1. Find $p_0(t)$ and $q_0(t)$ satisfying (i) the partial differential equation

$$F(x_0, y_0, u_0, p_0, q_0) = 0$$

along l, and (ii) the so-called "strip condition"

$$\frac{du_0}{dt} = p_0 \frac{dx_0}{dt} + q_0 \frac{dy_0}{dt}$$

along l.

2. Check that the Jacobian "noncharacteristic" condition

$$F_{p_0} \frac{dy_0}{dt} - F_{q_0} \frac{dx_0}{dt} \neq 0$$

is satisfied along l.

3. Integrate the five ordinary differential equations written above as an initial value problem.

Then under general conditions one is guaranteed a unique (i.e., for each given $(p_0 \ q_0)$ pair) solution $u(x, y)$ of the partial differential equation (2) in a neighborhood of the initial data curve l. For example, it is sufficient that the equation $F \in C^2$ with respect to x, y, u, p, q, that $F_p^2 + F_q^2 \neq 0$ so that the characteristic curves are never perpendicular to the x, y plane, that the initial curve l is C^2 for $t \in [a, b]$, and that C^1 quantities $p_0(t)$ and $q_0(t)$ have been established satisfying along l the equation F, the strip condition, and the noncharacteristic condition.

Let us illustrate this discussion with a simple example:

$$\begin{cases} u_x u_y = 1, \\ l: x_0(t) = 2t, \quad y_0(t) = 2t, \quad u_0(t) = 5t. \end{cases}$$

Letting $p = u_x$ and $q = u_y$, from $F = pq - 1$ we have $F_p = q$, $F_q = p$, and $F_u = F_x = F_y = 0$. To find initial conditions $p_0(t)$ and $u_0(t)$ we use the strip condition combined with the given equation,

$$\begin{cases} 5 = 2p + 2q \\ pq = 1 \end{cases},$$

from which we obtain the 2 initial condition sets (i) $p_0 = 2$, $q_0 = 1/2$, or (ii) $p_0 = 1/2$, $q_0 = 2$. The 2 corresponding ODE systems are, respectively, with solutions shown:

(i) $\quad x_s = 1/2 \Rightarrow x = s/2 + x(0) = s/2 + 2t,$

$\quad y_s = 2 \Rightarrow y = 2s + y(0) = 2s + 2t,$

$\quad u_s = 2 \Rightarrow u = 2s + u(0) = 2s + 5t,$

$\quad p_s = 0 \Rightarrow p = p_0 = 2,$

$\quad q_s = 0 \Rightarrow q = q_0 = 1/2,$

from which $u(x, y) = 2x + y/2$; and

(ii) $\quad x_s = 2 \Rightarrow x = 2s + 2t$

$\quad y_s = 1/2 \Rightarrow y = s/2 + 2t$

$\quad u_s = 2 \Rightarrow u = 2s + 5t$

$\quad p_s = 0 \Rightarrow p = 1/2$

$\quad q_s = 0 \Rightarrow q = 2$

from which $u(x, y) = 2y + x/2$. This second solution could have been obtained from the first by the symmetry of the equation and the initial conditions. Further information on first-order equations may be found in the cited literature.*

---

* See for example Courant and Hilbert, *Methods of Mathematical Physics*, Vol. II (Wiley-Inter-science, New York, 1962) for a further treatment of first-order equations. A good introductory view may be found in Garabedian, *Partial Differential Equations* (Wiley, New York, 1964). For a concise treatment in an interesting context see Courant and Friedrichs, *Supersonic Flow and Shock Waves* (Wiley-Interscience, New York, 1948).

Problem A.3

Solve the following first-order equations:

(a) $\begin{cases} u_x u_y - 2 = 0, \\ l: x_0(t) = t, \quad y_0(t) = t, \quad u_0(t) = 3t, \quad 0 \le t \le 1. \end{cases}$

(b) $\begin{cases} u_x u_y - u = 0, \\ l: x_0(t) = t, \quad y_0(t) = 1, \quad u_0(t) = t, \quad 0 \le t \le 1. \end{cases}$

(c) $\begin{cases} (u_x)^2 + (u_y)^2 = 1, \\ l: x_0(t) = \sin t, \quad y_0(t) = \cos t, \quad u_0(t) = 0, \quad 0 \le t \le 1. \end{cases}$

A.2 Self-Similar Solutions

The theory of self-similar solutions of a partial differential equation, although incomplete, originated in engineering design problems. When creating a large structure such as a ship or airplane by use of a small model (e.g., in a tank or wind tunnel) it is quite important that the observed results be scale invariant. That is, if flow be understood about a model, one wants the flow about the actual large object to be geometrically similar.

To secure this effect, dimensional analysis is performed on the governing structure or flow partial differential equations to render those equations dimensionless. This amounts to scale changes of variable. These methods go far back* and have been applied to pipe flow, frictional torque, aerodynmamic lift, shock decay, nozzle condensation, atomic explosion, boundary layer theory, turbulence, vortex theory, hypersonic flow, flame theory, porous media equations, etc. Although it is impossible to catalogue all of its applications, this theory is less widely known than many theories of smaller importance.

Here we can only provide a few insights. We will stress some elementary viewpoints that are quite useful for keeping the theory in bounds. In particular in this section we will emphasize the idea that the right change of variables can reduce a second-order partial differential equation in two independent variables, say x and t, to an ordinary differential equation in one independent variable η, say a judicious combination of x and t, thereby providing from the solutions of the ordinary differential equation some special solutions of the partial differential equation. Moreover, we will assert that the judiciously chosen combined variable η should often be thought of in terms of the characteristics of the partial differential equation.

An example to impress this view, and already available to us, is the wave equation initial value problem as studied in Sections 1.1, 1.9.2, and 2.5. We recall the equations

$$\begin{cases} u_{tt} - u_{xx} = 0, \text{ all } x, t, \\ u(x, 0) = f(x), t = 0, \\ u_t(x, 0) = 0, t = 0. \end{cases}$$

* Some say, to Galilei. H. Helmholtz, *Mcnatsber. der preuss. Akademic der Wiss. zu Berlin* (1873) developed a systematic similarity theory for the equations of fluid dynamics.

In converting the equation to the other canonical form $u_{\zeta\eta} = 0$ we employed the linear change of variable

$$\zeta = x + t$$
$$\eta = x - t$$

Was the new η variable judiciously chosen? Along the η direction, i.e., setting η constant, e.g., as in Fig. A.4,

$$\eta = x - t = x_0$$

the d'Alembert solution

$$u(\zeta, \eta) = \tfrac{1}{2}[f(\eta) + f(\zeta)]$$

was found by integrating the equation $u_{\zeta\eta} = 0$ as an ordinary differential equation in the (orthogonal) direction ζ. Notice that $\eta = $ constant is a characteristic curve along which initial data propagates. If all data is concentrated at the point x_0, then along the curve $x - t = x_0$ the solution is constant. Reversing the roles of η and ζ in the above gave us a second ordinary differential equation of $\zeta = $ constant in the η direction. For this example the special solutions of these two ordinary differential equations combined to give us the complete solution to the partial differential equation.

Another "right" change of variables may be found from those which leave a partial differential equation invariant. Under what changes of independent variable $\zeta = \zeta(x, t)$, $\eta = \eta(x, t)$, is the wave equation invariant:

$$u_{\zeta\zeta}(\zeta, \eta) - u_{\eta\eta}(\zeta, \eta) = 0 ?$$

Clearly the above classification linear change of variable, useful as it was to convert the equation to an exact derivative, was not intended to leave the form of the equation invariant. However, the uniform scale changes of variable

$$\zeta = \alpha x$$
$$\eta = \alpha t$$

do. Moreover, they form a (one parameter) group* of transformations. The partial reflections

$$x \to x, \qquad x \to -x$$
$$t \to -t, \qquad t \to t$$

also leave the equation invariant. The uniform scale transformation suggests the "velocity coordinate"

$$r = \zeta/\eta = x/t,$$

and phase plane reasoning (see for example Section 1.9.8 and Problem 8 of Section 1.9.9) suggests the "singularity coordinate"

$$s = \log t.$$

---

* Recall a group must satisfy axioms of closure, associativity, identity, and inverse. One would not have to insist on full group structure for the changes of variable. On the other hand, to transform back to the original variables once the problem is solved, the inverse transformation is needed.

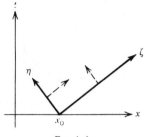

FIG. A.4

In these the wave equation is transformed to

$$((1 - r^2)u_{rr} - 2ru_r) - (u_{ss} - u_s) + 2ru_{rs} = 0.$$

From this we see that a function $u(r)$ of r only will be a solution if it satisfies the ordinary differential equation

$$(1 - r^2)u_{rr} - 2ru_r = 0,$$

and similarly a function $u(s)$ will be a solution if it satisfies the ordinary differential equation

$$u_{ss} - u_s = 0$$

Problem A.4

(a) Determine all of the nonsingular linear transformations

$$\begin{pmatrix} \zeta \\ \eta \end{pmatrix} = \begin{bmatrix} c_{11} & c_{12} \\ c_{21} & c_{22} \end{bmatrix} \begin{pmatrix} x \\ t \end{pmatrix}$$

which leave the wave equation invariant.
(b) Solve the two ordinary differential equations resulting from the "velocity" and "singularity" changes of variable above.
(c) Prove that the uniform scale changes form a transformation group. A similarity solution* is a solution of the form

$$u = u(x, t, \eta, g(\eta))$$

where $g(\eta)$ is an explicitly postulated but *a priori* unknown function of η, its existence arrived at for example by dimensional analysis or transformation group considerations. For example, $g(\eta)$ can be thought of as the solution of the ordinary differential equation from which the similarity solution $u(\eta)$ will be found.

For a second example we consider the heat equation initial value problem

$$\begin{cases} u_t - u_{xx} = 0, & \text{all } x, \ t > 0, \\ u(x, 0) = f(x), & \text{all } x, \ t = 0. \end{cases}$$

In Section 1.7 we derived this equation from physical principles. An examination of the physical units will show that there is no way to functionally combine the independent variables

---

* The exact definition is sometimes hard to come by in the literature, and in fact varies.

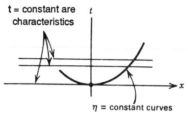

t = constant are characteristics

t

x

η = constant curves

Fig. A.5

t and x into a single new independent variable representing a new length or time scale. This is the usual first step in a dimensional analysis: isolate the important physical parameters that will determine the solution behavior. The heat equation has no such characteristic space or time parameters.

Accepting then possibly nonphysical new variables, a second-dimensional analysis procedure for homogeneous equations is to take as changed variable the relative differential order of the independent variables. For the heat equation this means we should try something like (see Fig. A.5)

$$\eta = \alpha \frac{x}{\sqrt{t}} \text{ or } \eta = \alpha \frac{x^2}{t},$$

α a yet to be determined constant depending, for example, on the local propagation speed. Let us see how this variable can be arrived at by dimensional reasoning.

Going back to Section 1.7 and the linearized heat equation

$$cu_t - ku_{xx} = 0$$

we recall that c had units cal/deg $\cdot$ cm^3 and k had units cal/deg $\cdot$ sec $\cdot$ cm. From the ratio

$$\frac{c}{k} = \frac{u_{xx}}{u_t} = \frac{\text{sec}}{\text{cm}^2}$$

we see that $\eta = x^2 c/tk$ is dimensionless and hence a good candidate for a similarity variable.

Letting $u = u(\eta)$ with $\eta = \alpha x^2/t$, if u satisfies the heat equation then $u_t = (-\alpha x^2/t^2)u_\eta$, $u_x = (2\alpha x/t)u_\eta$, $u_{xx} = (2\alpha/t)u_\eta + (4\alpha^2 x^2/t^2)u_{\eta\eta}$. These substituted into the heat equation, after multiplying through by t, give the ordinary differential equation

$$4\eta u_{\eta\eta} + (2 + \eta)u_\eta = 0.$$

This is easily solved with solution

$$u(\eta) = \int_{\eta_0}^{\alpha x^2/t} s^{-1/2} e^{-s/4\alpha} \, ds.$$

Let us next try a general scale change of both the independent variables and the dependent variable

$$\zeta = \alpha x$$

$$\eta = \beta t$$

$$v = \gamma u.$$

Substitution into the heat equation gives

$$\frac{\beta}{\gamma}v_\eta - \frac{\alpha^2}{\gamma}v_{\eta\eta} = 0$$

and hence for invariance of the equation we must take

$$\beta = \alpha^2 = \gamma.$$

Because this equation is linear homogeneous γ plays no role and we may as well take $\gamma = 1$ and $u = v$. Formally differentiating

$$u(x, y) = u(\zeta, \eta) = u(\alpha x, \alpha^2 t)$$

with respect to the scale parameter α we have

$$0 = xu_\zeta(\zeta, \eta) + 2\alpha t u_\eta(\zeta, \eta)$$

from which*

$$\alpha\frac{dt}{dx} = \frac{d(\alpha^2 t)}{d(\alpha x)} = -\frac{u_\zeta(\zeta, \eta)}{u_\eta(\zeta, \eta)} = \frac{2\alpha t}{x}.$$

This is a separable first-order equation with solution $t = cx^2$ where c is an arbitrary constant. Thus the scale change of variables brought us back to the same similarity variable $\eta = x^2/t$ found by dimensional analysis.

Problem A.5

(a) Solve the ordinary differential equation in the similarity independent variable $\eta = \alpha x^2/t$. Then verify that the resulting solution $u(\eta) = u(x, t)$ satisfies the heat equation expressed in x and t.

(b) Remembering that s is a dummy integration variable for $\eta = \alpha x^2/t$, the integrand exponential resembles the Poisson Kernel in the Green's function representation of the solution given in Section 1.3, Problem 2. Try to make more precise this connection.

(c) Nonlinear diffusion equations were mentioned at the end of Section 2.9.8. Show that the porous media equation

$$u_t = (u^m)_{xx}, \qquad 0 < x < \infty, \qquad t > 0$$

has a similarity variable $\eta = x(t + 1)^{-1/2}$ for which solutions of the form $u(x, t) = g(\eta)$ must satisfy the ordinary differential equation

$$(g^m)'' + \tfrac{1}{2}\eta g' = 0.$$

For the case of $m = 1$ (heat equation) show how this solution reduces to the one referred to in (b) above.

Recently there has been an upsurge in the theoretical development of (Lie) group methods to determine all "symmetries" of nonlinear differential equations. We will inspect this theory in the next section. However, the importance of the use of scaling arguments from dimensional

---

* Remember the implicit differentiation formula $y'(x) = -F_x/F_y$ from calculus.

analysis to find similarity solutions should not be underestimated. Not only do such methods often pick out the most important (e.g., geometrically scale invariant) physical solutions but they also can lead to important simplified differential equations.

We close this section with two important examples, the derivation of the Prandtl* boundary layer approximation to the Navier–Stokes equation, and a similarity solution analysis of an atomic bomb blast.†

The two-dimensional Navier–Stokes equations for viscous incompressible flow (see Section 1.7) are the momentum equations

$$\begin{cases} uu_x + vu_y = -p_x + \nu(u_{xx} + u_{yy}) \\ uv_x + vv_y = -p_y + \nu(v_{xx} + v_{yy}) \end{cases}$$

along with the continuity equation

$$u_x + v_y = 0.$$

We select the scaling region (Fig. A.6 below) with incoming flow velocity U_∞ with the thin boundary layer assumption $\delta \ll L$. Scaling arguments usually assume that differentiations do not change the orders of magnitude involved. This restricts the oscillations that may be accounted for.† Proceeding then with the scaling, we have

$$\begin{cases} \Delta u \sim U_\infty, & x \sim L \\ \Delta v \sim ?, & y \sim \delta, \end{cases}$$

which from the continuity equation gives

$$\frac{U_\infty}{L} \sim \frac{v}{\delta}$$

and hence v, $\Delta v \ll U_\infty$. The first momentum equation yields

$$\frac{U_\infty^2}{L} + \frac{U_\infty^2}{L} \sim (?)_x + \nu\left(\frac{U_\infty}{L^2} + \frac{U_\infty}{\delta^2}\right)$$

from which the smaller terms $\nu U_\infty/L^2$ will be dropped. The second momentum equation gives

$$\frac{\delta U_\infty^2}{L} + \frac{\delta U_\infty^2}{L^2} \sim (?)_y + \nu\left(\frac{\delta U_\infty}{L^3} + \frac{U_\infty}{\delta L}\right)$$

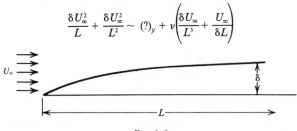

FIG. A.6

---

* L. Prandtl, Verhandlg 3rd Math. Congress, Heidelberg, 1904.

† G. Taylor, *Proc. Roy. Soc.* A 200, A 201, 1950, although done 8 years earlier. Sedov, *Prikl. Mat. Mekh.* **10**, 1946, had independently found a similarity solution during the same period. Apparently so had John Von Neumann in the United States.

† A similar restriction is built into the group methods of the next section, although its extent is somewhat obscured.

from which all terms will be dropped. The largest of those is the last one, and from it and the scaled first momentum equation we deduce

$$p_y \sim \nu \frac{U_\infty}{\delta L} \qquad\qquad p_x \sim \frac{U_\infty^2}{L} - \nu \frac{U_\infty}{\delta^2} \sim \nu \frac{U_\infty}{\delta^2}$$

From these we see that $p_y/p_x \sim \delta/L$. An argument similar to this led Prandtl to therefore assume p_y to be negligible within the boundary layer.

In summary, we have obtained the Prandtl equations

$$\begin{cases} uu_x + vu_y = -p_x + \nu u_{yy} \\ u_x + v_y = 0 \end{cases}$$

We see that scaling arguments have eliminated the second momentum equation. Furthermore, all second x derivatives have vanished so that the first momentum equation has changed type, from elliptic to parabolic. The latter is more amenable to solution and the model now contains fewer boundary conditions to satisfy. The pressure $p = p(x)$ now may be regarded as a known function, imposed from the outer flow.

These equations are of well-known importance and a wide literature exists. It turns out that by a change of variables they may be converted to porous media equations such as those mentioned in Problem A.5 above. See the following Problem A.6. From the porous media equations many of the Prandtl equation's mathematical properties may be investigated.*

Finally, we mention the interesting story of how the amount of energy released in an atomic bomb blast may be surmised from a similarity analysis of sequential photos of such an explosion. For an explosion with a spherical symmetry let $r = r(E, t, p_0, \rho_0, \eta)$ be the position of the ensuing wave front. Here E denotes the initial energy of the blast, and p_0 and ρ_0 the pressure and density before the blast, η the ratio of specific heats before and after. Treating η as the new independent similarity variable, and treating E and r as fundamental parameter and basic dependent variable, one may deduce dimensionally the dependence (where g is an as yet unknown function)

$$\frac{\rho_0 r^5}{E t^2} = g(\eta, p_0 r^3/E)$$

Thus the wave front position is thus given by

$$r(t) = \left(\frac{E}{\rho_0}\right)^{1/5} t^{2/5} g(\eta, p_0 r^3/E).$$

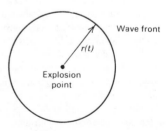

FIG. A.7

* O. Oleinik, *Seminari 1962/63 Anal. Alg. Geom. e. Topol. Vol. 1*, Ist. Naz. Alta. Mat., Ediz. Cremonese, Rome, 1965.

One may simplify this by the physical assumption that a gasdynamic shock has much higher pressure behind it: $p \gg p_0$. Dropping p_0, the function $g = g(\eta)$ is now dependent only on the similarity variable η. Basic gas dynamics then yields an explicit specific heat dependence $g(\eta)$ and hence the position of the wave front $r(t)$. Conversely, photographs of the latter at two or more times yield the energy E initially released.

Problem A.6

(a) Show that the boundary layer thickness δ is proportional to $Re^{-1/2}$, where Re denotes the Reynolds number

$$Re = \frac{U_\infty L}{\nu} .$$

(b) For the Prandtl equations written as

$$\begin{cases} uu_x + Vu_Y = -p_x + u_{YY} \\ u_x + V_Y = 0, \end{cases}$$

where $V = Re^{1/2}v$, $Y = Re^{1/2}y$, use the Von Mises stream function transformation

$$u = \psi_Y, \qquad v = -\psi_x$$

to arrive at the porous media equations

$$u(uu_\psi)_\psi = uu_x + p_x.$$

(c) The Prandtl equations can be described as setting up a two-deck structure describing a fluid near a wall: very near the wall the boundary layer equations, outside that small region the full Navier–Stokes equations. This has been extended to the so-called triple-deck model* and even more "decks." These theories contain many interesting mathematical and physical questions. Read about these.

A.3. Local Transformation Groups

The transformation of a partial differential equation under a class of variable changes was evident right at the beginning of this book, in the classification procedures of Section 1.1, 1.9.1 for rendering a linear second equation to its elliptic, parabolic, or hyperbolic canonical form. For nonlinear equations, classification was still formally possible but a suitable transformation of independent variables to canonical form (now that the coefficients depend on the dependent variable) becomes less tangible. Permitting the dependent variable u to enter into the class of allowable transformations thus makes sense.

The occurrences of the self-similar solutions of part A.2 above and other preserved entities of a partial differential equation were seen to be related to invariance properties of the equation with respect to transformations of the underlying space of independent variables and even to transformations involving both the independent and the dependent variables. The latter case describes the approach taken by Lie long ago,† principally for ordinary differential equations but with some application to partial differential equations. An instance

* K. Stewartson, *SIAM J. Appl. Math.* **28** (1975).

† S. Lie, Verhandlungen der Gesellschaft der Wissenshaften zu Christiania (Oslo before Norway's independence from Sweden) (1874).

of those techniques are the Legendre transformations mentioned earlier in Problem 5 of Section 1.9.9. As seen there, they permit one to "linearize" a given nonlinear partial differential equation by introducing additional variables. Because linear problems are easier to handle, these methods have recently enjoyed greatly renewed interest for use in solid-state physics, engineering control theory, transonic flow, and elsewhere.*

We have also seen in part A.1 above the introduction of new variables, e.g., $p = \partial u/\partial x$, $q = \partial u/\partial y$ as a useful, even essential, device to enable a geometrical picture of the solution locus of a general first-order partial differential equation. Earlier, in Section 1.9.4, we saw how by assigning new variables to each partial derivative, any analytic partial differential initial value problem could be transformed to a system of quasilinear partial differential equations of first order. The latter was then guaranteed a (local) solution by the Cauchy–Kowalewski Theorem.

Recall the exact equation rule of elementary ordinary differential equations: given the expression

$$M dx + N dy = 0,$$

if $\partial M/\partial y = \partial N/\partial x$, the expression is an exact differential

$$d\phi = M dx + N dy = 0$$

and the differential equation

$$M(x, y) + N(x, y)y'(x) = 0$$

is easily solved by first integrating one of $M = \partial\phi/\partial x$ or $N = \partial\phi/\partial y$ to obtain $\phi(x, y)$, for which the partial constant of integration is then determined by substituting into the other of M and N.

For example, the differential equation

$$x + yy'(x) = 0$$

satisfies $M_y = N_x = 0$ and hence has solution

$$\phi(x, y) = \int^x M = x^2/2 + c(y)$$

which by

$$\phi_y = c'(y) = y$$

provides the solution

$$\phi(x, y) \equiv x^2 + y^2 = c,$$

where c is the arbitrary constant of integration.

---

* Group theory for the analysis of linear partial differential equations has been used extensively in mathematical and theoretical physics for some time now. The use of infinitesimal transformation groups and their relation to finite transformation groups for the analysis of nonlinear ordinary and partial differential equations has recently seen a resurgence, with a number of excellent mathematical references. See for example G. Bluman and J. Cole, *Similarity Methods for Differential Equations* (Springer, New York, 1974), L. Ovsiannikov, *Group Analysis of Differential Equations* (Academic Press, New York, 1982), and the references cited therein.

Under what change of variables does this solution locus $x^2 + y^2 = c$ remain invariant? As the solution locus consists of circles, clearly a rotation of coordinates x and y through any angle will not affect it. Remember that, in contrast to the situation in the classification procedure of Section 1.1.1 wherein we wanted to alter the equation and hence solution form, here we are interested in *not altering at all* the form of the solution. If that is to be the case, the solution must be invariant under even a very small portion of the envisioned change of variables $\zeta = \zeta(x, y)$, $\eta = \eta(x, y)$. Therefore we restrict attention to the very small movement

$$\tilde{x} = x + \varepsilon X(x, y) + 0(\varepsilon^2)$$

$$\tilde{y} = y + \varepsilon Y(x, y) + 0(\varepsilon^2).$$

To do so amounts to assuming that the envisioned transformation is analytic. Dropping all terms of order $0(\varepsilon^2)$, we are then linearly approximating it for very small

$$\Delta x = \tilde{x} - x = \varepsilon X,$$

$$\Delta y = \tilde{y} - y = \varepsilon Y.$$

For x and y in the solution locus $x^2 + y^2 = c$ we then have

$$(\tilde{x} - \varepsilon X)^2 + (\tilde{y} - \varepsilon Y)^2 = c,$$

and upon requesting the invariance $\tilde{x}^2 + \tilde{y}^2 = c$ we obtain

$$-2\varepsilon(X\tilde{x} + Y\tilde{y}) = 0.$$

The very small increments $\varepsilon X = \Delta x$ and $\varepsilon Y = \Delta y$ may be approximated by their differentials dx and dy, from which we have

$$x\, dx + y\, dy = 0.$$

We recognize this as the differential equation from which we started. From the assumed analyticity of the transformation we have $X(x, y) = -y$ and $Y(x, y) = x$ and the envisioned change of variables can be only a rotation.

Generally, in this approach one quickly writes down the above *characteristic equation*

$$\frac{dx}{X} = \frac{dy}{Y} = \frac{du}{U} \, .$$

Here we have also allowed a change in the dependent variable. When X and Y do not depend on u, the solutions curves $\phi(x, y) = c$ of the first equality of the characteristic equation are called the *invariant curves* (of the given differential equation), and $\eta \equiv \phi$ is called the *similarity variable* and regarded as a new independent variable. In the example above, $\phi \equiv \eta = r^2$ is just the radial direction squared. Seen in polar coordinates, the characteristic equation

$$\frac{\cos\theta\, dr - \sin\theta\, d\theta}{-r\sin\theta} = \frac{\sin\theta\, dr + \cos\theta\, d\theta}{r\cos\theta}$$

becomes $2r\, dr = 0$. Thus in the polar similarity variable the differential equation in two variables x and y has been reduced to the one-variable equation $\eta' = 0$.

Problem A.7

(a) Above we asserted that an expansion $\tilde{x} = x + \varepsilon X + 0(\varepsilon^2)$, $\tilde{y} = y + \varepsilon Y + 0(\varepsilon^2)$ led to the characteristic equation $dx/X = dy/Y$. Show more generally that the expansions

$$\begin{cases} \hat{x} = x + \varepsilon X(x, y, u) + 0(\varepsilon^2) \\ \tilde{y} = y + \varepsilon Y(x, y, u) + 0(\varepsilon^2) \\ \tilde{u} = u + \varepsilon U(x, y, u) + 0(\varepsilon^2) \end{cases}$$

leads to the characteristic equation

$$\frac{dx}{X} = \frac{dy}{Y} = \frac{du}{U}.$$

(b) Ruminate on and relate the following notions.

(i) Divergence Theorem. See Section 1.6.1. The tangent plane to a surface $u(x, y)$ is given by

$$u - u_0 = (x - x_0)u_x + (y - y_0)u_y.$$

(ii) Characteristics. See Section 1.9.2. When $u = u(x, y)$ is requested to contain an implicit relation $y = y(x)$ between the independent variables, along which u is constant, then from implicit differentiation

$$0 = du = dx u_x + dy u_y$$

(iii) Legendre transformation. See Problem 1.9.9(4). Regarding the solution surface $u(x, y)$ of a partial differential equation as the envelope of its tangent planes, note that the Legendre transformation is of the form (extended, old literature; prolonged, new literature)

$$u(x, y) + \omega(u_x, u_y) = xu_x + yu_y.$$

(iv) First-order equations. See Appendix A.1. The solution of the equation

$$c(x, y, u) = a(x, y, u)u_x + b(x, y, u)u_y$$

may be obtained from the ODE system

$$\begin{cases} \dfrac{dx}{ds} = a(x, y, u) \\[2mm] \dfrac{dy}{ds} = b(x, y, u) \\[2mm] \dfrac{du}{ds} = c(x, y, u) \end{cases}$$

integrated in the s direction.

(v) Cauchy–Kowalewski Theorem. See Appendix A.1 and Problem 1.9.4. The solution of the first-order equation

$$F(x, y, u, u_x, u_y) = 0$$

may be obtained from the ODE system

$$\begin{cases} \dfrac{dx}{ds} = F_{u_x} \\[2mm] \dfrac{dy}{ds} = F_{u_y} \\[2mm] \dfrac{du}{ds} = F_{u_x}u_x + F_{u_y}u_y \\[2mm] \dfrac{du_x}{ds} = -F_x - u_x F_u \\[2mm] \dfrac{du_y}{ds} = -F_y - u F_u . \end{cases}$$

(vi) Group Invariance. See this Appendix. The infinitesimal change of variables

$$\begin{cases} \Delta x = \tilde{x} - x = \varepsilon X(x, y, u) \\ \Delta y = \tilde{y} - y = \varepsilon Y(x, y, u) \\ \Delta u = \tilde{u} - u = \varepsilon U(x, y, u) \end{cases}$$

plus equation invariance requires the *invariant surface condition*

$$U(x, y, u) = Xu_x + Yu_y$$

which is solved by the characteristics equation

$$\frac{dx}{X} = \frac{dy}{Y} = \frac{du}{U} .$$

(c) How do you solve the simple case of the first characteristic equation

$$\frac{dx}{X(x)} = \frac{dy}{Y(y)}$$

to always obtain a similarity variable?

For an important example of the application of this method of local transformation groups, let us consider the transonic flow equation

$$u_{xx} - u_y u_{yy} = 0$$

This is a type of ''full potential'' approximation to flow over a wing, where u_x is the velocity $(v - v_\infty)/v_\infty$ relative to an assumed far-field steady velocity and where we have normalized to one other gas dynamic constants.

Proceeding as above, we consider the transformation

$$\tilde{x} = x + \varepsilon X(x, y, u) + 0(\varepsilon^2)$$
$$\tilde{y} = y + \varepsilon Y(x, y, u) + 0(\varepsilon^2)$$
$$\tilde{u} = u + \varepsilon U(x, y, u) + 0(\varepsilon^2)$$

The method of equation invariance under such infinitesimal transformations involves three steps:

1. By use of the chain rule, replace the equation by its version in the $\bar{x}$, $\bar{y}$, $\bar{u}$ variables.
2. Use the original relation $u_y u_{yy} = u_{xx}$ and drop all $0(\varepsilon^2)$ terms to simplify the expression of l.
3. Set all coefficients of the remaining dependent variables equal to zero to obtain the determining equations for the coefficients X, Y, and U.

For the equation at hand, a very horrendous chain rule calculation leads to

$$
\begin{aligned}
& U_{xx} + (2U_{xu} - X_{xx})\theta_x + (U_{uu} - 2X_{xu})\theta_x^2 \\
& - X_{uu}\theta_x^3 + X_y\theta_x\theta_{yy} - X_u\theta_x\theta_y\theta_{yy} \\
& + (Y_{xx} - U_{yy})\theta_y + (Y_{yy} - 2U_{yu})\theta_y^2 \\
& + (2Y_{yu} - U_{uu})\theta_y^3 + Y_{uu}\theta_y^4 + (3Y_y - U_u - 2X_x)\theta_y\theta_{yy} \\
& + 2X_y\theta_y\theta_{xy} - 2Y_x\theta_{xy} - 2Y_u\theta_x\theta_{xy} + 3Y_u\theta_y^2\theta_{yy} \\
& + 2X_u\theta_{xy}\theta_y^2 + U_y\theta_{yy}
\end{aligned}
$$

as the coefficient of ε that must therefore vanish. All $0(\varepsilon^2)$ terms have been dropped and θ denotes the sought solution. Setting all coefficients of the dependent variables to zero produces the simultaneous equations (some are redundant) below, where we have also indicated their implications in the same fashion as in part A.1:

$$
\left.\begin{aligned} X_y &= 0 \\ -X_u &= 0 \end{aligned}\right\} \Rightarrow X = X(x)
$$

$$
\left.\begin{aligned} -2Y_x &= 0 \\ -2Y_u &= 0 \end{aligned}\right\} \Rightarrow Y = Y(y)
$$

$$
\left.\begin{aligned} U_y &= 0 \\ Y_{yy} - 2U_{yu} &= 0 \end{aligned}\right\} \Rightarrow Y = \alpha y + \beta
$$

$$
\left.\begin{aligned} U_{xx} &= 0 \\ 2Y_{yu} - U_{uu} &= 0 \end{aligned}\right\} \Rightarrow U = \tau xu + \lambda u + \gamma x + \sigma
$$

$$
\left.3Y_y - U_u - 2X_x = 0 \;\right\} \Rightarrow X = \frac{-\tau}{4}x^2 + \left(\frac{3\alpha}{2} - \frac{\lambda}{2}\right)x + \kappa
$$

$$
\left.2U_{xu} - X_{xx} = 0 \;\right\} \Rightarrow \tau = 0.
$$

Hence

$$
\begin{cases}
X = \left(\dfrac{3\alpha}{2} - \dfrac{\lambda}{2}\right)x + \kappa, \\
Y = \alpha y + \beta, \\
U = \lambda u + \gamma x + \sigma.^*
\end{cases}
$$

As the α, β, θ, κ, λ, σ are all arbitrary constants, one has a wide range of freedom in choosing specific changes of variable that will leave the form of the equation invariant.

The more important goal being to reduce the PDE to an ODE from which at least one solution may be extracted, let $\alpha = \lambda = 1$ and the other constants be 0. Then we are back to the characteristic equation

* This important invariance group for the transonic flow equation was first obtained in G. Nariboli, *Appl. Sci. Res.* **22** (1970).

$$\frac{dx}{x} = \frac{dy}{y} = \frac{du}{u}$$

Integrating the first equation $dx/x = dy/y = ds$ we have

$$\ln x - \ln x_0 = \ln y - \ln y_0 = s$$

hence

$$\frac{x}{x_0} = \frac{y}{y_0} = e^s.$$

Integrating the second equation $dy/y = du/u = ds$ gives

$$\frac{y}{y_0} = \frac{u}{u_0} = e^s.$$

Hence the choice $\alpha = \lambda = 1$, others $= 0$ gave us the stretching transformation group

$$\begin{cases} x = x_0 e^s \\ y = y_0 e^s \\ u = u_0 e^s. \end{cases}$$

Each of the two integrations above produces an arbitrary constant of integration. We could choose as new similarity independent variable η any constant of integration of the first equation

$$\ln x = \ln y + \ln c$$

and we choose $\eta = c = x/y$. Then the constant of integration of the second equation depends on η and becomes a new dependent variable $g(\eta)$. Thus

$$\ln y = \ln u + g(\eta)$$

from which

$$u = e^{-g(\eta)} y = \frac{e^{-g(\eta)}}{\eta} x.$$

Had we used the $dx/x = du/u$ equation for the second integration we would have arrived at the same place

$$u = f(\eta)x$$

where f is an (essentially) arbitrary function of η.

Note that $u = f(\eta)x$ is now, by use of the similarity change of variables, in a separated variables form. This is, as always in separation of variables procedures, a special choice of solution. From it we may now find an ordinary differential equation satisfied by f. Substituting $u = f(\eta)x$ with $\eta = x/y$ into the transonic flow equation yields

$$\eta^5 f'' f' + 2\eta^4 (f')^2 + \eta f'' + 2f' = 0$$

where $'$ denotes derivative with respect to η. Letting h denote f' and trying solutions of the form $h = a\eta^b$ yields a solution with $a = -1$ and $b = -4$, from which

$$u = x(\eta^{-3/3} + c_1) = y^3/3x^2 + c_1 x$$

is a solution.

The stipulating of invariance under a local transformation group has thus enabled us to find solutions of a nonlinear partial differential equation by separation of variables.

Problem A.8

(a) Perform the chain rule calculations and collections of terms to obtain the coefficient of ε above.

(b) Investigate other possible solutions to the ordinary differential equation found above.

(c) For other α, β, γ, κ, λ, σ parameter choices, solve the corresponding characteristics equations to find the local group represented and the resulting similarity variables, ordinary differential equations, and resultant special separated variables solutions.

As three last topics in this appendix, we want to just mention the methods of prolongation, generalized symmetries, and conservation laws.

Each arbitrary group constant α_i, such as the six parameters α, β, γ, κ, λ, σ for the invariance group for the transonic flow equation, may be associated with an infinitesimal operator V_i. The latter are sometimes called the symmetries, or symmetry generators, or better, the group generators of the invariance group,* and each may be obtained explicitly from the expression

$$V = X\frac{\partial}{\partial x} + Y\frac{\partial}{\partial y} + U\frac{\partial}{\partial u}$$

with $\alpha_i = 1$, all other $\alpha_j = 0$.

In searching for the group generators V_i in cases where the invariance group is not yet known, one tries the above expression with unknown coefficients $X(x, y, u)$, $Y(x, y, u)$, $U(x, y, u)$. Note that by letting these coefficients depend on u we are envisioning a change of variables not only in the independent variables x and y but also in the dependent variable u.

Very much in the spirit of both the Legendre transformation (see Section 1.9.9.4) which transforms not only variables but also surface elements, (x, y, u, u_x, u_y) to $(\omega_\zeta, \omega_\eta, \omega, \zeta, \eta)$, and the Cauchy–Kowalewski quasilinearization procedure in which derivatives are regarded as new variables, one may consider coefficients X, Y, and U depending also on the additional derivative variables $p = u_x$ and $q = u_y$, and even selected higher derivatives. This especially makes sense for partial differential equations possessing nonlinearities explicitly in those derivatives. Once you allow new variables such as $p = u_x$ and $q = u_y$ it is also natural to consider (extended, old literature; prolonged, new literature) expressions such as

$$V = X\frac{\partial}{\partial x} + Y\frac{\partial}{\partial y} + U\frac{\partial}{\partial z} + P\frac{\partial}{\partial p} + Q\frac{\partial}{\partial q}.$$

---

* The invariance group is defined as the largest local group of transformations acting on the independent and dependent variables which transform solutions of the equation to other solutions of the equation. From our point of view those would be the changes of variable that leave the form of the equation unaltered. The use of local infinitesimals to find the group means that there may be other "far away" variable changes (e.g., reflections) which leave the equation invariant. Also, of course, there can be other changes of variable which alter the equation while simplifying it.

These prolongations (generalized symmetries) allow for investigation of further changes of variables and provide the possibility of further simplification (e.g. , linearization) of the differential equation.

There results a very extensive, still developing theory. Let us, solely for informational purposes, portray it as follows, in a hierarchy of increasingly general settings for change of variables. We admit to over-emphasizing the latter viewpoint rather than more theoretical ones. The names given below are meant to be indicative rather than precise, complete, or exhaustive in any way whatsoever.

Coefficients of V	Theory	Associated Names
X_i (x_j only)	Similarity Variables	Galileo, Helmholtz
X_i (x_j, u)	Banal Case	Galois, Lie
X_i (x_j, u, u_{x_j}, $u_{x_j k}$)	Contact Transformations	Cartin, Vessiot
X_i (x_j, u, u_{x_j}, $u_{x_j k}$ . . . .)	Conservation Laws	Lax, Zakharov

We conclude with a comment about conservation laws for partial differential equations. As is well-known, many important partial differential equations are derived from physical conservation laws (see Section 1.7). Often these are associated with a variational principle. For partial differential equations which are of the Euler equations (see Section 1.9.5(3)) of a variational extremum there is an important result, sometimes called a generalized Noethers theorem,[*] under which the generalized symmetries V_i found above may be seen to be in a 1-1 correspondence with generalized conservation laws.[+] The latter may be regarded as generalizations of the fact that a similarity variable η is constant along a similarity curve.

Problem A.9

(a) Write down the six group generators V_i for the transonic flow equation. If you wish to approach the algebraic aspect of this theory, show that they form a Lie algebra under the commutator operation.

(b) Accepting a "universal conservation law" as epitomized for example by the saying "you don't get something for nothing," what really underlies this theory?

(c) Connect this theory to an important scheme in the numerical analysis of partial differential equations.

FIFTH PAUSE: EXAMPLES, EXPLANATIONS, EXERCISES

In Appendix A. 1 we considered the single first order quasilinear partial differential equation $a(x, y, u)u_x + b(x, y, u)u_y = c(x, y, u)$ in a rather general setting. Considerable interest attaches to systems of such equations of the form

$$u_t + (f(u))_x = 0$$

called *hyperbolic systems of conservation laws,* coming from fluid and gas dynamics, for example. But before we sketch that theory, it will be instructive to gain intuition about solution behavior from the scalar case.

---

[*] For E. Noether, *Nachr. Ges. Göttingen (math. phys.* KI. 1981).
[+] Some examples for the Korteweg-deVries equation were given in Section 2.9.8.

Example

The conservation law $u_t + (f(u))_x = 0$ can be written, assuming that the flux function f is differentiable, in the quasilinear form $u_t + f_u u_x = 0$. Here we want to consider three special scalar equations, which we call

$$u_t + cu_x = 0 \qquad \textit{translation equation,}$$
$$u_t + uu_x = 0 \qquad \textit{convection equation,}$$
$$u_t + c(u) u_x = 0 \qquad \textit{advection equation.}$$

For a discussion of our choice of names, see the Exercise 2 below.

For each of these equations, consider qualitatively the solution behavior for the pure initial value problem

$$\begin{cases} u_t + cu_x = 0 & -\infty < x < \infty, t > 0 \\ u(x, 0) = u_0(x) \text{ given} & -\infty < x < \infty. \end{cases}$$

Solution. The key to this theory is the chain rule from calculus, specifically, the total derivative

$$\frac{du}{dt} = \frac{\partial u}{\partial t} \frac{dt}{dt} + \frac{\partial u}{\partial x} \frac{dx}{dt}.$$

If we can find $x(t)$ curves in the $x - t$ plane such that we may set $dx/dt = c$ on those curves, then on those curves, (called *characteristics* of the conservation law)

$$0 = \frac{du}{dt} = \frac{\partial u}{\partial t} \frac{dt}{dt} + \frac{\partial u}{\partial x} \frac{dx}{dt} = u_t + cu_x = (u_t, u_x) \cdot \begin{pmatrix} 1 \\ c \end{pmatrix},$$

which we may interpret as follows.

See Fig. 5Pa, and we consider the constant coefficient translation equation first, with c taken positive. Fix an initial point x_0. From $dx/dt = c$ we have the curve $x(t) = ct + x_0$ emanating from $(x_0, 0)$ on which the total derivative du/dt of the solution $u(x,t)$ vanishes. Therefore $u(x, t)$ is constant on that curve, and therefore the solution retains the initial value u_0 along that curve. Doing this for all $-\infty < x_0 < \infty$ reveals the solution $u(x,t)$ as nothing more than the initial profile translating to the right with speed c. Thus $u(x,t) = u_0(x - ct)$. Let us also note that the relationship the characteristics must have with the solution gradient is that of normality. We have illustrated this in Fig. 5Pb for the initial profile and translation equation of Fig. 5Pa.

Does this reasoning work as well for the convection and advection equations? For the convection equation, the characteristic curves $x(t)$ are set to satisfy $dx/dt = u(x, t)$. But because the total derivative du/dt must vanish on such a curve, $u(x,t)$ will be constant, and hence just u_0, along it. Thus, once we fix the initial point x_0, we have fixed the initial value u_0, which propagates along the characteristic. Hence $dx/dt = u_0$ there and the characteristic is again a straight line, $x(t) = u_0 t + x_0$. Note now, however, that a shock discontinuity is produced if an initial value u_1 at some x_1 greater than x_0 is smaller than u_0, see Fig. 5Pc.

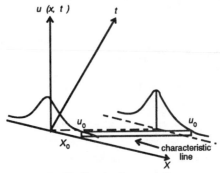

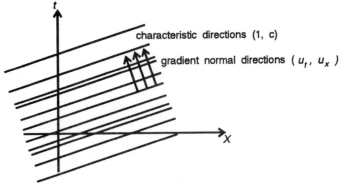

FIG 5P*a*. Translated wave solution

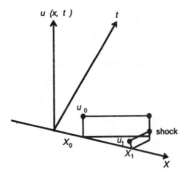

FIG 5P*b*. Topographic normality

FIG 5P*c*. Convection induced shock

For the advection equation, the same reasoning holds, producing characteristic lines again, $x(t) = c_0 t + x_0$ where $c_0 = c(u_0)$. Note that the $c(u_0)$ will now determine the fanning out or fanning in of the characteristics.

Exercises

1. Use the shock initial profile $u_0(x) = 1$ for $x \leq 0$, $u_0(x) = 0$ for $x > 0$ for the translation equation initial value problem. What does the solution look like?

2. We could also have used the names 'Burger's equation' or 'Material or Substantial' equation for what we called the convection equation? Why do we prefer the latter? Why do we call $u_t + c(u) u_x = 0$ the advection equation? What other names could be considered?

3. Reconcile this presentation of $u_t + u u_x = 0$ with that of Appendix A.1, especially, for example, with the worked example $u u_x + u_y = 1$ given there.

4. What are the jump, entropy, criteria for uniqueness of (weak) solutions to these equations? Consult the literature about the *shock dynamics* of advection equations.

5. Observe a connection to the similarity solution of Appendix A.2. State a connection to the group point of view of Appendix A.3.

Now we want to consider hyperbolic systems of conservation laws,

$$u_t + (f(u))_x = 0,$$

where u, f, and x are vectors.

Example

Consider the one space variable case, with $u(x, t)$ and $f(u)$ n-vectors. Convert from conservation law form to quasilinear form

$$u_t + A(u) u_x = 0$$

and analyze by the characteristics methods used above in the scalar case.

Solution. The key will again be the chain rule and conversion to total derivatives on characteristics. First, however, let us define what is meant by *hyperbolic* in this context, and why it is useful to us.

If at each point $(x, t, u_1, \ldots, u_n)$ there is an invertible matrix Q such that

$$Q^{-1} A Q = D = \begin{bmatrix} \lambda_1 & & & 0 \\ & \cdot & & \\ & & \cdot & \\ 0 & & & \lambda_n \end{bmatrix}$$

where the eigenvalues λ_i of A are all real, then the quasilinear first order system of partial differential equations is said to be *hyperbolic* at that point. If the eigenvalues λ_i are all distinct, the system is called strictly hyperbolic there. The interest in the quasilinear form of the system and its hyperbolicity lies in the fact that we can then decouple the system. Indeed, letting $v = Q^{-1}u$, we have $u_t + A(u)u_x = 0 \Rightarrow$ $u_t + QDQ^{-1}u_x = 0 \Rightarrow Q^{-1}u_t + DQ^{-1}u_x = 0 \Rightarrow v_t + Dv_x = 0$, a diagonal and hence decoupled system, *provided* that in the last step we ignore the fact that Q depends on the point (x, t, u).

Let us turn then to the conversion from conservation law form to quasilinear form. By the chain rule applied to the first component of $(f(u))_x$ we have

$$f_1(u)_x = \frac{\partial f_1(u)}{\partial u_1} \cdot \frac{\partial u_1}{\partial x} + \ldots + \frac{\partial f_1(u)}{\partial u_n} \cdot \frac{\partial u_n}{\partial x}$$

so that by doing the same to the other components of f we arrive at the quasilinear system

$$u_t + A(u)u_x = 0$$

where $A(u)$ is the $n \times n$ Jacobian matrix whose ij-th element is $\partial f_i / \partial u_j$. We will assume that A is hyperbolic, i.e., real diagonalizable, at all points $(x, t, u_1, \ldots u_n)$. This is the case, for example, when A is a symmetric matrix, and also in important applications in which A is not symmetric.

Remembering that an $n \times n$ matrix is diagonalizable if and only if both A and its transpose $A*$ possess n linearly independent eigenvectors, and also remembering that the eigenvalues λ_i of A and $A*$ are the same when real, we form the inner products of the equation with the n eigenvectors v_j of $A*$ to get n *orthogonality* (or normal) *equations:* $0 = (u_t + Au_x, v_j) = (u_t, v_j) + (u_x, A* v_j) = (u_t, v_j) + (u_x, \lambda_j v_j) = (u_t + \lambda_j u_x, v_j)$, ie,

$$0 = (u_t + \lambda_j u_x, v_j), \quad j = 1,\ldots,n$$

To now bring in characteristic directions, we would like to express each component of the vector $u_t + \lambda_j u_x$ as a total derivative

$$\frac{du_i}{dt} = \frac{\partial u_i}{\partial t}\frac{dt}{dt} + \frac{\partial u_i}{\partial x}\frac{dx}{dt}.$$

Thus, just as in the scalar case above, we assume that we can find characteristic curves $x_j(t)$ in the $x - t$ plane satisfying

$$\frac{dx_j(t)}{dt} = \lambda_j (x, t, u_1, \ldots, u_n).$$

Given such curves $x_j(t)$ in the $x - t$ plane, the n orthogonality equations above become the n ordinary differential equations

$$\frac{du_1}{dt}v_{11}(u) + \frac{du_2}{dt} v_{12}(u) + \ldots + \frac{du_n}{dt} v_{1n}(u) = 0 \quad j = 1$$
$$\vdots \qquad\qquad \vdots \qquad\qquad \vdots$$
$$\frac{du_1}{dt} v_{n1}(u) + \frac{du_2}{dt}v_{n2}(u) + \ldots + \frac{du_n}{dt} v_{nn}(u) = 0 \quad j = n,$$

each holding only on the j-th characteristic curve $x_j(t)$.

The PDE system has thus been reduced to ODE's along characteristics, just as in the scalar case. These ODE's are sometimes called the *normal* equations or normal forms for the original system. As ODE's they, along with the ODE's defining the characteristics, enable (in principle) the integration of the PDE system via ODE solvers.

Exercises

6. Review some linear algebra about diagonalizability.

7. What are the eigenvalues and eigenvectors in the three scalar equations above?

8. Integrate the vector conservation law and interpret.

Important first order hyperbolic systems of conservation laws occur in fluid dynamics when one ignores viscous effects.

Example

Show that the one dimensional Euler equations of inviscid compressible gas dynamics are a hyperbolic system of conservation laws, and find the eigenvalues λ_1, λ_2, and λ_3 of the system.

Solution. These equations can be written

$$\begin{bmatrix} \rho \\ \rho u \\ e \end{bmatrix}_t + \begin{bmatrix} \rho u \\ \rho u^2 + p \\ (e+p)u \end{bmatrix}_x = \begin{bmatrix} 0 \\ 0 \\ 0 \end{bmatrix}$$

where ρ is gas density, $m = \rho u$ is the mass, u is the velocity, p is the pressure, $e = \rho\varepsilon + \rho u^2/2$ is the total energy per unit volume, ε being a postulated internal energy per unit mass. The three equations in the system represent conservations of mass, momentum, and energy, respectively. Additionally* we assume a perfect gas law $p = (\gamma - 1)\rho\varepsilon$, where $\gamma > 1$ is a material dependent constant called the ratio of specific heats: for definiteness one can just take $\gamma = 1.4$ for air, for example. This additional equation of state permits the elimination (note that $p = (\gamma - 1)(e - \rho u^2/2)$) of both the total energy e and the internal energy ε from the equations, in favor of the primitive variables, ρ, u, and p (see Exercise 9).

Still written in terms of density, mass, and energy variables, using the additional state equation, we have the conservation system

---

*We refer to any book on gas dynamics for the derivation of these equations. However, as an interesting sidelight, recall the discussion in Section 1.7.1, where the postulated internal energy function U played a similarly somewhat disconcerting, or at least, intangible role. There, we simplified U_u to $c(u)$ to an assumed constant specific heat c throughout the body and temperature range under consideration, in order to produce linearity. Here, it is embarrassingly convenient to simplify the internal energy ε to be proportional to the ratio of pressure to density, in order to produce three equations in three unknowns.

$$\begin{bmatrix} \rho \\ m \\ e \end{bmatrix}_t + \begin{bmatrix} m \\ (\gamma - 1)e + (3 - \gamma)m^2/2\rho \\ \gamma em/\rho - (\gamma - 1)m^3/2\rho^2 \end{bmatrix}_x = \begin{bmatrix} 0 \\ 0 \\ 0 \end{bmatrix}$$

The Jacobian matrix $A = [\partial f_i/\partial u_j]$ may be easily computed to convert to quasilinear form

$$\begin{bmatrix} \rho \\ m \\ e \end{bmatrix}_t + \begin{bmatrix} 0 & 1 & 0 \\ (\gamma - 3)u^2/2 & (3 - \gamma)u & \gamma - 1 \\ (\gamma - 1)u^3 - \gamma eu/\rho & \gamma e/\rho - 3(\gamma - 1)u^2/2 & \gamma u \end{bmatrix} \begin{bmatrix} \rho \\ m \\ e \end{bmatrix}_x = \begin{bmatrix} 0 \\ 0 \\ 0 \end{bmatrix}$$

from which, less easily, the eigenvalues $\lambda_1 = u$, $\lambda_2 = u + c$, $\lambda_3 = u - c$ of A are found. Here $c = (\gamma p/\rho)^{1/2}$ may be regarded acoustically as a local sound speed. Because γ, p, and ρ are all assumed to be positive, the eigenvalues are real and distinct. Hence the system is strictly hyperbolic and diagonalizable.

Exercise

9. A somewhat simpler quasilinear form of the gasdynamics equations is possible when writing them in terms of the primitive variables ρ, u, p, resulting in the system

$$\begin{bmatrix} \rho \\ u \\ p \end{bmatrix}_t + \begin{bmatrix} u & \rho & 0 \\ 0 & u & 1/\rho \\ 0 & \gamma p & u \end{bmatrix} \begin{bmatrix} \rho \\ u \\ p \end{bmatrix}_x = \begin{bmatrix} 0 \\ 0 \\ 0 \end{bmatrix}.$$

Find its eigenvalues, eigenvectors, and decompose the system to ODE's as above.

APPENDIX B: COMPUTATIONAL METHODS

Computational methods, namely, those using computers, have recently come to the forefront. Although the development of computational methods for the solution of partial differential equations traces back to at least the beginning of this century, [+] and for ordinary differential equations much earlier, [§] their widespread development and application was restricted by technology: hand calculation or mechanical calculator. The advent of high-speed, large-memory, low-cost (per operation) electronic computers has changed all that, and in so doing, has changed the role of mathematics and the way it is perceived, not only

---

[+] For example C. Runge, *Z. Math. Phys.* **56** (1908), and L. Richardson, *Trans. R. Soc.* **A 210** (1910), employed finite difference methods for problems in two-space dimensions.

[§] L. Euler, Institutiones Calculi Integralis, St. Petersburg(1768).

within mathematics but to an even larger extent within science.

Because each of the subjects

computational methods for elliptic equations

computational methods for parabolic equations

computational methods for hyperbolic equations

is a large one, we shall restrict ourselves here to the presentation of some basic and recent methodology for elliptic equations, less for parabolic equations, and very little for the hyperbolic ones. Space and time limitations preclude any systematic numerical analysis (eg, error estimation, operation counts, algebraic features, and the like).

B.1 Finite Difference Methods

These methods have already been introduced in Sections 2.6 and 2.9.6. They are at the heart of all numerical approaches to solving partial differential equations and are sued extensively. A valuable key to both the theory and the implementation of finite difference methods is the notion of *stencil*.* Werecall the problem of Section 2.6: find numerically a good approximate solution to $\Delta u = 0$ in the 3×3 square portrayed in Figure B.1, given the values of u at the indicated points on $\partial\Omega$. Reasoning from the mean value theorem for harmonic functions we arrived at the set of 4 equations in 4 unknowns

$$\begin{bmatrix} 4 & -1 & -1 & 0 \\ -1 & 4 & 0 & -1 \\ -1 & 0 & 4 & -1 \\ 0 & -1 & -1 & 4 \end{bmatrix} \begin{bmatrix} u_{1,1} \\ u_{2,1} \\ u_{1,2} \\ u_{2,2} \end{bmatrix} = \begin{bmatrix} u_{1,0} + u_{0,1} \\ u_{2,0} + u_{3,1} \\ u_{1,3} + u_{0,2} \\ u_{2,3} + u_{3,2} \end{bmatrix}$$

Notice that we have shifted notation here from that of Section 2.6. Here $u_{i,j}$ denotes the ith row and the j column of the grid in Figure B.1. This notation is commonly used in the numerical literature. Also when a numerical scheme is being finalized for implementation

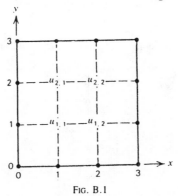

FIG. B.1

* Sometimes called the computational molecule.

capital letters U are often used to denote the discretized solution as contrasted to the use of lower case letters for the analytic solution u, and we shall sometimes do this.

The stencil of our method is the graph of Figure B.2. This is called the five-point stencil for discretizing the Laplacian operator Δ. It catalogues which points are used in the scheme. By including more points in the averaging (i.e., larger stencils) one can obtain higher-order approximations to u, but at a higher cost in computing time. Five-point stencils, perhaps with weights, remain in wide use for second-order elliptic partial differential equations. For fourth-order equations one may be forced to a more extensive stencil. The accuracy of a scheme, usually given in terms of an error guaranteed to be $0(h^n)$ where h denotes the grid spacing, can always be determined in principle from a Taylor series expansion.

Below in Figure B.3 are three other important stencils, two for parabolic equations and one for hyperbolic equations.

Figure B.3a is the stencil for the forward-Euler treatment of the heat equations, it being understood from the stencil that one takes a forward first-order time difference and a centered second-order space difference to approximate the equation. The discretized partial differential equation is thus

$$\frac{U_{i,j+1} - U_{i,j}}{\Delta t} = \frac{U_{i+1,j} - 2U_{i,j} + U_{i-1,j}}{(\Delta x)^2}$$

from which the algorithm of the scheme is seen to be

$$U_{i,j+1} = U_{i,j} + \frac{\Delta t}{(\Delta x)^2} [U_{i+1,j} - 2U_{i,j} + U_{i-1,j}].$$

This scheme is said to be *explicit* because the solution U on the next $j+1$st level may be explicitly calculated from known values of U on the jth level. Writing the algorithm as

$$U_{i,j+1} = (1 - 2r)U_{i,j} + r(U_{i+1,j} + U_{i-1,j})$$

reveals the ratio $r = (\Delta t)/(\Delta x)^2$ as an important one. Clearly if $r = 0$ the solution does not advance in time and is hence uninteresting. On the other hand, there would be little hope of convergence of the algorithm if the ratio $U_{i,j+1}/U_{i,j}$ were to exceed 1. As this ratio, if we ignore the two-side contributions, is exactly $(1 - 2r)^{-1}$, it would appear that we should

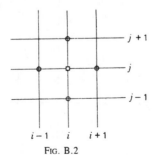

FIG. B.2

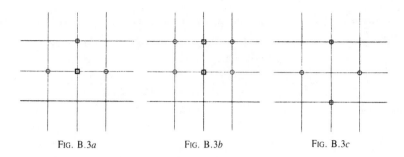

FIG. B.3a   FIG. B.3b   FIG. B.3c

take $r < 1/2$ to avoid any divergence. In fact a more careful stability analysis shows that for

$$0 < r \leqq 1/2$$

the finite difference solution $U_{i,j}$ converges to the solution $u_{i,j}$ of the differential equation as Δt and Δx go to zero. Recall that in Section 2.6 it was demonstrated that for consistent schemes, convergence and stability often go together.

There are two customary approaches to showing stability of a scheme. Numerical stability is usually taken to mean that a bounded initial error remains bounded. This is stability in the maximum norm as we have used the concept in this book. Both approaches are already implicit in the argument we gave above. The only effect that we have not taken into account is that of the coupled side terms.

The Von Neumann method, also called the Fourier method or the method of mode analysis, expresses the initial data errors in a (finite) Fourier series, separates variables just as we did in the continuous case, and then assures that no Fourier components grow exponentially. Thus, for the forward Euler scheme, if $u_{k,0} \equiv u_k$, $k = 0, \ldots, n$ denotes an initial error distribution, it may be represented by a Fourier sum

$$U(0) = \sum_{k=0}^{n} c_k e^{\frac{ik\pi x}{l}}$$

where the c_k are the discrete Fourier coefficients found from the $(n + 1) \times (n + 1)$ linear equations $U(x_k) = u_k$. In other words, one constructs a continuous fit to the initial errors, which is exact at the grid points. Linearity then enters strongly. First, we need only consider errors U superimposed on any solution. Secondly, we may set all $u_k = 0$ except one, solve, and then construct the total error from each of its $n + 1$ constituent parts. As far as growth or nongrowth of an error is concerned, the numerical c_k initial value is irrelevant. Thus one need only look at a single mode $e^{\frac{ik\pi x_k}{l}} = e^{\frac{ik\pi m}{n}}$. This is why the method is sometimes called mode analysis.

Assuming, therefore, just as in continuous separation of variables for the heat equation, that the solution error is of the form

$$U(x, t) = e^{\frac{ik\pi m}{n}} e^{\lambda t}$$

we substitute into the forward Euler discretization of the heat partial differential equation to obtain

$$\frac{e^{\frac{ik\pi m}{n}} e^{\lambda(t + \Delta t)} - e^{\frac{ik\pi m}{n}} e^{\lambda t}}{\Delta t} = \frac{e^{\frac{ik\pi(m + 1)}{n}} e^{\lambda t} - 2e^{\frac{ik\pi m}{n}} e^{\lambda t} + e^{\frac{ik\pi(m - 1)}{n}} e^{\lambda t}}{(\Delta x)^2}$$

from which

$$e^{\lambda t} - 1 = r(e^{\frac{ik\pi}{n}} - 2 + e^{\frac{-ik\pi}{n}})$$

$$= r\left(2 \cos \frac{k\pi}{n} - 2\right)$$

$$= -4r \sin^2 \left(\frac{k\pi}{2n}\right).$$

For stability the time step must be taken small enough to guarantee $|e^{\Delta t}| = |1 - 4r \sin^2 (k\pi/2n)| \leqq 1$, i.e., $0 \geqq -4r \sin^2 (k\pi/2n) \geqq -2$, which is assured when

$$r \leqq 1/2.$$

The above mode analysis of stability takes no note of boundary conditions. These are usually treated by a second method, that of the amplification matrix. For Dirichlet boundary conditions $u(0, t) = u(l, t) = 0$ the discrete Euler approximation is advanced the next time step by

$$U_{N+1} = AU_N$$

where

$$A = \begin{bmatrix} 0 & 0 & & & & \\ 1-2r & r & 0 & \cdots & & \\ r & 1-2r & r & \cdots & & \\ 0 & r & \cdots & & & \\ & & & & & r \\ & & & & r & (1-r) \\ & & & & 0 & 0 \end{bmatrix}$$

The vectors $\{U_N\}_{N=1}^{\infty}$ will remain bounded provided that all eigenvalues of A are of magnitude no greater than 1. For if some $|\lambda_k|$ exceeds 1, we may use its corresponding eigenvector as an initial error which will blow up. Conversely if each mode cannot blow up, then neither can any combination of them.

The eigenvalues of A may be found to be $\lambda_0 = \lambda_n = 0$ (due to the boundary conditions) and then for $k = 1, \ldots, n - 1$,

$$\lambda_k = 1 - 4r \sin^2 \frac{k\pi}{2n}.$$

Note that this is the same amplification factor found in the above Von Neumann analysis. Hence to guarantee $|\lambda_k| \leqq 1$ we need $r \leqq 1/2$. Different boundary conditions will change at least the first and last rows of A and hence can change the largest $|\lambda_k|$ and thus the permissible range of r.

Problem B.1.

(a) Show from Taylor Series that the forward Euler scheme is $0(h^2)$ in x and $0(k)$ in t, where h denotes Δx and k denotes Δt.

(b) Find good weights for the nine point stencil for the Laplacian.

(c) Show that an $n \times n$ tridiagonal matrix

$$
A = \begin{bmatrix}
a_0 & a_1 & 0 & \\
a_{-1} & a_0 & & a_1 \\
0 & a_{-1} & &
\end{bmatrix}
$$

has eigenvalues $a_0 + 2(a_1 a_{-1})^{1/2} \cos (k\pi/n + 1)$, $k = 1, \ldots, n$. Thus (ignoring the first and last zero rows) the forward Euler interior amplification matrix has eigenvalues $(1 - 2r) + 2r \cos k\pi/n = 1 - 4r \sin^2 (k\pi/2n)$.

The stencil of Figure B.3b is that of the Crank–Nicolson scheme for the heat equation

$$
\frac{U_{i,j+1} - U_{i,j}}{\Delta t} = \frac{[U_{i+1,j+1} - 2U_{i,j+1} + U_{i-1,j+1}] + [U_{i+1,j} - 2U_{i,j} + U_{i-1,j}]}{2(\Delta x)^2}.
$$

Notice that this scheme amounts to averaging the space central differences at both the jth and $j+1$st levels. In so doing it becomes *implicit*: U_{j+1} must be solved for implicitly rather than being given explicitly in terms of j values as in the forward Euler scheme above. Letting $r = \Delta t/(\Delta x)^2$ the Crank–Nicolson equations may be written as

$$
- ru_{i-1,j+1} + (2 + 2r)u_{i,j+1} - ru_{i+1,j+1} = ru_{i-1,j} + (2 - 2r)u_{i,j} + ru_{i+1,j}
$$

where the three unknowns are grouped on the left side. For Dirichlet boundary conditions the interior approximation U_N is advanced by

$$
U_{j+1} = A^{-1}BU_j
$$

$$
= \begin{bmatrix}
(2 + 2r) & -r & 0. \\
-r & (2 + 2r) & \\
0. & &
\end{bmatrix}^{-1}
\begin{bmatrix}
(2 - 2r) & r & 0. \\
r & (2 - 2r) & \\
0. & &
\end{bmatrix}
\begin{bmatrix}
U_{1,j} \\
\vdots \\
U_{n-1,j}
\end{bmatrix}.
$$

It can be shown that the eigenvalues of $A^{-1}B$ are all of magnitude less than one and hence the Crank–Nicolson scheme for this problem is unconditionally stable (i.e., stable for any value of r).

Here are some numerical problems that may be done by hand.

Problem B.2.

(a) Use the forward Euler scheme to advance two time-steps the problem

$$\begin{cases} u_t - u_{xx} = 0, & 0 < x < 1, \quad t > 0, \\ u(0, t) = u(1, t) = 0, & t > 0, \\ u(x, 0) = -x(x - 1), & 0 < x < 1, \end{cases}$$

with $h = \Delta x = 1/4$, $k = \Delta t = 1/64$, to determine $U(1/2, 1/32)$.

(b) Use the Crank–Nicolson scheme to advance one time-step the problem

$$\begin{cases} u_t - u_{xx} = 0, & 0 < x < 1, \quad t > 0, \\ u(0, t) = u(1, t) = 0, & t > 0, \\ u(x, 0) = 1, & 0 < x < 1, \end{cases}$$

with $h = \Delta x = 1/4$, $k = \Delta t = 1/32$.

(c) Use the Crank–Nicolson scheme to advance one time-step the problem

$$\begin{cases} u_t - u_{xx} = 0, & 0 < x < \pi, \quad t > 0, \\ u(0, t) = u(\pi, t) = 0, & t > 0, \\ u(x, 0) = \sin x, & 0 < x < \pi, \end{cases}$$

with $\Delta x = \pi/4$ and $\Delta t = \pi^2/32$. Compare to the analytic solution.

The third stencil above, Figure B.3c, is that of a so called "leap frog" scheme

$$\frac{U_{i,j+1} \pm U_{i,j-1}}{\Delta t} = \frac{U_{i+1,j} \pm U_{i-1,j}}{\Delta x}$$

useful for hyperbolic partial differential equations. The ratio $\Delta t/\Delta x$ should be adjusted to the coefficients (propagation speeds) of the particular equation, and the choice of $\pm$ depends on whether the equation is second- or first-order. These methods* are related to the theories of characteristics and conservation laws treated in Appendix A and are best understood in that context.

Here are some numerical problems that may be done on a machine.

Problem B.3.

(a) Solve Problem 1 of Section 2.1 by use of the discretization

$$\frac{U_{i,j+1} + U_{i,j-1}}{\Delta t} = \frac{U_{i+1,j} + U_{i-1,j}}{\Delta x}$$

---

* We will not elaborate them here. Examples were touched on briefly in Section 2.9.6 (Buckley–Leverett equation) and Problem 2.9.8(a) (Traffic flow equation). An especially good treatment of numerical theory for hyperbolic systems may be found in Kreiss and Oliger, *Methods for the Approximate Solution of Time Dependent Problems*, Global Atmospheric Research Program Publications No. 10 (World Meteorological Organization, 1973).

To get started, introduce the fictitious values

$$U_{i,-1} = U_{i,1} - 2g(x_i)\Delta t$$

from which

$$U_{i,1} = \frac{U_{i-1,0} + U_{i+1,0}}{2} + g(x_i)\Delta t.$$

(b) Solve (b) of Problem B.2 computationally. Run time out a long way.
(c) Compare the analytical and numerical solutions of (b).

B.2 Finite Element Methods

As finite element methods were discussed in Section 2.6 and Section 2.9.6 and because the subject is becoming so large, this section will be very brief. A number of references were given in Section 2.6. Finite element methods were pioneered principally by civil engineers for use in structural design. More recently they have been applied to problems in fluid dynamics.* Finite elements may be placed in increasing density around corners and obstacles and therefore possess great advantages in modeling partial differential equations in irregular regions. Implementing the input and output can be tedious.

Table B.1 presents a comparison of the Fourier Series, Finite Difference, and Finite Element methods applied to the problem of Section 2.6. The Fourier series converges with 10 terms, the Finite Differences approximate to about 5 percent with a 10 × 10 grid, and the Finite Elements approximate to about 10 percent with 10 × 10 nodes. The latter is not as good here because a linear plate element code with only 12 fixed degrees of freedom permitting only deflection and rotation was used. A quadratic element would presumably produce much better results. The boundary function of Figure 2.6b is really too sharp for the use of a plate code.

Table B.1

(x, y)	Fourier series 10 terms	Finite difference 10 × 10 (9 × 9 interior)	Finite element 10 × 10 (9 × 9 interior)
1, 0	0.988	1.0	1.0
2, 0	1.913	2.0	1.99
1, 1	0.431	0.434	0.47
2, 1	0.517	0.535	0.55
1, 2	0.143	0.146	0.16
2, 2	0.153	0.159	0.17

* Good recent references, among many, are F. Thomasset, *Implementation of Finite Element Methods for Navier–Stokes Equations* (Springer, New York, 1981), R. Peyret and T. Taylor, *Computational Methods for Fluid Flow* (Springer, New York, 1983), and A. Baker, *Finite Element Computational Fluid Mechanics* (McGraw-Hill, New York, 1983).

When an irregular domain Ω is triangulated into tetrahedral or rectangular elements, and the approximating functions thereupon are only loosely connected to one another, very wild matrices in the resulting linear system can result. This can even happen in a square when the approximating spline structure over the triangulation is complicated, (e.g., by the need to approximate vector valued rather than scalar functions). Without going into the details, here is (Fig. B.4) a sparse matrix from a quadratic finite element basis for incompressible flow.[*] There are many interesting combinatorial and algebraic questions that arise in the process of assembling finite elements.[†] Interesting combinatorial and analytic "Green's Function" pole and dipole placement questions arise in using (method of panels) combined finite element/finite difference schemes for irregular regions about aerodynamic structures.[‡]

Rather than delving into such multidimensional element combination problems, we offer some further practice with variational quadratic approximation, which, stretching the definitions, could be regarded as approximation on a single element.

Problem B.4.

(a) Solve $u'' + u' = 4$ on $0 < x < 1$ with boundary conditions $u(0) = u(1) = 0$ using an approximating function $\phi(x) = x(x - 1)$ and the Galerkin method of Section 2.6.
(b) Solve it analytically.
(c) Compare at $x = 0.2, 0.5$, and 0.8.

Problem B.5.

(a) Solve the Airy's equation problem of Sections 1.3 and 2.9.9(9) $u'' + 4xu = 16$ on $0 < x < 2$ with boundary conditions $u(0) = 0$, $u(2) = -1$, using quadratic Galerkin functions $\phi_0 = 1/2x(1 - x)$ and $\phi_1 = x(2 - x)$.
(b) Solve it also by Finite differences at $x = 0, 0.5, 1.0, 1.5$, and 2.0.
(c) Compare.

Problem B.6.

(a) Solve $u'' + u = x^2$ on $0 < x < 1$ with boundary conditions $u(0) = 0$ and $u'(1) = 1$, using first $\phi = x$ and then $\phi_1 = x$ and $\phi_2 = x^2$ as approximants.
(b) Solve it analytically.
(c) Compare at $x = 0.2, 0.5$, and 0.8.

B.3 Finite Spectral Methods

The terminology *spectral methods* has come into the numerical solution of differential equations as a rather general lumping together of a number of methods employing Fourier or other transforms. Their increasing popularity is based for example on the great speed with

---

* See K. Gustafson and R. Hartman, "Graph theory and fluid dynamics," *SIAM J. Alg. Disc. Math.* 6 (1985). The V^1, V^2, V^3 are discrete versions of H_p, H_c, H_d of Appendix C.1.

† These are also being pursued by the French School, see for example F. Hecht, *RAIRO Anal. Numer.* (1981).

‡ See K. Gustafson, K. Halasi, D. P. Young, *Intern. J. Num. Meth. in Fluids* 5 (1985).

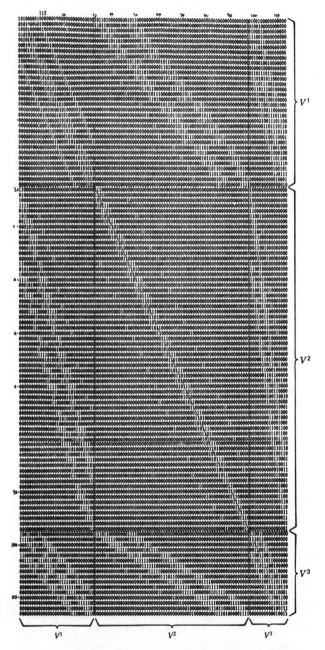

FIG. B.4. Sparseness matrix, with Helmholtz components indicated.

which one can take a Fast Fourier Transform (FFT). These methods include the Galerkin and *Rayleigh-Ritz* methods discussed in Section 2.6

From our point of view expressed at the beginning of Chapter 2, that of generalized Fourier expansions (e.g., eigenfunction expansions), spectral methods could as well be called *Fourier methods*. In their implementation one posits and approximation such as

$$u_N(x, t) = \sum_{n=1}^{N} T_N(t)X_n(x)*$$

where the $X_n(x)$ expansion functions are specified. Although these need not be eigenfunctions, that is, they could be pulled arbitrarily out of the air, experience with a few examples will indicate the desirability that they satisfy the boundary conditions and hopefully have other properties common to the given problem. Indeed, as already seen in Section 2.6, in most cases the exact eigenfunctions $X_n(x)$, when known, provides the optimal choice. ¶

Let us consider the example

$$\begin{cases} u_t - u_{xx}, & 0 < x < \pi, \quad t > 0, \\ u(x, 0) = f(x), & 0 < x < \pi, \\ u(0, t) = u(\pi, t) = 0, & t > 0. \end{cases}$$

Were we to proceed analytically by separation of variables, we would be led to the expansion

$$u(x, t) = \sum_{n=1}^{\infty} T_n(t)X_n(x)$$

where $X_n(x) = \sin nx$, the eigenfunctions of the space part of the heat operator with the given Dirichlet boundary condition. Rather than then solving directly from the separation of variables relation for the $T_n(t)$, the spectral numerical procedure truncates the Fourier series at N terms and then asks that this approximation satisfy the partial differential equation

exactly	Fourier
in projected sense	Galerkin
at specified points	Collocation
orthogonally	Tau
in weighted sense	Weighted residual
weighted, locally	Finite element

We have taken some liberties but the matchings just given are essentially correct.

---

*The spectral method literature appears to prefer the notation $a_n(t)$ for $T_n(t)$, perhaps to connote that the $a_n(t)$ are coefficients approximatin the exact coefficients

$$a_n(t) = \frac{2}{\pi} \int_0^{\pi} u(x, t) \sin nx \, dx.$$

The latter coefficients $\{a_n\}_{n=1}^{\infty}$ are sometimes called the (Fourier) Sine Transform of the solution $u(x, t)$.

¶ In Section 2.6 we emphasized the point of view that the numerical approximations u_N correspond to Projections in a Hilbert space. For other (e.g, Tchebycheff) basis functions $X_n(x)$ or ϕ_n they may more naturally be regarded as Projections in a Banach space. The main point is, however, whether one be a functional analyst or numerical analyst, to get Projections as naturally related to the given partial differential operator as possible. Then good results will follow, in either context.

For the example at hand, for the "exact" solution we would use the relations of $T_n(t)$ to $X_n(x)$ by the separation of variables procedure, from which $T_n(t) = c_n e^{-n^2 t}$. Of course then fitting the initial data $f(x)$ exactly would require that $f(x)$ happen to lie in the subspace spanned by the $X_n(x)$, $n = 1, \ldots, N$. Otherwise we would take the c_n to be the first N Fourier coefficients in the $X_n(x)$ expansion of $f(x)$.

The Galerkin procedure, see Section 2.6, approximates the solution by a projection

$$u_N(x, t) = \sum_{n=1}^{N} a_n(t)\phi_n(x)$$

in the subspace spanned by a prechosen set $\{\phi_n\}$. That the equation also be approximated within this subspace is implemented by the set of N equations,

$$\left\langle \left(\frac{\partial}{\partial t} - \frac{\partial^2}{\partial x^2}\right) u_N, \phi_m \right\rangle = 0, \qquad m = 1, \ldots, N.$$

This becomes the set of N equations in N unknowns a_n

$$a_1'(t)\langle \phi_1, \phi_1 \rangle + \cdots + a_N'(t)\langle \phi_N, \phi_1 \rangle = a_1(t)\langle \phi_1'', \phi_1 \rangle + \cdots + a_N(t)\langle \phi_N'', \phi_1 \rangle$$

$$\vdots \qquad\qquad \ddots \qquad\qquad \vdots$$

$$a_1'(t)\langle \phi_1, \phi_N \rangle + \cdots + a_N'(t)\langle \phi_N, \phi_N \rangle = a_1(t)\langle \phi_1'', \phi_N \rangle + \cdots + a_N(t)\langle \phi_N'', \phi_N \rangle$$

which is to hold for each time t. This is a first-order ODE system of the form

$$Aa'(t) = Ba(t)$$

with constant coefficient matrices and with the initial condition $a(0)$ taken from the system

$$Aa(0) = \begin{pmatrix} \langle f, \phi_1 \rangle \\ \vdots \\ \langle f, \phi_N \rangle \end{pmatrix}.$$

Notice how convenient it is when the $\{\phi_n\}$ are chosen to be orthonormal. Then not only is A invertible but it is the identity. By choosing the $\{\phi_n\}$ to be the natural eigenfunctions $\phi_n(x) \equiv X_n(x) = \sin nx$ the eigenfunction reproducing property $\phi_n'' = -n^2\phi_n$ renders B diagonal and the ODE system uncouples. The fast numerical approximation of the Fourier sine coefficients $2/\pi \int_0^\pi f(x) \sin nx\, dx$ then sets up the initial value $a(0)$, from which we recover the above "exact" solution. The only error is in the spectral transform $c_n = a_n(0) = \langle f, \phi_n \rangle$.

Generally one is not so lucky as to know the exact eigenfunctions. Then to minimize the errors uniformly in x, and especially near the boundary, Tchebycheff polynomials $\{\phi_n\}$ are often employed for the Galerkin spectral expansion. Care must be taken to meet the boundary conditions of the problem as well as possible. Other special functions (e.g., Laguerre, Legendre, see Section 2.1) are sometimes chosen, depending on the geometry of the problem and the availability of fast spectral transforms to provide the coefficients.

Collocation combined with Fourier eigenfunction expansion approximants is often called a pseudospectral method. This, because collocation, by itself, obviates the need to take any spectral transforms. In collocation one requires that the equation residual be identically zero, that is,

$$\left(\frac{\partial}{\partial t} - \frac{\partial^2}{\partial x^2}\right) u_N(x_m, t) = 0$$

at specified points x_m, (e.g., $m = 1, \ldots, N$). For a trial approximation u_N in the $\{\phi_n\}$ subspace this sets up a first order $N \times N$ ODE system

$$Aa'(t) = Ba(t)$$

similar to that above, except that now the coefficients are values of the ϕ_n and ϕ_n'' at the collocation points x_m. This can be a powerful method, faster than Galerkin, and avoiding problems at the boundary.

The so-called Tau (or Lanczos*) method may be regarded as a variation of the Galerkin method which offers a way to satisfy M boundary condition constraints by allowing the approximation M additional degrees of freedom while no longer demanding that each of the $\{\phi_n\}$ satisfy all of the boundary conditions. For such

$$u_{N+M}(x, t) = \sum_{n=1}^{N} a_n(t)\phi_n(x) + \sum_{n=N+1}^{N+M} b_m(t)\phi_n(x)$$

the residual turns out to be

$$\left(\frac{\partial}{\partial t} - \frac{\partial^2}{\partial x^2}\right) u_{N+M}(x, t) = \sum_{n=N+1}^{\infty} \tau_n(t)\phi_n(x)$$

where $\{\phi_n\}$ is assumed here to be a complete orthonormal set. Recalling Section 2.3, this means that the residual lies in the orthogonal complement of the space spanned by the $\{\phi_n\}_{n=1}^{N}$. As such, this method is also sometimes referred to as a method of orthogonal collocation: the part of the residual in the basic approximating subspace exactly vanishes.

The name Tau is obviously evidenced in the coefficients $\tau_n(t)$ above. The implementation of this algorithm goes as follows. As in the Galerkin method, to effect a projected approximation of the solution and the equation, one forms scalar products with the $\{\phi_n\}$. This leads to the $N + M$ equations in $N + M$ unknowns

$$a_n'(t) = \left(\frac{\partial^2}{\partial x^2} u_{N+M}, \phi_n\right), \qquad n = 1, \ldots, N,$$

and, for the case at hand of $M = 2$ boundary constraints,

$$u_{N+M}(0) = u_{N+M}(\pi) = 0.$$

It turns out that the τ coefficients in the residual are given by $\tau_n = -\left(\frac{\partial^2}{\partial x^2} u_{N+M}, \phi_n\right)$ for $n = N + M + 1, \ldots$ and with the intermediate τ_n for $n = N + 1, \ldots, N + M$ determined implicitly by the above system. Rather than getting more involved in this method, we will content ourselves with the example and exercises below.

There are other weighted residual methods† into which we shall not delve. These are all variational methods in the sense that the equation is multiplied by a finite set of weighting functions $\{\phi_i\}$ and then integrated over the full domain Ω or some subdomain such as the support of the weighting function. The latter case includes the finite element method discussed

---

* C. Lanczos, *J. Math. Physics* **17** (1938); *Applied Analysis*, Prentice Hall (1956).

† For example, see Ames, Finlayson, *op. cit.* in Section 2.6.

in Sections 2.6, 2.9.6, and in part B.2 above. Whether or not the method should also be considered a finite spectral method will turn on whether the coefficient integrals need be obtained by spectral for example, Fast Fourier, transform techniques.

Problem B.7.

(a) For the heat equation example, write out the Galerkin equations $Aa'(t) = Ba(t)$ for ϕ_n the orthogonal functions $\sin nx$, $n = 1, \ldots, N$.

(b) Write out the Collocation equations $Aa'(t) = Ba(t)$ in general and then for the ϕ_n as in (a).

(c) Read further about spectral methods, where other examples and problems may be found.

A discussion of computational methods would not be complete without remarks on linear algebra and graphics output. Unfortunately these are subjects in themselves. Therefore we content ourselves here with a few examples of the value of the latter in understanding solutions.

Consider Problem 1 of Section 1.2. With the data x^2 on two sides, the graphics at first shows very little difference between the solution for the Laplacian operator and that for the Minimal Surface operator (Fig. B.5). However, the linearization error multiplied by 100 shows that the minimal surface solution sags considerably below that of the Laplacian.*

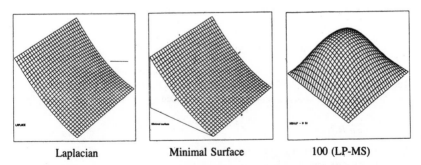

Laplacian   Minimal Surface   100 (LP-MS)

FIG. B.5. Computational Solution of Problem 1 of Section 1.2.

* The linear algebra was done with SOR with $\omega = 1.816$ in most of these examples.

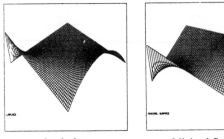

| Laplacian | Minimal Surface | 100 (LP-MS) |

FIG. B.6. Computational Solution of Problem 2 of Section 2.1.

More interesting is a comparison of the numerical solution of Problem 2 of Section 2.1 first with the Laplacian approximation and then with the full minimal surface equation. The views here (Fig. B.6) are from a perspective from the left of Figure 2.1*l*.

The Fourier series "spectral method" approximation of the square wave was requested in Problem 1.9.6(2)(c) to illustrate the Gibbs effect at initial data discontinuities. The square wave $f(x) = -1$ on $(-\pi, 0)$ and $f(x) = +1$ on $(0, \pi)$ has Fourier series $\sum_{n=1}^{\infty} \dfrac{4}{(2n-1)}$ $\sin (2n - 1)x$. The maximum value occurs one-fourth of the shortest wave length from the discontinuity and for the square wave at a jump from -1 to $+1$ one expects the series to reach above 1 to the value

$$1 + 2(0.0895) \approx 1.179$$

according to the discussion of Section 1.9.6(2). This series was evaluated to a large number of terms to find

Terms	Gibbs Effect
$N = 20,000$	1.17897974530
$N = 200,000$	1.17897974448
$N = 2,000,000$	1.17897974447

as graphically portrayed for $N = 100$, 10,000, 100,000, and 1,000,000 terms in Fig. B.7. The effect is instantly smoothed by parabolic equations but persists in hyperbolic ones.

Some additional computational practice is provided by the following problems.

Problem B.8.

(a) Solve computationally Problem 3 of Section 2.1.

(b) Solve computationally

$$\begin{cases} u_t - u_{xx} = 0, & 0 < x < 1, \quad t > 0, \\ u(0, t) = u(1, t) = 0, & t > 0, \\ u(x, 0) = 2x, & \text{for } 0 < x \le 0.5, \\ \qquad\quad = 2 - 2x, & \text{for } 0.5 \le x \le 1. \end{cases}$$

(c) Read about the recently developed important Multigrid Approximation Schemes.

Problem B.9.

Solve computationally the traffic flow problem of Section 2.9.8, namely,

$$\rho_t + c(\rho)\rho_x = 0$$

with initial profile

$$\rho(x, 0) = 200 + 100e^{-0.5\left(\frac{x - 0.5}{0.04}\right)^2}$$

and with propagation coefficient

$$c(\rho) = 76.184 - 17.2 \, ln \, \rho$$

by means of (a) a Leapfrog scheme, (b) a Lax–Wendroff scheme, and (c) a Method of Characteristics scheme. Which is more natural to the problem?

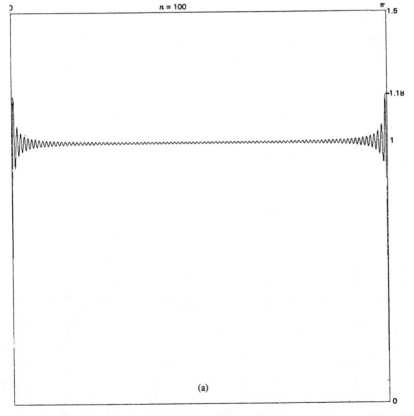

(a)

FIG. B.7. Computational Solution of Problem (c) of Section 1.9.6(2). (a) 100 Fourier terms (b) 10,000 Fourier terms (c) 100,000 Fourier terms (d) 1,000,000 Fourier terms.

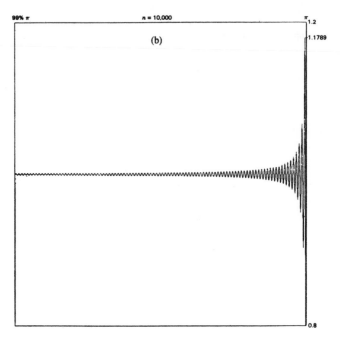

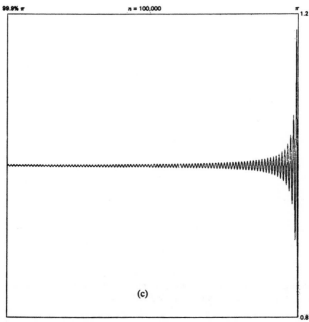

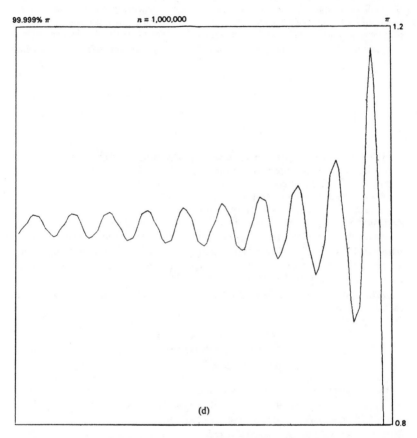

FIG. B.7 Computational Solution of Problem (c) of Section 1.9.6(2). (a) 100 Fourier terms (b) 10,000 Fourier terms (c) 100,000 Fourier terms (d) 1,000,000 Fourier terms.

SIXTH PAUSE: EXAMPLES, EXPLANATIONS, EXERCISES

In the Fifth Pause we noticed that shocks could develop in solutions of advection equations or advective systems, even when starting from smooth initial values. This has led to the development of shock capturing schemes, computational methods which can find and resolve the shock dynamics of an evolving hyperbolic system.

Example

The most used test problem for such schemes in the *Riemann Problem*, also called the Shock Tube Problem,* Some shock capturing schemes actually use the exact analytical solutions of Riemann Problems at each time step. Discuss this test problem ** and the principal types of schemes used for numerical solution of hyperbolic conservation laws.

Solution. We recall the Euler equations of inviscid gas dynamics from the Fifth Pause, namely

$$\begin{pmatrix} \rho \\ \rho u \\ e \end{pmatrix}_t + \begin{pmatrix} \rho u \\ \rho u^2 + p \\ (e + p)u \end{pmatrix}_x = \begin{pmatrix} 0 \\ 0 \\ 0 \end{pmatrix},$$

with the same ideal gas state equation assumptions as before. *Physically*, the Riemann problem imagines a thin tube in which gas on the left at constant high pressure is separated from gas on the right at constant low pressure, separated by a diaphragm. At $t = 0$ the diaphragm is punctured and the flow is viewed as a piston moving to the right, ahead of which is an induced *compressive* wave. Trailing behind the "piston" is an *expansive* wave. *Analytically*, the problem is treated as a pure initial value problem, the equations to determine the flow of the physical primitive variables density ρ, velocity u, and pressure p from their given initial values $\rho(x, 0), u(x, 0), p(x, 0)$. *Computationally*, because shocks are expected to develop, it is customary to start with a shock, to test the robustness of the numerical scheme being developed.

Although the position of the origin is arbitrary, a standard Riemann Solver test problem uses the following initial data:

density	$e_l = 1.000$	$c_r = 0.125$
velocity	$u_l = 0.000$	$u_r = 0.000$
pressure	$p_l = 1.000$	$p_r = 0.100$

with origin placed at $x = 0.500$. The flow is then computed for a short time in the interval $0 \leq x \leq 1$ and compared to the exact analytical solution. The most used stopping time is $t = 0.140$, short enough that no dynamics has reached the "ends" of the tube $0 \leq x \leq 1$ so that no reflections, etc., need be considered.

The exact analytical solution at $t = 0.140$ is shown in Fig. 6Pa. In Fig. 6Pb we have sketched the evolution of the density from its initial values to its final values at $t = 0.140$. Notice that the flow evolves forward (in time) in 5 regions. In the farthest left region I,

---

*Since this problem involves the gasdynamics equations in only one space dimension, it is helpful to realize that this is not a "tube" problem at all. In a "tube" one could envision all kinds of highly complicated two and three dimensional wave interactions. This test problem is a "shock ray" problem, analogous to the "rod" setting of the First Pause. Movement is tracked only in a fixed single space direction. The resulting *shock dynamics* are usually called *wave dynamics*, but we will seek some conceptual simplification by thinking in terms of *ray dynamics* for this one space dimensional setting.

Although analytically the extension to two or three space dimensions is a large one, the computational schemes discussed here have been successfully applied to complicated aerodynamics and other gasdynamics applications in two and three space dimensions.

** An excellent reference is G. Sod, *J. Computational Physics* 27 (1978). Like many others, we use Sod's initial values and terminal time $t = 0.14$. Sod also describes how to find the analytical solutions.

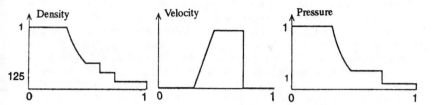

FIG 6P*a*. Exact Solution of Riemann Test Problem at $t = 0.14$

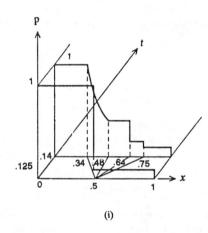

(i)

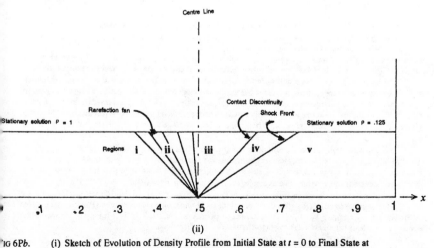

(ii)

FIG 6P*b*. (i) Sketch of Evolution of Density Profile from Initial State at $t = 0$ to Final State at $t = 0.14$. (ii) The Five Evolution Regions for the Riemann Test Problem

nothing has yet changed, the initial constant e_l, u_l, p_l are not yet affected. The next region II represents a *rarefaction action* of high pressure and high density on the left being reduced by the generally rightward flow. In the middle region III a maximum velocity behind the "piston" has been achieved. A particle initially at the data discontinuity (the piston) is now at the rightmost point of region III. This point is called a *contact discontinuity* and represents a *translation action* (e.g., of the initial density discontinuity). Region IV separates the "piston" front from the *advection-induced shock* ahead of it. This *compressive action*, due to decreasing solution profiles, induces discontinuous shocks in all flow variables, in the manner explained in Pause 5. Ahead of the shock is the as yet undisturbed region V of right initial values ρ_r, u_r, p_r Computational simulations of this Riemann Test Problem are given in the last Example and Fig. 6Pc in pages 329 to 330.

Let us at this point observe an analytical fact of some importance: the invariance of the Euler equations under the uniform stretching group. Consider the Euler equations in primitive variable (nonconservative) form

$$\rho_t + u\rho_x = - \rho u_x$$

$$u_t + uu_x = -\frac{1}{\rho}p_x$$

$$p_t + up_x = -\gamma p u_x$$

If we replace x and t with new variables λx and λt, λ a fixed constant, we see that if $\rho(x, t), u(x, t), p(x, t)$ is a solution, then $\rho_\lambda(x, t) = \rho(\lambda x, \lambda t), u_\lambda(x, t) = u(\lambda x, \lambda t), p_\lambda (x, t) = p(\lambda x, \lambda t)$ is also a solution: the λ comes out as λ^{-1} in each derivative. In other words, a classical solution may be continued out on rays $x/t = \eta =$ constant, and more generally, on rays $(x - x_0)/(t - t_0) = \eta =$ constant.[*]

Thus for the Riemann Solver test problem, because the initial data is classical (flat, in fact) except at the origin $x = 0.5$, we may view the evolving solution as mostly classical (flat, in fact) continued along rays from the initial profile. The four possible exceptions occur at the interior boundaries of the five regions described above, where the ray solutions may be weak solutions. If one sits at a point x to the right or left of the origin, there are

---

[*] All first order conservation systems $u_t + (f(u))_x = 0$ in one space dimension enjoy this property, as may be varified by the chain rule. We (and others) have used it, for example, in scalar equations when choosing the rarefaction fan solutions $u(x,t) = u(x/t)$, see Exercise 4 of Pause 5. But we have not seen it emphasized as a general analytical principle for the selection of the (unique) physical solutions, which we might call (see previous footnote) *ray solutions*. More to the point: we may call them particle solutions.

The same scale invariance will hold for conservation laws in higher space dimensions. solutions would be viewed as plane, sphere, solutions but we are not prepared to go into that here. This observation is just scale invariance, using the arguments of dimensional analysis or local transformation groups, as one prefers. The continued solutions are similarity solutions on rays. See Appendix A. When one uses the group viewpoints, one is ignoring initial and boundary value constraints, and imagining infinite domains $- \infty < x < \infty$.

This analysis depends on the solutions being classical, and breaks down at solution or derivative discontinuities, e.g., at shocks, contact discontinuities, and fan boundaries. But by suitable reinitialization, e.g., along the lines of our discussion of shock dynamics in Pause 5, one can in principle continue ray solutions even from those points. For example, for the convection equation $u_t + uu_x = 0$, at a shock point we "reinitialized" the ray direction to the average of the two incoming characteristic directions. If one views this as a fast particle overtaking a slow one, both of the same mass, our various criteria (Rankine-Hugoniot, Weak Solution, Regularity) just produce a "lumped" particle whose speed is the average of the incoming speeds, and whose direction is the average of the incoming directions.

Roe's 2nd order Time=0.035, # of G.P=1001, dt=0.00014

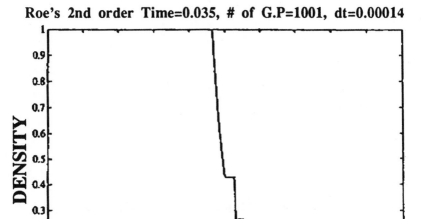

Roe's 2nd order Time=0.035, # of G.P=1001, dt=0.00014

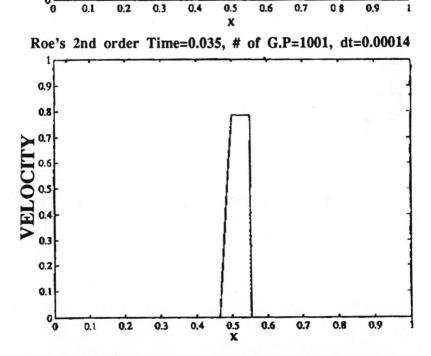

Fig 6Pc.   (i) High Resolution fine-grid ($\Delta_x = 0.001$) small time-step ($\Delta_t = 0.00014$)
Computational Solution of Riemann Shock Tube test problem for the Euler Flow
Equations Hyperbolic System

Roe's 2nd order Time=0.035, # of G.P=1001, dt=0.00014

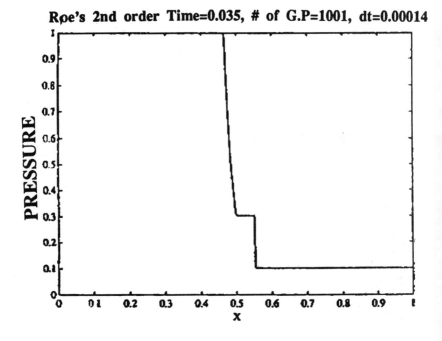

Fig 6Pc. (i) High Resolution fine-grid ($\Delta_x = 0.001$) small time-step ($\Delta_t = 0.00014$)
Computational Solution of Riemann Shock Tube test problem for the Euler Flow
Equations Hyperbolic System

Roe's 2nd order Time=0.070, # of G.P=1001, dt=0.00014

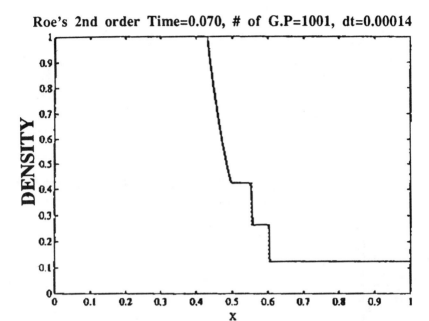

Roe's 2nd order Time=0.070, # of G.P=1001, dt=0.00014

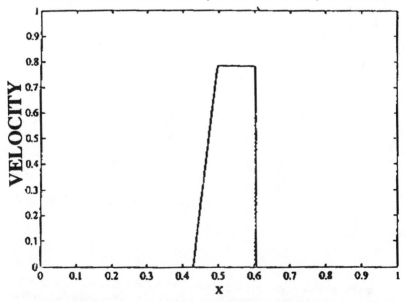

Fig 6Pc. (i) High Resolution fine-grid ($\Delta_x = 0.001$) small time-step ($\Delta_t = 0.00014$)
Computational Solution of Riemann Shock Tube test problem for the Euler Flow
Equations Hyperbolic System

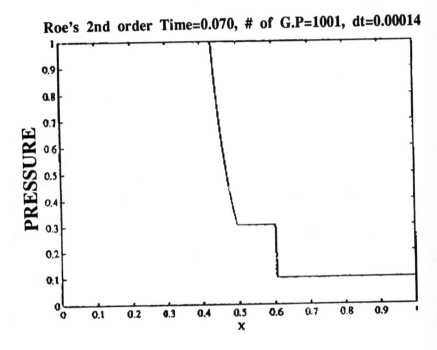

Fig 6Pc. (i) High Resolution fine-grid ($\Delta_x = 0.001$) small time-step ($\Delta_t = 0.00014$)
Computational Solution of Riemann Shock Tube test problem for the Euler Flow
Equations Hyperbolic System

Roe's 2nd order Time=0.105, # of G.P=1001, dt=0.00014

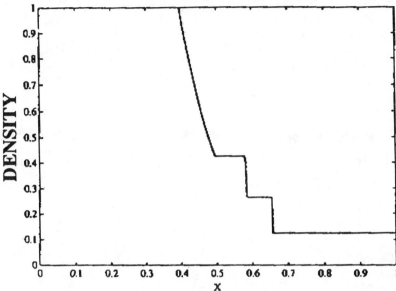

Roe's 2nd order Time=0.105, # of G.P=1001, dt=0.00014

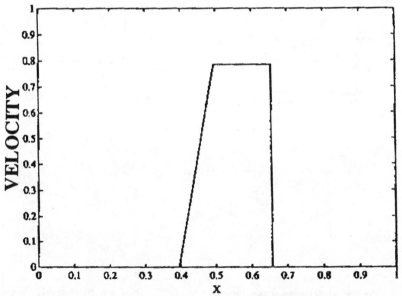

Fig 6Pc. (i) High Resolution fine-grid ($\Delta_x = 0.001$) small time-step ($\Delta_t = 0.00014$)
Computational Solution of Riemann Shock Tube test problem for the Euler Flow
Equations Hyperbolic System

Roe's 2nd order Time=0.105, # of G.P=1001, dt=0.00014

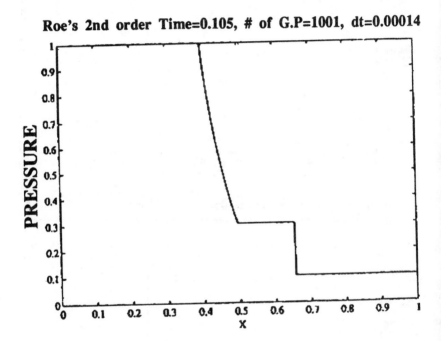

Fig 6Pc. (i) High Resolution fine-grid ($\Delta_x = 0.001$) small time-step ($\Delta_t = 0.00014$)
Computational Solution of Riemann Shock Tube test problem for the Euler Flow
Equations Hyperbolic System

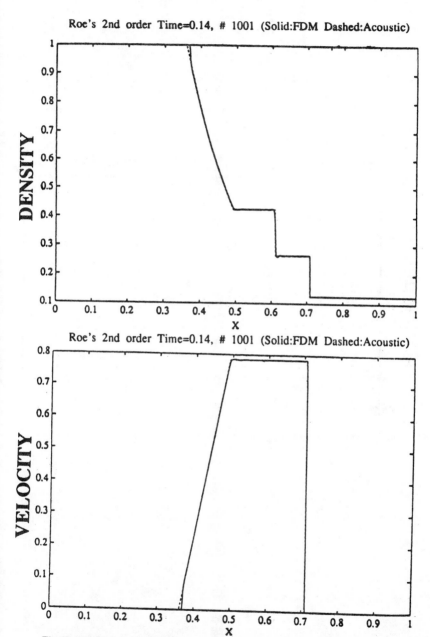

Roe's 2nd order Time=0.14, # 1001 (Solid:FDM Dashed:Acoustic)

Roe's 2nd order Time=0.14, # 1001 (Solid:FDM Dashed:Acoustic)

Fig 6Pc. (i) High Resolution fine-grid ($\Delta_x = 0.001$) small time-step ($\Delta_t = 0.00014$) Computational Solution of Riemann Shock Tube test problem for the Euler Flow Equations Hyperbolic System

Roe's 2nd order Time=0.14, # 1001 (Solid:FDM Dashed:Acoustic)

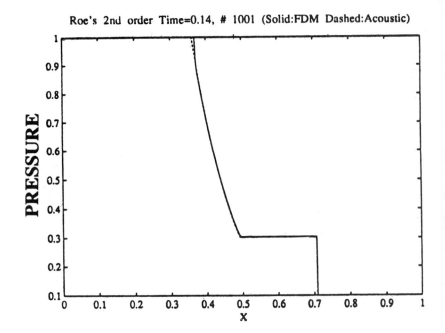

Fig 6Pc. (i) High Resolution fine-grid ($\Delta_x = 0.001$) small time-step ($\Delta_t = 0.00014$)
Computational Solution of Riemann Shock Tube test problem for the Euler Flow
Equations Hyperbolic System

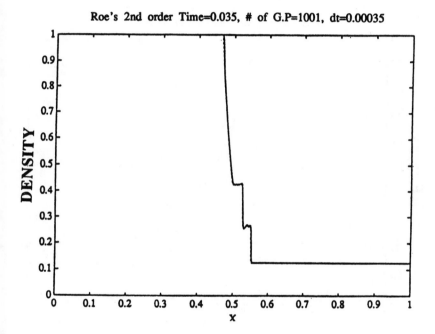

Roe's 2nd order Time=0.035, # of G.P=1001, dt=0.00035

Fig 6Pc.(ii)   Gibbs Effect when ($\Delta t = 0.00035$) too large

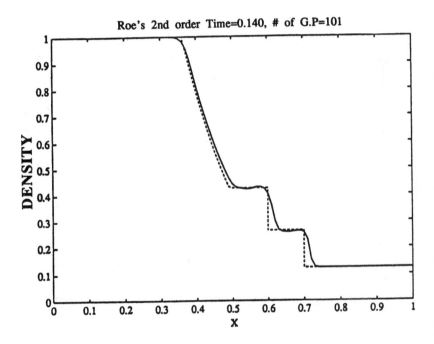

Fig 6Pc.(iii)   Smearing Effect when ($\Delta x = 0.01$) too large

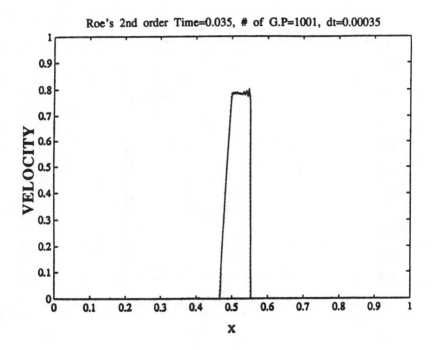

Roe's 2nd order Time=0.035, # of G.P=1001, dt=0.00035

Fig 6Pc.(ii) Gibbs Effect when ($\Delta_t = 0.00035$) too large

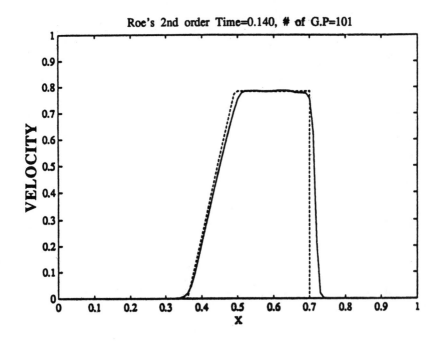

Fig 6Pc.(iii)   Smearing Effect when ($\Delta_x = 0.01$) too large

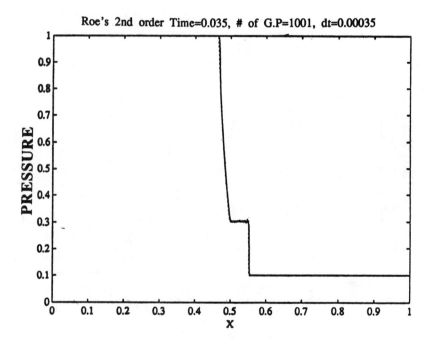

Roe's 2nd order Time=0.035, # of G.P=1001, dt=0.00035

Fig 6Pc.(ii) Gibbs Effect when   ($\Delta_t = 0.00035$) too large

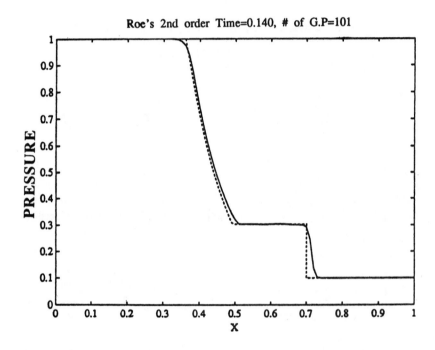

Fig 6Pc.(iii)   Smearing Effect when ($\Delta x = 0.01$) too large

discontinuities passing through at just two instants. Once they have passed, the solutions are steady forever.

See Exercises 1 and 2 for further qualitative discussion.

Turning now to quantitative methods, the computational schemes in use may be categorized into three classes:

Upwinding (Godunov, Glimm, Roe, Van Leer, Steger, Osher,...)

Centered (Lax, Wendroff, MacCormack, Yee,...)

Total Variation Diminishing (Harten, Boris, Book, Colella, Jameson,...)

See, respectively, Exercises 3, 4, 5 below. Limited space and time permit only a very brief description of the essentials of these schemes, there is some overlap in the categories, many authors' names are missing.

Exercises

1. Provide some further qualitative insight into the nature of Riemann Solver solutions.
2. Discuss qualitatively the contact discontinuity, the so-called Riemann Invariants, and the roles of the eigenvalues and eigenvectors of the system.
3. Give a brief description of some upwind schemes.
4. Give a brief description of some centered schemes.
5. Give a brief description of some TVD schemes.

Overall, the computational picture for hyperbolic systems can be held together conveniently by a brief discussion of *flux-splitting**. Consider a general hyperbolic system in quasilinear form, then diagonalized:

$$U_t + F(U)_x = 0 \rightarrow U_t + A(U)U_x = 0 \rightarrow u_t + Du_x = 0$$

where U and $F(U)$ are the physical quantities and fluxes of interest, A is the Jacobian $\partial F / \partial U$, $D = Q^{-1}AQ$ is the diagonal eigenvalue matrix, Q is the eigenvector (columns) matrix, and $u = Q^{-1}U$.

Example

Given such a diagonalized system, examine the stability of one-sided differencing. Relate these considerations to flux splitting and to the Euler equation system.

Solution. The uncoupled general hyperbolic system consists of n equations

$$\frac{\partial u_i}{\partial t} + \lambda_i \frac{\partial u_i}{\partial x} = 0 \quad i = 1, \ldots, n,$$

where the eigenvalues λ_i may be assumed to be real and distinct. If we use a backward spatial difference (see Appendix B)

$$\frac{\partial u_i}{\partial x}\bigg|_{x_j = j\Delta x} = \frac{u_j - u_{j-1}}{\Delta x} + O(\Delta x) \quad i = 1, \ldots, n$$

then the uncoupled PDE system becomes the semidiscrete ODE system

$$\frac{du_j}{dt} + \lambda \frac{u_j - u_{j-1}}{\Delta x} = 0 \quad j = 1, \ldots, m$$

---

*See R. Warming and R. Beam, *Symposium on Computational Fluid Dynamics*, SIAM-AMS Proceedings **11** (1978), pp. 85-129, for a good exposition of Flux Splitting and its application to the Euler gasdynamic equations in one and two space dimensions.

By the Von Neumann method (see Appendix B.1), the stability of this scheme may be tested by the single Fourier mode $u_j(t) = v(t)e^{i k j \Delta x}$, which when substituted yields

$$\frac{dv}{dt} + \frac{\lambda(1 - e^{-i k \Delta x})}{\Delta x} v = 0.$$

The solution $v(t)$ is bounded for all $t > 0$ iff $\lambda \geq 0$ (see Exercise 6).

In the same way the forward spatial semidiscrete ODE system

$$\frac{du_j}{dt} + \lambda \frac{u_{j+1} - u_j}{\Delta x} = 0 \quad j = 1, \ldots, m,$$

for stability needs $\lambda \leq 0$.

For the Euler equations we know (Pause 5, Exercise 9) that the eigenvalues are $\lambda_1 = u - c$, $\lambda_2 = u$, $\lambda_3 = u + c$. When the speed u is greater than the sound speed $c = (\gamma p / \rho)^{1/2}$, all three eigenvalues are positive and backward differencing would be favored. This is *upwinding*: the upstream slope $(u_j - u_{j-1})/\Delta x$ is used to predict the next downstream values on the right-moving wave. Similarly, when the speed u is less than c, downwinding, i.e., forward differencing, would be preferred. But generally, the eigenvalues are of mixed sign.

The way out of this apparent incompatibility is to "split" the flux: decompose the flux in the conservation law so that

$$F = F^+ + F^- \rightarrow A = A^+ + A^-, \text{ all } \lambda_i (A^+) \geq 0, \text{ all } \lambda_i (A^-) < 0.$$

Then the explicit conservative first order difference scheme

$$\frac{U_j(t_n + \Delta t) - U_j(t_n)}{\Delta t} + \frac{(\nabla_x F_j^+(t_n) + \Delta_x F_j^-(t_n))}{\Delta x} = 0,$$

where ∇_x means forward differencing, Δ_x means backward differencing, will be stable whenever the Courant-Friedrichs-Levy stability condition $\left| \lambda^{\pm} \right| \Delta t / \Delta x \leq 1$ is satisfied for all of the eigenvalues.

The Euler gasdynamic equations possess an important analytical property: the flux vectors $F(U)$ are homogeneous functions of degree one in U, i.e., $F(\lambda U) = \lambda F(U)$.* This property implies that $F(U) = F'(U)U$, $F'(U)$ being the Jacobian $A(U)$ we have seen in the quasilinear formulation of a conservation system. Thus in conservation form the system is

$$U_t + (AU)_x = 0$$

*It is interesting to combine this property with the invariance under uniform space-time scaling which we observed earlier. Letting $\bar{x} = \lambda x$, $\bar{t} = \lambda t$, and $\bar{u} = \Lambda u$, we see that the Euler equations are invariant under this enlarged Lie stretching group. This gives a further dimension to our view of solutions as 'ray' solutions. Both invariances hold as well for two and three space dimensions. But now one needs to visualize the solution in four or five dimensions, which is harder.

whereas in quasilinear formulation we had

$$U_t + AU_x = 0$$

In other words, the $A(U)$ can move in or out of the spatial derivative, as if it were constant. Use of this property enables the flux splitting above according to $F^+ = A^+U, A^+ = QD^+Q^{-1}$, where D^+ is D with the $\lambda_i < 0$ entries set equal to 0. Similarly $F^- = A^-U, A^- = QD^-Q^{-1}$, (see Exercise 7).

Any upwinded scheme introduces in a subtle way a numerical viscosity. Let us illustrate this with the simple scalar translation equation

$$u_t + cu_x = 0, \quad -\infty < x < \infty, t > 0,$$

with the eigenvalue $\lambda = c > 0$. The forward-time backward-spatial differencing

$$\frac{u_j^{n+1} - u_j^n}{\Delta t} + \lambda \frac{(u_j^n - u_{j-1}^n)}{\Delta x} = 0$$

can be rewritten

$$u_j^{n+1} = \frac{\Delta t}{\Delta x} \lambda u_{j-1}^n + \left(1 - \frac{\Delta t}{\Delta x} \lambda\right) u_j^n.$$

Under the CFL condition $\lambda \Delta t / \Delta x \leq 1$, this scheme is stable and *monotone*: in the maximum norm,

$$\max_j \left|u_j^{n+1}\right| \leq \max_j \left|u_j^n\right|.$$

Rewritten again as

$$u_j^{n+1} = u_j^n - \Delta t\lambda \frac{(u_{j+1}^n - u_{j-1}^n)}{2\Delta x} + \Delta t\lambda \frac{(u_{j+1}^n - 2u_j^n + u_{j-1}^n)}{2\Delta x}$$

this stable upwind monotone scheme can be viewed as a second order discretization of the continuous viscous equation

$$u_t + \lambda u_x - \left(\lambda \frac{\Delta x}{2}\right) u_{xx} = 0.$$

Exercises

6. Complete the stability discussion for one-sided backward and forward differences.

7. Carry out the details of the flux-splitting argument for a general hyperbolic system. Look at it also for the Euler equation example.

8. Reflect upon viscosity, physically, analytically, or computationally as mentioned above.

Having discussed shock capturing schemes and the Riemann Shock Tube test problem, a computational simulation is now in order.

Example

Code* and implement a numerical solution of the Riemann test problem for the Euler hyperbolic system.

Solution. A Roe's second order scheme produced the best result. The initial values for density, velocity, and pressure as described above were prescribed at $t = 0$ for $0 \le x \le 1$ and the flow was allowed to evolve until terminal time 0.14. The schematics of the expected flow were given above in Figs. 6Pa and 6Pb. We have plotted the evolution at $t = 0.035$, 0.070, 0.105, and 0.140 in Figure 6Pc(i). Note that on this fine grid ($\Delta x = 0.001$) and small time step ($\Delta t = 0.00014$) it is almost impossible to distinguish the computational solution (solid lines) from the analytic solution (dashed lines). The velocity in the first few time steps (not shown here) shoots up to its maximum value with a (finite) delta-function-like profile to begin driving the flow.

In order to see better the discrepancies between numerical and analytic solution profiles, in Figure 6Pc(ii) we show a small Gibbs oscillation occurring for a larger time step, and in Figure 6Pc(iii) some smearing when the space discretization is coarser.

Exercise

9. Summarize or comment on the usual three questions (existence, uniqueness, stability) and the other three questions (construction, regularity, approximation) for the hyperbolic systems of conservation laws treated in Pause 5 and Pause 6. To complete this "confirmation" exercise, reflect upon the other "trinities" of Chapter I in this context.

APPENDIX C: ADVANCED FLUID DYNAMICS

By "advanced" fluid dynamics we mean advanced in time: that is, what is going on currently in mathematical fluid dynamics research. Many of the older difficult problems in fluid dynamics, both theoretical and experimental, remain unresolved. But the advent or the computer has allowed a third window, beyond those of pure theory and laboratory physical experiment, to enable knowledge of fluid behavior. The combination of the analytical, experimental, and computational methods has stimulated advances in each.

We can only touch on this subject here, with a brief look at the Navier-Stokes equations recent views as to the makeup of final fluid states and turbulence and a whole new field of mathematics called computational fluid dynamics.

---

* Speaking here about all computational solutions shown in this book, some code was written by the author, some code was written by coworkers during research, and some code was written by students at the author's request. It is a pleasure to thank everyone who helped. In particular, the code for this simulation was written by Kisa Matsushima, a visiting student from Japan who took my PDE course 1990-91.

C.1 Navier–Stokes Equations

The Navier–Stokes equations* correctly describe many fluid flows. Their validity extends to very high speeds, even higher than one might first imagine.† Assuming the body forces to be zero and the density to be uniform, consider the general viscous incompressible Navier–Stokes equations given in Section 1.7, namely, written out, the momentum equation

$$\begin{cases} u_t - v(u_{xx} + u_{yy} + u_{zz}) + uu_x + vu_y + wu_z = -p_x \\ v_t - v(v_{xx} + v_{yy} + v_{zz}) + uv_x + vv_y + wv_z = -p_y \\ w_t - v(w_{xx} + w_{yy} + w_{zz}) + uw_x + vw_y + ww_z = -p_z, \end{cases}$$

and the mass continuity equation

$$u_x + v_y + w_z = 0.$$

Here $\vec{u} = (u, v, w) = (u_1, u_2, u_3) = u$ in different notations is the velocity and p is the (unknown) pressure whose effect is felt only through its gradient. For simplicity we shall consider only a closed container Ω, and Dirichlet boundary and initial conditions: $\vec{u}(\vec{x}, t)$ known on $\partial\Omega$ for all time, $\vec{u}(\vec{x}, 0)$ given initially. The viscosity $v = 1/Re$ where Re denotes the Reynolds number which briefly described is a specified average flow velocity.

Problem C.1.

  (a) It is always instructive** to look at the one-dimensional versions of important physical partial differential equations. Write down the one-dimensional version of the Navier–Stokes equations.
  (b) For the one-, two-, or three-dimensional equation, apply the incompressibility constraint to the momentum equation to arrive at a Neumann–Poisson Problem for the pressure.
  (c) What does (b) mean for the one-dimensional version?

From this problem we may induce the importance of dimension for the Navier–Stokes equations. For two dimensions the basic existence, regularity, and uniqueness theory is in reasonably good shape.‡ For three dimensions the important question of when the solution can develop singularities remains unresolved.

When body forces are present (e.g., for gravity effect), the nonhomogeneous Navier–Stokes equations

$$u_t - \frac{1}{Re}\Delta u + (u \cdot \nabla)u = -\nabla p + f$$

are generally treated by the methods for parabolic equations, essentially thinking first of the heat-like (vector) equation

$$u_t - \frac{1}{Re}\Delta u = f$$

---

* Originally obtained by C. Navier, *Memoire sur les lois du mouvement des fluides* (*Mem. Acad. Sci.* **6**, 1823), and corrected by G. Stokes (*Cambridge Phil. Soc. Trans.* **8**, 1849).

† See D. Tritton, *Physical Fluid Dynamics* (Van Nostrand, 1977).

** We sometimes must remind ourselves of this.

‡ See R. Temam, *Navier–Stokes Equations and Nonlinear Functional Analysis* (SIAM, Philadelphia, 1983).

and then trying to accommodate the (nonlinear) perturbation terms $(u \cdot \nabla)u$. An important orthogonality relation between the pressure gradient and the incompressible fluid velocity arises here. We call this Helmholtz's theorem and it may be regarded as a type of generalization of the Dirichlet Orthogonality Projection Theorem of Section 2.6, extended now to the projection of vector fields, i.e., vector valued functions $u = (u_1, u_2, u_3)$. One may write this Helmholtz Projection Theorem for such functions u with n components defined on a region Ω as

$$(L^2(\Omega))^n = H_{\text{pot}} \oplus H_{\text{sol}} \oplus H_{\text{irrot}}$$
$$= H_p \oplus H_c \oplus H_d$$

where

$$H_p = \{u \mid \text{div } u = 0, \quad \text{curl } u = 0\}$$
$$H_c = \{u \mid \text{div } u = 0, \quad \text{curl } u \neq 0\}$$
$$H_d = \{u \mid \text{div } u \neq 0, \quad \text{curl } u = 0\}$$

denote subspaces which we call here the potential subspace, the solenoidal subspace, and the divergence subspace. The last subspace can also be called the irrotational subspace but we have taken the subscript d because it is shorter and to remind us that for any u decomposed into its three orthogonal constituent parts $u = u_p \oplus u_c \oplus u_d$ one has div $u = $ div u_d. Similarly, one has curl $u = $ curl u_c.*

The following algorithm establishes the above theorem in an operational way.

Problem C.2.

(a) Solve the Dirichlet–Poisson problem

$$\begin{cases} \Delta\phi_3 = \text{div } u \text{ in } \Omega, \\ \phi_3 = 0 \text{ on } \partial\Omega; \end{cases}$$

let $u_d = \nabla\phi_3$.

(b) Solve the Neumann–Poisson problem

$$\begin{cases} \Delta\phi_1 = 0 \text{ in } \Omega, \\ \dfrac{\partial\phi_1}{\partial n} = n \cdot u - \dfrac{\partial\phi_3}{\partial n} \text{ on } \partial\Omega; \end{cases}$$

let $u_p = \nabla\phi_1$.

(c) Let $u_c = u - u_p - u_d$. Then $u = u_p \oplus u_c \oplus u_d$.

Taking advantage of the Helmholtz decomposition we may eliminate the pressure† from the momentum equation by noting that gradients are othogonal to solenoidal functions v. That is, differentiating by parts

---

* Discrete versions of Helmholtz's theorem also hold. See Fig. B.4, which exhibits a basis support matrix for projected quadratic elements. Also see the paper cited there for further details.

† It must be admitted that this slights the pressure, which may be rather important to find in real applications.

$$(v_1 p)_x = v_1 p_x + v_{1_x} p$$

$$(v_2 p)_y = v_2 p_y + v_{2_y} p$$

$$(v_3 p)_z = v_3 p_z + v_{3_z} p$$

we have $v \cdot \nabla p = \mathrm{div}(p(v_1, v_2, v_3)) - p \, \mathrm{div} \, v$ and hence

$$\int_\Omega v \cdot \nabla p = \oint_{\partial\Omega} p v \cdot n = 0$$

when $\mathrm{div} \, v = 0$. We have also used its (immiscibility) form $v \cdot n = 0$ on $\partial\Omega$, a special case being the often assumed no-slip condition $v = 0$ on $\partial\Omega$. Continuing, we may now cast the momentum equation into the weak (Galerkin) form by multiplying by an arbitrary v in the divergence-free subspace $H_s = H_p \oplus H_c$ and integrating over Ω

$$\frac{d}{dt} \int_\Omega v \cdot u - \frac{1}{Re} \int_\Omega v \cdot \nabla u + \int_\Omega v \cdot (u \cdot \nabla u) = \int_\Omega v \cdot f$$

This form is very useful not only for theory but also for numerical approximation.

Assuming that there exists a unique or at least attainable long-time steady limit of the flow, the first term drops out and we have the weak-form steady flow equations. For low Reynolds numbers (highly viscous flows) the nonlinear term $u \cdot \nabla u$ is often neglected, and absorbing Re in the data f we arrive at the weak form Dirichlet–Poisson Problem

$$\int_\Omega (-\Delta u - f) \cdot v = 0.$$

It may be shown that this weak form is equivalent to the steady Stokes problem

$$\begin{cases} -\Delta u + \nabla p = f & \text{in } \Omega, \\ \mathrm{div} \, u = 0 & \text{in } \Omega, \\ u = 0 & \text{on } \partial\Omega. \end{cases}$$

Problem C.3.

(a) It would be very convenient if the nonlinear term $\int_\Omega (u \cdot \nabla u)v$ would vanish in the weak formulation. Show $\int (u \cdot \nabla v)v$ does.

(b) Argue essentially how one gets from the weak form to the strong Stokes formulation above.

(c) Verify the following solution to the full three-dimensional Navier–Stokes equations in the steady case.

$$\begin{cases} u = \dfrac{-6}{Re}\dfrac{x}{y^2} - 3c_1 y^2 + \dfrac{c_2}{y^2} + \dfrac{c_3}{y^3}, \\[2ex] v = \dfrac{-6}{Re} \cdot \dfrac{1}{y}, \\[2ex] w = \dfrac{-12}{Re}\dfrac{x}{y^3} + 6c_1 y + \dfrac{2c_2}{y^3} + \dfrac{3c_3}{y^4}, \\[2ex] p = \dfrac{-60}{Re}c_1 x - \dfrac{12}{(Re)^2} \cdot \dfrac{1}{y^2}. \end{cases}$$

C.2 Turbulence and Attractors

Many physical flows display a turbulent behavior for sufficiently high parameter values. Fluid turbulence has long been an important scientific problem and there are many approaches to its study. *Ad hoc* models for simulating turbulent effects have been developed for airplane design and other applications. However, a satisfactory mathematical foundation has not yet been found.

Recent advances in the theory of dynamical systems have shown that solution trajectories can exhibit wildly chaotic yet deterministic behavior. This behavior is often controlled by an attracting set to which the trajectories periodically, often, intermittently, or eventually return. There is currently much study of these (strange) attractors to see how well they can be used as a foundation for understanding turbulent behavior.[*]

Let us consider swirling flow in a unit square or cube. Kolmogorov[†] postulated that, although viscosity will not affect the larger eddies of the flow, there is a critical small length l_d, dependent on viscosity, below which the eddies cannot persist. This length may be seen on a physical basis to be

$$l_d = (\varepsilon \, Re^3)^{-1/4}$$

where ε is a local average rate of energy dissipation. From this it may be deduced that a flow will go turbulent or at least very chaotic provided that it has a sufficiently large but finite number of degrees of freedom $N = 0(l_d^{-n})$, for $n = 2$ or 3 dimensions. Assuming that we can get enough (unit) energy dissipation ε into some small part of Ω, this means that a flow of

$$N \sim Re^{3n/4}$$

degrees of freedom might be necessary before one could achieve a flow with true turbulence. As Re must be large itself, e.g., $Re \sim 10^3$ to 10^6 depending on the situation, the estimated $N \sim 10^9$ is beyond the capabilities of the largest computers available today.

We content ourselves here with a brief consideration of the so-called space periodic model: Ω is all space R^2 or R^3 latticed into unit squares or cubes with corners at the multi-integer point Z^n, and the velocity vectors u are laterally periodic on these squares or cubes in each coordinate direction. Then we may write u in its Fourier series expansion

$$u = \sum_k c_k e^{2\pi i k \cdot x} \qquad k \in Z^n$$
$$\text{div } u = \sum_k k \cdot c_k (2\pi i) e^{2\pi i k \cdot x}$$

from which div $u = 0$ iff $k \cdot c_k = 0$ for all $k \in Z^n$. Thus, for the simplified case of the Stokes equation, the real divergence-free eigenfunctions are

$$c_k e^{2\pi i k \cdot x} + \bar{c}_k e^{-2\pi i k \cdot x}$$

provided $k \cdot c_k = 0$. For example, for $k = (1, 2, 4)$ in R^3, the eigenfunctions are all of the above functions for which $1 \cdot c_1 + 2c_2 + 4c_4 = 0$. The eigenvalues are the numbers $4\pi^2(k_1^2 + k_2^2 + k_3^2)$.

---

[*] For surveys of this work see G. Barenblatt, G. Iooss, D. Joseph, *Nonlinear Dynamics and Turbulence* (Pitman. Boston, 1983).

[†] See the account in L. Landau, E. Lifschitz, *Fluid Dynamics* (Addison-Wesley, New York, 1953).

The idea now is that, if we think of the Kolmogorov critical length as the side of a small cube within which eddy structure no longer maintains itself, then l_d small means that $k_d = l_d^{-1}$ is a large frequency of the motion, at or above which disorderly motion will set in. The number of frequencies to monitor to describe the flow will be the number of integers k_1, k_2, and k_3 such that

$$k_1^2 + k_2^2 + k_3^2 \leqq k_d^3 = l_d^{-3}$$

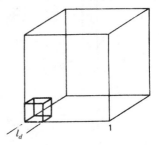

FIG. C.1. Kolmogorov Box

and thus the required number of degrees of freedom to obtain a turbulent flow is $N \sim l_d^{-n}$ for $n = 2$ or 3 dimensions.

Moreover, the fact that it is more and more difficult to satisfy the Pythagorean relation as k_d increases should be taken as an indication that it is more difficult to compose an orderly flow at the higher frequencies.

Problem C.4.

(a) Given the Kolmogorov postulate of a critical small length l_d below which eddies lose structure, deduce on a physical basis that the number of degrees of freedom needed for a flow to become "turbulent" is $0((l_0/l_d)^n)$ for $n = 2$ or 3 dimensions. Here l_0 is the physical scale of the whole flow (e.g., the diameter of the containing vessel).

(b) Recently it has been shown rigorously that the number of degrees of freedom of a turbulent flow $\sim$ the fractal dimension of a corresponding universal attractor. How can this be done?

(c) Why in the answer to (b) did we denote the limiting average of the energy dissipation rate $\varepsilon(x, t) = \nu|\text{grad } u(x, t)|^2$ by $\nu\lambda_m^M$?

We have been speaking above of an attracting set to which all flows tend. Because we have no intention of going into the theory of such sets, let us simply describe them here as generalizations of the Poincaré limit cycles one finds in the phase plane analyses of nonlinear second-order ordinary differential equations. An example of such was the Van der Pol limit cycle given in Section 1.9.8. In two dimensions the theory of all possible limit sets of second-order equations has a good structure. The problem with three dimensions is that the extra degree of freedom eliminates such a simple final limit set structure.

A classical fluid dynamics physical experiment has been the Taylor Problem of flow between two long cylinders, the outer fixed and the inner turning at a controlled steady angular acceleration ω. Physically one sees a steady Couette flow for an interval of small ω, for the next ω interval bands of horizontal counter-rotating Taylor vortices superimposed on the steady rotating flow, then a next interval in which the horizontal bands of Taylor vortices develop a wavyness or helical upward pattern. The appearance of the Taylor vortices represents a primary bifurcation off the basic Couette flow, in the manner described in Section 1.8. The wavy solutions at higher Reynolds number would appear to be a secondary bifurcation, called a Hopf bifurcation which means a loss of steady flow to a periodic flow.

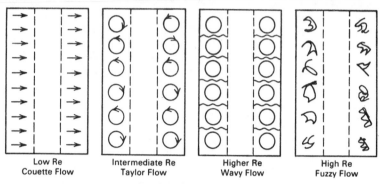

| Low Re | Intermediate Re | Higher Re | High Re |
| Couette Flow | Taylor Flow | Wavy Flow | Fuzzy Flow |

FIG. C.2. Taylor Cylinders

Problem C.5.

(a) Has the existence of a Hopf bifurcation been proven mathematically analytically for the Navier–Stokes equations?
(b) Has a Hopf bifurcation been demonstrated mathematically numerically for the Navier–Stokes equations?
(c) Give an example of a Hopf bifurcation for ordinary differential equations.

After a possible Hopf bifurcation to a wavy pattern (i.e., at higher Reynolds numbers) the flow in physical experiments appears to become chaotic. There are two views toward the mathematical understanding of this. The classical view pictures a rapid transition through an infinite sequence of higher-order bifurcations to a final chaotic state. A modern view predicts a chaotic state right after a wavy one.

Problem C.6.

(a) How could a chaotic state occur right after two or three bifurcations?
(b) Give an example of a system of differential equations producing very chaotic trajectories.
(c) Is it necessary to have three dimensions to produce real turbulence?

C.3 Computational Fluid Dynamics

For flows of a reasonable number of degrees of freedom, computational methods for the numerical treatment of the Navier–Stokes equations have greatly increased our understanding of fluid dynamics. This is a rapidly expanding field, in some sense becoming an entity in itself, and we give here only one example—flow within a driven cavity. This flow contains within a simple geometry a sufficient richness of separation, bifurcation, and vortex dynamics so as to become a fundamental model for understanding the laminar and recirculation patterns of flow along surfaces and near geometrical discontinuities such as corners. Although it is far from being understood, the fluid features found easily transfer to other geometries such as flow against marine obstructions or over aerodynamic structures (e.g., see Fig. C.5).

For brevity we consider only two cases, the unit cavity and the cavity of aspect ratio A = depth/width = 2. In each case a no-slip condition $(u, v) = (0, 0)$ will be imposed on the sides and bottom, and the top lid will move steadily to the left with $(u, v) = (-1, 0)$ there. We will consider only the two-dimensional case, in which we still cannot resolve all features, and will use formulations in both the primitive variables velocity (u, v) and pressure p and also in the stream function, vorticity variables (ψ, ω).

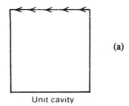

(a)

Unit cavity

FIG. C.3. Cavity Flow Problems (a) $A = 1$ (b) $A = 2$.

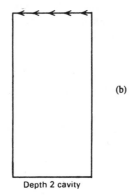

(b)

Depth 2 cavity

Problem C.7.

(a) Derive from the steady Stokes equations

$$\left.\begin{array}{l} -\Delta u = -\nabla p \\ \nabla \cdot u = 0 \end{array}\right\} \text{ in } \Omega$$

$$\left.\begin{array}{l} u = 0 \text{ on sides, base} \\ u = (-1, 0) \text{ on top} \end{array}\right\} \text{ on } \partial\Omega$$

in primitive variables, the stream function-vorticity formulation

$$\left.\begin{array}{l} \Delta\psi = -\omega \\ \Delta\omega = 0 \end{array}\right\} \text{ in } \Omega$$

$$\left.\begin{array}{l} \partial\psi/\partial n = 0 \text{ on sides, base} \\ \partial\psi/\partial n = -1 \text{ on top} \\ \psi = 0 \text{ on all boundaries} \end{array}\right\} \text{ on } \partial\Omega$$

for the cavity flow problem. Here $(u, v) = (\psi_y, -\psi_x)$, ψ being the stream function, ω denoting the vorticity $\omega = v_x - u_y$.

(b) In using discrete finite difference Poisson solvers to solve for the stream function and vorticity one needs boundary conditions for the vorticity. Show that using

$$\begin{cases} \omega(x, 0) = -2\psi(x, \Delta y)/(\Delta y)^2 \\ \omega(0, y) = -2\psi(\Delta x, y)/(\Delta x)^2 \\ \omega(1, y) = -2\psi(1 - \Delta x, y)/(\Delta x)^2 \\ \omega(x, 1) = -2\psi(x, 1 - \Delta y)/(\Delta y)^2 + 2/\Delta y \end{cases}$$

are $0(h)$ accurate, where $\Delta x = \Delta y = h$ is the grid size.

(c) For large (e.g., 256 × 256) grids, one is led for efficiency to employ iterative matrix solvers rather than straight Gauss elimination. These go under the names of Gauss–Seidel, Jacobi, Successive-overrelaxation, Conjugate Gradient, Multigrid methods. In using them, for computing the vorticity how can you assure that they will converge?

Very early Lord Rayleigh* analyzed the steady motion of viscous incompressible flow in a corner:

The general problem thus represented is one of great difficulty, and all that will be attempted here is the consideration of one or two particular cases. We inquire what solutions are possible such that ψ, as a function of r (the radius vector from the corner), is proportional to r^m.

Assuming the slow motion (Stokes) linearized equations

$$\begin{cases} \Delta^2\psi = 0 \text{ in } \Omega, \\ \psi = \partial\psi/\partial n = 0 \text{ on } \partial\Omega, \end{cases}$$

in a flow region Ω near the corner, Rayleigh was unable to fit the boundary conditions with an assumed stream function of the form

$$\psi(r, \theta) = r^m f(\theta)$$

because he restricted these trial solutions to $m = n\pi/\alpha$ where α was the corner angle. For a 90° corner this would require m to be an integer.

The reader should immediately recognize the separated trial form of ψ as that of a candidate similarity solution.† It is the fourth-order nature of the Stokes operator which makes the attempted separation of variables difficult, just as it did for plate problems (see the Second Pause, and Section 1.7.2 Problem 1).

Later studies tried similarity solutions of the form

$$\psi(r, \theta) = r^\lambda f(\theta)$$

for noninteger λ and found the rather remarkable fact that there is an infinite sequence of vortices descending into the corner for corner angles $\alpha < 146.3°$. Two or three of these

---

* Lord Rayleigh, *Phil. Magazine* **21** (1911). For a full history and recent results see K. Gustafson, *Applied Numerical Analysis* 3 (1987)

† The guessing of self-similar solutions has been a practically important way of finding them, and has the advantage of providing explicitly the functional relation $(\eta, g(\eta))$ on which their existence depends.

have been recently verified experimentally.* Because their intensity drops off $O(10^{-4})$ from one to the next, finding them is a delicate task.

In a demonstration of the great power of adaptive iterative relaxation methods, we recently found ¶ more than 20 of these corner eddies. Here are the first 10 stream function intensities ψ_i and the locations of the zeros z_i between the stream function relative extrema measured along the 45° corner bisector, z_1 being the zero between the first main central vortex ψ_1 in the cavity and the first corner vortex ψ_2.

Local Maximum Stream Function Intensity	Stream Function Zero Measured Along Diagonal
1.0006×10^{-1}	6.97×10^{-2}
-2.232×10^{-6}	4.205×10^{-3}
6.155×10^{-11}	2.534×10^{-4}
-1.703×10^{-15}	1.535×10^{-5}
4.71×10^{-20}	9.247×10^{-7}
-1.30×10^{-24}	5.602×10^{-8}
3.59×10^{-29}	3.370×10^{-9}
-9.93×10^{-34}	2.040×10^{-10}
2.75×10^{-38}	1.236×10^{-11}
-7.59×10^{-43}	7.421×10^{-13}

Problem C.8.

(a) Show that Rayleigh's attempted trial solution leads to the fourth-order ordinary differential equation

$$\frac{d^4\psi}{d\theta^4} + [(m - 2)^2 + m^2]\frac{d^2\psi}{d\theta^2} + (m - 2)^2 m^2\psi = 0$$

and that one obtains real solutions only for angles $\alpha = \pi$ or 2π.

(b) Show that from the more general similarity trial solution $\psi(r, \theta) = r^\lambda \cdot (A \cos \lambda\theta + C \cos(\lambda - 2)\theta)$ one may determine the existence of an infinite progression of vortices descending into the corner.

(c) Show that the Stokes flow in the cavity is uniquely determined (hint, not needed, but of possible related interest: see the plate equations discussed in the Second Pause of Chapter 1, and Section 1.7.2 Problem 1).

Returning to the primitive variables formulation of the full nonlinear unsteady Navier–Stokes flow equations, by taking 360,000 time steps (with $\Delta t = .001$) in a staggered mesh scheme we were able to follow the full dynamics of the flow evolution at the relatively high Reynolds number $Re = 10,000$ in a depth 2 cavity § See Fig. C.4, where velocities have been normalized in the flow portrait. As can be seen, these contain transient off-wall eddies,

* S. Taneda, *J. Phys. Soc. Japan* **46** (1979).

¶ K. Gustafson, R. Leben, *Applied Mathematics and Computation* **19** (1986) and *Proc. 1st International Symp. on Domain Decomposition of Partial Differential Equations* (SIAM Philadelphia, 1988). These intensities and locations agree very closely to those predicted by theory.

§ K. Gustafson, K. Halasi, *J. Computational Physics*, **64**(1986) and *J. Computational Physics*; **70** (1987).

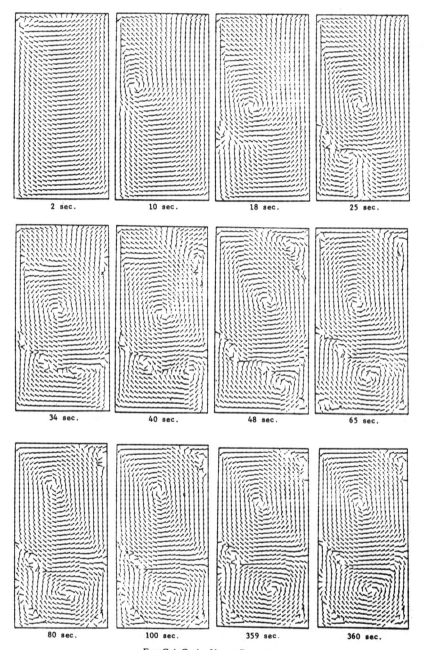

2 sec.   10 sec.   18 sec.   25 sec.

34 sec.   40 sec.   48 sec.   65 sec.

80 sec.   100 sec.   359 sec.   360 sec.

FIG. C.4 Cavity Vortex Dynamics

recirculation structures near a separation point, vortex fission after collision with a wall, sublayer bursting, multiple vortex shedding and fusion sequences, and a final persistent (Hopflike) wavyness. These patterns would appear to be related to the simulated fluid's absolute adherence to certain vortex parity matching rules for flow in a closed region.[*]

The cavity model provides a foundation for the investigation of a variety of other interesting domains exhibiting subvortical structure. By mapping techniques (grid generation) a cavity multigrid scheme has been adapted to wedges, multiple-corner regions, and flow over an airfoil.[+] As Fig. C.5 shows, the agreement of the resulting numerical simulation (on the right) with physical flow visualization[++] (on the left) is remarkable.[§]

Problem C.9.

(a) Analytically, there are no mathematical proofs yet for the flow features found above computationally. Nonetheless, argue the existence of an infinite sequence of corner eddies for the final state of the full Navier-Stokes flow.

(b) Assign "+" to counterclockwise rotating vortices and "—" to clockwise rotating vortices and examine a parity rule which asserts that opposite signed states, once generated, cannot merge.

(c) The crux of all such large computations are the linear solvers. A number of these were mentioned in Problem C.7(c). Read about them. How well will the relative speed advantage of the iterative ones persist into the coming days of Parallel Computations?

---

[*] A rather general vortex theory is put forth in K. Gustafson, "Four Principles of Vortex Motion," *Vortex Methods* and *Vortex Motion* (SIAM, Philadelphia, 1991). Rather conclusive (numerical) proof that the (depth 2) driven cavity has a periodic solution for Re as low as 5000 may be found in J. Goodrich, K. Gustafson, K. Halasi, *J. Computational Physics* **90** (1990). See also K. Gustafson, *Mathl. Comput. Modelling* **22** (1995) for a survey of the cavity flow problem results, both theoretical and computational.

[†] See K. Gustafson, R. Leben, "Robust Multigrid Computational Visualization of Separation and Vortex Evolution in Aerodynamic Flows," *Proceedings 1st National Fluid Dynamics Congress* (AIAA/ASMF/SIAM/APS, 1988), and K. Gustafson, R. Leben, "Computation of Dragonfly Aerodynamics," *Computer Physics Communication* **65** (1991). See also K. Gustafson, R. Leben, J. McArthur, M. Mundt, *Theor. and Comput. Fl. Dyn.* **8** (1996).

[††] P. Freymuth, *Prog. Aerospace Science* **22** (1985).

[§] And the end of all our exploring
Will be to arrive where we started
And know the place for the first time. . . . T. S. Eliot

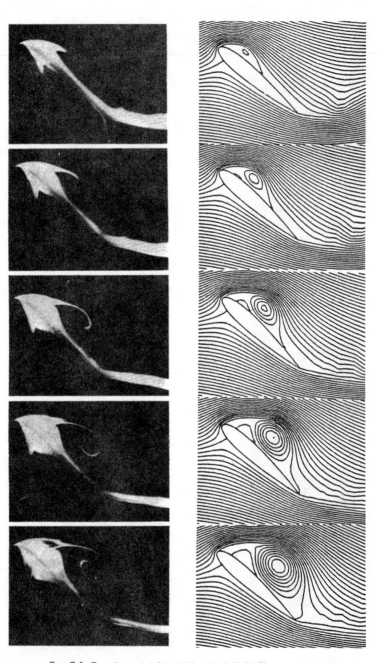

Fig. C.5. Experimental (left) and Numerical (right) Flows over an airfoil.

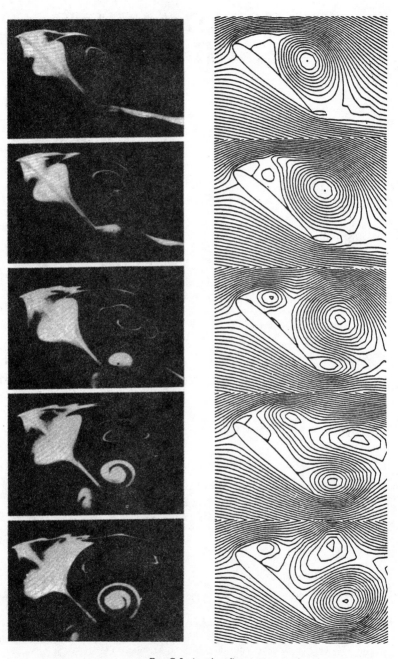

FIG. C.5. (continued)

SELECTED ANSWERS, HINTS, AND SOLUTIONS

1.0 EXERCISES

1. (a) $u(x) = c_1 x + c_2$, c_1 and c_2 arbitrary constants of integration, $-\infty < x < \infty$.
   (b) $u(x) = c_1 x + 1$, $x \geq 0$.
   (c) $u(x) = c_1 \sin 2x + c_2 \cos 2x$, c_1 and c_2 arbitrary constants of integration, $0 < x < \pi$.

2. (a) $u(x) = -x + 1$, $0 \leq x \leq 1$.
   (b) $u(x) = e^{-2x}$, $x \geq 0$.
   (c) $u(x) = c_1 \sin 2x$, c_1 arbitrary, $0 \leq x \leq \pi$.

3. (a) $u(x, y) = c_1(y)$, c_1 an arbitrary function of y.
   (b) $u(x, y) = c_1(y)x + c_2(y)$, c_2 an arbitrary function of y.
   (c) $u(x, y) = c_1(y)x + 1$.

SECTION 1.1

1. (a) $d = y$. Thus the operator is elliptic for $y > 0$, parabolic for $y = 0$, and hyperbolic for $y < 0$.
   (b) $d = 1 + (u_x)^2 + (u_y)^2 > 0$.

2. (a) $C = \begin{bmatrix} 1 & -1 \\ 1 & 1 \end{bmatrix}$, among others.

   (b) $C^{-1} = \begin{bmatrix} \frac{1}{2} & \frac{1}{2} \\ -\frac{1}{2} & \frac{1}{2} \end{bmatrix}$, for C as given in (a).

3. (a) $u = ax + by + cxy + d$, $x^2 - y^2$, $e^y \sin x, \ldots$
   (b) $u = ax + b$, $x^2 + 2t$, $e^{-t} \sin x, \ldots$
   (c) $u = ax + bt + cxt + d$, $x^2 + t^2$, $e^{ax + bt}$ with $a^2 = b^2, \ldots$
   (d) $u = ax + by + cxy + d$, $3x^2 - y^3, \ldots$
   (e) $u = ax + by + c$

1.1 EXERCISES

1. (a) $d = AC - B^2 = 8$. Hence the equation is elliptic. Note that the lower-order derivatives are ignored.
   (b) $d = -8$, hyperbolic equation.
   (c) $d = 0$, parabolic equation. Note that the discriminant rule given here applies only to the case of two independent variables.

2. (a) Note that $\ln r = \frac{1}{2}\ln(x^2 + y^2)$ and differentiate.

   (b) Substitute and verify. Note that $\sqrt{2t}$ may be interpreted as the spread (standard deviation σ) of the solution, and that this spread increases with t.

   (c) Substitute and verify. Note that the symmetry in x and t guarantee this function to be a solution.

3. (a) Direct computation (chain rule) yields

$$u_{xx} = u_{\zeta\zeta} + 2ru_{\zeta\eta} + r^2 u_{\eta\eta}$$

$$u_{xy} = u_{\zeta\eta} + ru_{\eta\eta}$$

$$u_{yy} = u_{\eta\eta}$$

   which upon substitution into the equation gives

$$Au_{xx} + 2Bu_{xy} + Cu_{yy} = Au_{\zeta\zeta} + (Ar^2 + 2Br + C)u_{\eta\eta} + 2(rA + B)u_{\zeta\eta}$$

$$= Au_{\zeta\zeta}.$$

   Note that $r = -B/A$ is a root of $Ar^2 + 2Br + C = 0$ because the equation was parabolic and hence $d = AC - B^2 = 0$

   (b) Direct computation, substitution, and collecting terms yield the canonical form. Partial integration yields the general solution form. See Problem 1.9.1 for further ramifications of the discriminant classification rule, and see Section 2.5 for a full integration of the wave equation.

   (c) As $d = AC - B^2 = y^2 - x^2 = (y + x)(y - x)$, this equation has type as indicated below:

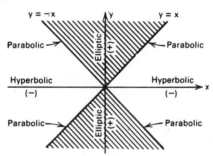

SECTION 1.2

3. (a) $u = \sin x \cos t$

1.2 EXERCISES

1. (a) Solutions to the Dirichlet Problem in one dimension are straight lines $u(x) = c_1 x + c_2$, with the c_1 and c_2 determined by the boundary data. For the interior problem in which Ω is a finite interval, c_1 and c_2 are uniquely determined by the two given values $f(-1)$ and $f(1)$.

   (b) Again, as in (a), the coefficients c_1 and c_2 of the general solution of the differential equation, $u(x) = c_1 x + c_2$, are determined by the boundary data. In particular,

$c_2 = f(0)$, but now we must take care on the interpretation of the condition at $x = \infty$. In applications usually we need $u(x) \to 0$ as $x \to \infty$, and in that case there is a solution only if $f(0) = 0$. More generally, if $f(\infty)$ is any finite quantity, there will be a solution (again, a horizontal line) iff $f(0) = f(\infty)$. Should we allow the boundary condition: $f(\infty) = \infty$, $-\infty$, or $\pm\infty$, there are an infinite number of solutions whose determination and meaning have become rather vague.

(c) The considerations are the same as in (b), except now we have both a left unbounded domain, $\Omega_- = (-\infty, -1)$, a right unbounded domain, $\Omega_+ = (1, \infty)$, on which the solutions are independent. For further discussion of unbounded exterior domain implications, see Section 2.7.

2. (a) For each fixed x, the solution $u(x, t)$ is a line horizontal in t of height $f(x)$. The solution set is thus an $f(x)$ cylinder.

(b) $u(x, t) = f(x)e^{-3t}$.

(c) $u(x, t) = c_1(x)t + c_2(x) = c_1(x)t + f(x)$ where $c_1(x)$ is arbitrary for each x. From ordinary differential equations, we know we need another initial condition to determine the solution uniquely.

3. (a) If $Lu = \lambda u$, then $L(cu) = cLu = c\lambda u = \lambda(cu)$, remembering the properties of linear transformations put forth on p. 1.

(b) The function $v_5(x) = \sin 5x$ is the fifth principal eigenfunction, corresponding to the fifth eigenvalue $\lambda_5 = (5)^2 = 25$. Because the ordinary derivative operator d^2/dx^2 is linear, any multiple of v_5 is also an eigenfunction solution.

(c) Recalling from ODE the variable change $v = du/dx$, the expression becomes the separable one

$$v\,dv = \lambda u^3\,du$$

from which

$$\left(\frac{du}{dx}\right)^2 = \frac{\lambda}{2}u^4 + c_1$$

Without integrating this equation, it suffices to note that substituting $u = cu_\lambda$ will not retain the eigensolution property at λ, although it does generate an eigensolution at λc^{-2}. See Section 1.8 for more details on nonlinear eigenvalue problems.

SECTION 1.3

1 (a) Let u_1 and u_2 be $C^0(\overline{\Omega}) \cap C^2(\Omega)$ solutions to the Dirichlet problem for the same data f. Then $u_1 - u_2$ is harmonic in Ω with zero boundary value, from which $u_1 = u_2$ by the maximum principle.

(b) The $C^2(\Omega)$ is necessary to the validity of the maximum principle. The counter-example $u(x, y) = 2xy/(x^2 + y^2)^2$ for $(x, y) \neq (0, 0)$ $u(0, 0) = 0$, shows this.

A straightforward computation yields $u_{xx} = 24xy(x^2 - y^2)/(x^2 + y^2)^4$, and hence $u_{yy} = 24xy(y^2 - x^2)/(x^2 + y^2)^4$ by symmetry. Thus $\Delta u = 0$ at all points except possibly $(0, 0)$. From the difference quotient limits one also verifies that $\Delta u(0, 0) = 0$. But along $y = x$, $u = x^{-2}$ which blows up as $x \to 0$. The

$C^0(\overline{\Omega}) \cap C^2(\Omega)$ requirement along with $\Delta u = 0$ guarantees in fact that a function u is analytic.

(c) For $\Delta u_1 = 0$ in Ω, $u_1 = f_1$ on $\partial\Omega$, and $\Delta u_2 = 0$ in Ω, $u_2 = f_2$ on $\partial\Omega$, we have from the Maximum Principle

$$\max_\Omega |u_1 - u_2| \leq \max_{\partial\Omega} |f_1 - f_2|.$$

Hence if the data f_1 is close to f_2, the solutions are close, an instance of continuous dependence of the solutions on the data.

2. Differentiate under the integral, in the classical way. Denoting the integral kernel of the integrand by P, for $t \neq 0$,

$$P = (4\pi t)^{-1/2} e^{[-(x-y)^2/4t]}$$

$$P_x = (4\pi t)^{-1/2} e^{[-(x-y)^2/4t]} \cdot [-2(x-y)/4t]$$

$$P_{xx} = (4\pi t)^{-1/2} e^{[-(x-y)^2/4t]} \cdot [-(1/2t) + (x-y)^2/4t^2]$$

$$P_t = (4\pi t)^{-1/2} e^{[-(x-y)^2/4t]} \cdot [(x-y)^2/4t^2]$$
$$\quad + e^{[-(x-y)^2/4t]} \cdot [(1/4\pi)^{1/2} (-t^{-3/2}/2)]$$
$$= (4\pi t)^{-1/2} e^{[-(x-y)^2/4t]} \cdot [(x-y)^2/4t^2 - (1/2t)]$$
$$= P_{xx}$$

Thus $u_t - u_{xx} = 0$. Thus, granting the gift of this solution, we have obtained (1) existence of a solution and (1') construction of a solution. In just setting $u = f$ at $t = 0$, the solution could well be spurious as discussed in the text for the Dirichlet problem. Later we will establish that it continuously attains the initial data f as $t \to 0$ and that it possesses the related regularity (2'), stability (3), and uniqueness (2) properties. Numerical approximation (3') methods are discussed in Chapter 3.

3. (1) Existence of a solution to the differential equation is guaranteed by the analytic theory of linear ordinary differential equations. One needs only that the coefficients be continuous. However, the existence question is not settled until the boundary conditions are obtained. We will come back to this question. For construction (1') we will use the familiar power series method from ODE. Assume $u(x) = \sum_{k=0}^\infty a_k x^k$.

It is convenient to first consider the uniqueness (2) of such solutions. Substituting the assumed series solution into

$$\begin{cases} u_{xx} + 4xu = 0, & 0 < x < 2, \\ u(0) = u(2) = 0, \end{cases}$$

yields

$$0 = 2a_2 + \sum_{k=0}^\infty ((k+3)(k+2)a_{k+3} + 4a_k)x^{k+1}.$$

Thus $a_2 = 0$ and $a_{k+3} = -4a_k/(k+3)(k+2)$, $k = 0, 1, 2, \ldots$. The boundary condition $u(0) = 0$ makes $a_0 = 0$. Induction then gives $a_{3k} = a_{3k+2} = 0$ and $a_{3k+1} = (-4)^k a_1 / \prod_{j=1}^k (3j+1)(3j)$. Hence the formal series solution is

$$u(x) = a_1 \left[x + \sum_{k=1}^{\infty} \frac{(-4)^k}{\prod_{j=1}^{k}(3j + 1)(3j)} x^{3k + 1} \right]$$

For convenience call this $u = a_1 \sum_{0}^{\infty} c_k x^{3k+1}$. The radius of convergence is

$$\lim_{k \to \infty} \left| \frac{c_{k+1}}{c_k} \cdot \frac{x^{3k+4}}{x^{3k+1}} \right| = |x|^3 \lim_{k \to \infty} \frac{4}{(3k+4)(3k+3)} = 0$$

so that the series converges absolutely for all x. Hence u is real analytic.

Now apply the other boundary condition $u(2) = 0$ to obtain $a_1 \sum_{0}^{\infty} c_k 2^{3k+1} = 0$. To show that the series $\sum_{0}^{\infty} c_k 2^{3k+1}$ is nonvanishing, we can for example proceed as follows:

$$\sum_{k=0}^{6} c_k 2^{3k+1} = -0.454088581$$

whereas

$$\left| \sum_{k=7}^{\infty} c_k 2^{3k+1} \right| \leq 2^{22} |c_7| = 0.000256096.$$

Hence $a_1 = 0$ and thus $u = 0$. Since the original problem was linear, this guarantees uniqueness.

We now return to the construction of the solution. From

$$\begin{cases} u_{xx} + 4xu = 16, \\ u(0) = 0, \end{cases}$$

we have

$$u(x) = a_1 \sum_{k=0}^{\infty} b_k x^{3k+1} + 8 \sum_{k=0}^{\infty} c_k x^{3k+2}$$

where now

$$b_k = (-4)^k / \prod_{j=1}^{k}(3j + 1)(3j), \quad c_k = (-4)^k / \prod_{j=1}^{k}(3j + 2)(3j + 1)$$

and in which we made use of $2a_2 = 16$. Applying the right boundary condition $u(2) = -1$ gives

$$a_1 = \frac{-1 - 8\sum_{0}^{\infty} c_k 2^{3k+2}}{\sum_{0}^{\infty} b_k 2^{3k+1}},$$

Let us summarize. Existence (1), uniqueness (2), regularity (analyticity in this problem) (2′), and construction (1′) have been accomplished. Approximation (3′) by truncating the series solution to a finite number of terms is in fact implicit in the construction (1′) in cases such as this where one has taken a series solution. Even if summable, the closed form solution generally involves elementary functions that need to be approximated when calculating actual solution values.

Which kind of stability (3) one wants will depend on the problem or application. Following our use of the power series method, suppose ε_0, ε_2, and

ε_f are errors in $u(0)$, $u(2)$, and $f(x) = 16$ for $0 < x < 2$, respectively. Then if $\bar{u}$ is the solution of the perturbed equation with slightly erroneous data, we have (roughly) that

$$\bar{u} - u \cong \varepsilon_0\alpha(x) + \varepsilon_2\beta(x) + \varepsilon_f(x)\gamma(x)$$

where

$$\alpha = \sum_0^\infty a_k x^{3k} \qquad \beta = \sum_0^\infty b_k x^{3k+1}, \qquad \gamma = \sum_0^\infty c_k x^{3k+2}$$

where $a_k = (-4)^k/\pi_1^k(3k)(3k - 1)$ and b_k and c_k are as above.

One could go further with this example, e.g., in terms of the Sturm–Liouville theory, but this suffices to illustrate the six $(3 + 3)$ questions here.*

1.3 EXERCISES

1. The existence (a) and uniqueness (b) of the solution $u(x) = ax + b$ were guaranteed by its construction (by methods of elementary calculus) and by the fitting of the boundary conditions. The degree of stability (c) of the solution under the varying of the data a and b is evident geometrically regarding the solution as a line dependent on those two parameters.

2. As pointed out in 1, construction (a) depended on our knowledge of calculus (although in this simple example, geometry would have sufficed), the regularity (b) of the solution is its analyticity, as is so often the case in ordinary differential equations, and its approximation (c) is unnecessary.

3. Consult any ODE book to find that the differential equation possesses a two-parameter solution family under general conditions (e.g., the coefficients $a_i(x)$ be continuous). However, the existence (a), which depends on the ability to satisfy the boundary conditions, and similarly the uniqueness (b), e.g., in the case $a_0 = 1$, $a_1 = 0$, $a_2 = \lambda$, depends on the nature of the coefficients, and the construction (c) has still not been resolved in all equations.

SECTION 1.4

1. (b) No solution.

2. (c) For $u''(x) = 0$ on an interval $[a, b]$ with left boundary condition $u(a) = 0$, there exists a unique solution for any of the three boundary conditions $u(b) = c$, $u'(b) = d$, or $u'(b) + ku(b) = 0$ with $k \neq (a - b)^{-1}$. For the latter case the solution is $u(x) = c(b - a)^{-1}(x - a)$, where c may be chosen arbitrarily.

3. (a) The solution for $k > 0$ is $u = 0$ unless $k = 2$. Recall uniqueness fails at $k = 0$ also.

   (c) For $f = 0$ the problem is called a Steklov eigenvalue problem (with eigenvalue k in the boundary condition). Nontrivial solutions can exist. Recall the ODE versions of parts (a) and (b). See also Problem 3 of Section 1.6.1.

---

* Its solutions are in fact important special functions (see 2.9.9(9)).

1.4 EXERCISES

1. (a) Dirichlet,   (b) Neumann,   (c) Robin.

2. (a) $u_1(x) \equiv c$ and $u_2(x) = c/r$, from the radial equation $u'' + 2r^{-1}u' = 0$.
   (b) $u(x) \to 0$ as $x \to \infty$ leaves only $u_2(x)$.
   (c) $\partial u/\partial r = 0$ leaves only $u_1(x)$, $c \neq 0$.

3. By linearity of the problem, the difference $u(x) = u_1(x) - u_2(x)$ of any two possibly different solutions $u_1(x)$ and $u_2(x)$ satisfies

$$\begin{cases} u''(x) = 0, & 0 < x < 1, \\ u(0) = u(1). = 0. \end{cases}$$

Recognizing the solution geometrically (a) to be the line passing through $x = 0$ and 1, we see that the difference $u(x)$ is identically zero. (b) The maximum principle of Section 1.3 asserts that both $u(x)$ and its negation $-u(x)$, as harmonic (1-dimensional, here) functions, take their maxima on the boundary $x = 0$ and $x = 1$. Therefore, since both the maximum and minimum values of u are zero, so is the whole function. (c) Multiplying $u''(x) = 0$ by $u(x)$ and integrating by parts, we obtain

$$0 = \int_0^1 u(x)u''(x)\,dx = u(x)u'(x)\Big|_0^1 - \int_0^1 (u'(x))^2\,dx.$$

Because $u(0) = u(1) = 0$, we may conclude that $u'(x) = 0$ throughout the interval $0 \leqq x \leqq 1$. Hence $u(x)$ is constant and by its boundary values, identically zero.

SECTION 1.5.1

1. (a) $u(x, y) = \sum\limits_{n=1}^{\infty} c_n \sin nx \sinh n(\pi - y)$,

$c_n = \dfrac{2}{\pi \sinh n\pi} . \int_0^{\pi} f(s) \sin ns\,ds.$

2. (a) $u_{rr} + r^{-1}u_r + r^{-2}u_{\theta\theta}$.

   (b) $u(r, \theta) = \frac{1}{2}a_0 + \sum\limits_{n=1}^{\infty} r^n(a_n \cos n\theta + b_n \sin n\theta)$,

   $a_n = \dfrac{1}{\pi} \int_{-\pi}^{\pi} f(s) \cos ns\,ds, \qquad b_n = \dfrac{1}{\pi} \int_{-\pi}^{\pi} f(s) \sin ns\,ds.$

3. (a) $u(x, y) = \sum\limits_{n=1}^{\infty} \dfrac{2(-1)^{n+1}}{n \sinh n\pi} \sin nx \sinh n(\pi - y)$

   $u(r, \theta) = \sum\limits_{n=1}^{\infty} \dfrac{2(-1)^{n+1}r^n}{n} \sin n\theta.$

1.5(1) EXERCISES

1. (a) All c_n vanish except for $c_1 = 1$. A Fourier series expansion thus perfectly preserves any one of its fundamental modes.

(b) An interesting identity*

$$\sin^3 x = \frac{3}{4} \sin x - \frac{1}{4} \sin 3x$$

represents the given data in terms of the two fundamental modes $\sin x$ and $\sin 3x$. Hence $c_1 = 3/4$, $c_3 = -1/4$, and all other c_n vanish.

(c) After some work (integration by parts) one arrives at $c_n = 0$ for n even, $c_n = 8/n^3\pi$ for n odd. The even-numbered coefficients vanish because the even-numbered modes are odd about the interval midpoint $x = \pi/2$, whereas the given function is even about that point. The same remark applies to parts (a) and (b).

2. One needs only assert the Fourier coefficients calculated in Exercise 1 above. Thus
   (a) $u(x, t) = \sin x \cos t$
   (b) $u(x, t) = \frac{3}{4} \sin x \cos t - \frac{1}{4} \sin 3x \cos 3t$
   (c) $u(x, t) = \displaystyle\sum_{n=1}^{\infty} \frac{8}{(2n - 1)^3 \pi} \sin(2n - 1)x \cos(2n - 1)t.$

3. The separation of variables procedure yields the general solution formula

$$u(x, t) = \sum_{n=1}^{\infty} c_n e^{-n^2 t} \sin nx$$

and thus the solutions are, as in Exercise 2 above,
   (a) $u(x, t) = e^{-t} \sin x$
   (b) $u(x, t) = \frac{3}{4} e^{-t} \sin x - \frac{1}{4} e^{-9t} \sin 3x$
   (c) $u(x, t) = \dfrac{8}{\pi} \displaystyle\sum_{n=1}^{\infty} \dfrac{e^{-(2n-1)^2 t}}{(2n - 1)^3} \sin(2n - 1)x.$

SECTION 1.5.2

1. (b) $G(x, s) = \begin{cases} s, & 0 \leq s \leq x, \\ x, & x \leq s \leq \pi. \end{cases}$

3. (b) There are several ways to do this, including the following:
   (i) We recall that $\int_{-\infty}^{\infty} e^{-s^2} ds = \sqrt{\pi}$. One way to establish that is from

---

* We would like to note here an instance of how interpersonal communication often plays a vital role in the development and application of mathematics. This identity, when pointed out by the author to Professor N. Bazley over fondue in Geneva, Switzerland one day in the early 1970s, led to a theory of "reproducing nonlinearities." See N. Bazley, *Manuscripta Math.* 18 (1976), P. Rutkowski, *J.A. Math. and Physics* (ZAMP) 34 (1983), for an account and applications of this theory.

The identity follows immediately from De Moivres Theorem:

$$[r(\cos \theta + i \sin \theta)]^n = r^n[\cos n\theta + i \sin n\theta].$$

Nonetheless one could speculate that it reveals a possibly far-reaching "theory of nonlinear Fourier analysis" which has not yet, to the author's knowledge, been fully developed. Other instances of such a theory would include for example the Clebsch–Gordan relations of elementary particle theory and other fundamental relations within classes of special functions.

$$\left[\int_{-\infty}^{\infty} e^{-s^2}\, ds\right]^2 = \int_{-\infty}^{\infty}\int_{-\infty}^{\infty} e^{-(x^2+y^2)}\, dx\, dy = \int_0^{2\pi}\int_0^{\infty} e^{-r^2} r\, dr\, d\theta = \pi.$$

(ii) By letting $s = -\dfrac{x-y}{2t^{1/2}}$ in the Poisson kernel we then have

$$\int_{x-\delta}^{x+\delta} \frac{1}{\sqrt{4\pi t}} e^{-(x-y)^2/4t}\, dy = \pi^{-1/2}\int_{-\delta/2t^{1/2}}^{\delta/2t^{1/2}} e^{-s^2}\, ds \to 1 \quad\text{as}\quad t \to 0^+.$$

(iii) It is easily checked that

$$\int_{-\infty}^{x-\delta} \frac{1}{\sqrt{4\pi t}} e^{-(x-y)^2/4t}\, dy \to 0$$

and

$$\int_{x+\delta}^{\infty} \frac{1}{\sqrt{4\pi t}} e^{-(x-y)^2/4t}\, dy \to 0 \quad\text{as}\quad t \to 0^+,$$

for fixed $\delta > 0$.

(iv) Now let $\varepsilon > 0$ and let f be continuous and bounded (i.e., $|f(y)| \leq M$ for all $-\infty < y < \infty$). Given ε, there exists a $\delta > 0$ such that $|f(x_1) - f(x_2)| < \varepsilon/4$ provided that $|x_1 - x_2| < \delta$. If also $|y - x_2| < \delta$, then $|f(y) - f(x_1)| < \varepsilon/2$. Since by (i) above,

$$f(x_1) - \int_{-\infty}^{\infty} \frac{1}{\sqrt{4\pi t}} e^{-(x_2-y)^2/4t} f(y)\, dy$$

$$= \frac{1}{\sqrt{4\pi t}}\left[\int_{x_2-\delta}^{x_2+\delta} (f(x_1) - f(y))e^{-(x_2-y)^2/4t}\, dy\right.$$

$$\left. + \int_{(-\infty,\, x_2-\delta)}^{(x_2+\delta,\, \infty)} (f(x_1) - f(y))e^{-(x_2-y)^2/4t}\, dy\right],$$

we thus have, by (iii) above, for all $0 < t \leq T$ for T chosen sufficiently small, that

$$\left| f(x_1) - \int_{-\infty}^{\infty} \frac{1}{\sqrt{4\pi t}} e^{-(x_2-y)^2/4t} f(y)\, dy\right|$$

$$< \frac{\varepsilon}{2} + 2M \int_{(-\infty,\, x_2-\delta)}^{(x_2+\delta,\, \infty)} \frac{1}{\sqrt{4\pi t}} e^{-(x_2-y)^2/4t}\, dy < \frac{\varepsilon}{2} + \frac{\varepsilon}{2} = \varepsilon.$$

1.5(2) EXERCISES

1. The discriminant $b^2 - 4ac$ determines the cases
   (a) $d > 0$, (b) $d = 0$, (c) $d < 0$.
2. (a) $y_1(x) = e^{r_1 x}$, $y_2(x) = e^{r_2 x}$; $y_1(x) = e^{rx}$, $y_2(x) = xe^{rx}$; $y_1(x) = e^{\alpha x}\cos\beta x$, $y_2(x) = e^{\alpha x}\sin\beta x$, for roots $r = \alpha \pm i\beta$.
   (b) Let $y_p(x) = c_1(x)y_1(x) + c_2(x)y_2(x)$ and require that

$$\begin{cases} c_1' y_1 + c_2' y_2 = 0 \\ c_1' y_1' + c_2' y_2' = f/a. \end{cases}$$

(c) Solving (b) by, for example, Cramer's rule.

3. (a) Yes, for each fixed P.   (b) Yes, since

$$\delta_p(u(Q) + v(Q)) = u(P) + v(P) = \delta_p(u(Q0) + \delta_p(v(Q)).$$

(c) $u(x) = \int_0^\pi G(x, s)\, f(s)ds$

$$= \int_0^\pi G(x, s)\underbrace{(-\frac{d^2}{ds^2})\, u(s)ds}_{\delta(x,\, s)}$$

$$\underbrace{\phantom{= \int_0^\pi G(x, s)(-\frac{d^2}{ds^2})\, u(s)ds}}_{\delta_x(s)}$$

where the ds is understood to be there in both.

SECTION 1.5.3

2. (a) $R(v) = 10\pi^{-2} \approx 1.013$.

(b) $\| v - v_1 \| \approx 0.038$.

1.5(3) EXERCISES

1. (a) The solution of $u''(x) = 0$ for $a < x < b$ is obtained by minimizing $\int_a^b |v'(x)|^2 dx$ over all functions $v(x)$ which take on the prescribed Dirichlet Boundary conditions $f(a)$ and $f(b)$.

(b) Clearly 0 is a lower bound, hence an infimum exists. Later (Poincaré Inequality) we will see that the lower bound can be taken to be greater than zero.

(c) An infimum is merely a limit which need not actually be attained.

2. (a) Integration by parts, especially after multiplication of the equation by u or some other relevant expression in u.

(b) From $u''(x) = 0$, $uu'' = 0$, $\int_a^b uu'' = 0$, hence $D(v) = vv'|_a^b - \int_a^b vv''$ would appear to be minimized by requiring the second term to vanish, i.e., $v'' = 0$

(c) Roughly, there are two variation, first in v' on the boundary, and then in v'' over the domain.

3. (a) Stipulating too much regularity can "lose" the solution, see Problem 3 above.

(b) Weakening regularity allows proofs of the existence of at least weak solutions.

(c) Exactly the right amount of regularity will be evident once the problem is fully understood.

FIRST PAUSE

1. $c_n = 4A/n\pi$ for n odd, $c_n = 0$ for n even. Hence the solution is

$$u(x, t) = \frac{4a}{\pi} \sum_{n=1}^{\infty} \frac{1}{(2n-1)} \sin \frac{(2n-1)\pi x}{l} e^{\frac{-k(2n-1)^2\pi^2}{l^2} t}$$

2. Letting $u(x, t) = e^t v(x, t)$ reduces the problem to $v_t - 2v_{xx} = 0$, with the same boundary conditions and initial condition. The calculation of the Fourier coefficients of the initial value is more laborious but finally one arrives at the solution

$$u(x, t) = \frac{8a}{\pi} e^t \sum_{n=0}^{\infty} \frac{1}{4n + 2} \sin \frac{(4n + 2)\pi x}{10} e^{-\frac{2(4n+2)^2\pi^2}{10^2}t}.$$

3. After much chain rule

$$\Delta u = r^{-2}(r^2 u_r)_r + r^{-2}(\sin\theta)^{-1}(\sin\theta u_\theta)_\theta + (r\sin\theta)^{-2} u_{\phi\phi}$$

Then insert $u(r, \theta, \phi) = R(r)\Theta(\theta)\Phi(\phi)$.

4. $u(x, t) = \dfrac{8A}{\pi^2} \displaystyle\sum_{n=1}^{\infty} \dfrac{(-1)^{n+1}}{(2n - 1)^2} \sin \dfrac{(2n - 1)\pi x}{l} \cos \dfrac{k(2n - 1)\pi}{l} t.$

5. Find $g(x)$ so that v satisfies

$$v_t - v_{xx} = 0, \qquad 0 < x < 1, \qquad t > 0,$$
$$v(0, t) = v(1, t) = 0, \qquad t \geq 0,$$
$$v(x, 0) = f(x) - g(x),$$

and then solve the latter.

6. $u(x) = c_1 + c_2 x + \int_0^x (x - s)f(s)\, ds$ for $\lambda = 0$,

$$u(x) = c_1 e^{(-\lambda)^{1/2}x} + c_2 e^{-(-\lambda)^{1/2}x} + (-\lambda)^{-1/2} \int_0^x \sinh((-\lambda)^{1/2}(x - s))f(s)\, ds$$

for $\lambda < 0$. Then do the boundary conditions, algebra, and trigonometry.

7. Either directly or by plugging into the above $G(x, s)$ found for $\lambda > 0$,

$$G(x, s) = \begin{cases} -\sin s \cos x, & 0 \leq s \leq x, \\ -\sin x \cos s, & x \leq s \leq \pi/2. \end{cases}$$

The solution for $f(x) = x$ is $u(x) = x - \dfrac{\pi}{2}\sin x$.

8. $G(x, s) = \begin{cases} [\cos(x + s) + \sin(x - s)]/2, & 0 \leq s \leq x, \\ [\cos(x + s) - \sin(x - s)]/2, & x \leq s \leq \pi/2, \end{cases}$

$$u(x) = \frac{\pi}{4}(\cos x - \sin x) - \sin x + x.$$

9. (a) ODE gives $2 - 2\cos(\Lambda^{1/2}(b - a)) - \Lambda^{1/2}(b - a)\sin(\Lambda^{1/2}(b - a)) = 0.$
   (b) Integration by parts shows Dirichlet orthogonality: $D(u_n, u_m) = 0.$
   (c) All shorter length trial functions are trial functions for the longer length variational quotients when extended.

SECTION 1.6.1

1. It suffices by linearity of the operators involved to show that zero data implies necessarily the zero solution. (Write this out as a useful general lemma for linear operators.) Thus the regularity per se of f and g does not matter. We assume the

domain to be a regular or divergence domain and functions u to be in $C^2(\Omega) \cap C^1(\overline{\Omega})$. Then from Green's first identity and zero data we have

$$\int_\Omega |\operatorname{grad} u|^2 \, dx = \oint_{\partial\Omega} u \frac{\partial u}{\partial n} ds - \int_\Omega u \, \Delta u = 0$$

and hence (write out why) u is constant over Ω. Since u was assumed continuous on $\overline{\Omega}$, $u = 0$ there.

1.6(1) EXERCISES

1. (a) By the divergence theorem

$$\frac{1}{3} \oint_{\partial\Omega} x \cdot n = \frac{1}{3} \int_\Omega \operatorname{div}(x) dV = \int_\Omega dV = \operatorname{vol}(\Omega)$$

(b) As in part (a).

(c) $\operatorname{vol}(\Omega) = \dfrac{1}{m} \oint_{\partial\Omega} x \cdot n = \dfrac{1}{m} \oint_{\partial\Omega} x \cdot \dfrac{x}{|x|} = \dfrac{r}{m} \oint_{\partial\Omega} dS = \dfrac{r}{m} \operatorname{Area}(\partial\Omega)$

2. (a) and (b). The relation

$$\int_0^{\pi/2} \cos^m(\theta) \, d\theta = \frac{m-1}{m} \int_0^{\pi/2} \cos^{m-2}(\theta) \, d\theta$$

is helpful here.

(c) $\operatorname{Vol}(B_4(r)) = 2r \operatorname{Vol}(B_3(r)) \int_0^{\pi/2} \cos^4\theta \, d\theta = \pi^2 r^4/2$,

$$\operatorname{Area}(S_4(r)) = \frac{4}{r} \operatorname{Vol}(B_4(r)) = 2\pi^2 r^3.$$

3. (a) Solutions of $u''(x) = 0$ are of the form $u(x) = ax + b$,

$$\frac{1}{\omega_1} \oint_{S_{PQ}=r} u(Q) dS_Q = \frac{1}{2}[u(-r) + u(r)] = b = u(0) \text{ for any } r.$$

(b) Let $\sigma(t) = (r \cos(t), r \sin(t))$, $t \in (0, 2\pi)$, be a parameterization of $S_{PQ} = r$. Then

$$\frac{1}{\omega_2 r} \oint_{S_{PQ}=r} u(Q) dS_Q = \frac{1}{2\pi r} \int_0^{2\pi} u(\sigma(t)) |\sigma'(t)| dt$$

$$= \frac{1}{2\pi} \int_0^{2\pi} [a \cos(t) + b \sin(t) + c] = c = u(0).$$

(c) The mean value theorem implies $u(x) = \dfrac{u(x + r) + u(x - r)}{2}$. Use Taylor's theorem to conclude

$$u''(x) = \lim_{r \to 0} \frac{u(x + r) - 2u(x) + u(x - r)}{r^2} = 0.$$

SECTION 1.6.2

2. (a) $\lambda_n = \inf \int_\Omega |\text{grad } u|^2 \, dx / \int_\Omega u^2 \, dx$, the infimum taken over all sufficiently regular u that are orthogonal to all eigenfunctions $u_1, \ldots, u_{n-1}$ corresponding to the previous eigenvalues $\lambda_1, \ldots, \lambda_{n-1}$, that is, $\int_\Omega uu_k \, dx = 0$ for all such u_k.

   (b) $\lambda_{mn} = m^2 + n^2$, $m = 1, 2, 3, \ldots$, $n = 1, 2, 3, \ldots$, and the $u_{mn} = c_{mn}$ $\sin mx \times \sin ny$, c_{mn} arbitrary.

3. (a) $a \leq c_1 b_1 + c_2 b_2 \Rightarrow a^2 \leq (1 + \varepsilon)c_1^2 b_1^2 + (1 + \varepsilon^{-1})c_2^2 b_2^2$.
   $a^2 \leq c_1 b_1^2 + c_2 b_2^2 \Rightarrow a \leq \sqrt{c_1} b_1 + \sqrt{c_2} b_2$.

   (b) $\|u\|_{\hat\Omega} \leq \|u\|_{\partial\Omega} + c_3 \|\Delta u\|_{\hat\Omega}$, where $c_3 = \max_{Q \in \partial\Omega} d^2 (0, Q)/2n$, $\|\cdot\|_S$ denoting the maximum over the set S, n the dimension. The origin 0 should be chosen most advantageously within Ω.

   (c) Indirectly, and for sharp bounds, it involves the geometry of the domain Ω and working with the relationship between the gradient and the tangential derivatives of u on $\partial\Omega$.*

   One may obtain such a bound for $|u(P)|^2$, rather than for the requested $\int_\Omega u^2$, at each P in Ω from the Poisson integral formula given in Remark 3 of Section 1.6.1 as follows. As there, one has (it holds for an arbitrary nice bounded domain Ω)

$$u(P) = -\oint_{\partial\Omega} \frac{\partial G}{\partial n} u \, ds$$

from which (by Schwarz's inequality)

$$u^2(P) \leq \oint_{\partial\Omega} r_{PQ} \frac{\partial r_{PQ}}{\partial n} \left(\frac{\partial G}{\partial n}\right)^2 ds_Q \cdot \oint_{\partial\Omega} \left(r_{PQ} \frac{\partial r_{PQ}}{\partial n}\right)^{-1} u^2(Q) \, ds_Q.$$

From a calculation, using the fact that

$$\frac{\partial r_{PQ}}{\partial n} \cdot \frac{\partial G}{\partial n} = \frac{\partial G}{\partial r_{PQ}},$$

the first factor equals ω_n^{-1}, where ω_n is the area of the unit sphere in n-dimensions (see 1.6.1 Remark 1). Thus

$$u^2(P) \leq \frac{1}{m_P \omega_n} \oint_{\partial\Omega} u^2 \, ds$$

where m_P is a positive lower bound for $r_{PQ} \, \partial r_{PQ}/\partial n$ on $\partial\Omega$. The latter exists for all P for example if Ω is convex. One can see now the difficulty in trying to generate the just obtained inequality over all P in Ω to obtain a bound for $\int_\Omega u^2$, because the m_P is not uniform and may go to zero as P approaches $\partial\Omega$.

The existence however of the desired bound may be seen as follows. As shown in 1.6.2(1'') the obtaining of the desired bound $\int_\Omega u^2 \leq c \oint_{\partial\Omega} u^2$ for $\Delta u = 0$ in Ω was equivalent to obtaining a bound $\oint_{\partial\Omega} (\partial\psi/\partial n)^2 \leq k \int_\Omega (\Delta\psi)^2$ for $\psi = 0$ on $\partial\Omega$. For the latter we consider the clamped plate Steklov eigenvalue problem

---

* See, for example, L. Payne and H. Weinberger, *Pac. J. Math.* **8** (1958).

$$\begin{cases} \Delta^2\psi = 0 \text{ in } \Omega, \\ \psi = 0 \text{ on } \partial\Omega, \\ \Delta\psi = \lambda\dfrac{\partial\psi}{\partial n} \text{ on } \partial\Omega. \end{cases}$$

From Green's second identity we then have

$$\oint_{\partial\Omega} \psi\frac{\partial(\Delta\psi)}{\partial n} - \oint_{\partial\Omega} \lambda\frac{\partial\psi}{\partial n}\cdot\frac{\partial\psi}{\partial n} = -\int_\Omega \Delta\psi\,\Delta\psi$$

and thus

$$\lambda = \int_\Omega (\Delta\psi)^2 \Big/ \oint_{\partial\Omega} \left(\frac{\partial\psi}{\partial n}\right)^2.$$

The eigenvalues are discrete and as just shown nonnegative. It is immediate to check (we leave this as an exercise for the student) that the first eigenvalue λ_1 is not zero, and hence the desired k above may be taken to be λ_1^{-1}.

1.6(2) EXERCISES

1. (a) For $x = (x_1, x_2, x_3)$ and $y = (y_1, y_2, y_3)$, we have

$$x_1y_1 + x_2y_2 + x_3y_3 \leq (x_1^2 + x_2^2 + x_3^2)^{1/2}(y_1^2 + y_2^2 + y_3^2)^{1/2}.$$

(b) For functions u and v we have

$$\int_\Omega \text{grad } u(x)\cdot\text{grad } v(x) \leq \left(\int_\Omega(\text{grad } u(x))^2\right)^{1/2}\left(\int_\Omega(\text{grad } v(x))^2\right)^{1/2}$$

(c) For vectors $x = (x_1, x_2, x_3, \ldots)$ and $y = (y_1, y_2, y_3, \ldots)$, we have

$$x_1y_1 + x_2y_2 + x_3y_3 + \cdots$$
$$\leq (x_1^2 + x_2^2 + x_3^2 + \cdots)^{1/2}(y_1^2 + y_2^2 + y_3^2 + \cdots)^{1/2}$$

Note that for both (b) and (c) we must assume that the quantities used all exist.

2. (a) $((x_1 - y_1)^2 + (x_2 - y_2)^2 + (x_3 - y_3)^2)^{1/2} \leq ((x_1 - z_1)^2 + (x_2 - z_2)^2 + (x_3 - z_3)^2)^{1/2} + ((z_1 - y_1)^2 + (z_2 - y_2)^2 + (z_3 - y_3)^2)^{1/2}.$

(b) $\left(\int_\Omega(\text{grad }(u + v))^2\right)^{1/2} \leq \left(\int_\Omega(\text{grad } u)^2\right)^{1/2} + \left(\int_\Omega(\text{grad } v)^2\right)^{1/2}$

(c) Same as (a) except append the $\cdots$.

3. (a) The Dirichlet integral presumes that we are considering functions u and v over some region Ω in some n-dimensional space. Then

$$\text{grad } v\cdot\text{grad } u = \left(\frac{\partial v}{\partial x_1}, \cdots, \frac{\partial v}{\partial x_n}\right)\cdot\left(\frac{\partial u}{\partial x_1}, \cdots, \frac{\partial u}{\partial x_n}\right), \text{ the } R^n \text{ dot product.}$$

(b) The l.h.s. squared is 14884, the rhs squared is 14938.

(c) Close.

SECTION 1.6. 3

2. (a) $\displaystyle\sum_{n=1}^{\infty} \frac{2(-1)^{n+1}}{n} \sin nx.$

(b) Converges pointwise to $f(x) = x$ for $0 \le x < \pi$, to 0 at $x = \pi$ is not uniformly convergent on $[0, \pi]$, converges in root mean square to $f(x)$.

1.6(3) EXERCISES

1. (a) Assuming we can do termwise differentiation yields

$$u'(x) = \frac{d}{dx} \sum_{n=0}^{\infty} a_n x^n = \sum_{n=1}^{\infty} n a_n x^{n-1} = \sum_{n=0}^{\infty} a_n x^n = u(x)$$

Equating like coefficients yields $na_n = a_{n-1}$ and $u(0) = 1$ implies $a_0 = 1$

Hence $a_n = \dfrac{1}{n!}$ and $u(x) = \displaystyle\sum_{n=0}^{\infty} \frac{1}{n!} x^n = e^x.$

(b) $u(x, t) = x^2 + \displaystyle\sum_{n=1}^{\infty} \frac{1}{n!} \frac{\partial^n u}{\partial t^n} (x, 0) t^n$ by initial assumptions.

$$\frac{\partial u}{\partial t}(x, 0) = \frac{\partial^2 u}{\partial x^2}(x, 0) = 2 + \sum_{h=1}^{\infty} \frac{1}{n!} \frac{\partial^{n+2} u}{\partial t^n \partial x^2}(x, 0) t^n \Big|_{t=0} = 2.$$

Hence $u(x, t) = x^2 + 2t + \displaystyle\sum_{n=2}^{\infty} \frac{1}{n!} \frac{\partial^n u}{\partial t^n}(x, 0)\, t^n$

$$\frac{\partial^2 u}{\partial t^2} = \frac{\partial}{\partial t}\left[\frac{\partial^2 u}{\partial x^2}\right] = \frac{\partial}{\partial t} \sum_{n=2}^{\infty} \frac{1}{n!} \frac{\partial^{n+2} u}{\partial t^n \partial x^2}(x, 0)\, t^n$$

$$= \sum_{n=2}^{\infty} \frac{1}{(n-1)!} \frac{\partial^{n+2} u}{\partial t^n \partial x^2}(x, 0)\, t^{n-1}$$

Hence $\dfrac{\partial^2 u}{\partial t^2}(x, 0) = 0$ and in similar fashion $\dfrac{\partial^n u}{\partial t^n}(x, 0) = 0$ for $n \ge 2$. Therefore $u(x, t) = x^2 + 2t$.

(c) Using techniques as in (b) one gets $u(x, t) = \displaystyle\sum_{n=0}^{\infty} \frac{1}{n!}(-1)^n \sin(x) t^n$ or $u(x, t)$

$$= \sin(x) e^{-t}.$$

2. (a) $\displaystyle\int_0^x e^s\, ds = x + x^2/2! + \ldots = e^x - 1.$

(b) $\displaystyle\lim_{n\to\infty} \left|\frac{u_{n+1}}{u_n}\right| = \lim_{n\to\infty} \frac{n}{n+1} |x| = |x|.$

(c) Use Theorem D to show $u'(x) = \displaystyle\sum_{n=1}^{\infty} \frac{d}{dx} \frac{1}{n} x^n = \sum_{n=0}^{\infty} x^n = \frac{1}{1-x}$, (Geometric Series) .

Then solve

$$\begin{cases} u'(x) = \dfrac{1}{1-x} \\[2mm] u(0) = 0 \end{cases}$$

to obtain $u(x) = -\ln(1 - x).$

3. (a) Using $x = r \cos \theta$ and $y = r \sin \theta$ we have by the chain rule

$$\frac{\partial}{\partial r} = \frac{\partial x}{\partial r}\frac{\partial}{\partial x} + \frac{\partial y}{\partial r}\frac{\partial}{\partial y} = \cos \theta \frac{\partial}{\partial x} + \sin \theta \frac{\partial}{\partial y},$$

$$\frac{\partial}{\partial \theta} = \frac{\partial x}{\partial \theta}\frac{\partial}{\partial x} + \frac{\partial y}{\partial \theta}\frac{\partial}{\partial y} = -r \sin \theta \frac{\partial}{\partial x} + r \cos \theta \frac{\partial}{\partial y}.$$

Written in matrix notation, this becomes

$$\begin{bmatrix} \dfrac{\partial}{\partial r} \\[2mm] \dfrac{\partial}{\partial \theta} \end{bmatrix} = \begin{bmatrix} \cos \theta & \sin \theta \\ -r \sin \theta & r\cos \theta \end{bmatrix} \begin{bmatrix} \dfrac{\partial}{\partial x} \\[2mm] \dfrac{\partial}{\partial y} \end{bmatrix}$$

Inverting, we have

$$\begin{bmatrix} \dfrac{\partial}{\partial x} \\[2mm] \dfrac{\partial}{\partial y} \end{bmatrix} = \frac{1}{r}\begin{bmatrix} r \cos \theta & -\sin \theta \\ r \sin \theta & \cos \theta \end{bmatrix} \begin{bmatrix} \dfrac{\partial}{\partial r} \\[2mm] \dfrac{\partial}{\partial \theta} \end{bmatrix}$$

$$= \langle \cos \theta, \sin \theta \rangle \frac{\partial}{\partial r} + \langle -\sin \theta, \cos \theta \rangle \frac{1}{r}\frac{\partial}{\partial \theta}.$$

(b) $\Delta_2 r^m = \dfrac{\partial^2}{\partial r^2}r^m + r^{-1}\dfrac{\partial}{\partial r}r^m + r^{-2}\dfrac{\partial^2}{\partial \theta^2}r^m = m(m-1)r^{m-2} + mr^{m-2}$

$\qquad = m^2 r^{m-2},$

and

$$\text{grad } r^m = (\cos \theta, \sin \theta)\partial(r^m)/\partial r = mr^{m-1}(\cos \theta, \sin \theta).$$

(c) From (a) and (b) we have by termwise differentiation

$$u = 1 + r + r^2 + \cdots + r^m + \cdots$$

$$\Delta u = 0 + r^{-1} + 4 + 9r + 16r^2 + \cdots + m^2 r^{m-2} + \cdots$$

$$\frac{\partial u}{\partial n} = \text{grad}(1 + r + r^2 + \cdots + r^m + \cdots) \cdot n$$

$$= (0 + 1 + 2r + \cdots + mr^{m-1} + \cdots)(\cos \theta, \sin \theta) \cdot n.$$

On $r = R < 1$ the latter integrates termwise as

$$\oint_{r=R<1} \frac{\partial u}{\partial n}ds = 2\pi R(1 + 2R + \cdots + mR^{m-1} + \cdots)$$

$$= 2\pi(R + 2R^2 + \cdots + mR^m + \cdots)$$

where we noted that on the sphere, $n = (\cos \theta, \sin \theta)$. On the other hand, the volume integral is

$$\iint_{r \leq R} \Delta_2 u \, r \, dr d\theta = 2\pi \int_0^R (1 + 4r + 9r^2 + \cdots + m^2 r^{m-1} + \cdots) \, dr$$

$$= 2\pi(R + 2R^2 + \cdots + mR^m + \cdots)$$

SECOND PAUSE

1. That $\Lambda_n = \int_\Omega |\Delta u_n|^2 / D(u_n)$ follows from Green's identities. The variational characterization can then be written down formally from experience, although its proof requires more care. See Sections 1.9.5(3) and 2.6 for relevant discussion.

2. Use both $u = 0$ and $\partial u/\partial n \equiv \text{grad } u \cdot n = 0$ on $\partial\Omega$. The former guarantees that the tangential derivative $\partial u/\partial \tau = 0$ on $\partial\Omega$.

3. $\Lambda_1 = \lambda_2$ iff Ω is an n-sphere, and then the trial functions are just the u_{x_i}.

4. It is equivalent to the fixed membrane problem. From $\Delta(\Delta u + \Lambda u) = 0$ and the boundary conditions, a plate eigenfunction is also one for the membrane at the same frequency. The vice versa statement needs only sufficient regularity of a membrane eigenfunction.

5. By Green's second identity the Steklov eigenvalues are given by

$$\mu = \int_\Omega (\Delta \psi)^2 \bigg/ \oint_{\partial\Omega} \left(\frac{\partial \psi}{\partial n} \right)^2.$$

These eigenvalues are positive, for if not, $v = \Delta\psi$ is zero on both Ω and $\partial\Omega$ and hence ψ is identically zero. Hence the constant $k(\Omega)$ in the Dirichlet–Poisson bound can be taken to be $k = \mu_1^{-1}$.

6. By Schwarz's inequality,

$$\int_a^b |h(x)| \cdot \chi_{(a,b)}(x) \, dx \leq \left(\int_a^b 1 \, dx \right)^{1/2} \left(\int_a^b |h(x)|^2 dx \right)^{1/2},$$

so that

$$\int_a^b |h| \, dx \leq (b - a) \left(\int_a^b |h|^2 \right)^{1/2} \quad \text{for any } h \, \varepsilon \, L^2(a, b).$$

7. This follows from the interesting property that the Fourier coefficients provide the best least squares $\{\phi_N\}$ fit to h. That is, for any partial sum $S_M = \sum_{N=1}^M c_N \phi_N$, one has

$$\int_a^b (h - S_M)^2 dx = \int_a^b h^2 \, dx - 2\sum_{N=1}^M c_N \int_a^b h \phi_N \, dx + \sum_{N=1}^M c_N^2$$

$$\geq \int_a^b h^2 \, dx - \sum_{N=1}^M c_N^2$$

where the latter minimum is obtained by taking $c_N = \int_a^b h \phi_N$, $N = 1, \ldots, M$. Then

$$\sum_{N=1}^M c_N^2 \leq \int_a^b h^2 \, dx < \infty$$

independent of M so that the series of Fourier coefficients squared converges and hence necessarily $c_N \to 0$ as $N \to \infty$. In fact (see Section 2.3) the series $\Sigma_{N=1}^{\infty} c_N^2$ sums to exactly $\int_a^b h^2\, dx$ when the $\{\phi_N\}$ are a complete orthonormal set. This result is called Parseval's Equation or Theorem.

8. The classical ϕ_N have in fact the additional property that their L^2 integrals $\int_{-\pi}^{\pi} \phi_N^2(x)\, dx$ are independent of N. For the Dini ϕ_N use the hint in 1.9.6(2).

9. (a) We consider only the case of a smooth $\partial\Omega$ here.* Recall the calculus formulae for arc length, length, and area when $\partial\Omega$ is represented parametrically by $x = x(s)$, $y = y(s)$, s running from 0 to L:

$$(ds)^2 = (dx)^2 + (dy)^2, \qquad L = \int_0^L \sqrt{(\dot{x}(s))^2 + (\dot{y}(s))^2}\, ds,$$

$$A = \frac{1}{2}\int_0^L (x\dot{y} - y\dot{x})\, ds.$$

For use of Fourier series representations of the functions $x(s)$ and $y(s)$ and their derivatives it is useful to reparametrize by $t = (2\pi s/L) - \pi$. A trick to obviate the need to integrate the square root in L is to note that $(dx/ds)^2 + (dy/ds)^2 = 1$ so that $(dx/dt)^2 + (dy/dt)^2 = (L/2\pi)^2$. Upon integrating this constant expression we have

$$2\pi\left(\frac{L}{2\pi}\right)^2 = \int_{-\pi}^{\pi}\left[\left(\frac{dx}{dt}\right)^2 + \left(\frac{dy}{dt}\right)^2\right] dt = \sum_{n=1}^{\infty} n^2(a_n^2 + b_n^2 + \bar{a}_n^2 + \bar{b}_n^2)$$

where we have used the Fourier representations (see Section 1.5.1)

$$x(t) = a_0/2 + \sum_{n=1}^{\infty} (a_n \cos nt + b_n \sin nt)$$

$$y(t) = \bar{a}_0/2 + \sum_{n=1}^{\infty} (\bar{a}_n \cos nt + \bar{b}_n \sin nt)$$

$$\dot{x}(t) = \sum_{n=1}^{\infty} (nb_n \cos nt - na_n \sin nt)$$

$$\dot{y}(t) = \sum_{n=1}^{\infty} (n\bar{b}_n \cos nt - n\bar{a}_n \sin nt).$$

The area formula in the t parametrization becomes

$$A = \frac{1}{2}\int_{-\pi}^{\pi} (x(t)\dot{y}(t) - y(t)\dot{x}(t))\, dt = \sum_{n=1}^{\infty} n(a_n\bar{b}_n - b_n\bar{a}_n)$$

$$\le \frac{1}{2}\sum_{n=1}^{\infty} n(a_n^2 + b_n^2 + \bar{a}_n^2 + \bar{b}_n^2) \le \frac{1}{2}\sum_{n=1}^{\infty} n^2(a_n^2 + b_n^2 + \bar{a}_n^2 + \bar{b}_n^2)$$

$$= L^2/4\pi,$$

---

* For full rigor in the statement and proof of the isoperimetric inequality it would appear advisable to begin in the context of the Jordan curve theorem of complex analysis, which delineates all regions homeomorphic to the disc.

thereby establishing the isoperimetric inequality. Note the use of the arithmetic–geometric mean inequality as a key element in this proof of the isoperimetric inequality.

We remark that in the above, the summations of Fourier coefficients obtain formally directly from the orthogonality relations of the trigonometric functions. More generally they should be regarded as an instance of the Parseval's expression for the inner product of any two Fourier representations $u = \Sigma_n c_n \phi_n$, $v = \Sigma_n d_n \phi_n$, namely, $\langle u, v \rangle = \Sigma_n c_n d_n$. See Section 2.3 and the next Pause.

(b) Denote the left-hand side by $\sqrt{\cdot}$ Then by Schwarz's inequality

$$\dot{x}(s) \int_a^b \dot{x}(t)\,dt + \dot{y}(s) \int_a^b \dot{y}(t)\,dt \leqq (\sqrt{\cdot}) \sqrt{(\dot{x}(s))^2 + (\dot{y}(s))^2}$$

which provides the desired inequality upon integration with respect to s and division by $\sqrt{\cdot}$

(c) The left-hand side of b is zero in the situation of (a), so it is not sharp enough to provide an interesting lower bound for L.

SECTION 1.7.1

2. $\dfrac{\text{cal}}{\text{cm deg}} \cdot \dfrac{\text{deg}}{\text{sec}} = \dfrac{\text{cal} \cdot \text{cm}}{\text{deg} \cdot \text{sec}} \cdot \dfrac{\text{deg}}{\text{cm}^2}$.

SECTION 1.7.2

1. Directly from Green's second identity one has

$$\mu = \frac{\int_\Omega (\Delta u)^2}{\int_\Omega u^2} = \frac{\int_\Omega (\Delta u)^2}{D(u)} \frac{D(u)}{\int_\Omega u^2} \geqq \Lambda_1 \lambda_1.$$

The first eigenvalues Λ_1 and λ_1 for the buckled clamped plate and the fixed membrane, respectively, are both positive.

2. (a) Note first by symmetry that $u'(0) = u'(\frac{1}{2}) = u'(1) = 0$.

(b) $v''(x) + (v''(x) - 2\lambda)v(x + \frac{1}{2}) = 0$.

3. You should fail. The main difficulty is in the term representing the electron–electron interaction.

SECTION 1.7.3

1. Yes. It is known that the first eigenfunction may change sign on a nonsimply connected Ω.*

2. For practical purposes, the most important unresolved question is (3'), the approximation (e.g., numerically, in 3 dimensions) of the flow. This problem has many

* See R. J. Duffin and D. H. Shaffer, *Bull. Amer. Math. Soc.* 58 (1952).

important applications (e.g., to aerodynamic design) and great gains have been made recently. See Appendix C. For theoretical purposes, a longstanding problem has been Leray's conjecture* that a singularity "always" develops somewhere in a three-dimensional flow history. This is a question of regularity (2'), which in turn affects our understanding of existence (1) and uniqueness (2).

3. It is a three-body problem.

1.7 EXERCISES

1. This is just Green's second identity with the boundary terms vanishing. See Section 1.9.7(1) for the more precise meaning of self-adjointness of an operator.

2. Let $w = \Delta u + \Lambda u$. Then

$$\begin{cases} \Delta w = 0 \text{ in } \Omega, \\ \quad w = 0 \text{ on } \partial\Omega \end{cases}$$

and hence $w = 0$ in Ω. Therefore the problem simplifies (modulo regularity conditions) to the vibrating membrane problem

$$\begin{cases} -\Delta u = \Lambda u \text{ in } \Omega, \\ \quad u = 0 \text{ on } \partial\Omega. \end{cases}$$

3. Not so far. If you could, you would also be able to solve analytically a number of related problems in fluid dynamics. See Appendix C.

SECTION 1.8

1. (a) $x = \pm \int_{\phi_0}^{\phi} [2\lambda(\cos(s) - \cos \phi_0)]^{-1/2} ds$.

3. (b) Let $u(x) = \beta + \gamma \ln \upsilon(x)$, which eventually leads to $u(x) = \beta - 2 \ln (\cosh((\lambda e^\beta/16)^{1/2} x) + c)$. The boundary conditions then constrain β, c, and λ.

PROBLEM 1.9.1

(i) $d = AC - B^2 = -5/4$ implies hyperbolic. The transformation

$$\begin{pmatrix} \zeta \\ \eta \end{pmatrix} = \begin{bmatrix} c_{11} & c_{12} \\ c_{21} & c_{22} \end{bmatrix} \begin{pmatrix} x \\ t \end{pmatrix} = \begin{bmatrix} 1 & 0 \\ 2 & 4 \end{bmatrix} \begin{pmatrix} x \\ t \end{pmatrix}$$

converts the equation to

$$u_{\zeta\zeta} - 20u_{\eta\eta} - u_\zeta - 6u_\eta = 0.$$

(ii) $d = 15/4$ implies elliptic. The rotation

$$\begin{pmatrix} \zeta \\ \eta \end{pmatrix} = \begin{bmatrix} \cos\theta & \sin\theta \\ -\sin\theta & \cos\theta \end{bmatrix} \begin{pmatrix} x \\ t \end{pmatrix}$$

with $\theta = 45$ degrees transforms to

$$(2.5)u_{\zeta\zeta} + (1.5)u_{\eta\eta} + u = 0.$$

---

* J. Leray, *Acta Math.* 63 (1934).

(iii) $d = -16$ implies hyperbolic. The transformation

$$\begin{pmatrix} \zeta \\ \eta \end{pmatrix} = \begin{bmatrix} \tfrac{1}{2} & \tfrac{1}{2} \\ -\tfrac{1}{4} & \tfrac{1}{4} \end{bmatrix} \begin{pmatrix} x \\ t \end{pmatrix}$$

transforms the equation to

$$u_{\zeta\zeta} - u_{\eta\eta} + 3\,u_{\zeta} - 2\,u_{\eta} + u = 0.$$

PROBLEM 1.9.2

(a) $d = AC - B^2 = \sin^2 x - 1 = -\cos^2 x$. Hence for all $x \neq (2n-1)\,\pi/2$, the equation is hyperbolic. Along the lines $((2n-1)\pi/2, t)$ it is parabolic. From the characteristics equation $dx/dt = 1 \pm |\cos x|$ the characteristic curves are

$$t = -\cot x - \csc x + c_1 = -\cot(x/2) + c_1$$
$$t = -\cot x + \csc x + c_2 = \tan(x/2) + c_2$$

(c) $dx = \pm\sqrt{-y}\,dy$ for $y \le 0$, a separable ordinary differential equation, yielding the characteristics $x = \pm\tfrac{2}{3}(-y)^{3/2} + c$

PROBLEM 1.9.3

(b) Let $v = u + \tfrac{1}{4}(x^2 + y^2)\,\max_{\overline{\Omega}} |\Delta u|$. Then $\Delta v \ge 0$ so that $v \le \max_{\partial\Omega} u +$

$\dfrac{R^2}{4}\,\max_{\overline{\Omega}} |\Delta u|$ in Ω. Hence $u \le \max_{\partial\Omega} u + \dfrac{R^2}{4}\,\max_{\overline{\Omega}} |\Delta u|$ in Ω. Doing the same for

$-u$ gives the result. For f_1 and f_2 close, F_1 and F_2 close, this implies the stability

$$|u_1 - u_2| \le \max_{\partial\Omega} |f_1 - f_2| + \frac{R^2}{4}\,\sup_{\Omega} |F_1 - F_2|$$

of two Dirichlet-Poisson problems

$$\begin{cases} \Delta u_1 = F_1 \text{ in } \Omega \\ u_1 = f_1 \text{ on } \partial\Omega \end{cases} \qquad \begin{cases} \Delta u_2 = F_2 \text{ in } \Omega \\ u_2 = f_2 \text{ on } \partial\Omega \end{cases}$$

with close data.

(d) See for example the book by Protter and Weinberger.[*]

PROBLEM 1.9.4

(b) The contraction mapping theorem is easily proved itself. One may start with any point x_0 in the space X. Let $F(x)$ denote the contractive map and let $x_n = F(x_{n-1})$ $= F^n(x_0)$ be the iterates of x_0 under the mapping, $n = 1,2,3,.....$We recall that F is said to be contractive (more precisely, strongly contractive) if there exists some $0 \le r \le 1$ such that (and let us consider here just complete normed spaces,

---

[*] M.H. Protter and H.F. Weiberger, *Maximum Principles in Differential Equations* (Prentice-Hall, Englewood Cliffs, New Jersy, 1967).

the proof being identical for all complete metric spaces by just replacing the norm distance $\|x - y\|$ by a general distance function $d(x, y)$)

$$\|F(x) - F(y)\| \leq r\|x - y\| \quad \text{all } x, y \in X.$$

Thus we have for all $m \geq n > N$,

$$
\begin{aligned}
\|x_m - x_n\| &= \|F^N(F^{m-N}(x_0)) - F_N(F^{n-N}(x_0))\| \\
&\leq r^N[\|F^{m-N}(x_0) - F^{m-N-1}(x_0)\| + \|F^{m-N-1}(x_0) - F^{m-N-2}(x_0)\| \\
&\quad + \cdots + \|F^{n-N+1}(x_0) - F^{n-N}(x_0)\|] \\
&\leq r^N[(r^{m-N-1} + r^{m-N-2} + \cdots + r + 1)\|x_1 - x_0\|] \\
&= r^N\left(\frac{1}{1-r} - \frac{r^{m-N}}{1-r}\right)\|x_1 - x_0\| \\
&< r^N\frac{\|x_1 - x_0\|}{1-r},
\end{aligned}
$$

having used the triangle inequality and the sum of a geometric series. As $N \to \infty$, $\|x_m - x_n\| \to 0$, and from the assumed completeness of the space X, there exists a (unique) limit point $x = \lim_{n\to\infty} x_n$. By the continuity of the map F (it is easily shown that a contractive map is continuous), then

$$F(x) = \lim_{n\to\infty} F(x_{n-1}) = \lim x_n = x$$

and thus x is the (unique) fixed point for F on X.

Two remarks are in order here:

First, it should be remembered that there is an extremely simple example of the contraction mapping theorem, namely, the lines

$$F(x) \equiv y(x) = ax + b$$

of positive slope a. For $a \equiv r < 1$ the line eventually crosses the fixed point line $y \equiv x$, but for $a \equiv r \geq 1$ this need not be the case.

The second remark is that proofs of the contraction mapping theorem, the successive approximations method, the Picard theorem, and even the Cauchy–Kowalewski theorem, can all be given by means of a majorization by a convergent geometric series.

PROBLEM 1.9.5(1)

(a) Dirichlet's lectures on the Dirichlet Problem were published posthumously in P. G. Lejeune–Dirichlet, *Vorlesungen über die im umgekehrten Verhältnis des Quadrats der Entfernung wirkenden Kräfte*, Ed.: P. Grube, Leipzig, 1887.

(b) A father-son team was involved in this problem, although it appears to be named for the son, Carl.

PROBLEM 1.9.5(2)

(a) Divergence theorem domains are rather general—roughly, more general than can be described. One can for example tolerate a finite number of discontinuities in the tangent planes to $\partial\,\Omega$, provided that each boundary point is not too inaccessible. One builds up the general acceptable domains roughly as follows:

normal domains $\subset$ regular domains $\subset$ divergence domains

| (as in 1.6.1) | (aproximable by normal domains) | (approximable by regular domains, and beyond) |

Dirichlet domains are also very general and can permit (for example in two dimensions) slits and cuts that are not acceptable for a divergence theorem domain.

(b) For example Hadamard and Zaremba* and not Lebesgue who first showed that the Dirichlet problem is not solvable for arbitrary domains.

PROBLEM 1.9.5(3)

(b) There can be more than one Lagrangian for a given Euler equation.**

(c) $L = \dfrac{1}{2}\,(u_t^2 - u_x^2 - u_y^2)$.

PROBLEM 1.9.6(1)

(a) Although the divergence theorem, taken abstractly, is the fundamental theorem of calculus generalized to n-dimensions, its historical origins are in the physical theories of electricity and magnetism, viewed as fluids and studied by Green, Gauss, and others in the first half of the nineteenth century. George Green's work, barely recognized during his life, later inspired the Cambridge school of mathematical physicists, including Sir William Thomson, Sir Gabriel Stokes, Lord Rayleigh, and Clerk Maxwell.

PROBLEM 1.9.6(2)

(b) As an example of another version, consider the following Dini test and proof (sketched): Let $f \in C^0[-\pi, \pi] \cap P\,C^1\,[-\pi, \pi]$ such that $f(-\pi) = f(\pi) = 0$. Then $s_N^f \to f$ uniformly pointwise.

Proof: (i) f as postulated is absolutely continuous so that the Fourier coefficients a_n' and b_n' for f are $n\,b_n$ and $-\,n\,a_n$, respectively, by integration by parts. (ii) Thus

$$\left| (s_N^f - s_M^f)\,(x) \right| \le \sum_{M+1}^{N} (a_n^2 + b_n^2)^{1/2}$$

$$\le \left[\sum_{M+1}^{N} \frac{1}{n^2} \right]^{1/2} \left[\sum_{M+1}^{N} ((a_n')^2 + (b_n')^2) \right]^{1/2}$$

$$\le \frac{\|f'\|^2}{M^{1/2}}$$

---

* S. Zaremba, *Acta Mathematica* **34** (1911).

** An interesting recent book concerned with such questions is that of Santilli, *Foundations of Theoretical Mechanics I : The inverse problem of Newtonian mechanics,* Springer (1978)

(iii) The partial sums s_N^f for f are thus uniformly Cauchy and $|f(x) - s_M(x)| \leq M^{-1/2}\|f'\|_{L^2(-\pi,\pi)}$, independent of x.

(c) The Gibb's effect to 2,000,000 terms is calculated near the end of Appendix B.3.

PROBLEM 1.9.6(3)

(a) Given that $\partial f(x, y)/\partial y$ exists and is bounded almost everywhere in the given interval, then from the convergence of the difference quotients

$$U_n(x, y) \equiv \frac{f(x, y + \Delta_n y) - f(x, y)}{\Delta_n y} \to \frac{\partial f(x, y)}{\partial y}$$

follows the convergence of the integrals

$$\frac{d}{dy} \int_a^b f(x, y)\, dx \equiv \lim_{n \to \infty} \int_a^b U_n(x, y) \to \int_a^b \frac{\partial f(x, y)}{\partial y}\, dx$$

and thus the Leibnitz formula of Section 1.6.3.

PROBLEM 1.9.7(1)

(a) and (b). Integration by parts and, although not necessarily in this terminology, the Riesz representation theorem for linear functionals.

PROBLEM 1.9.7(2)

(b) $0 = \int_\Omega u\, \Delta\varphi = \int_\Omega \varphi\, \Delta u$ and the denseness in $L^2(\Omega)$ of the test functions $\{\varphi\}$.

PROBLEM 1.9.7(3)

(c) Whereas earlier (see Exercise 2(b) of Section 1.1) we noted that $\sqrt{2t}$ plays the role of the standard deviation σ in the normal probability distribution, here we should note that x plays the role of the mean. In other words, the Green's function spreads initial heat data in a way taking into account only first and second moments.

PROBLEM 1.9.8

(b) $u(t) = a \sin(t + b)$ substituted into the equation for $\lambda \neq 0$ yields

$$(a^2 \sin^2(t + b) - 1)a \cos(t + b) = 0,$$

which is not possible unless $a = 0$.

PROBLEM 1.9.9

2. (a) *Hint*: Use data $f_n(x) = n^{-1} \sin nx$, $n = 1, 2, 3, \ldots$.
   (b) *Hint*: Use data $f_n(x) = e^{-\sqrt{n}} \sin(4n + 1)x$, $n = 1, 2, 3, \ldots$, solutions via separation of variables, and stability in the sense of maximum norms.

3. (a) $u(x, t) = \sum_{n=1}^{\infty} \frac{2(-1)^{n+1}}{n} e^{-3n^2 t} \sin nx$.

(b) $u(r, \theta) = \dfrac{1}{2} + \dfrac{2}{\pi} \displaystyle\sum_{n=1}^{\infty} \dfrac{1}{(2n-1)} r^{2n-1} \sin(2n-1)\theta$.

(c) $u(x, t) = -\dfrac{4}{\pi^2} \displaystyle\sum_{n=1}^{\infty} \dfrac{(-1)^n}{(2n-1)^2} \sin(2n-1)x \cos(2n-1)t$.

7. (c) (i) yes; (ii) yes; (iii) yes; (iv) yes for nonnegative solutions. A proof including (iii) goes as follows. If $\Delta u_i - u_i^3 = F$ in Ω, $u_i = f$ on $\partial\Omega$, $i = 1, 2$, then $\Delta(u_1 - u_2) = u_1^3 - u_2^3$ in Ω and $u_1 - u_2 = 0$ on $\partial\Omega$. By Green's identity

$$0 \geq -\int_\Omega (\operatorname{grad}(u_1 - u_2))^2 = \int_\Omega (u_1 - u_2)\Delta(u_1 - u_2)$$

$$= \int_\Omega (u_1 - u_2)^2 (u_1^2 + u_1 u_2 + u_2^2) \geq 0$$

using the arithmetic–geometric mean inequality. Thus $u_1 = u_2$.

Similarly for (ii), if $\Delta u_i = u_i \int u_i^2 \, dy = F$ in Ω, $u_i = f$ on $\partial\Omega$, then $\Delta(u_1 - u_2) = u_1 \int_\Omega u_1^2 - u_2 \int_\Omega u_2^2$ in Ω and $u_1 - u_2 = 0$ on $\partial\Omega$. By Green's identity

$$0 \leq \int_\Omega (\operatorname{grad}(u_1 - u_2))^2 = \int_\Omega u_1 u_2 \left[\int_\Omega u_1^2 + \int_\Omega u_2^2 \right]$$

$$- \left[\left(\int_\Omega u_1^2 \right)^2 + \left(\int_\Omega u_2^2 \right)^2 \right]$$

$$= [\|u_1\|^2 + \|u_2\|^2](u_1, u_2) - [\|u_1\|^4 + \|u_2\|^4]$$

$$\leq \tfrac{1}{2}[\|u_1\|^2 + \|u_2\|^2]^2 - [\|u_1\|^4 + \|u_2\|^4]$$

$$= \|u_1\|^2 \|u_2\|^2 - \tfrac{1}{2}[\|u_1\|^4 + \|u_2\|^4]$$

$$= -\tfrac{1}{2}(\|u_1\|^2 - \|u_2\|^2)^2 \leq 0$$

again using the arithmetic–geometric mean inequality. Thus $u_1 = u_2$.

8. This is an interesting equation going under the names Bratu, Frank–Kamenetskii, Poisson–Boltzman, Chandrasekhar, and Liouville. It occurs in Arrhenius models of reactions, theories of isothermal gas stars, and electrolytics, among others.*

PROBLEM 1.9.9(9)

1. (a) elliptic (b) parabolic (c) elliptic (d) hyperbolic (e) elliptic (f) parabolic (g) elliptic for $t > 0$, parabolic for $t = 0$, hyperbolic for $t < 0$ (h) elliptic when $u_t > 0$, parabolic when $u_t = 0$, hyperbolic when $u_t < 0$.

2. (e) $C = \begin{bmatrix} 1 & 1 \\ 1 & 0 \end{bmatrix}$ (f) $C = \begin{bmatrix} 1 & 2 \\ 1 & 1 \end{bmatrix}$

* For more information see K. Gustafson, "Combustion and Explosion Equations and Their Calculation," *Computational Techniques in Heat Transfer* (Pineridge, Swansea, 1985), and E. Ash, B. Eaton, K. Gustafson, *J. Appl. Math. and Physics* **41** (1990).

3. (a) $-2 \sum_{n=1}^{\infty} \frac{(-1)^n}{n} \sin nx$

   (b) $2 \sum_{n=1}^{\infty} \frac{1}{n} \sin nx$

   (c) $-4 \sum_{n=1}^{\infty} \frac{(1 + 2(-1)^n)}{n^3} \sin nx$

   (d) $\frac{\pi}{2} - \frac{4}{\pi} \sum_{n=1}^{\infty} \frac{1}{(2n-1)^2} \cos(2n-1)x$

4. (a) $N = 3$ (b) $N = 7$

5. (a) $\sum_{n=1}^{\infty} \frac{1}{(2n-1)} \sin(2n-1)x$ (b) Convergence in $L^2(-\pi, \pi)$ to f and pointwise to f

   except at points $x_0 = -\pi, 0, \pi$, where the series converges to $(f(x_0^-) + f(x_0^+))/2$. (c)
   Direct from the representation (d)Direct from the Fourier series for $f(x) = x(\pi - x)$ on

   $-\pi \le x \le \pi$ which is $\frac{\pi^2}{6} - \sum_{n=1}^{\infty} \frac{\cos 2nx}{n^2}$

6. Given (i), for $u = u_1 - u_2$ the difference of any two solutions, one has $L(u_1 - u_2)$
   $= f - f = 0 \Rightarrow u_1 = u_2$. Given (ii), because $u \equiv 0$ is a solution, it is the only solution.

7. (a) For 0 data (see Excercise 6 above), one has by Green's identity and the boundary
   condition

   $$D(u) = \int_{x^2+y^2 < 1} |\text{grad } u|^2 = \oint_{x^2+y^2=1} u \frac{\partial u}{\partial n} = -7 \oint_{x^2+y^2=1} u^2.$$

   Hence u is constant and zero on $x^2 + y^2 = 1$ and hence zero everywhere in $x^2 + y^2 \le 1$,
   which implies uniqueness.

   (b) For 0 data, $D(u) = \oint_{\partial\Omega} u \frac{\partial u}{\partial n} = 0$ so that u is constant. Uniqueness follows provided the

   Dirichlet part of the mixed boundary condition holds on at least, say, a small interval.

   (c) All constraints are linear. Then $D(u) = 0$ implies is u constant, and either homogeneous
   boundary constraint implies $u = 0$.

   (d) Letting L denote the uniformly elliptic operator on the left, by the divergence theorem
   we may modify Green's first identity to

   $$-\int_{\Omega} u L u + \oint_{\partial\Omega} (ue^x u_x n_x + ue^y u_y n_y) = \int_{\Omega} e^x (u_x)^2 + e^y (u_y)^2.$$

   Then 0 data implies the right side is 0, hence u is constant and 0.

8. (a) $u(0, 0) = (2\pi)^{-1} \int_0^\pi 1 \, d\phi = 1/2$ by the mean value theorem.

   (b) $u(0, 0) = (4\pi)^{-1} \oint\limits_{x^2+y^2+z^2=1} 1 \, ds = 1$ by the mean value theorem.

   (c) $u(0, 0) = (2\pi)^{-1} \int_{-\pi}^\pi \sin^2\phi \, d\phi = (4\pi)^{-1} \int_{-\pi}^\pi (1 - \cos 2\phi) \, d\phi = 1/2$.

   (d) $u(0, 0) = 1/4$ from symmetry by letting $u = u_1 + u_2 + u_3 + u_4$ where $\Delta u_i = 0$ and $u_i = 1$ on side i, 0 on the other sides, and noting that $u \equiv 1$.

9. We must first check the compatibility condition $\int_\Omega \Delta u = \oint_{\partial\Omega} \dfrac{\partial u}{\partial n}$. We have

$$\int_0^2 \int_0^4 x^2 \, y \, dx \, dy = 128/3, \quad \oint_{\partial\Omega} g = \int_0^2 (y - 1) \, dy + \int_4^0 (x - 2) \, dx = 0.$$

Hence the constant $c = -\frac{16}{3}$ is the correct one for compatibility. As $\partial\Omega$ is piecewise smooth and the data analytic, solutions will exist, and be unique up to an arbitrary constant.

To construct a solution u by separation of variables, let it be composed of the three parts $u = u_1 + u_2 + u_3$ where

$$\begin{cases} \Delta u_1 = x^2 y - \frac{16}{3} \text{ in } \Omega \\ \dfrac{\partial u_1}{\partial n} = 0 \text{ on } \partial\Omega \end{cases} \qquad \begin{cases} \Delta u_2 = 0 \text{ in } \Omega \\ \dfrac{\partial u_2}{\partial y} = \dfrac{x}{2} \text{ on top} \\ \dfrac{\partial u_2}{\partial n} = 0 \text{ elsewhere} \end{cases} \qquad \begin{cases} \Delta u_3 = 0 \text{ in } \Omega \\ \dfrac{\partial u_3}{\partial x} = y \text{ on right side} \\ \dfrac{\partial u_3}{\partial n} = 0 \text{ elsewhere} . \end{cases}$$

Expand $u_1 = c_0 + \sum\limits_{n=1}^\infty \sum\limits_{m=1}^\infty c_{nm} \cos \dfrac{n\pi x}{4} \cos \dfrac{m\pi y}{2}$ in the Neumann eigenfunctions.

Similarly expand $x^2 y - \frac{16}{3} = d_0 + \sum\limits_{n=1}^\infty \sum\limits_{m=1}^\infty d_{nm} \cos \dfrac{n\pi x}{4} \cos \dfrac{m\pi y}{2}$. Now equate coefficients of $\Delta u_1 = x^2 y - \frac{16}{3}$ as expanded (Hint: look ahead to p.147), from which $c_{nm} = d_{nm}/\lambda_{nm}$ where $\lambda_{nm} = (n^2\pi^2 + 4m^2\pi^2)/16$ are the eigenvalues of this Neumann problem. Note that $d_0 = 0$ as required, thanks to the compatibility condition. Since

$$d_{nm} = \frac{1}{2} \int_0^2 \int_0^4 (x^2 y - \frac{16}{3}) \cos\frac{n\pi x}{4} \cos\frac{m\pi y}{2} dx \, dy = \cdots = \frac{256(-1)^{n+1}}{(nm\pi^2)^2}[(-1)^n - 1]$$

we find

$$u_1(x, y) = \sum_{n=1}^\infty \sum_{m=1}^\infty \frac{4096[(-1)^{n\cdot m} + (-1)^n]}{n^2 m^2 \pi^4 (n^2\pi^2 + 4m^2\pi^2)} \cos\frac{n\pi x}{4} \cos\frac{m\pi y}{2}.$$

By separation of variables we find $u_2(x, y) = \sum\limits_{n=1}^\infty c_n \cos\dfrac{n\pi x}{4} \cosh\dfrac{n\pi y}{4}$, where c_n as computed from the boundary condition is

$$c_n = \frac{4}{n\pi \sinh\left(\frac{n\pi}{2}\right)} \int_0^4 (x - 1) \cos\frac{n\pi x}{4} dx = \cdots = \left(\frac{4}{n\pi}\right)^3 \left[\frac{(-1)^n - 1}{\sinh\left(\frac{n\pi}{2}\right)}\right].$$

By separation of variables we find $u_3(x, y) = \sum\limits_{n=1}^\infty c_n \cos\dfrac{n\pi y}{2} \cosh\dfrac{n\pi x}{2}$, where

$$c_n = \frac{2}{n\pi \sinh(2n\pi)} \int_0^2 (y - 1) \cos\frac{n\pi y}{2} dy = \cdots = \left(\frac{2}{n\pi}\right)^3 \left[\frac{(-1)^n - 1}{\sinh(2n\pi)}\right].$$

Combining these three solutions gives us the overall solution in the case that $c = -16/3$. The general solution is this solution plus an arbitrary constant. For any other value of c, solutions do not exist.

It is interesting to consider the same problem with data $y - 1$ and $x - 2$ replaced by y and x, respectively, which produces a more difficult question.

SECTION 2.1

1. $u(x, t) = \sum\limits_{n=1}^{\infty} c_n \sin nx \sin nt$,

$$c_n = \frac{2}{n\pi} \int_0^{\pi} g(s) \sin ns \, ds.$$

2. (a) $u(x, y) = \sum\limits_{n=1}^{\infty} c_n \sin nx \sinh n(\pi - y) + \sum\limits_{n=1}^{\infty} d_n \sin ny \sinh n(\pi - x)$,

$$c_n = \begin{cases} 0 \text{ for } n \text{ even} \\ \dfrac{8}{n^3 \pi \sinh n\pi} \text{ for } n \text{ odd}, \end{cases} \qquad d_n = \frac{2(-1)^{n+1}}{n \sinh n\pi}.$$

3. (a) $u(x, t) = \sum\limits_{n=1}^{\infty} c_n e^{-n^2 t} \sin nx$, $\qquad c_n = \begin{cases} \dfrac{2}{n\pi} \text{ for } n \text{ odd}, \\ -\dfrac{4}{n\pi}, \ n = 2, 6, 10, \ldots, \\ 0, \ n = 4, 8, 12, \ldots. \end{cases}$

(b) $u(0, t) = u(\pi, t) = 0$, $\qquad u(\pi/2, t) = \dfrac{2}{\pi} \sum\limits_{n=1}^{\infty} \dfrac{e^{-(2n-1)^2 t}(-1)^{n+1}}{2n - 1}$,

and in particular $u(\pi/2, 0) = \dfrac{2}{\pi}(1 - \tfrac{1}{3} + \tfrac{1}{5} - \tfrac{1}{7} + \cdots) = \dfrac{2}{\pi} \cdot \dfrac{\pi}{4} = \dfrac{1}{2}$.

(c) *Hint:* Generally one has the estimate (reasonably good for large t)

$$|u(x, t)| \le \frac{2}{\pi} \frac{e^{-t}}{1 - e^{-t}}$$

which shows that $u(x, t) \to 0$ uniformly in x as $t \to \infty$. To see what to expect for small t one may look at the solution for several x_i in the interval $[0, \pi]$. For example, at $x = 0.6\pi$ one may sketch the solution from the following data:

$u(0.6\pi, t) \cong \left(\dfrac{2}{\pi}\right) \sin(0.6\pi)e^{-t} +$	$\left(\dfrac{-2}{\pi}\right) \sin(1.2\pi)e^{-4t} +$	$\left(\dfrac{2}{3\pi}\right) \sin(1.8\pi)e^{-9t}$	
0.8549	0.6055	0.3742	-0.1247	$t = 0$
0.7480	0.5478	0.2508	-0.0527	0.1
0.6432	0.4957	0.1681	-0.0206	0.2
0.5529	0.4485	0.1127	-0.0084	0.3
0.4165	0.3672	0.0506	-0.0014	0.5
0.2296	0.2227	0.0069	-0.0000	1.0

SECTION 2.2

1. (a) To begin, for the solution $u = \Sigma\, u_n$ note that for applications of Weierstrass M-test one has

$$|u_n(x, t)| \le \frac{2}{\pi} \int_0^\pi |f(s)|\, ds \cdot e^{-n^2 t}, \quad \left|\frac{\partial^2 u_n(x, t)}{\partial x^2}\right| \le \frac{2}{\pi} \int_0^\pi |f(s)|\, ds \cdot n^2 e^{-n^2 t},$$

$$\left|\frac{\partial u_n}{\partial x}(x, t)\right| \le \frac{2}{\pi} \int_0^\pi |f(s)|\, ds \cdot n e^{-n^2 t}, \quad \left|\frac{\partial u_n(x, t)}{\partial t}\right| \le \frac{2}{\pi} \int_0^\pi |f(s)|\, ds \cdot n^2 e^{-n^2 t}.$$

By the ratio test and l'hospital's rule $\Sigma_{n=1}^\infty\, n^k e^{-n^2 t_0}$ is a convergent series for any integer $k > 0$ and any $t_0 > 0$, so that by the Weierstrass M-test all series of derivatives converge uniformly.

2. (a) Two things are wrong with it: (i) The M-test will not apply; (ii) violation of symmetry. (c) Yes.

3. For example, $\|u(x, t)\|_{L^2} \le \dfrac{\pi}{2}\,(E(0))^{1/2}$, that is,

$$\int_0^\pi u^2(s, t)\, ds \le \frac{\pi^2}{8} \int_0^\pi [(u_s(s, t))^2 + (u_t(s, t))^2]\, ds,$$

for s equal to any x.

SECTION 2.3

2. By eliminating the overlapping elements, it suffices to show $\mathbb{Z} \times \mathbb{Z}$ is countable, where $\mathbb{Z} = \{1, 2, 3, \ldots\}$. One usual way to do the latter is to count $\mathbb{Z} \times \mathbb{Z}$ via the diagram

3. (b) (i) The following basis for $C[0, 1]$ was given by J. Schauder in 1927. Let $\chi_{(a,b)}$ denote the characteristic function χ of the interval (a, b), so that $\chi(x) = 1$ for x in (a, b), $\chi(x) = 0$ elsewhere. Then $\varphi_0(x) = \chi_{[0,1]}(x)$, $\varphi_1(x) = x\varphi_0(x)$, $\varphi_2(x) = \varphi_1(2x) + \chi_{(0,1]}(2x - 1) - \varphi_1(2x - 1), \ldots, \varphi_{2^n + k}(x) = \varphi_2(2^n x - k + 1)$, $n = 1, 2, 3, \ldots, k = 1, \ldots,$ is a Schauder basis for $C[0, 1]$.

SECTION 2.4

1. (a) *Hint:* Note that $y'(x) = (r(x)p^{-1}(x))^{1/2} = v^2(y)u^{-2}(x)p^{-1}(x)$ so that $2vv'y' = 2uu'py' + u^2p'y' + u^2py''$, the prime denoting differentiation with respect to x. From this one eventually arrives at

$$c = \frac{q}{r} + \frac{p''}{4r} - \frac{(p')^2}{16pr} + \frac{p'r'}{8r^2} + \frac{pr''}{4r^2} - \frac{5p(r')^2}{16r^3},$$

which becomes $c(y)$ upon substitution of

$$x(y) = \int_{y_0}^{y} (p(x)r^{-1}(x))^{1/2} \, dx + x(y_0).$$

3. Eigenvalues are simple (i.e., of multiplicity one).

SECTION 2.5

1. (a) Let

$$v(\xi, \eta) = u(x(\xi, \eta), t(\xi, \eta)) = u\left(\frac{\xi + \eta}{2}, \frac{\xi - \eta}{2}\right)$$

so that the wave equation becomes

$$v_{\xi\eta} = -F\left(\frac{\xi + \eta}{2}, \frac{\xi - \eta}{2}\right).$$

With (x_0, t_0) fixed as shown in Figure 2.5e, integrate with respect to ξ from $\xi = \eta$ to $\xi = x_0 + t_0$, as indicated in the figure. Then integrate with respect to η from $x_0 - t_0$ to $x_0 + t_0$, thereby completing the integration over the characteristic triangle. The result follows upon changing back to the x and t variables and upon the insertion of the data. (b) Assume $F \, \varepsilon \, C^2$, $f \, \varepsilon \, C^2$, $g \, \varepsilon \, C^2$, and differentiate under the integral. (c) From the d'Alembert formula.

2. (c) Differentiate under the integral to show $Lu = LL^{-1}F = F$. The Green's function is thus the kernel of an integral operator inverting the Laplacian L with zero boundary conditions. In fact L^{-1} is a "compact" operator, a condition stronger than just being bounded. See the discussion in Section 2.7.2.

3. (a) With a propagation speed now of c rather than normalized to 1, the d'Alembert solution is

$$u(x, t) = \frac{1}{2}(f(x - ct) + f(x + ct)) + \frac{1}{2c} \int_{x-ct}^{x+ct} g(s)ds$$

$$+ \frac{1}{2c} \int_0^t \int_{x-c(t-s)}^{x+c(t-s)} F(y, s) \, dy \, ds.$$

Thus $u(x, t) = x^2 + c^2t^2$ and $u(100, 100) = 10^4(1 + c^2)$.

(b) $u\left(\frac{1}{2}, \frac{3}{2}\right) = \frac{1}{2}(2 + 2) + \frac{1}{4}\int_{-5/2}^{7/2} ds + \frac{1}{4}\int_0^{3/2} \int_{2s-5/2}^{7/2-2s} e^z \, dz \, ds$

$$= \frac{7}{2} - \frac{1}{8}[2e^{1/2} - e^{7/2} - e^{-5/2}].$$

(c) Because of the free end boundary conditions we extend the initial data evenly as far as needed. Then

$$u(1/2, 2) = \frac{1}{2} [1 + 1] + \frac{1}{2} \int_{-3/2}^{5/2} |sin_e \, \pi s| \, ds$$

$$= 1 + \frac{4}{2} \int_0^1 sin\pi x \, dx = 1 + 4/\pi$$

Here $sin_e \pi s$ signifies the extension of $sin \, \pi s$ to the other intervals.

THIRD PAUSE

1. $\int_\Omega (u - u_N)^2 \, r \, dx = \int_\Omega u^2 \, r \, ds + \sum_{n=1}^N \left(\frac{\int_\Omega u\phi_n r \, dx}{\int_\Omega \phi_n^2 \, r \, dx} - c_n \right)^2$

$$- \sum_{n=1}^N \left(\frac{\int_\Omega u\phi_n r \, dx}{\int_\Omega \phi_n^2 \, r \, dx} \right)^2.$$

Hence choose $c_n = \int_\Omega u\phi_n r \, dx / \int_\Omega \phi_n^2 r \, dx = (u, \phi_n)/(\phi_n, \phi_n)$.

2. The $\phi_n = (2/\pi)^{1/2} cos \, nx$ are a maximal orthonormal set in $L^2 (0, \pi)$ by the Sturm–Liouville Theorem applied to the problem $-u'' = \lambda u$ on $(0, \pi)$, $u'(0) = u'(\pi) = 0$. Thus $\int_0^\pi \ln x \, cos \, nx \, dx = \left(\frac{\pi}{2} \right)^{1/2} (\ln x, \phi_n) = \left(\frac{\pi}{2} \right)^{1/2} c_n \to 0$ provided that $\ln x$ is square integrable. The latter may be seen from calculus and l'hospital's rule:

$$\int_0^1 (\ln x)^2 \, dx = \lim_{\varepsilon \to 0} [x(\ln x)^2 - 2x(\ln x - 1)]|_\varepsilon^1$$

$$= 2 - \lim_{\varepsilon \to 0} [\varepsilon(\ln\varepsilon)^2 - 2\varepsilon\ln\varepsilon + 2\varepsilon]$$

$$= 2 - \lim_{\varepsilon \to 0} \frac{2(\ln\varepsilon)(\varepsilon^{-1})}{(-\varepsilon^{-2})} - 2 \lim_{\varepsilon \to 0} \left(\frac{\varepsilon^{-1}}{-\varepsilon^{-2}} \right)$$

$$= 2$$

3. $u(x) = |x|^{-1/2}$ is not square integrable so we cannot apply Parseval's equality. However, we may apply the Riemann–Lebesgue Lemma (see the Second Pause) provided that $u(x)$ is just integrable, noting that in this instance the ϕ_n are uniformly bounded. Thus the result follows from

$$\int_{-\pi}^\pi |x|^{-1/2} \, dx = 2 \lim_{\varepsilon \to 0} \int_\varepsilon^\pi x^{-1/2} \, dx = 4 \lim_{\varepsilon \to 0} [x^{1/2}]|_\varepsilon^\pi = 4\pi^{1/2}.$$

4. This follows from the continuity of the inner product which is implicit in the Fourier representation of Theorem H. One has

$$u = \lim_{N \to \infty} \sum^N \frac{(F, \phi_{nm})}{n^2 + m^2} \phi_{nm} = \lim_{N \to \infty} \left(\sum^N \frac{\phi_{nm} \phi_{nm}}{n^2 + m^2}, F \right).$$

Alternately, one can use the inner product form of Parseval's Equality in Exercise 5.

5. From $\|u - v\|^2 = \|u\|^2 + \|v\|^2 - 2(u, v)$ we have from the norm squared version of Parseval's Equation that

$$\sum (c_n - d_n)^2 = \sum c_n^2 + \sum d_n^2 - 2(u, v)$$

from which the inner product version follows.

6. Recall the normal form, Section 2.4, Problem 1. Rather than using the general recipes given there, it is faster to proceed directly. Letting $\hat{x} = (\sqrt{m\omega/\hbar})x$, $\hat{x}$ is a dimensionless new independent variable because $\sqrt{m\omega/\hbar}$ has units (length)$^{-1}$ because m is mass, ω is angular frequency, and $\hbar$ is in units of energy · sec. The chain rule and substitution yields

$$\frac{-\hbar^2}{2m} \frac{m\omega}{\hbar} \frac{d^2}{d\hat{x}^2} + \frac{1}{2} m\omega^2 \frac{\hbar}{m\omega} (\hat{x})^2 u\left(\sqrt{\frac{\hbar}{m\omega}} \hat{x}\right) = Eu\left(\sqrt{\frac{\hbar}{m\omega}} \hat{x}\right)$$

which after division by $\hbar\omega/2$ becomes

$$\left(-\frac{d^2}{d\hat{x}^2} + \hat{x}^2\right) u\left(\sqrt{\frac{\hbar}{m\omega}} \hat{x}\right) = 2\varepsilon u\left(\sqrt{\frac{\hbar}{m\omega}} \hat{x}\right)$$

where $\varepsilon = E/\hbar\omega$. Letting $\hat{u}(\hat{x}) = u(\sqrt{\hbar m\omega}\, x)$ brings us to normal form

$$-\hat{u}''(\hat{x}) + c(\hat{x})\hat{u}(\hat{x}) = \lambda\hat{u}(\hat{x})$$

with $c(\hat{x}) = \hat{x}^2$ and $\lambda = 2\varepsilon = 2E/\hbar\omega$.

7. Put the equation obtained above into self-adjoint form. For ease of notation let u and x denote $\hat{u}$ and $\hat{x}$, respectively. The substitution $u(x) = e^{-x^2/2} h(x)$ into the normal form

$$-u'' + x^2u = \lambda u$$

gives

$$(1 - x^2)e^{-x^2/2} h + 2xe^{-x^2/2} h' - e^{-x^2/2} h'' + x^2 e^{-x^2/2} h - \lambda e^{-x^2/2} h = 0$$

which becomes

$$-h'' + 2xh' + (1 - \lambda)h = 0$$

which has the alternate form of Section 2.1

$$h'' - 2xh' + \mu h = 0$$

where $\mu = \lambda - 1 = 2(\varepsilon - 1/2)$. We have called this the ODE form: $a_0y'' + a_1y' + a_2y = 0$ with $a_0 = 1$. It is clearly equivalent to the self-adjoint form

$$-(e^{-x^2}h')' + \mu e^{-x^2}h = 0.$$

Beyond this symmetric form of the equation, for self-adjointness and hence completeness of the eigenfunctions one also needs correct boundary conditions (e.g., see Section 2.4, the earlier discussion in Section 1.9.7(1), and the later discussion in Section 2.9.4). For the Hermite equation above, $p = r = e^{-x^2}$, the Hilbert space is $L^2(-\pi, \pi, e^{-x^2})$, and hence we need to specify that the eigenfunctions h to vanish at infinity appropriately relative to the weight e^{-x^2}.

We mention that one finds such important equations written in different ways depending on context. For example the two normal forms

$$-u'' + x^2 u = \lambda u, \qquad u'' - \frac{x^2}{4} u + \lambda u = 0$$

in which the harmonic oscillator Hermite equation is often found are related by setting $\bar{x} = \alpha x$ from which one finds $\alpha = \sqrt{2}$.

8. Let $h(x) = \sum_{n=0}^{\infty} a_n x^n$ be a general solution, necessarily analytic because the coefficients of the equation are. Substituting into the ODE form we have

$$\sum_{n=0}^{\infty} n(n-1)a_n x^{n-2} - 2 \sum_{n=0}^{\infty} n a_n x^n + \mu \sum_{n=0}^{\infty} a_n x^n = 0.$$

Shifting index this requires that the coefficients

$$n(n-1)a_n - 2(n-2)a_{n-2} + \mu a_{n-2} = 0$$

for $n \geqq 2$. Thus the recursion

$$a_{n+2} = 2a_n \frac{[n - (\varepsilon - 1/2)]}{(n+1)(n+2)}$$

will give all odd and even coefficients respectively from the values of a_0 and a_1. Moreover, for $\varepsilon - 1/2 = n$ an integer, the series will terminate.

Taking $a_0 = 1$ and $a_1 = 0$ and $\varepsilon - 1/2 = 0$ yields the first eigenfunction $H_0(x) = 1$. From $\varepsilon - 1/2 = 2$ comes $H_2(x) = 1 - 2x$. Continuing with n an even integer one gets all of the Hermite polynomials of even index.

Taking $a_0 = 0$ and $a_1 = 1$ and $\varepsilon - 1/2 = 1$ yields $H_1 = x$, with the other odd index Hermite polynomials $H_3 = x - (2/3)x^3, \ldots$ resulting from n an odd integer.

There can be no others, for if $[n - (\varepsilon - 1/2)]$ does not vanish, all terms will be present in the power series, which will eventually overwhelm the decay factor e^{-x^2} and force the solution out of the Hilbert space.

9. (a) Getting into the details of the Haar functions would take us too far afield. Their orthogonality is clear from their signed support sets, their orthonormality then given by their weights. That they are maximal is less immediate but may be shown (e.g., as in Section 2.4 by using the densities of the C_0 functions). (b) Haar's point was that they do not come from a regular Sturm–Liouville equation. (c) The Haar system has been very useful in functional analysis and ergodic theory. Haar's second and later contribution, by far the most well-known, was to generalize Lebesgue measure to the construction of an invariant measure on any locally compact topological group. A third contribution, the least well-known but the most important for differential equations, is the rather remarkable fact that a large class of Sturm–Liouville series expansions converge or diverge pointwise along with the Fourier trigonometric expansion. That is, Dini's tests and conclusions apply to these other eigenfunction expansions as well.*

---

* A. Haar, *Math Annalen* **69** (1910), **71** (1911). The Haar system and the more general conclusion from the Dini tests constituted the main portions of Haar's dissertation.

SECTION 2.6

1. (b) $u(x, y) = \sum_{n=1}^{\infty} \dfrac{18 \sin(2n\pi/3)}{n^2\pi^2\sinh n\pi} \sin\left(\dfrac{n\pi}{3} x\right) \sinh(n\pi(1 - y/3))$. Additionally it
is somewhat interesting to see how well this rather gross finite difference works. By computer we can get a good approximation to the solution. For example, summing to 50 terms, one has the following comparisons.

$u(x_i, y_i)$	$s_{50}(x_i, y_i)$	Finite difference
$u(1, 1) \cong \displaystyle\sum_{n=1}^{50} \dfrac{18 \sin 2n\pi/3}{n^2\pi^2\sinh n\pi} \sin \dfrac{n\pi}{3} \sinh\left(\dfrac{2n\pi}{3}\right)$	$= 0.43051$	$\sim\ 0.45833$
$u(1, 2) \cong \displaystyle\sum_{n=1}^{50} \dfrac{18 \sin 2n\pi/3}{n^2\pi^2\sinh n\pi} \sin \dfrac{n\pi}{3} \sinh\left(\dfrac{n\pi}{3}\right)$	$= 0.14285$	$\sim\ 0.16667$
$u(2, 1) \cong \displaystyle\sum_{n=1}^{50} \dfrac{18 \sin 2n\pi/3}{n^2\pi^2\sinh n\pi} \sin \dfrac{2n\pi}{3} \sinh\left(\dfrac{2n\pi}{3}\right)$	$= 0.51732$	$\sim\ 0.66667$
$u(2, 2) \cong \displaystyle\sum_{n=1}^{50} \dfrac{18 \sin 2n\pi/3}{n^2\pi^2\sinh n\pi} \sin \dfrac{2n\pi}{3} \sinh\left(\dfrac{n\pi}{3}\right)$	$= 0.15310$	$\sim\ 0.20833$

A slightly finer finite difference grid would yield very close solutions.

2. (a) Note that for any n-vectors x and y one has

$$x \cdot x = y \cdot y + (x - y) \cdot (x - y) + 2x \cdot y - 2y \cdot y.$$

With $x = \text{grad } v$ and $y = \text{grad } u$ this gives

$$|\text{grad } v|^2 = |\text{grad } u|^2 + |\text{grad}(v - u)|^2 + 2 \text{ grad } v \cdot \text{grad } u$$
$$- 2 \text{ grad } u \cdot \text{grad } u$$
$$= |\text{grad } u|^2 + |\text{grad}(v - u)|^2 + 2 \text{ grad } u \cdot \text{grad}(v - u)$$

This may also be seen directly and in fact the one dimensional version was used in Section 1.5.3. Thus integrating over the region Ω we have

$$D(v) = D(u) + D(v - u) + 2D(u, v - u)$$
$$= D(u) + D(v - u) \geqq D(u)$$

because by Green's first identity

$$D(u, v - u) = \oint_{\partial\Omega} (v - u) \frac{\partial u}{\partial n} - \int_{\Omega} (v - u)\, \Delta u = 0,$$

since $v = u$ on $\partial\Omega$ and $\Delta u = 0$ in Ω.

3. $R(u) - R(v) = 2\displaystyle\oint_{\partial\Omega} u \frac{\partial u}{\partial n} - \int_{\Omega} |\text{grad } u|^2 - 2\oint_{\partial\Omega} f \frac{\partial v}{\partial n} + \int_{\Omega} |\text{grad } v|^2$

$$= 2D(u) - D(u) - 2D(u, v) + D(v)$$
$$= D(u - v).$$

Note that $R(u) = D(u)$ so that the approximations are from below in the Dirichlet norm.

FOURTH PAUSE

1.
$$
\begin{bmatrix}
4 & -1 & 0 & -1 & 0 & 0 & 0 & 0 & 0 \\
-1 & 4 & -1 & 0 & -1 & 0 & 0 & 0 & 0 \\
0 & -1 & 4 & 0 & 0 & -1 & 0 & 0 & 0 \\
-1 & 0 & 0 & 4 & -1 & 0 & -1 & 0 & 0 \\
0 & -1 & 0 & -1 & 4 & 0 & 0 & -1 & 0 \\
0 & 0 & -1 & 0 & -1 & 4 & -1 & 0 & -1 \\
0 & 0 & 0 & -1 & 0 & 0 & 4 & -1 & 0 \\
0 & 0 & 0 & 0 & -1 & 0 & -1 & 4 & -1 \\
0 & 0 & 0 & 0 & 0 & -1 & 0 & -1 & 4
\end{bmatrix}
\begin{bmatrix}
u(.75, .75) \\
u(.75, 1.5) \\
u(.75, 2.25) \\
u(1.5, .75) \\
u(1.5, 1.5) \\
u(1.5, 2.25) \\
u(2.25, .75) \\
u(2.25, 1.5) \\
u(2.25, 2.75)
\end{bmatrix}
=
\begin{bmatrix}
.75 \\
0 \\
0 \\
1.5 \\
0 \\
0 \\
1.5 \\
0 \\
0
\end{bmatrix}
$$

yields the solution

$$(.416, .209, .126, .707, .295, .294, .618, .263, .139).$$

Notice that this grid is not a refinement of the original coarse 3×3 grid and hence offers poor comparisons. Also when we linearly interpolated the data f, this grid failed to take note of the peak point of f at $x = 2$.

2.
$$
\begin{bmatrix}
4 & -1 & -1 & 0 \\
-1 & 4 & 0 & -1 \\
-1 & 0 & 4 & -1 \\
0 & -1 & -1 & 4
\end{bmatrix}
\begin{bmatrix}
u(\pi/3, \pi/3) \\
u(\pi/3, 2\pi/3) \\
u(2\pi/3, \pi/3) \\
u(2\pi/3, 2\pi/3)
\end{bmatrix}
$$
$$
=
\begin{bmatrix}
\pi/3 + (\pi/3)(\pi - \pi/3) \\
2\pi/3 + 0 \\
0 + (2\pi/3)(\pi - 2\pi/3) \\
0 + 0
\end{bmatrix}
=
\begin{bmatrix}
3.24 \\
2.09 \\
2.19 \\
0
\end{bmatrix}
$$

yields the solution

$$(1.302, 0.471, 0.996, 0.492).$$

This problem is small enough to be done on a programmable hand calculator. For example a Hewlett-Packard 41 C yields

$$
\begin{bmatrix}
4 & -1 & -1 & 0 \\
-1 & 4 & 0 & -1 \\
-1 & 0 & 4 & -1 \\
0 & -1 & -1 & 4
\end{bmatrix}^{-1}
=
\begin{bmatrix}
.2917 & .0833 & .0833 & .0417 \\
.0833 & .2917 & .0417 & .0833 \\
.0833 & .0417 & .2917 & .0833 \\
.0417 & .0833 & .0833 & .2917
\end{bmatrix}
$$

These numbers are certainly adequate for data given on such a coarse grid.

3. $AU = b$, where A is the 9×9 matrix of (1.) above, and

$$b = (2.636, 1.571, 2.356, 2.467, 0, 0, 1.850, 0, 0)$$

yields, with points ordered as in (2.) above,

$$u = (1.32, 1.21, 1.03, 1.42, 0.93, 0.56, 0.95, 0.54, 0.27).$$

4. From the separation of variables solution obtained previously (Section 2.1, Problem 2(a), see Answer Section) summed to 10 terms we have

$$u = (1.29, 1.21, 1.02, 1.39, 0.90, 0.52, 0.92, 0.51, 025).$$

5. For a graphics output and a comparison with the solution of the minimal surface solution with the same boundary conditions, see Appendix B, Fig. B.6.

6. Use fact $H_0^1[0, 1]$ is the completion of the test functions $C_0^\infty[0, 1]$. Let u_m in C_0^∞ converge in H^1 to u in H^1 as in the example above, and call $w = u'$ the weak derivative of u. Let $\phi \in C_0^\infty(0, 1)$ be any test function and $[a, b]$ an interval in $[0, 1]$. Then

$$\int_a^b \phi'(u_m - u)\,dx \leq \|u_m - u\|_2 \|\phi'\|_2 \to 0,$$

$$\int_a^b \phi(u_m' - w)\,dx \leq \|u_m' - w\|_2 \|\phi\|_2 \to 0$$

by Schwarz's inequality. Hence

$$\phi u_m|_a^b = \int_a^b (\phi u_m)' = \int_a^b (\phi' u_m + u_m' \phi) - \int_a^b (\phi' u + w\phi) \to 0.$$

Letting $a = 0$ and $b = x$ we thus have

$$\lim_{m \to \infty} \phi(x)u_m(x) = \int_0^x (\phi'(s)u(s) + w(s)\phi(s))\,ds$$

for all x. Calling this limit $h^\phi(x)$, from the fundamental theorem of (Lebesgue) calculus h^ϕ is absolutely continuous. As $\phi(x)u_m(x) \to h^\phi(x)$ pointwise and in L^2, we have $\phi(x)u(x) = h^\phi(x)$ almost everywhere. Taking ϕ to be a smoothed characteristic function (see Section 2.9.3) we may conclude therefore that u is absolutely continuous.

7. Let

$$f_n(x) = \begin{cases} 1 & 1/n \leq x \leq 1 \\ \sin\dfrac{n\pi x}{2} & -1/n \leq x \leq 1/n \\ -1 & -1 \leq x \leq -1/n \end{cases}$$

We check that $f_n'(x)$ is continuous so that each f_n is in $C^1(-1, 1)$. Their antiderivatives are

$$F_n(x) = \begin{cases} x - 1 \\ \dfrac{1}{n} - 1 - \dfrac{2}{n\pi}\cos\dfrac{n\pi x}{2} \\ -1 - x \end{cases}$$

which converge in the H^1 norm to

$$F(x) = \begin{cases} x - 1 & 0 \leq x \leq 1 \\ -1 - x & -1 \leq x \leq 0 \end{cases}$$

which is not in $C^1(-1, 1)$ due to the discontinuity in its derivative at $x = 0$.

8. Letting S_N denote the partial sum $\sum_1^N n^{-1}(w, \phi_n)\phi_n$, it suffices to show that $|S_N(x) - S_M(x)| \to 0$ uniformly in x. This follows from the Schwarz and Bessel inequalities

$$|S_N(x) - S_M(x)| \le \sum_{N+1}^M \frac{1}{n} |(w, \phi_n)| (2/\pi)^{1/2}$$

$$\le (2/\pi)^{1/2} \left(\sum_{N+1}^M \frac{1}{n^2} \right)^{1/2} \cdot \left(\sum_{N+1}^M |(w, \phi_n)|^2 \right)^{1/2}.$$

We mention that one may roughly interpret w as the weak derivative of u, the given weighted Fourier series performing the integration back to a continuous function.

9. (a) Taking for example $\phi = 1, x, x^2, x^3, \ldots$, and cutting them off by mollification to compact support very close to the whole interval $\Omega = (0, 1)$, the condition $\int_0^1 u\phi'' = 0$ requires that u be orthogonal to all of them except possibly 1 and x.

   (b) All derivatives exist and are continuous at $x = 0$ (by l'hospital's rule) but the Taylor series is identically 0.

   (c) This follows from (b). These are bell-shaped cutoff functions important for use as molifier kernels. See Section 2.9.3.

SECTION 2.7.1

1. (b) For example, to stay in $L^2(\Omega)$ it is reasonable to impose that $u \to 0$ as $|x| \to \infty$. This now gives uniqueness for the Neumann problem in which one had the condition $\partial u/\partial n = 0$ at $|x| = 1$. But the problem was then really a mixed Neumann–Dirichlet problem.

3. (d) Not in general.

SECTION 2.7.2

1. (a) L^{-1} is 1–1 but not onto $L^2(0, \pi)$. (c) *Hint*: The variation of parameters formula.

SECTION 2.7.3

2. (a) The convolution theorem follows from (i) the unitarity of the Fourier transform, that is, in particular, that it preserves angles, and by (ii) observing that the convolution product $f * h = \int_{-\infty}^{\infty} f(x - y)h(y) \, dy$ is an inner product. To show that $f * h = \hat{f}\hat{h}$, we may show instead that $f * h = \widehat{\hat{f}\hat{h}}^{-1}$. The latter follows from

$$\int_{-\infty}^{\infty} f(x-y)h(y) \, dy = \langle f(x-y), \overline{h(y)} \rangle = \langle \widehat{f(x-y)}, \widehat{h(y)} \rangle$$

$$= (2\pi)^{-n/2} \langle e^{ikx} \hat{f}(-k), \overline{\hat{h}(-k)} \rangle$$

$$= (2\pi)^{-n/2} \int_{-\infty}^{\infty} e^{-ikx} \hat{f}(k)\hat{h}(k) \, dk.$$

SECTION 2.8

1. (a) Remember that even though $G_0(P, Q) = r^{-1}$, $r = |P - Q|$, the element of integration is $r^2 d(\text{angle})$. See Section 1.6 and Green's third identity.

   (c) One way in which to proceed is to consider the Helmholtz equation $\Delta u + (\lambda + i\varepsilon)u_\varepsilon = f$ for $\varepsilon > 0$, which has a solution

$$u_\varepsilon(P) = \int_{R^3} \frac{e^{ia_\varepsilon r}}{4\pi r} \cdot e^{-ib_\varepsilon r} f(Q) \, dV_Q$$

   with the property that u decays well at ∞. Here a_ε and b_ε come from $\sqrt{\lambda + i\varepsilon}$ and are given by $a_\varepsilon = 2^{-1/2}((\lambda^2 + \varepsilon^2)^{1/2} + \lambda^2)^{1/2}$ and $b_\varepsilon = 2^{-1/2}((\lambda^2 + \varepsilon^2)^{1/2} - \lambda^2)^{1/2}$. Then show that $u_\varepsilon \to u$ as $\varepsilon \to 0^+$ for appropriate f.

2. (a) By a consideration of x small and x large one is led in both cases to $W_+^* W_- \phi = \varphi$.

PROBLEM 2.9.1

(a) The point here is that the solution by separation of variables corresponds to tensoring the partial differential equation in such a way that the solution u is a diagonal of a tensor represented in $L^2(0, \pi)^* \otimes L^2(0, \pi)^*$. It is perhaps easier to write down the tensor machinery first for the Poisson problem $-\Delta u = F$ in Ω.

PROBLEM 2.9.2

(a) The method of proof is indicated directly after the problem.

(b) By Problem 3 of Section 1.7 one has $\|Vu\| \le a\|u\| + b\|H_0 u\|$ for some $b < 1$. From this, $(1 - b)\|H_0 u\| \le \|(H_0 + V)u\| + a\|u\|$. From the latter, that $H_0 + V$ is closed follows from H_0 closed and the definition of a closed operator.

PROBLEM 2.9.3 ·

(d) That functions such as $\delta_\varepsilon(x)$ exist and provide C_0^∞ "cutoffs" is rather important in the abstract theory of partial differential operators and moreover in the more modern theory of pseudodifferential operators (see Problem 2.9.9, Exercise 6). Consider

$$f(x) = \begin{cases} e^{-(1-x^2)^{-1}}, & |x| < 1, \\ 0, & |x| \ge 1. \end{cases}$$

Clearly f has compact support and is C^∞ except possibly at $x = \pm 1$. Consider the case $x = 1$; clearly the right-hand derivatives all exist there and are zero. Let us show that the left-hand derivative is also zero there, the vanishing of the higher order left-hand derivatives following in the same way. The left-hand difference quotient is indeterminant as $h \to 0^-$,

$$\frac{\Delta f}{\Delta h} = \frac{f(1 - h) - f(1)}{h} = \frac{e^{-(2h - h^2)^{-1}}}{h} \to \frac{0}{0},$$

so that by l'hospital's rule we may proceed to

$$\frac{\Delta f}{\Delta h} \rightarrow \frac{h^{-1}}{e^{(2h - h^2)^{-1}}} \rightarrow \frac{-h^{-2}}{e^{(2h - h^2)^{-1}}[(2h - 2)/(2h - h^2)^2]}$$

$$= \frac{(2 - h)^2/(2h - 2)}{e^{(2h - h^2)^{-1}}} \rightarrow \frac{-2}{\infty} = 0.$$

PROBLEM 2.9.4

(c) In the regular case the solutions of the Sturm–Liouville equations are continuous on bounded intervals and hence square integrable.

PROBLEM 2.9.5

(a) By linearity of A it suffices to show that A is bounded iff A is continuous at 0. We recall from the ε, δ definition of a continuous function in a metric space that A continuous at 0 means that for every $\varepsilon > 0$ there exists a $\delta = \delta(\varepsilon) > 0$ such that $A(B_\delta) \subset B_\varepsilon$, where B_γ denotes the ball $\|x\| \leq \gamma$. (i) Let A be bounded, that is, $\|Ax\| \leq M$ for $\|x\| \leq 1$. Then taking $\delta = \varepsilon/M$ we have for $\|x\| \leq \delta$ that $\|Ax\| \leq (\varepsilon/M)\|A(M/\varepsilon)x\| \leq \varepsilon$. (ii) Let A be continuous. Then taking $\varepsilon = 1$ we have $\|Ax\| \leq 1$ for $\|x\| \leq \delta$ for some $\delta > 0$, from which $\|Ax\|\delta^{-1}$ for $\|x\| \leq 1$.

PROBLEM 2.9.6(1)

(a) This is a special case of the Friedrichs–Poincaré Lemma shown in the Second Pause.

PROBLEM 2.9.6(2)

(a) In matrix form the system of differential equations becomes

$$\underbrace{\begin{bmatrix} du(x_1, t)/dt \\ du(x_2, t)/dt \\ \vdots \\ du(x_{N-1}, t)/dt \end{bmatrix}}_{\dfrac{dY}{dt}} = \underbrace{\begin{bmatrix} -2/h^2 & 1/h^2 & 0 & \cdots \\ 1/h^2 & -2/h^2 & 1/h^2 & 0 & \cdots \\ 0 & 1/h^2 & -2/h^2 & \ddots \\ \vdots & & \ddots & \ddots \\ & & 0 & 1/h^2 & -2/h^2 \end{bmatrix}}_{A} \underbrace{\begin{bmatrix} u(x_1, t) \\ u(x_2, t) \\ \vdots \\ u(x_{N-1}, t) \end{bmatrix}}_{Y}$$

with initial condition

$$Y(0) = C = [f(x_1), f(x_2), \cdots, f(x_{N-1})].$$

(b) Approximate $du(x_i, t)/dt$ by the forward difference $(u(x_i, t_{j+1}) - u(x_i, t_j))/k$. This now becomes the forward Euler scheme. See Appendix B.

PROBLEM 2.9.6(3)

(a) *Hint*: Use Green's identities and the "energy" method.

(b) For example,

$$\frac{u_{i,j}^{\gamma+1/2} - u_{i,j}^{\gamma}}{\frac{1}{2}k_{\gamma}} = \frac{1}{h_x^2}[A_{i+1/2,j}^n(u_{u+1,j}^{\gamma+1/2} - u_{i,j}^{\gamma+1/2}) - A_{i-1/2}^n(u_{i,j}^{\gamma+1/2} - u_{i-1,j}^{\gamma+1/2})]$$

$$+ \frac{1}{h_y^2}[A_{i,j+1/2}^n(u_{i,j+1}^{\gamma} - u_{i,j}^{\gamma}) - A_{i,j-1/2}^n(u_{i,j}^{\gamma} - u_{i,j-1}^{\gamma})].$$

(c) *Hint:* See Chapter 3, Appendices A and B.

PROBLEM 2.9.7(1)

(b) (ii) Let $z = re^{i\theta}$, then Ω is found to be the lens-shaped domain prescribed by $-\pi/3 < \theta < \pi/3$ and $r - 4r^{1/2}\cos(\theta/2) + 3 < 0$.

(c) (ii) After a somewhat long calculation one arrives at

$$u(r, \theta) = -\oint_{\partial\Omega} \frac{\partial G(r, \theta, \rho, \psi)}{\partial n} u(\rho, \psi)\,ds$$

$$= \int_{-\pi/3}^{\pi/3} \frac{[4r^{1/2}\cos(\theta/2) - r - 3]\rho^{1/2}}{4\pi[r + \rho - 2r^{1/2}\rho^{1/2}\cos((\theta - \psi)/2)](2\cos\psi - 1)^{1/2}}$$
$$\cdot u(\rho, \psi)\,d\psi$$

where it is understood that the integral is to be evaluated twice as $Q = (\rho, \psi)$ runs over both parts of the $\partial\Omega$. Part (i) is much easier and is left to the student.

PROBLEM 2.9.7(2)

(c) *Hint:* See the answer to Problem 1(c) of Section 2.8.

PROBLEM 2.9.7(3)

(b) This reduces to showing the following (a Stoke's or Duhamel method):

$$\begin{cases} \Box u = 0 \\ u(0) = 0 \\ u_t(0) = f \end{cases} \Rightarrow \begin{cases} \Box v = 0 \\ v(0) = f \\ v_t(0) = 0. \end{cases}$$

With some continuity assumptions (e.g., u_{tt} continuous at 0) one shows that $\Box v = 0$ and $v(0) = f$. There are several ways to then show $v_t(0) = 0$, one employing the divergence theorem.

PROBLEM 2.9.8

(a) See Problem B-9.

(b) The stream function φ and variable change $\ln w = -\phi/2\nu$ may be combined in the (Hopf–Cole) transformation $u = -2\nu w_x/w$ which converts $u_t + uu_x - \nu u_{xx} = 0$ to $w_t - \nu w_{xx} = 0$. Thus the nonlinear and linear initial value problems are related by

$$\begin{cases} u_t + uu_x - \nu u_{xx} = 0, \quad t > 0 \\ u(0) = f(x) \end{cases} \leftrightarrow \begin{cases} w_t - \nu w_{xx} = 0, \quad t > 0, \\ w(0) = \exp(-(2\nu)^{-1}\int_0^x f(s)\,ds). \end{cases}$$

One may feed various w and f into this relation.

(e) See Problem B-9(c).

PROBLEM 2.9.9

1. (a)

$$u(x, y, t) = 4\sum_{n=0}^{\infty} \sum_{m=0}^{\infty} \frac{(-1)^{n+m}}{(2n+1)^2(2m+1)^2}$$

$$\times \sin(2n+1)x \, \sin(2m+1)y \, \cos(m^2+n^2)^{1/2}t.$$

(b) $u(x, y, t) = 4\pi^2 \sum_{n=1}^{\infty} \sum_{m=1}^{\infty} \left(\frac{(-1)^{n+1}}{n} - \frac{2(1 - (-1)^n)}{n^3 \pi^2} \right)$

$$\cdot \left(\frac{(-1)^{m+1}}{m} - \frac{2(1 - (-1)^m)}{m^3 \pi^2} \right)$$

$$\times \sin nx \, \sin my \, \cos(m^2+n^2)^{1/2}t.$$

(d) $u(x, y, t) = \displaystyle\sum_{n=1}^{\infty} \sum_{m=1}^{\infty} c_{nm} e^{-k(m^2+n^2)t} \sin nx \, \sin my,$

$$c_{nm} = 4\pi^{-2} \int_0^{\pi} \int_0^{\pi} f(s, \tau) \sin ns \, \sin n\tau \, ds \, d\tau.$$

2. (c) No.
   (d) Self-adjoint.
   (f) *Hint:* $u = c_1 e^{i\sqrt{\lambda}x} + c_2 e^{-i\sqrt{\lambda}x}$ is square integrable when $\int_0^{\infty} e^{-2\beta x} dx < \infty$, $\sqrt{\lambda} = \alpha + i\beta$.

3. (b) d'Alembert's formula.
   (c) $E(t) = \frac{1}{2}\int_{-\infty}^{\infty} ((u_t)^2 + (u_x)^2)\, dx, \qquad t \geq 0.$
   (d) *Hint:* See Problem 2.9.7(3).

7. (b) Those of radius R_n satisfying the equation $\tan \lambda^{1/2} R_n = \lambda^{1/2} R_n,$ $n = 1, 2, 3, \ldots .$

9. 1. (a) is easily solved by separation of variables or even inspection with solution $u(x, t) = 1 + e^{-t} \cos x$. (b) leads to the transcendental equations

$$\tan \sqrt{-\lambda}\, \pi = -\sqrt{-\lambda}/2, \qquad \cot \sqrt{-\lambda}\pi = \sqrt{-\lambda}/2$$

for the eigenvalues λ_n. Note also that the initial conditions and boundary conditions in (a) are compatible at $(\pm\pi, 0)$ whereas they are not in (b).

2. The Hermite equation $-u'' + x^2 u = \lambda u$ has variational Rayleigh quotient $\lambda = \int u(-u'' + x^2 u)/\int u^2$ and hence we seek

$$\min_l \int_{-l}^{l} (1 - x^2/l^2)\left(-\frac{d^2}{dx^2} + x^2\right)(1 - x^2/l^2)dx \bigg/ \int_{-l}^{l} (1 - x^2/l^2)\, dx$$

$$= \min_l \int_{-l}^{l} \left(\frac{2}{l^2} + x^2 - \frac{x^4}{l^2} - \frac{2x^2}{l^4} - \frac{x^4}{l^2} + \frac{x^6}{l^4}\right)dx \bigg/ \int_{-l}^{l} \left(1 - \frac{2x^2}{l^2} + \frac{x^4}{l^4}\right)dx$$

$$= \min_l \left(\frac{2}{l} + \frac{l^3}{3} - \frac{2l^5}{5l^2} - \frac{2l^3}{3l^4} + \frac{l^7}{7l^4}\right) \bigg/ \left(l - \frac{2l^3}{3l^2} + \frac{l^5}{5l^4}\right)$$

$$= \min_l \frac{15}{8}\left(\frac{4}{3}\left(\frac{1}{l^2}\right) + \frac{8}{105}(l^2)\right).$$

The derivative set to zero gives $l^2 = \sqrt{35/2}$ which upon substitution yields $\lambda = (2/7)\sqrt{35/2} \approx 1.2$, reasonably close to the true value $\lambda_1 = 1$.

3. Recalling $S_N(r, \theta) = \dfrac{a_0}{2} + \Sigma_1^N a_n r^n \cos n\theta + \Sigma_1^N b_n \sin n\theta$, where

$$a = \frac{1}{\pi}\int_{-\pi}^{\pi} f(\theta)\cos n\theta\, d\theta, \qquad b = \frac{1}{\pi}\int_{-\pi}^{\pi} f(\theta)\sin n\theta\, d\theta,$$

we assume the boundary function $f(\theta)$ is piecewise $C^1[-\pi, \pi]$ and continuous. By the maximum principle

$$|S_M(r, \theta) - S_N(r, \theta)| \le \max_{-\pi \le \theta \le \pi} |S_M(1, \theta) - S_N(1, \theta)|.$$

Because $S_N(1, \theta) \xrightarrow{\text{unif}} f$ by the Dini test, $S_N(r, \theta)$ converges uniformly to its limit, called $u(r, \theta)$, for all $r \le 1$.

For termwise differentiation let $c_n = na_n r^{n-1}\cos n\theta$. Then

$$|c_n| \le \frac{n}{\pi}\int_{-\pi}^{\pi}|f(\theta)|\, d\theta r^{n-1} \le M_n$$

where $M_n = (n/\pi)\int_{-\pi}^{\pi}|f(\theta)|\, d\theta\,(1-\varepsilon)^{n-1}$ for all $r < 1 - \varepsilon$. The ratio $M_{n+1}/M_n = ((n+1)/n)(1-\varepsilon) \to (1-\epsilon)$, $n \to \infty$. Hence $\Sigma\, M_n < \infty$ by the ratio test and $\Sigma\, nr^{n-1} a_n \cos n\theta$ converges uniformly for $-\pi \le \theta \le \pi$, $0 \le r \le 1 - \varepsilon$ for any positive ε. Similarly $\Sigma\, nr^{n-1} b_n \sin n\theta$ so converges. Consequently $\Sigma^\infty nr^{n-1}(a_n \cos n\theta + b_n \sin n\theta)$ equals $u_r(r, \theta)$ for any $r < 1$. Likewise $u_{rr}(r, \theta)$, u_θ, and $u_{\theta\theta}$ equal their termwise differentiation series in Ω.

4. With P, Q, Ω_ε, etc. as in Section 1.6.1, we first establish that the fundamental singularity $-\ln r_{PQ}$ is harmonic for $Q \ne P$. This is seen from $P = (0, 0)$ first. In this case $-\ln r = -\frac{1}{2}\ln(x^2 + y^2)$ from which $(-\ln r)_{xx} = (x^2 - y^2)/(x^2 + y^2)^2$ and $(-\ln r)_{yy} = (y^2 - x^2)/(x^2 + y^2)^2$. Change of variable and the chain rule establish the general result.

Remembering Green's second identity for the region Ω_ε,

$$\int_{\Omega_\varepsilon} u\Delta v - \int_{\Omega_\varepsilon} v\Delta u = \oint_{\partial\Omega} u\frac{\partial v}{\partial n} - \oint_{\partial\Omega} v\frac{\partial u}{\partial n} + \oint_{S_\varepsilon} u\frac{\partial v}{\partial n} - \oint_{S_\varepsilon} v\frac{\partial u}{\partial n},$$

we consider the terms one by one when v is taken as $v = -(2\pi)^{-1}\ln r_{PQ}$. The first term $\int_{\Omega_\varepsilon} u\Delta v = 0$ and hence its limit as $\varepsilon \to 0$ vanishes also. The second $\int_{\Omega_\varepsilon} v\Delta u$ is regular and converges for $u \in C^2(\Omega)$ as $\varepsilon \to 0$ because the integration element dA is of the form $dA = r dr d\theta$, so that $-r_{PQ}\ln r_{PQ}$ is bounded at P. The third and fourth terms are independent of ε. The fifth term involves, putting in all the details this time,

$$\frac{\partial v}{\partial n_Q} = \lim_{t \to 0}\frac{v_{P,Q+tn_Q} - v_{P,Q}}{t}$$

$$= \frac{1}{2\pi}\lim_{t \to 0}\ln(r_{PQ}) - \ln(r_{P,Q+tn_Q})$$

$$= \frac{1}{2\pi} \lim_{t \to 0} \frac{\ln(r_{PQ}) - \ln(r_{PQ} - t)}{t}$$

$$= -\frac{1}{2\pi} \lim_{t \to 0} \frac{\ln((r_{PQ} - t)/r_{PQ})}{t}$$

$$= -\frac{1}{2\pi} \lim_{t \to 0} (r_{PQ}/r_{PQ} - t)(-1/r_{PQ}) = \frac{1}{2\pi}\left(\frac{1}{r_{PQ}}\right),$$

by l'hospitals rule. Hence the fifth term

$$\oint_{S_\varepsilon} u \frac{\partial v}{\partial n_Q} = \frac{1}{2\pi\varepsilon} \oint_{S_\varepsilon} u \, ds = \frac{1}{2\pi\varepsilon} u(Q_\varepsilon) \cdot 2\pi\varepsilon = u(Q_\varepsilon),$$

the existence of a Q_ε guaranteed by the calculus mean value theorem applied to the interval $[-\pi, \pi]$. By continuity of u, $u(Q_\varepsilon) \to u(P)$ as $\varepsilon \to 0$. The sixth term vanishes because on the S_ε circle

$$\oint_{S_\varepsilon} v \frac{\partial u}{\partial n} = -\frac{1}{2\pi} \ln\varepsilon \oint_{S_\varepsilon} \frac{\partial u}{\partial n_Q} = -\frac{\ln\varepsilon}{2\pi} \frac{\partial u}{\partial n}(Q_\varepsilon) \cdot 2\pi\varepsilon \to 0$$

as $\varepsilon \to 0$. Here we used the mean value theorem together with $u \in C^1(\Omega)$, and then l'hospitals rule.

5. (a) A square of side length n has eigenvalues $(l^2 + m^2)\pi^2/n^2$, $l, m = 1, 2, \ldots$. Letting u be an eigenfunction for Ω_n, we may extend u by defining it to be 0 on all other squares. Thus u remains an eigenfunction for Ω with the same eigenvalue. On the other hand, any eigenfunction u of all of Ω must be nonzero on some Ω_n, and restricted to that Ω_n, is a local eigenfunction there. Hence the set of all eigenvalues is the set of numbers $\{(l^2 + m^2)\pi^2/n^2: l, m, n \text{ are positive integers}\}$.

   (b) There is no smallest eigenvalue, hence 0 is in the spectrum of the operator. This should raise a doubt about a variational characterization of the eigenvalues.

6. Let $u(P_0) > 0$, put a ball B_{r_0} about P_0, then in polar coordinates, using Schwarz's inequality

$$\int_\Omega u^2 \geq \int_{B_{r_0}} u^2 = \int_0^{r_0} \oint_{\partial B_r} u^2 \, dr \geq \int_0^{r_0} \left[\oint_{\partial B_r} u\right]^2 \cdot \frac{dr}{l(\partial B_r)}$$

where $l(\partial B_r)$ denotes the length or surface measure of the boundary of B_r, this quantity being ωr^{n-1}, see Section 1.6.1. By the mean value theorem applied to the ball B_r we know that $\oint_{\partial B_r} u = u(P_0)\omega_n r^{n-1}$. Hence

$$\int_\Omega u^2 \geq u^2(P_0)\omega_n \int_0^{r_0} r^{n-1} \, dr = u^2(P_0)\omega_n r_0^n/n$$

which exceeds any bound by taking r_0 sufficiently large.

7. See Figure 2.9h. For a Dirichlet boundary condition, extending the initial position f oddly about the boundary will guarantee that the value of the solution $u(x, t)$ will stay 0 on the boundary when d'Alembert's formula is used. This gives rise to "ray-tracing" diagrams such as that below to keep track of the correct extensions of f. Similarly for the Neumann boundary condition, one extends f oddly

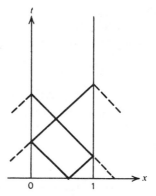

about the boundary points. Data g and F may be similarly handled. This method is useful theoretically but is not so practical unless one is only interested in one or a few points $u(x_0, t_0)$.

8. The equation $u_{xx} + 4xu = 16$ becomes

$$u_{zz} - zu = 2^{8/3}$$

under the variable change $z = \alpha x$ with $\alpha = (-4)^{1/3}$. Thus its solution may be expressed in terms of the Airy functions $Ai((-4)^{1/3}x)$ and $Bi((-4)^{1/3}x)$. Not all such linear second-order equations have known special function solutions, and not all have been solved.*

9. (a) The helicoid is given by $u(x, y) = \tan^{-1}(y/x)$ and the catenoid by $u = \cosh^{-1}((x^2 + y^2)^{1/2})$. For the helicoid from the quotient rule

$$u_x = -y/(x^2 + y^2)$$
$$u_{xx} = 2xy/(x^2 + y^2)^2$$

and so on to the verification. Similarly for the catenoid, which is a bit more involved.

(c) The suggested "additively separated" solution yields

$$\frac{X''}{1 + (X')^2} = -\frac{Y''}{1 + (Y')^2} = c$$

from which $X(x) = -c^{-1} \ln|\cos cx| + c_2$ and $Y(y) = c^{-1} \ln|\cos cy| + c_3$ and hence

$$u(x, y) = c^{-1} \ln \left| \frac{\cos y}{\cos x} \right|.$$

---

* The expression $u_{xx} + 4xu$ was just written down randomly in class and only afterward seen to be of Airy's type.

We left out some constants of integration. Notice that this solution exists on a checkerboard domain in the plane.*

(d) For the Born–Infeld equation it works, giving

$$\frac{X''}{1 + (X')^2} = \frac{T''}{1 - (T')^2} = c,$$

which may be integrated similarly to the above. For the slicing equation it produces only a special case of the quadratic solutions.

APPENDIX PROBLEMS

A.1. (a)

$$u(x) = u_0 \exp\left(-\int_{x_0}^{x} p(s) \, ds\right)$$

$$+ \exp\left(-\int_{x_0}^{x} p(s) \, ds\right) \int_{x_0}^{x} \exp\left(\int_{x_0}^{s} p(\tau) \, d\tau\right) q(s) \, ds.$$

(b) The solution would have to be $u(x) = x + u_0$ but then $u'(0) \neq 0$.

(c) Let $u(x) = a_0 + a_1 x + a_2 x^2 + \cdots$, then $a_0 = 0, a_1 = 1, a_2 = 0, \ldots$.

A.2. (a) $u(x, y)$ remains a solution wherever defined, i.e., except at $y = 2$.

(b) (i) $a(y_0(t))_t - b(x_0(t))_t = t \cdot 1 - 1 \cdot t = 0$. l is characteristic because the ratio

$$\frac{dx_0(t)}{dt} : \frac{dy_0(t)}{dt} : \frac{du_0(t)}{dt}$$

agrees with that of the characteristic direction numbers $a : b : c$. One may now proceed to the general solution in a number of ways, but not by integrating the ordinary differential equations. Note that $uu_x = (\frac{1}{2}u^2)_x$, from which $u^2(x, y) = 2x - 2\int^x u_y + w(y)$, w arbitrary but such that $-2\int^x u_y + w(y)$ must vanish on the initial curve.

A solution $u = s + t, y = s + t, x = (s + t)^2$ may be found by solving the ODE initial value problem as in the example preceding this problem. From the discussion following this problem, a complete solution could be determined from the two auxiliary equations

(i) $u_0 p_0 + q_0 = 1$

and

(ii) $u_0' = p_0 x_0' + q_0 y_0'$.

But in this case in which l is characteristic, these two equations are the same, namely,

$$tp_0 + q_0 = 1$$

---

* It carries the name Scherks surface, although we did not know that when we used the trick of additive separation of variables to find it.

and hence there are an infinite number of p_0, q_0 pairs, to generate an infinite number of solutions.

(ii) l is not characteristic but the Jacobian vanishes identically. Integrating the ODE system

$$\begin{cases} dx/ds = u \Rightarrow x = s^2/2 + ts + t^2 \\ dy/ds = 1 \Rightarrow y = s + 2t \\ du/ds = 1 \Rightarrow u = s + t \\ x(0) = t^2 \\ y(0) = 2t \\ u(0) = t \end{cases}$$

yields the surfaces

$$u(x, y) = y/2 \pm \sqrt{x - y^2/4}$$

which indeed satisfy the given partial differential equation outside of l. However, along l, i.e., as one approaches l, we have on the surfaces

$$u_x = \pm \frac{1}{2}(x - y^2/4)^{-1/2} \rightarrow \pm \frac{1}{2}(t^2 - 4t^2/4)^{-1/2}$$

which blows up.

(iii) The interesting point is clearly $t = 2$.

(c) The main point is that integral surfaces $u = u(x, y)$ of equation (1) are themselves generated by constructing characteristic strips from such points x_0, y_0, u_0, p_0, and q_0.

A.3. (a) From the partial differential equation and the strip condition we have the two equations $p_0 q_0 = 2$ and $p_0 + q_0 = 3$, from which results the pair of initial directions $p_0 = 1$, $q_0 = 2$, and $p_0 = 2$, $q_0 = 1$. The Jacobian "noncharacteristic" condition $F_{p_0} y_0' - F_{q_0} x_0' = q_0 - p_0 \neq 0$ is satisfied. The characteristic ordinary differential equations are then (for the first pair $p_0 = 1$, $q_0 = 2$)

$$dx/ds = 2 \Rightarrow x(s, t) = 2s + t,$$
$$dy/ds = 1 \Rightarrow y(s, t) = s + t,$$
$$dy/ds = 4 \Rightarrow u(s, t) = 4s + 3t,$$
$$dp/ds = 0 \Rightarrow p = p_0,$$
$$dq/ds = 0 \Rightarrow q = q_0,$$

the solutions thereof as indicated, and hence $u(x, y) = x + 2y$. For $p_0 = 2$, $q_0 = 1$, $u(x, y) = 2x + y$.

(b) From the two auxiliary equations we have $p_0 = 1$ and $q_0 = t$.

(c) $(p_0, q_0) = (\sin t, \cos t)$ and $(-\sin t, -\cos t)$ and the solution locus is given by $(u \pm 1)^2 = x^2 + y^2$.

A.4. (a) From the classification procedure of Problem 1.9.1 the wave equation invariance requires

$$b^{11} = c_{11}^2 - c_{12}^2 = 1 \qquad \text{(or } -1\text{)}$$

$$b^{12} = c_{11}c_{21} - c_{12}c_{22} = 0$$

$$b^{22} = c_{21}^2 - c_{22}^2 = -1 \qquad \text{(or } 1\text{)}.$$

The first and third equations are hyperbolas. To couple them we use the second equation substituted into the discriminant relation (see Problem 1.9.1)

$$c_{11}^2 c_{22}^2 + c_{12}^2 c_{21}^2 - 2c_{11}c_{22}c_{12}c_{21} = 1,$$

that is,

$$(c_{11}^2 - c_{12}^2)c_{22}^2 + (c_{21}^2 - c_{22}^2)c_{12}^2 = 1,$$

from which, using the first and third equations above,

$$c_{22}^2 - c_{12}^2 = 1.$$

The latter added to the third equation gives the coupling $c_{12} = \pm\, c_{21}$ which when substituted into the second equation gives the additional coupling $c_{11} = \pm\, c_{22}$. Thus the set of possible transformations has been reduced to those given by matrices of the form

$$\begin{bmatrix} a & b \\ \pm b & \pm a \end{bmatrix}$$

with $a^2 - b^2 = \pm\, 1$.

(b) The r equation is missing the u term so let $v = u_r$ and solve the first order equation $v_r - 2r(1 - r^2)^{-1}v = 0$ from which $v = (1 - r^2)^{-1}$ and by integration $u = (1/2)ln((1 + r)/(1 - r))$. Note that this solution can also be expressed in terms of the classification change of variables $\zeta = x + t$, $\eta = x - t$, $u(r) = u(\zeta, \eta) = (1/2)ln(\zeta/\eta)$. The s equation is less interesting with solution $u(s) = e^s = t$, a trivial instance of the fact that all linear functions of x and t are solutions of the wave equation.

(c) Check the four group axioms.

A.5. (a) Upon substitution, $(\alpha x^2/t)u_\eta + 2\alpha u_\eta + (4\alpha^2 x^2/t)u_{\eta\eta} = 0$, where we have multiplied through by $-t$. This is a first-order equation in u_η, from which $u_\eta = v = \eta^{-1/2}e^{-\eta/4\alpha}$.

(b) Instead of using the similarity variable $\eta = x^2/t$, its square root $\eta = x/\sqrt{t}$ yields the ordinary differential equation

$$g''(\eta) + \frac{\eta}{2}g'(\eta) = 0$$

for any solution $u(x, t) = g(\eta)$. Thus

$$g'(\eta) = c_1 e^{-\int \eta/2} = c_1 e^{-\eta^2/4}$$

and

$$g(\eta) = c_1 \int_{\eta_0}^{\eta} e^{-s^2/4}\, dx + c_2.$$

(c) Chain rule gives the equation. When $m = 1$ it is the equation of (b) above.

A.6. (a) Setting $p_x = 0$ the Prandtl momentum equation becomes $U_\infty^2/L \sim \nu U_\infty/\delta^2$ from which $\delta^{-2} = U_\infty L/\nu$.

(b) One also obtains $uY_\psi = 1$ and $uY_x = V$ on streamlines. An interpretation is that in the limit a boundary layer is just a streamline.

(c) See further work of the British school.*

A.7. (a) See an ODE book. Essentially, with Δx denoting $\bar{x} - x$, Δy denoting $\bar{y} - y$, and Δu denoting $\bar{u} - u$, the variable change is

$$\frac{\Delta x}{X} = \frac{\Delta y}{Y} = \frac{\Delta u}{U} = \varepsilon = ds$$

in the limit as the $0(\varepsilon^2)$ contributions disappear.

(b) Prolongations are mentioned in the sequel. The $\varepsilon = ds$ in part (a) above is a valuable clue in connecting the group infinitesimal arguments to those for first-order equations in Appendix 1.

(c) This is separable and hence in principle always yields a similarity variable $\eta(x, y)$. For example when $X(x) = x$ and $Y(y) = y$ the solution is $x/y = c$ which we set equal to the similarity independent variable, $\eta = c$.

A.8. (a) This can be very tedious with much room for error. Computer codes using symbolic manipulators such as MACSYMA have been developed for this purpose.†

(b) The quadratic term $h'h$ hints at another solution. Indeed we may factor the equation into

$$(\eta^4 h + 1)(\eta h' + 2h) = 0.$$

The first factor yields the previously found solution, $f_1(\eta) = \eta^{-3}/3 + c_1$, whereas the second factor yields a new solution $f_2(\eta) = c_2\eta^{-1} + c_3$, where $c_3 = -c$. As a solution to the transonic flow equation, the latter is not so interesting as it yields $u = c_2 y + c_3 x$, already a knowable solution.

(c) Alternately, do the heat equation.

A.9. (a) $V_1 = \dfrac{\partial}{\partial x}$, $V_2 = \dfrac{\partial}{\partial y}$, $V_3 = \dfrac{\partial}{\partial u}$, $V_4 = \dfrac{3}{2}x\dfrac{\partial}{\partial x} + y\dfrac{\partial}{\partial y}$, $V_5 = -\dfrac{x}{2}\dfrac{\partial}{\partial x} + u\dfrac{\partial}{\partial u}$, $V_6 = x\dfrac{\partial}{\partial u}$.

(b) Overregularity, in the sense introduced in Section 2.2. To further support this observation, which we advance here, note that one is in the above theory usually treating homogeneous (no data) equations. The functional analytic domain $D(L)$ of a differential equation $L(u) = f$ considered for all data f limits the general regularity of u to accommodate all such data functions f in the range of the operator. For a given particular data f and especially for homogeneous (data = 0) problems the solutions will be much more regular. In

* F. T. Smith, "On the high Reynolds number theory of laminar flows," *IMA J. of A. Math.* **28** (1982).

† See P. Rosenau and J. Schwarzmeier, *Courant Inst. Math. Sci. Report* C00-3077-160/MF-94 (1979).

practice many of the generalized symmetries may therefore be viewed as obtained by redifferentiating the equation one or many times.

(c) For the treatment of hyperbolic equations in conservation law form*

$$u_t + (f(u))_x = 0$$

the Lax-Wendroff schemes depend essentially on the redifferentiating of the given equation according to

$$u_n = -(f(u))_{xt} = -(f(u)_t)_x = -(A(u) u_t)_x = -(A(u)(f(u))_x)_x$$

where the matrix $A(u)$ denotes the Jacobian of $f(u)$ with respect to u. This observation supports that of (b).

FIFTH PAUSE

1. The solution $u(x, t)$ is the shock moving to the right with speed c.

2. The '*Material or Substantial Derivative*'

$$\frac{D}{Dt} = \frac{\partial}{\partial t} + u \cdot \nabla$$

is defined in mechanics to represent '*the derivative following the motion*', For example, in a fluid, it is the derivative of the fluid as computed by an observer 'floating downstream' with the fluid. Thus $u_t + uu_x = 0$ means that there is no relative acceleration going on. D/Dt is also called the *particle derivative*. The $\partial/\partial t$ measures the local rate of change of something at a point x, and the $u \cdot \nabla$ measures the *convected* (by u) rate of change due to transport of materials to or away from x. Thus $u_t + uu_x = 0$ means that the local change u_t equals the loss $- uu_x$ due to *convection*.

A better name would be *Euler equation*, as truncated from the Euler equations of gas and fluid dynamics, but *Euler* has been overused. Recently it has been fashionable to call $u_t + uu_x = 0$ in the conservation form $u_t + (u^2/2)_x = 0$ *Burger's equation*, but Burger's interest was definitely in having a viscosity term in the equation, e.g. , $u_t + uu_x + vu_{xx} = 0$.

A good name for the *advection* equation would be *normal* equation, from the sense apparent in $(u_t, u_x) \perp (1, c)$, as illustrated in Fig. 5Pb. Recall that the tangential derivative $\partial u/\partial \tau$ of u along the characteristic is just grad u dotted with that τ direction, i.e., $\partial u/\partial \tau = u_t + cu_x$. Thus $u_t + cu_x = 0$ says that there is no change in the tangential direction, all change is in the *normal* direction, the gradient of u is all in the normal direction, which therefore may be thought of as the direction of steepest (and only) descent**. Further, within the theory of first order systems of PDE's, there are *normal* equations and *normal* forms not inconsistent with calling $u_t + cu_x = 0$ the normal equation. But the name *normal equation* is overused, and already in many contexts.

---

* That is, in divergence form.
** Skiers who want to win ski the *fall line*.

A better name would be the *characteristics equation*, but that has already been reserved for the ODE $dx/dt = c$. Indeed, all of these three equations $u_t + cu_x = 0$ could have been called characteristics equations, indeed the convective derivative $D/DT = \partial/\partial t + c\partial/\partial x$ could be called the *characteristic derivative*, as a directional derivative in characteristic directions. But as we noted at the beginning of Appendix A, the word *characteristics* is enough used.

A very good name for all of these equations of the form

$$u_t + c(u, \text{other})u_x = g$$

is: *advection* equations. Advection connotes to *transport, to carry, convey*. The "other" means other variables but no derivatives, as, for example, in the vorticity transport equation $\omega_t + v\omega_x = v\omega_{xx}$ where $c=v$, a velocity which drives the advection of the vorticity ω, modulo the viscous righthand side. The "g" just means data, i.e., $g= g(x, t)$, so that we do not just end up with the full quasilinear equation of Appendix A.1. The name *transport* equations is equally good, but already much used in other contexts.

3. The variable correspondence is $y =t$, and let τ be the initial curve parameter to avoid confusion. The point is that Appendix A.1 is more general by treating also inhomogeneous equations. The latter become more important in Pause 6.

4. *Shock dynamics* is a subject in itself.* To keep Pause 5 just a pause, we have restricted our discussion of shock dynamics to this one Exercise. Some of the notions we give here will be needed to understand the numerical *shock capturing* schemes discussed in Pause 6.

As illustrated in Fig. 5Pc, any decreasing portion of the initial profile $u(x, 0)$ for the convection equation induces a shock shortly thereafter downstream. More precisely, at the point (x, t) at which the characteristics meet, one encounters a conflict between two possible solution values. More accurately, one has two initial values in conflict.**

How to uniquely continue the solution downstream beyond such a point introduces the notions of jump (Rankine-Hugoniot) and entropy criteria. Historically these were motivated by physical gas dynamic considerations (e.g., the jump relation $u_l/u_r = \rho_r/\rho_l$ between velocity and density across a shock, and that entropy must increase at a shock). Here we will motivate both notions by the idea of regularity: trying to keep the solution as regular as possible.

Turning first to the case of u_l (left value) $> u_r$ (right value) as in Fig. 5Pc, we give up the idea of a classical (e.g., C^1 i.e., continuously differentiable, which we assumed in the chain rule) solution and accept a weak solution. Recall (Pause 4) that weak solutions are really very good, except at certain points or curves, here to be the shock curve. Our object now is to assume a weak solution regularity condition and find the shock curve.

Proceeding exactly as in the last Example of Pause 4, we ask that a weak solution u be at least piecewise C^1, and multiply the conservation equation $u_t + (f(u))_x = 0$ by an

---

*See the book of Whitham referred to in problem 2.9.8. Another excellent and more recent treatment may be found in J. Smoller, Shock Waves and Reaction-Diffusion Equations (Springer, New York, 1983.)

** More correctly, one is paying the price for dropping the viscous terms in the fluid equations. In physical reality, the solution has a very steep but continuous shock. Here one finds quite opposing views by experts who prefer either the Euler equations or the full Navier-Stokes equations (with viscosity).

Many paradoxes, blowups, discontinuities found in solutions of model equations, are eventually understood as having been introduced by the modelers themselves. On the other hand, until one can resolve the full equations, the simpler model must be used. And, it must be admitted, a discontinuous shock front quite graphically depicts a physical effect.

arbitrary cutoff test function ϕ in C_0^1, all such ϕ assumed to be zero outside and one just inside a small neighborhood B of the shock point (x,t) in question. Assuming that the only discontinuity of u occurs on a shock curve $x(t)$ splitting B into a left portion B_l and a right portion B_r we have by the divergence theorem

$$\iint_B (\phi_t\, u + \phi_x f(u)) = -\iint_{B_l} (u_t + (f(u))_x\, \phi - \iint_{B_r} (u_t + (f(u)_x)\phi$$

$$+ \oint_{x(t)} (u^l\, n_t^l + f(u^l) n_x^l)\, \phi + \oint_{x(t)} (u^r\, n_t^r + f(u^r) n_x^r)\, \phi$$

where (n_t^l, n_x^l) is the left unit outer normal to $x(t)$, similarly (n_t^r, n_x^r) the right unit outer normal to the curve $x(t)$. But the first B integral vanishes for all such ϕ, by the definition of weak solutions, and the B_l and B_r integrals vanish since u is a classical solution on those pieces. Thus the total interface $x(t)$ must vanish. Now remember (draw it) that the variable order is t, x, that $n^l = (-\,dx, dt)$ suitably normalized, and $n^r = (dx, -\,dt)$. Thus

$$u^l\,(-\,dx) + u^r\,(dx) + f(u^l)dt + f(u^r)\,(-\,dt) = 0$$

all along the shock curve $x(t)$, because ϕ was arbitrary. This is the jump condition, usually written

$$s[u] = [f(u)]$$

where $[u] = u_l - u_r$, $[f(u)] = f(u_l) - f(u_r)$, and $s = dx/dt$. Thus we have found the jumps in u and $f(u)$ required for the regularity of a weak solution, and the required shock speed s. Moreover, we have found the shock itself: the curve $x(t)$ of slope

$$\frac{dx}{dt} = \frac{f(u^l) - f(u^r)}{u^l - u^r}$$

in terms of the left and right solution values.

As an important example, for the convection equation $u_t + (u^2/2)_x = 0$ the shock curve is given by

$$\frac{dx}{dt} = \frac{(u_l^2 - u_r^2)/2}{u_l - u_r} = \frac{u_l + u_r}{2}$$

the average of the slopes of the two characteristics colliding to form the shock. It is worth drawing this, to see that the shock bisects upward through the parallelograms formed by the intersecting characteristic grids.

This takes care of the case when $u_l > u_r$ the shock case. When $u_r > u_l$ the characteristics fan out rather than in. How do we fill the gap downstream, not provided for by propagating the initial data now? There are many ways we could interpolate, and in fact a little fooling around with the convection equation shows that there can be an infinite number of interpolations that are actually weak (e.g.,piecewise C^1) solutions. As mentioned above, here in the literature one usually finds a discussion of entropy conditions: entropy must increase across a shock, which happens when $u_r < u_l$. But in the present case of $u_r > u_l$, a shock would decrease entropy. So one wishes, on physical grounds, to avoid a shock in this case. The solution choice is again guided by the gas dynamics: one selects the fan-like solution called a *rarefaction wave*. These are solutions $u(x, t) = u(x/t)$ depending only on the ratio $\eta = x/t$, or, if the initial point is not $(0,0)$, then $u(x,t) = u((x - x_0)/(t - t_0))$. These are also called *centered simple waves*. It can be shown that they satisfy various conditions motivated by entropy considerations, such as the *entropy inequality*

$$f'(u_l) \leq f'(u(\eta)) \leq f'(u_r).$$

They could also be called *fan solutions*, or *ray solutions*. By definition they are constant on the rays x/t = constant.

Let's see what this means for the important example, the convection equation $u_t + (u^2/2)_x = 0$. Let the initial point be the origin $(0,0)$, let the initial profile be $u_l(x, 0) = 0$ for $x < 0$, $u_r(x, 0) = 1$ for $x > 0$. Then the characteristics $x(t)$ go straight up to the left of the t-axis, they go right at a $45°$ angle to the right of the t-axis. We fill the gap with the rarefaction wave

$$u(x, t) = x/t, \quad 0 < x < t.$$

It is easily seen to be a solution, for all x and $t > 0$:

$$u_t + uu_x = -x/t^2 + (x/t)(1/t) = 0$$

It forms, when joined to the constant solutions 0 to the left of the t-axis and 1 to the right of the $45°$ ray, a continuous weak solution $u(x,t)$ for all $x,t > 0$. Since $f'(u) = u$ for the convection equation, it satisfies the entropy inequality.

This is all we will say about *shock dynamics*. Our point of view was a little different than some: we stress that the preferred solution should be determined by making it as regular as possible. By so doing, the best approximation to (viscous) reality should obtain. Consider the convection equation example. From this regularity point of view, the *jump* condition should be thought of as "defining" a solution $u(x,t)$ exactly on the shock: let $u = (u_l + u_r)/2$ The shock curve $x(t)$ can then be regarded as an induced characteristic determined from $dx(t)/dt = u$ initially. Such an assigned u gives the "most regular" derivative at the discontinuity: a vertical drop from u_l through mid-value $\bar{u}$ to u_r. Similarly, the *entropy* requirements may be thought of as "smoothing" a solution $u(x,t)$ across a "characteristic gap" in the best (lowest order, nonoscillatory) physical way: the *rarefaction* fan. This linear *monotone* classical solution interpolates a discontinuity $u_l < u_r$ to induce a shock-free solution not only at the discontinuity but also as far downstream as possible. The rarefaction singularity at its vertex should be thought of as "defining" a "vertical" solution $u(x,t)$ exactly at the discontinuity: let $\bar{u}(0, 0, x/t) = x/t$ there. This vertical solution connects u_l to u_r and slants continuously into the fan as one departs the discontinuity. The fan lines $x(t)$ can then be regarded as a continuous family of induced characteristics determined from $dx(t)/dt = u$ initially. Such an assigned u initiates the rarefaction fan and provides the "most regular" solution u at the discontinuity. In both instances, the shock and fan speeds are intermediate to and best average the characteristic speeds at both sides of the discontinuity.

5.   The constant solutions $u(x,t) = u_0$ along the characteristics $\eta = x - ct = x_0$ are similarity solutions. The solution u to the translation equation is unchanged under translations in the characteristic $(1, c)$ direction, illustrating invariance under that group action. The rarefaction fans (see answer to Exercise 4) are both similarity solutions and also invariant under the dialation group $x \to \alpha x, t \to \alpha t$.

6.   A matrix A is diagonalizable $Q^{-1} A Q = D$ implies that the columns of Q are in fact the eigenvectors corresponding to the λ_i ; and vice versa. Thus the eigenvectors v_j of A^* are the rows of Q^{-1} seen from $Q^{-1}AQ = D \Rightarrow (Q^{-1}AQ)^* = D^* = D$ $= Q*A*(Q*)^{-1}$. The eigenvectors of A and those of A^* are sometimes called the right and left eigenvectors of A, respectively. In Pause 6 it will be interesting to use the fact that for a real diagonalizable $n \times n$ matrix A, the right eigenvectors r_i and the left eigenvectors l_i are *bi-orthogonal*:$(r_i, l_j) = 0$ if $i \neq j$ $(r_i, l_i) \neq 0$. This is easily seen from the relation $Q^{-1} Q = I$.

7. $\lambda = c, u$ and $c(u)$, respectively, with eigenvector $v = 1$.

8.   From the conservation law system $u_t + (f(u))_x = 0$, we have by the divergence theorem that

$$\frac{d}{dt}\int_\Omega u \, dx + \oint_{\partial\Omega} f(u) . n = 0$$

where Ω is some region of interest. This gives a general meaning to all such systems: the variation of the average of u is the total flux through the boundary of Ω The region Ω may be the total physical region, or, of numerical interest, some control volume or discretization volume.

9.   The eigenvalues of the matrix A are easily found:

$$\det (A - \lambda I) = \begin{vmatrix} (u - \lambda) & \rho & 0 \\ 0 & (u - \lambda) & 1/\rho \\ 0 & \gamma p & (u - \lambda) \end{vmatrix} = (u - \lambda)((u - \lambda)^2 - \gamma p/\rho) \cdot$$

from which $\lambda_1 = u - c, \lambda_2 = u, \lambda_3 = u + c$,

where $c = (\gamma p/\rho)^{1/2}$. The corresponding eigenvectors of A are $(1, - c/\rho, c^2), (1, 0, 0)$, and $(\rho/c, 1, \rho c)$, and those of A^* being $(0, 1, -1/\rho c), (1, 0, -1/c^2)$, and $(0, 1, 1/\rho c)$. Using the latter eigenvectors we have the decomposition to ODE's

$$\frac{du}{dt} - \frac{1}{\rho c}\frac{dp}{dt} = 0 \text{ on 1st characteristic (slope } u - c \text{).}$$

$$\frac{d\rho}{dt} - \frac{1}{c^2}\frac{dp}{dt} = 0 \text{ on 2nd characteristic (slope } u \text{)}$$

$$\frac{du}{dt} + \frac{1}{\rho c}\frac{dp}{dt} = 0 \text{ on 3rd characteristic (slope } u + c \text{)}$$

B.1 (a) $u(x, t + k) = u(x, t) + ku_t(x, t) + O(k^2)$ and $u(x \pm h, t) = u(x, t) \pm hu_x(x, t) + h^2 u_{xx}(x, t)/2 \pm h^3 u_{xxx}(x, t)/6 + O(h^4)$, which upon addition gives

$$\frac{u(x + h, t) - 2u(x, t) + u(x - h, t)}{h^2} + O(h^2)$$

$$= \frac{u(x, t + k) - u(x, t)}{k} + O(k)$$

(b)   The stencil below produces an $O(h^4)$ approximation to $\Delta u = 0$.

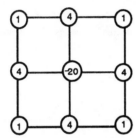

(c)   Tridiagonal, pentadiagonal, and other such constant diagonal matrices are studied under the theory of Toeplitz matrices.

B.2. (a) $U_{1, 1} = 5/32 = U_{3, 1}$, $U_{2, 1} = 7/32$, from which $U_{2, 2} = 3/16$.

(b)   Clearing fractions,

$$\begin{bmatrix} 6 & -1 & 0 \\ -1 & 6 & -1 \\ 0 & -1 & 6 \end{bmatrix} \begin{bmatrix} U_{1, 1} \\ U_{2, 1} \\ U_{3, 1} \end{bmatrix} = \begin{bmatrix} 2 & 1 & 0 \\ 1 & 2 & 1 \\ 0 & 1 & 2 \end{bmatrix} \begin{bmatrix} U_{1, 0} \\ U_{2, 0} \\ U_{3, 0} \end{bmatrix},$$

from which $(U_{1, 1}, U_{2, 1}, U_{3, 1}) = (11/17, 15/17, 11/17)$.

(c)   The same A and B as in (b) may be used, hence

$$\begin{bmatrix} 6 & -1 & 0 \\ -1 & 6 & -1 \\ 0 & -1 & 6 \end{bmatrix} \begin{bmatrix} U_{1, 1} \\ U_{2, 1} \\ U_{3, 1} \end{bmatrix} = \begin{bmatrix} 2 & 1 & 0 \\ 1 & 2 & 1 \\ 0 & 1 & 2 \end{bmatrix} \begin{bmatrix} \sqrt{2}/2 \\ 1 \\ \sqrt{2}/2 \end{bmatrix},$$

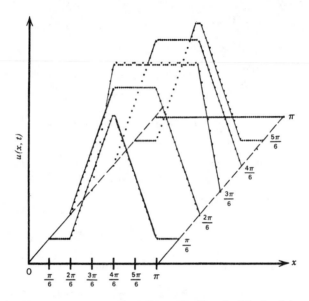

Solutions, Problem B.3(a). Numerical solutions of Problem 1 of Section 2.1.

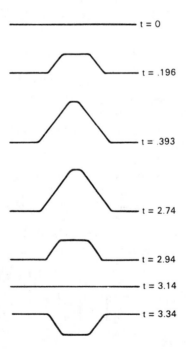

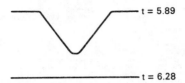

from which

$$\begin{bmatrix} U_{1,1} \\ U_{2,1} \\ U_{3,1} \end{bmatrix} = \begin{bmatrix} (7\sqrt{2} + 8)/34 \\ (4\sqrt{2} + 7)/17 \\ (7\sqrt{2} + 8)/34 \end{bmatrix} \cong \begin{bmatrix} 0.526 \\ 0.745 \\ 0.526 \end{bmatrix}$$

The analytic solution is $u(x, t) = e^{-t} \sin x$, giving

$$\begin{bmatrix} u(\pi/4, \pi^2/32) \\ u(\pi/2, \pi^2/32) \\ u(3\pi/4, \pi^2/32) \end{bmatrix} \cong \begin{bmatrix} 0.519 \\ 0.735 \\ 0.519 \end{bmatrix}$$

for a relative error of about 1.35%.

B.3. (a) Here are plots of the outputs of two different runs. The first extends only out to $T = \pi$ whereas the second shows sections at larger times. Notice the close agreement to the data spreading expected from the d'Alembert solution.

(b) Here is a code and output for arbitrary initial data and when the solution with UINIT $\equiv 1$, $h = 0.1$, $k = 0.01$. See the next pages.

(c) The analytical solution by separation of variables is seen to be

$$u(x, t) = \frac{4}{\pi} \sum_{n=0}^{\infty} \frac{1}{(2n + 1)} e^{-(2n + 1)^2 \pi^2 t} \sin (2n + 1)\pi x.$$

In Table B.3 we compare $u(0.1, t)$ to those obtained first by an explicit Euler run and then to the Crank–Nicholson run of this problem.

Table B.3.

Time	Analytical solution	Explicit Euler	Error (%)	Implicit C-N	Error (%)
0.001	0.9747	0.9000	7.7	0.9089	6.8
0.002	0.8862	0.8200	7.5	0.8337	5.9
0.005	0.6827	0.6566	3.8	0.6731	1.4
0.010	0.5205	0.5113	0.9	0.5234	−0.6
0.050	0.2442	0.2429	0.5	0.2443	negligible
0.100	0.1467	0.1460	0.4	0.1467	negligible
0.200	0.0547	0.0546	negligible	0.0551	−0.7

Note that Crank–Nicholson converged more quickly but then began to slightly overestimate the solution near $t = 0.200$.

UNIVERSITY OF MINNESOTA FORTRAN COMPILER (VERSION 5.4 - 79/03/01) ON THE 6400 UNDER
KRONOS 2.1.0 ON 81/09/30 AT 10.34 UNIVERSITY COMPUTING CENTER - UNIVERSITY OF COLORADO

```
                    MNF(I=CRANK1,PL,BL)

                C   ********************************************************************
                C   * THIS PROGRAM CALCULATES THE SOLUTION OF A PARABOLIC DIFFERENTIAL *
                C   * EQUATION IN A UNIT SPACE DOMAIN HAVING ZERO END CONDITIONS. THE   *
                C   * CALCULATIONS ARE PERFORMED BY MEANS OF A CRANK-NICOLSON FINITE    *
                C   * DIFFERENCE SCHEME. THE INITIAL CONDITION (AT TIME=0) IS SPECIFIED *
                C   * BY A FUNCTION (UINIT) WHICH CAN BE APPENDED TO THE MAIN PROGRAM.  *
                C   * THE MAXIMUM NUMBER OF INTERIOR POINTS IS 9. THE SYSTEM OF LINEAR  *
                C   * ALGEBRAIC EQUATIONS IS SOLVED BY GAUSS-JORDAN ELIMINATION METHOD  *
                C   ********************************************************************
                C
                C
  1.    000000B        REAL K
  2.    000000B        DIMENSION C(9,9),X(9),F(9),U(9),D(9)
                C
                C     ...READ DISCRETIZATION DATA
  3.    002131B        READ (5,102) H,K,TMAX
                C
                C     ...SET COEFFICIENTS
  4.    002325B        N=IFIX(1./H-1.)
  5.    002327B        R=K/(H*H)
  6.    002330B        TIME=0.0
  7.    002331B        N1=N-1
  8.    002333B        R2=2.-2.*R
                C
                C     ...SET COEFFICIENTS OF (C) MATRIX
  9.    002336B        DO 10 I=1,N
 10.    002340B          DO 10 J=1,N
 11.    002342B            C(I,J)=0.0
 12.    002342B   10   CONTINUE
 13.    002352B        DO 11 I=1,N1
 14.    002354B          J=I+1
 15.    002354B          C(I,J)=-R
 16.    002357B          C(J,I)=-R
 17.    002362B          C(I,I)=2.+2.*R
 18.    002365B   11   CONTINUE
 19.    002366B        C(N,N)=2.+2.*R
                C
                C     ...SET (X) ARRAY AND INITIALIZE (U)
 20.    002371B        DO 1 I=1,N
 21.    002373B          X(I)=FLOAT(I)*H
 22.    002374B          F(I)=UINIT(X(I))
 23.    002401B          U(I)=F(I)
 24.    002402B   1    CONTINUE
                C
                C     ...PRINT DISCRETIZATION DATA AND COEFFICIENT MATRIX
 25.    002405B        WRITE (6,105)
 26.    002410B        WRITE (6,100) R,H,K,TMAX
 27.    002417B        WRITE (6,106)
 28.    002422B        DO 40 I=1,N
 29.    002424B          WRITE (6,103) (C(I,J),J=1,N)
 30.    002441B   40   CONTINUE
                C
                C     ...PRINT HEADINGS AND INITIAL CONDITION
 31.    002443B        WRITE (6,101) (X(I),I=1,N)
 32.    002454B        WRITE (6,104) TIME,(U(I),I=1,N)
                C
                C     ...LOOP UNTIL DESIRED TIME (TMAX)
 33.    002465B   4    CONTINUE
                C        ...INCREMENT TIME
 34.    002465B          TIME=TIME+K
                C
                C     ...CALCULATE RIGHT-HAND VECTOR (D)
 35.    002466B        DO 2 I=2,N1
 36.    002471B          D(I)=R*(F(I-1)+F(I+1))+R2*F(I)
 37.    002471B   2    CONTINUE
 38.    002500B        D(1)=R2*F(1)+R*F(2)
 39.    002503B        D(N)=R2*F(N)+R*F(N1)
                C
                C     ...CALCULATE SOLUTION VECTOR (U)
 40.    002507B        CALL GAUSS(C,D,U,N)
                C
                C     ...PRINT SOLUTION VECTOR AND TIME
 41.    002513B        WRITE (6,104) TIME,(U(I),I=1,N)
                C
                C     ...RESET VALUES OF (F) FOR NEXT TIME STEP
 42.    002524B        DO 3 I=1,N
 43.    002526B          F(I)=U(I)
 44.    002526B   3    CONTINUE
                C
                C     ...CHECK CURRENT TIME AND STOP
 45.    002532B        IF (TIME.GE.TMAX) STOP
```

```
46.   002535B            GO TO 4
                    C
                    C
                    C      ************* FORMAT STATEMENTS *****************************
                    C
48.   002535B       100 FORMAT (3(/)35X,4HR = ,F4.1,10X,4HH = ,F4.2,10X,4HK = ,
                      1        F5.3,10X,7HTMAX = ,F5.3)
49.   002535B       101 FORMAT (1H1//////28X,5HX =   ,9(F3.1,6X)//)
50.   002535B       102 FORMAT (3F10.0)
51.   002535B       103 FORMAT ((/42X,9(F4.1,2X))
52.   002535B       104 FORMAT ((/10X,7HTIME = ,F5.3,8X,9(F7.4,2X))
53.   002535B       105 FORMAT (1H1,8(/),52X,5(1H*),21H  CALCULATION  DATA  ,5(1H*))
54.   002535B       106 FORMAT (8(/),51X,5(1H*),23H  COEFFICIENT  MATRIX  ,5(1H*)//)
54.   002535B           END




1.    000000B           SUBROUTINE GAUSS(C,B,X,N)
                    C  ***********************************************************
                    C  * SOLVE UP TO 9 SIMULTANEOUS LINEAR ALGEBRAIC EQUATIONS *
                    C  * BY GAUSS-JORDAN ELIMINATION METHOD                    *
                    C  ***********************************************************
                    C
2.    000000B           DIMENSION A(9,9),C(9,9),B(9),X(9)
                    C
                    C      ...RESET LOCAL COEFFICIENT MATRIX (A)
3.    000000B           DO 6 I=1,N
4.    000123B             DO 6 J=1,N
5.    000125B               A(I,J)=C(I,J)
6.    000125B       6 CONTINUE
                    C
                    C      ...PERFORM MANIPULATIONS
7.    000135B           DO 4 I=1,N
8.    000136B             DO 2 K=1,N
9.    000140B               IF (K.EQ.I) GO TO 2
10.   000142B               CONST=-A(K,I)/A(I,I)
11.   000147B               DO 1 J=1,N
12.   000152B                 A(K,J)=A(K,J)+CONST*A(I,J)
13.   000160B                 IF (J.EQ.I) A(K,J)=0.0
14.   000165B       1         CONTINUE
15.   000167B               B(K)=B(K)+CONST*B(I)
16.   000173B       2       CONTINUE
17.   000176B             CONST=A(I,I)
18.   000200B             DO 3 J=1,N
19.   000202B               A(I,J)=A(I,J)/CONST
20.   000202B       3       CONTINUE
21.   000210B             A(I,I)=1.0
22.   000212B             B(I)=B(I)/CONST
23.   000214B       4 CONTINUE
                    C
                    C      ...OBTAIN SOLUTION VECTOR
24.   000217B           DO 5 I=1,N
25.   000220B             X(I)=B(I)
26.   000220B       5 CONTINUE
27.   000224B           RETURN
28.   000226B           END




1.    000000B           FUNCTION UINIT(V)
                    C  ***********************************************************
                    C  * SPECIFY AND CALCULATE INITIAL CONDITION AT MESH POINTS (V) *
                    C  ***********************************************************
                    C
2.    000000B           UINIT=1.0
3.    000001B           RETURN
4.    000004B           END
      CAUTION  -----------------------------------DUMMY ARGUMENT V WAS NEVER USED




               ***** CALCULATION  DATA  *****


      R =  .1        H =  .10        K =  .001        TMAX =  .200
```

***** COEFFICIENT MATRIX *****

2.2	-.1	0	0	0	0	0	0	0
-.1	2.2	-.1	0	0	0	0	0	0
0	-.1	2.2	-.1	0	0	0	0	0
0	0	-.1	2.2	-.1	0	0	0	0
0	0	0	-.1	2.2	-.1	0	0	0
0	0	0	0	-.1	2.2	-.1	0	0
0	0	0	0	0	-.1	2.2	-.1	0
0	0	0	0	0	0	-.1	2.2	-.1
0	0	0	0	0	0	0	-.1	2.2

X =	.1	.2	.3	.4	.5	.6	.7	.8	.9
TIME = 0	1.0000	1.0000	1.0000	1.0000	1.0000	1.0000	1.0000	1.0000	1.0000
TIME = .001	.9089	.9959	.9998	1.0000	1.0000	1.0000	.9998	.9959	.9089
TIME = .002	.8337	.9848	.9990	.9999	1.0000	.9999	.9990	.9848	.8337
TIME = .003	.7709	.9695	.9971	.9998	1.0000	.9998	.9971	.9695	.7709
TIME = .004	.7181	.9514	.9940	.9994	.9999	.9994	.9940	.9514	.7181
TIME = .005	.6731	.9318	.9897	.9987	.9997	.9987	.9897	.9318	.6731
TIME = .006	.6345	.9115	.9843	.9978	.9995	.9978	.9843	.9115	.6345
TIME = .007	.6011	.8912	.9779	.9964	.9990	.9964	.9779	.8912	.6011
TIME = .008	.5719	.8710	.9707	.9946	.9984	.9946	.9707	.8710	.5719
TIME = .009	.5462	.8514	.9628	.9924	.9975	.9924	.9628	.8514	.5462
TIME = .010	.5234	.8323	.9544	.9897	.9963	.9897	.9544	.8323	.5234
TIME = .011	.5031	.8140	.9455	.9866	.9948	.9866	.9455	.8140	.5031
TIME = .012	.4848	.7965	.9363	.9831	.9930	.9831	.9363	.7965	.4848
TIME = .013	.4683	.7797	.9269	.9792	.9909	.9792	.9269	.7797	.4683
TIME = .014	.4533	.7636	.9174	.9750	.9884	.9750	.9174	.7636	.4533
TIME = .015	.4396	.7483	.9077	.9704	.9855	.9704	.9077	.7483	.4399
TIME = .016	.4271	.7338	.8981	.9655	.9823	.9655	.8981	.7338	.4271
TIME = .017	.4155	.7198	.8884	.9603	.9788	.9603	.8884	.7198	.4155
TIME = .018	.4048	.7066	.8788	.9548	.9749	.9548	.8788	.7066	.4048
TIME = .019	.3948	.6939	.8692	.9491	.9708	.9491	.8692	.6939	.3948
TIME = .020	.3856	.6818	.8597	.9432	.9663	.9432	.8597	.6818	.3856
TIME = .021	.3769	.6702	.8503	.9371	.9615	.9371	.8503	.6702	.3769
TIME = .022	.3688	.6591	.8410	.9307	.9565	.9307	.8410	.6591	.3688
TIME = .023	.3612	.6485	.8319	.9243	.9512	.9243	.8319	.6485	.3612
TIME = .024	.3540	.6383	.8228	.9177	.9457	.9177	.8228	.6383	.3540
TIME = .025	.3472	.6285	.8139	.9109	.9400	.9109	.8139	.6285	.3472
TIME = .026	.3408	.6191	.8052	.9041	.9341	.9041	.8052	.6191	.3408
TIME = .027	.3347	.6101	.7965	.8972	.9280	.8972	.7965	.6101	.3347
TIME = .028	.3289	.6013	.7880	.8901	.9218	.8901	.7880	.6013	.3289
TIME = .029	.3234	.5929	.7796	.8831	.9154	.8831	.7796	.5929	.3234

TIME									
= .030	.3181	.5848	.7713	.8759	.9088	.8759	.7713	.5848	.3181
= .031	.3131	.5769	.7632	.8687	.9022	.8687	.7632	.5769	.3131
= .032	.3083	.5693	.7552	.8615	.8955	.8615	.7552	.5693	.3083
= .033	.3036	.5619	.7473	.8543	.8886	.8543	.7473	.5619	.3036
= .034	.2992	.5547	.7395	.8470	.8817	.8470	.7395	.5547	.2992
= .035	.2949	.5477	.7318	.8397	.8747	.8397	.7318	.5477	.2949
= .036	.2908	.5409	.7243	.8324	.8677	.8324	.7243	.5409	.2908
= .037	.2868	.5343	.7168	.8251	.8606	.8251	.7168	.5343	.2868
= .038	.2829	.5279	.7094	.8179	.8535	.8179	.7094	.5279	.2829
= .039	.2792	.5216	.7022	.8106	.8464	.8106	.7022	.5216	.2792
= .040	.2756	.5155	.6950	.8033	.8392	.8033	.6950	.5155	.2756
= .041	.2721	.5095	.6879	.7961	.8320	.7961	.6879	.5095	.2721
= .042	.2687	.5037	.6810	.7889	.8248	.7889	.6810	.5037	.2687
= .043	.2653	.4980	.6741	.7817	.8177	.7817	.6741	.4980	.2653
= .044	.2621	.4924	.6673	.7746	.8105	.7746	.6673	.4924	.2621
= .045	.2590	.4869	.6606	.7674	.8033	.7674	.6606	.4869	.2590
= .046	.2559	.4815	.6539	.7604	.7961	.7604	.6539	.4815	.2559
= .047	.2529	.4763	.6474	.7533	.7890	.7533	.6474	.4763	.2529
= .048	.2500	.4711	.6409	.7463	.7819	.7463	.6409	.4711	.2500
= .049	.2471	.4660	.6345	.7393	.7748	.7393	.6345	.4660	.2471
= .050	.2443	.4610	.6282	.7324	.7677	.7324	.6282	.4610	.2443
= .051	.2416	.4561	.6219	.7255	.7607	.7255	.6219	.4561	.2416
= .052	.2389	.4513	.6157	.7187	.7536	.7187	.6157	.4513	.2389
= .053	.2363	.4465	.6096	.7119	.7467	.7119	.6096	.4465	.2363
= .054	.2337	.4418	.6036	.7052	.7397	.7052	.6036	.4418	.2337
= .055	.2312	.4372	.5976	.6985	.7329	.6985	.5976	.4372	.2312
= .056	.2287	.4327	.5917	.6919	.7260	.6919	.5917	.4327	.2287
= .057	.2262	.4282	.5858	.6853	.7192	.6853	.5858	.4282	.2262
= .058	.2238	.4238	.5800	.6788	.7124	.6788	.5800	.4238	.2238
= .059	.2215	.4195	.5743	.6723	.7057	.6723	.5743	.4195	.2215
= .060	.2191	.4152	.5687	.6659	.6991	.6659	.5687	.4152	.2191
= .061	.2168	.4110	.5631	.6595	.6925	.6595	.5631	.4110	.2168
= .062	.2146	.4068	.5575	.6532	.6859	.6532	.5575	.4068	.2146
= .063	.2124	.4027	.5520	.6469	.6794	.6469	.5520	.4027	.2124
= .064	.2102	.3986	.5466	.6407	.6729	.6407	.5466	.3986	.2102
= .065	.2080	.3946	.5413	.6345	.6665	.6345	.5413	.3946	.2080
= .066	.2059	.3906	.5359	.6284	.6601	.6284	.5359	.3906	.2059
= .067	.2038	.3867	.5307	.6224	.6538	.6224	.5307	.3867	.2038
= .068	.2017	.3828	.5255	.6164	.6475	.6164	.5255	.3828	.2017
= .069	.1997	.3790	.5203	.6104	.6413	.6104	.5203	.3790	.1997
= .070	.1976	.3752	.5152	.6045	.6352	.6045	.5152	.3752	.1976
= .071	.1957	.3715	.5102	.5987	.6291	.5987	.5102	.3715	.1957
= .072	.1937	.3678	.5052	.5929	.6230	.5929	.5052	.3678	.1937
= .073	.1917	.3641	.5003	.5872	.6170	.5872	.5003	.3641	.1917
= .074	.1898	.3605	.4954	.5815	.6111	.5815	.4954	.3605	.1898
= .075	.1879	.3570	.4905	.5759	.6052	.5759	.4905	.3570	.1879

TIME	=	.076	.1860	.3534	.4857	.5703	.5994	.5703	.4857	.3534	.1860
TIME	=	.077	.1842	.3499	.4810	.5648	.5936	.5648	.4810	.3499	.1842
TIME	=	.078	.1824	.3465	.4763	.5593	.5878	.5593	.4763	.3465	.1824
TIME	=	.079	.1805	.3431	.4716	.5539	.5822	.5539	.4716	.3431	.1805
TIME	=	.080	.1788	.3397	.4670	.5485	.5765	.5485	.4670	.3397	.1788
TIME	=	.081	.1770	.3364	.4625	.5432	.5709	.5432	.4625	.3364	.1770
TIME	=	.082	.1752	.3330	.4579	.5379	.5654	.5379	.4579	.3330	.1752
TIME	=	.083	.1735	.3298	.4535	.5327	.5599	.5327	.4535	.3298	.1735
TIME	=	.084	.1718	.3265	.4490	.5275	.5545	.5275	.4490	.3265	.1718
TIME	=	.085	.1701	.3233	.4447	.5224	.5491	.5224	.4447	.3233	.1701
TIME	=	.086	.1684	.3201	.4403	.5173	.5438	.5173	.4403	.3201	.1684
TIME	=	.087	.1668	.3170	.4360	.5123	.5385	.5123	.4360	.3170	.1668
TIME	=	.088	.1651	.3139	.4318	.5073	.5333	.5073	.4318	.3139	.1651
TIME	=	.089	.1635	.3108	.4276	.5024	.5281	.5024	.4276	.3108	.1635
TIME	=	.090	.1619	.3078	.4234	.4975	.5230	.4975	.4234	.3078	.1619
TIME	=	.091	.1603	.3048	.4193	.4927	.5179	.4927	.4193	.3048	.1603
TIME	=	.092	.1587	.3018	.4152	.4879	.5129	.4879	.4152	.3018	.1587
TIME	=	.093	.1572	.2988	.4111	.4831	.5079	.4831	.4111	.2988	.1572
TIME	=	.094	.1556	.2959	.4071	.4784	.5030	.4784	.4071	.2959	.1556
TIME	=	.095	.1541	.2930	.4031	.4738	.4981	.4738	.4031	.2930	.1541
TIME	=	.096	.1526	.2902	.3992	.4692	.4933	.4692	.3992	.2902	.1526
TIME	=	.097	.1511	.2873	.3953	.4646	.4885	.4646	.3953	.2873	.1511
TIME	=	.098	.1496	.2845	.3915	.4601	.4837	.4601	.3915	.2845	.1496
TIME	=	.099	.1481	.2817	.3877	.4556	.4790	.4556	.3877	.2817	.1481
TIME	=	.100	.1467	.2790	.3839	.4512	.4744	.4512	.3839	.2790	.1467
TIME	=	.101	.1453	.2763	.3801	.4468	.4697	.4468	.3801	.2763	.1453
TIME	=	.102	.1438	.2736	.3764	.4424	.4652	.4424	.3764	.2736	.1438
TIME	=	.103	.1424	.2709	.3728	.4381	.4606	.4381	.3728	.2709	.1424
TIME	=	.104	.1410	.2682	.3691	.4339	.4562	.4339	.3691	.2682	.1410
TIME	=	.105	.1397	.2656	.3655	.4296	.4517	.4296	.3655	.2656	.1397
TIME	=	.106	.1383	.2630	.3620	.4255	.4473	.4255	.3620	.2630	.1383
TIME	=	.107	.1370	.2605	.3584	.4213	.4430	.4213	.3584	.2605	.1370
TIME	=	.108	.1356	.2579	.3549	.4172	.4387	.4172	.3549	.2579	.1356
TIME	=	.109	.1343	.2554	.3515	.4132	.4344	.4132	.3515	.2554	.1343
TIME	=	.110	.1330	.2529	.3481	.4091	.4302	.4091	.3481	.2529	.1330
TIME	=	.111	.1317	.2504	.3447	.4051	.4260	.4051	.3447	.2504	.1317
TIME	=	.112	.1304	.2480	.3413	.4012	.4218	.4012	.3413	.2480	.1304
TIME	=	.113	.1291	.2456	.3380	.3973	.4177	.3973	.3380	.2456	.1291
TIME	=	.114	.1279	.2432	.3347	.3934	.4137	.3934	.3347	.2432	.1279
TIME	=	.115	.1266	.2408	.3314	.3896	.4096	.3896	.3314	.2408	.1266
TIME	=	.116	.1254	.2385	.3282	.3858	.4056	.3858	.3282	.2385	.1254
TIME	=	.117	.1242	.2362	.3250	.3820	.4017	.3820	.3250	.2362	.1242
TIME	=	.118	.1229	.2338	.3218	.3783	.3978	.3783	.3218	.2338	.1229
TIME	=	.119	.1217	.2316	.3187	.3746	.3939	.3746	.3187	.2316	.1217
TIME	=	.120	.1206	.2293	.3156	.3710	.3901	.3710	.3156	.2293	.1206
TIME	=	.121	.1194	.2271	.3125	.3674	.3863	.3674	.3125	.2271	.1194

TIME	=	.122	.1182	.2249	.3095	.3638	.3825	.3638	.3095	.2249	.1182
TIME	=	.123	.1171	.2227	.3065	.3603	.3788	.3603	.3065	.2227	.1171
TIME	=	.124	.1159	.2205	.3035	.3567	.3751	.3567	.3035	.2205	.1159
TIME	=	.125	.1148	.2184	.3005	.3533	.3715	.3533	.3005	.2184	.1148
TIME	=	.126	.1137	.2162	.2976	.3498	.3678	.3498	.2976	.2162	.1137
TIME	=	.127	.1126	.2141	.2947	.3464	.3643	.3464	.2947	.2141	.1126
TIME	=	.128	.1115	.2120	.2918	.3431	.3607	.3431	.2918	.2120	.1115
TIME	=	.129	.1104	.2100	.2890	.3397	.3572	.3397	.2890	.2100	.1104
TIME	=	.130	.1093	.2079	.2862	.3364	.3537	.3364	.2862	.2079	.1093
TIME	=	.131	.1082	.2059	.2834	.3331	.3503	.3331	.2834	.2059	.1082
TIME	=	.132	.1072	.2039	.2806	.3299	.3469	.3299	.2806	.2039	.1072
TIME	=	.133	.1061	.2019	.2779	.3267	.3435	.3267	.2779	.2019	.1061
TIME	=	.134	.1051	.1999	.2752	.3235	.3401	.3235	.2752	.1999	.1051
TIME	=	.135	.1041	.1980	.2725	.3203	.3368	.3203	.2725	.1980	.1041
TIME	=	.136	.1031	.1961	.2698	.3172	.3335	.3172	.2698	.1961	.1031
TIME	=	.137	.1021	.1941	.2672	.3141	.3303	.3141	.2672	.1941	.1021
TIME	=	.138	.1011	.1923	.2646	.3111	.3271	.3111	.2646	.1923	.1011
TIME	=	.139	.1001	.1904	.2620	.3080	.3239	.3080	.2620	.1904	.1001
TIME	=	.140	.0991	.1885	.2595	.3050	.3207	.3050	.2595	.1885	.0991
TIME	=	.141	.0982	.1867	.2570	.3021	.3176	.3021	.2570	.1867	.0982
TIME	=	.142	.0972	.1849	.2545	.2991	.3145	.2991	.2545	.1849	.0972
TIME	=	.143	.0962	.1831	.2520	.2962	.3115	.2962	.2520	.1831	.0962
TIME	=	.144	.0953	.1813	.2495	.2933	.3084	.2933	.2495	.1813	.0953
TIME	=	.145	.0944	.1795	.2471	.2905	.3054	.2905	.2471	.1795	.0944
TIME	=	.146	.0935	.1778	.2447	.2876	.3024	.2876	.2447	.1778	.0935
TIME	=	.147	.0926	.1760	.2423	.2848	.2995	.2848	.2423	.1760	.0925
TIME	=	.148	.0916	.1743	.2399	.2821	.2966	.2821	.2399	.1743	.0916
TIME	=	.149	.0908	.1726	.2376	.2793	.2937	.2793	.2376	.1726	.0908
TIME	=	.150	.0899	.1709	.2353	.2766	.2908	.2766	.2353	.1709	.0899
TIME	=	.151	.0890	.1693	.2380	.2739	.2080	.2739	.2380	.1693	.0890
TIME	=	.152	.0881	.1676	.2307	.2712	.2852	.2712	.2307	.1676	.0881
TIME	=	.153	.0873	.1660	.2285	.2686	.2824	.2686	.2285	.1660	.0873
TIME	=	.154	.0864	.1644	.2263	.2660	.2797	.2660	.2263	.1644	.0864
TIME	=	.155	.0856	.1628	.2240	.2634	.2769	.2634	.2240	.1628	.0856
TIME	=	.156	.0847	.1612	.2219	.2608	.2742	.2608	.2219	.1612	.0847
TIME	=	.157	.0839	.1596	.2197	.2583	.2716	.2583	.2197	.1596	.0839
TIME	=	.158	.0831	.1581	.2176	.2558	.2689	.2558	.2176	.1581	.0831
TIME	=	.159	.0823	.1565	.2154	.2533	.2663	.2533	.2154	.1565	.0823
TIME	=	.160	.0815	.1550	.2133	.2508	.2637	.2508	.2133	.1550	.0815
TIME	=	.161	.0807	.1535	.2113	.2484	.2611	.2484	.2113	.1535	.0807
TIME	=	.162	.0799	.1520	.2092	.2459	.2586	.2459	.2092	.1520	.0799
TIME	=	.163	.0791	.1505	.2072	.2435	.2561	.2435	.2072	.1505	.0791
TIME	=	.164	.0784	.1491	.2052	.2412	.2536	.2412	.2052	.1491	.0784
TIME	=	.165	.0776	.1476	.2032	.2388	.2511	.2388	.2032	.1476	.0776
TIME	=	.166	.0768	.1462	.2012	.2365	.2487	.2365	.2012	.1462	.0768
TIME	=	.167	.0761	.1447	.1992	.2342	.2462	.2342	.1992	.1447	.0761

TIME =									
.168	.0754	.1430	.1973	.2319	.2438	.2319	.1973	.1433	.0754
.169	.0746	.1419	.1954	.2297	.2415	.2297	.1954	.1419	.0746
.170	.0739	.1406	.1935	.2274	.2391	.2274	.1935	.1406	.0739
.171	.0732	.1392	.1916	.2252	.2368	.2252	.1916	.1392	.0732
.172	.0725	.1378	.1897	.2230	.2345	.2230	.1897	.1378	.0725
.173	.0718	.1365	.1879	.2208	.2322	.2208	.1879	.1365	.0718
.174	.0711	.1352	.1860	.2187	.2299	.2187	.1860	.1352	.0711
.175	.0704	.1338	.1842	.2166	.2277	.2166	.1842	.1338	.0704
.176	.0697	.1325	.1824	.2144	.2255	.2144	.1824	.1325	.0697
.177	.0690	.1312	.1806	.2124	.2233	.2124	.1806	.1312	.0690
.178	.0688	.1300	.1789	.2103	.2211	.2103	.1789	.1300	.0683
.179	.0677	.1287	.1771	.2082	.2190	.2082	.1771	.1287	.0677
.180	.0670	.1274	.1754	.2062	.2168	.2062	.1754	.1274	.0670
.181	.0663	.1262	.1737	.2042	.2147	.2042	.1737	.1262	.0663
.182	.0657	.1250	.1720	.2022	.2126	.2022	.1720	.1250	.0657
.183	.0651	.1238	.1703	.2002	.2105	.2002	.1703	.1238	.0651
.184	.0644	.1226	.1687	.1983	.2085	.1983	.1687	.1226	.0644
.185	.0638	.1214	.1670	.1964	.2065	.1964	.1670	.1214	.0638
.186	.0632	.1202	.1654	.1944	.2045	.1944	.1654	.1202	.0632
.187	.0626	.1190	.1638	.1926	.2025	.1926	.1638	.1190	.0626
.188	.0620	.1178	.1622	.1907	.2005	.1907	.1622	.1178	.0620
.189	.0614	.1167	.1606	.1888	.1985	.1888	.1606	.1167	.0614
.190	.0608	.1156	.1591	.1870	.1966	.1870	.1591	.1156	.0608
.191	.0602	.1144	.1575	.1852	.1947	.1852	.1575	.1144	.0602
.192	.0596	.1133	.1560	.1834	.1928	.1834	.1560	.1133	.0596
.193	.0590	.1122	.1545	.1816	.1909	.1816	.1545	.1122	.0590
.194	.0584	.1111	.1529	.1798	.1891	.1798	.1529	.1111	.0584
.195	.0579	.1100	.1515	.1780	.1872	.1780	.1515	.1100	.0579
.196	.0573	.1090	.1500	.1763	.1854	.1763	.1500	.1090	.0573
.197	.0567	.1079	.1485	.1746	.1836	.1746	.1485	.1079	.0567
.198	.0562	.1069	.1471	.1729	.1818	.1729	.1471	.1069	.0562
.199	.0556	.1058	.1456	.1712	.1800	.1712	.1456	.1058	.0556
.200	.0551	.1048	.1442	.1695	.1783	.1695	.1442	.1048	.0551

B.4. (a) Letting $u = cx(x - 1)$ we require that

$$0 = \int_0^1 \phi Lu = \int_0^1 \phi(u'' + u' - 4)$$

$$= -\int_0^1 (\phi'(u' + u) + 4\phi) + \phi(u' + u)\Big|_0^1$$

$$= -\int_0^1 [((2)(-1)^2 + x(x - 1)(2x - 1))c + 4x(x - 1)]$$

from which $c = 2$.

(b) $u(x) = (4e/1 - e) \cdot (1 - e^{-x}) + 4x$

(c) $x = .2, \ u_\phi = -.32, \ u = -.35$

$x = .5, \ u_\phi = -.50, \ u = -.49$

$x = .8, \ u_\phi = -.32, \ u = -.28$

B.5. (a) Letting $u = \frac{1}{2}x(1 - x) + cx(2 - x)$, from

$$0 = \int_0^1 \phi_1(u'' + 4xu - 16)$$

we arrive at $c = 89/6$. Note that ϕ_0 has been taken to satisfy the inhomogeneous boundary condition.

(b) By central differences $u(0.5) = 4$, $u(1.0) = 10$, $u(1.5) = 10$.

(c) From (a) $u_\phi(0.5) = 11.25$, $u_\phi(1.0) = 14.83$, $u_\phi(1.5) = 10.75$.

B.6. (a) $c = 9/8$, $u_\phi = 1.125x$.

$c_1 = 1.295$, $c_2 = -.1511$, $u_\phi \cong 1.295x - .1511x^2$

(b) $u(x) = 2\cos x + ((2\sin 1 - 1)/\cos 1)\sin x + x^2 - 2$.

(c) $x = .2$ $u_1 = .225$ $u_2 = .253$ $u = .251$

$x = .5$ $u_1 = .563$ $u_2 = .610$ $u = .611$

$x = .8$ $u_1 = .900$ $u_2 = .939$ $u = .940$.

B.7 (a) As the ϕ_n and their second derivatives are orthogonal, the equations uncouple to

$$\begin{cases} a_n'(t) = -n^2 a_n(t) & t > 0 \\ a_n(0) = \langle f, \phi_n \rangle & t = 0 \end{cases}$$

$n = 1, \ldots, N$.

(b) For collocation points $x_m = m\pi/(N + 1)$, $m = 1, \ldots, N$, the approximate solution $u(x_m, t)$ is to be of the form

$$u(x_m, t) = \sum_{n=1}^N a_n(t)\, \phi_n(x_m)$$

Use the discrete Wallis formula

$$\sum_{n=1}^N \sin(m\pi n/(N + 1))\sin(k\pi n/(N + 1)) = \begin{cases} (N + 1)/2 & k = m \\ 0 & k \neq m \end{cases}$$

when applying the collocation requirements.

(c) Some are given below.* Spectral methods are now possible because of the recent fast transform algorithms.

B.8 (a) How you discretize at $x = 0$, $\pi/2$, and π will create differing initial Gibb's effects, which will quickly disappear in this problem. Here is a Crank-Nicolson output.

T	U1	U2	U3	U4	U5	U6	U7	U8	U9
0	0	0	0	0	.500000	1.000000	1.000000	1.000000	1.000000
.100	.004912	.019518	.072653	.269201	.497131	.719399	.887779	.834645	.455110
.200	.021618	.066516	.165658	.304971	.483509	.640822	.696916	.597775	.357468
.300	.043913	.108553	.201929	.326730	.455256	.549239	.563662	.470286	.266422
.400	.060380	.132538	.223305	.324840	.418021	.475820	.467356	.376589	.211360
.500	.069618	.145692	.229124	.313273	.382045	.415208	.394092	.310661	.171553
.600	.073881	.149720	.226151	.295996	.347114	.364889	.337618	.261107	.142774
.700	.074423	.147957	.217420	.276234	.314765	.322786	.292743	.223281	.120975
.800	.072587	.142441	.205476	.255586	.285219	.287043	.256375	.193339	.104045
.900	.069277	.134755	.191881	.235097	.258345	.256354	.226275	.169150	.090534
1.000	.065147	.125942	.177692	.215342	.233966	.229706	.200933	.149183	.079511

* D. Gottlieb and S. Orszag, *Numerical Analysis of Spectral Methods* (SIAM, Philadelphia, 1977); P. Prenter, *Splines and Variational Methods* (Wiley, New York, 1975); R. Peyret and T. Taylor, *Computational Methods for Fluid Flow* (Springer, New York, 1983).

(b) Use for example the Crank–Nicolson code given in B.3(b), from which, for example, ensues the following output:

```
***** CALCULATION  DATA  *****

R = 1.0          H = .10          K = .010          TMAX = .100


***** COEFFICIENT  MATRIX  *****
```

4.0	-1.0	0	0	0	0	0	0	0
-1.0	4.0	-1.0	0	0	0	0	0	0
0	-1.0	4.0	-1.0	0	0	0	0	0
0	0	-1.0	4.0	-1.0	0	0	0	0
0	0	0	-1.0	4.0	-1.0	0	0	0
0	0	0	0	-1.0	4.0	-1.0	0	0
0	0	0	0	0	-1.0	4.0	-1.0	0
0	0	0	0	0	0	-1.0	4.0	-1.0.
0	0	0	0	0	0	0	-1.0	4.0

	X =	.1	.2	.3	.4	.5	.6	.7	.8	.9
TIME =	0	.2000	.4000	.6000	.8000	1.0000	.8000	.6000	.4000	.200#
TIME =	.01	.1989	.3956	.5834	.7381	.7691	.7381	.5834	.3956	.198#
TIME =	.02	.1936	.3789	.5397	.6461	.6921	.6461	.5397	.3789	.193#
TIME =	.03	.1826	.3515	.4902	.5843	.6152	.5843	.4902	.3515	.182#
TIME =	.04	.1683	.3218	.4461	.5267	.5555	.5267	.4461	.3218	.168#
TIME =	.05	.1538	.2932	.4047	.4770	.5019	.4770	.4047	.2932	.153#
TIME =	.06	.1399	.2664	.3672	.4321	.4546	.4321	.3672	.2664	.139#
TIME =	.07	.1270	.2418	.3330	.3916	.4119	.3916	.3330	.2418	.127#
TIME =	.08	.1153	.2193	.3019	.3550	.3733	.3550	.3019	.2193	.115#
TIME =	.09	.1045	.1989	.2738	.3219	.3385	.3219	.2738	.1989	.104#
TIME =	.10	.0948	.1803	.2482	.2918	.3069	.2918	.2482	.1803	.094#

(c) Originated in the 1960's in Russia, this theory has matured in the 1980's and 1990's.*

B.9 (a) The Leapfrog stencil (Figure B.3c) at first appears natural to this problem. However, there are difficulties in getting it started and for this reason it can generate spurious results, e.g., false subpeaks in the solution wave.

(b) Lax–Wendroff schemes often treat problems in conservation form

$$\rho_t + (F(\rho))_x = 0$$

by additional differentiation with respect to t.† Although better than Leapfrog, two codes tried here failed to produce adequate backward movement of the traffic bulge. **

(c) The method of characteristics is an important one for the numerical treatment of nonlinear hyperbolic wave problems. For the problem at hand, on curves in the x, t plane for which $dx/dt = c(\rho)$, the equation $\rho_t + c(\rho)\rho_x = 0$ means that the total derivative $d\rho/dt = 0$. On these curves, $c(\rho)$ remains constant, hence these curves are straight lines with slopes $c(\rho(x, 0))$ The solution $\rho(x, t)$ may therefore be constructed along these lines, producing a backward moving wave as shown below.

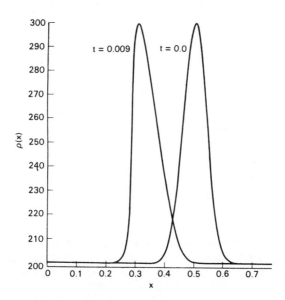

* Three good research treatment sources are Hackbusch and Trottenberg, *Multigrid Methods* (Springer, Berlin, 1982), and two special issues of the journal *Applied Mathematics and Computations* (1983, 1986). A recent excellent introduction is Briggs, *A Multigrid Tutorial* (SIAM, Philadelphia, 1987). A solid textbook treatment is that of W. Hackbusch, *Multi-Grid Methods and Applications* (Springer, Berlin, 1985).

† See the answer to Problem A.9(c), and the book by Ames cited in Section 2.6.

** But several of the shock capturing schemes, to be discussed in the following SIXTH PAUSE, produced excellent solutions to this traffic flow problem.

SIXTH PAUSE

1.   Writing once again the Euler equations

$$\rho_t + u\rho_x = -\rho u_x$$

$$u_t + uu_x = -\frac{1}{\rho}p_x$$

$$p_t + u\,p_x = -\gamma p u_x$$

we see that each is an *advection* equation of the type considered in Pause 5. Most analyses of the Riemann Problem show the existence of a unique solution for a conservation law $u_t + (f(u))_x = 0$ only among solutions restricted to constant states separated by shocks, fans, or contact discontinuities. Let us see why that is so.

Notice that on space intervals where u and p are constant, all three equations have zero righthand sides. By the analysis of Pause 5, they are then total differentials propagating their constant values along straight line characteristics of slope $dx/dt = u$. Thus in noninteraction regions, the solutions preserve their initial constant values. Moreover, when ρ is also constant, the solutions become steady, with no change in forward time.

This preference for regular (indeed, flat) solutions, given regular (indeed, flat) initial data, is entirely in accord with the physics of a system at rest preferring to stay that way. It is also the key to the success of numerical methods called cell or *finite volume* methods, which use an averaged (indeed, in many instances, flat) approximation at each time step.

2.   The notion of *contact discontinuity* (also called a slip line, or effective piston) mentioned in Pause 6 is a new *shock dynamic* which we should briefly discuss. It may be regarded as a degenerate, or one-sided, shock: its speed is not averaged according to gasdynamics physics, but instead is regarded as the pushing (effective piston) speed on the high pressure side.

Analytically, it occurs when a characteristic family $dx/dt = \lambda_k$ is *linearly degenerate*: $(\nabla \lambda, r_k) = 0$, where r_k is the kth right eigenvector of the Jacobian A of the quasilinear hyperbolic system. Rather than go further into the full theory, let us illustrate with the Riemann Problem. From Pause 5, Exercise 9, we know that $\lambda_2 = u$ and $r_2 = (1, 0, 0)$. Since the gradient ∇ in these variables (ρ, u, p) is $\nabla = (\partial/\partial\rho, \partial/\partial u, \partial/\partial p)$ we have $(\nabla \lambda_2, r_2) = 0 \cdot 1 + 1 \cdot 0 + 0 \cdot 0 = 0$. Thus the contact discontinuity occurs at the "middle" characteristic velocity $\lambda_2 = u$: it is just the "rushing" speed to the right created by the impulsive breaking of the diaphragm in the shock tube. Velocity and pressure are continuous across it, as indeed they constitute its "piston action". Again we see a physical " preference for the regular solution". Density and energy, however, may incur offsetting discontinuities, in order to accommodate the assumed equation of state (gas law).*

More generally, the right eigenvectors are used to define the *Riemann Invariants*: smooth functions $w(x, t)$ such that

$$(\nabla w, r_k) = 0$$

For example, for the case of $r_2 = (1, 0, 0)$ and the contact discontinuity just considered, we just noted that $w^1 = \lambda_2 = u$ is a Riemann invariant. It can be shown generally for a hyperbolic conservation system of n equations, that for each kth right eigenvector, there are $n - 1$ Riemann invariants with linearly independent gradients. What is the other one, for the 3 Euler equations at r_2? From $(\nabla w, r_2) = w_\rho \cdot 1 + w_u \cdot 0 + w_p \cdot 0 = 0$ we see that w should not depend on the density ρ. We have already chosen $w^1 = u$, so we now choose $w^2 = p$. Across a contact discontinuity, the Riemann invariants are generally continuous.

The Riemann invariants at the left moving rarefaction wave bounded by lines $dx/dt = \lambda_1 = u - c$, and at the right moving shock wave $dx/dt = \lambda_3 = u + c$, are entropy $s = c_v \ln (p/\rho^\gamma)$ and relative enthalpy $u \pm h$, respectively, where enthalpy $h = 2c/(\gamma - 1)$. These may be regarded as bringing into the picture favored (e.g., regular) energy and density "coordinates", just as the Riemann Invariants u and p are favored (e.g., regular) "coordinates" at $dx/dt = \lambda_2 = u$. Finding the Riemann invariants completes the construction of the exact analytic solution to the Riemann problem, see the cited references.

We close this answer with a nice way to geometrically view what is really going on, at the basic level of the eigenvalues $\lambda_1, \ldots, \lambda_n$ right eigenvectors $r_1, \ldots, r_n$, and left eigenvectors $l_1, \ldots, l_n$, for a general hyperbolic system, Recall (Pause 5, Exercise 6) that the right and left eigenvectors of any real diagonalizable matrix A are biorthogonal: $(r_i, l_j) = 0, i \neq j, (r_i, l_i) \neq 0$. The reader should check this for the Euler equations (see Pause 5, Exercise 9, for the r_i and l_j). If we think of $l_j = \nabla w^j$, this biorthogonality gives us $n-1$ equations $(r_i, l_j) = (r_i, \nabla w^j) = 0$ to solve for the $n-1$ Riemann invariants w^j with linearly independent gradients. For a system of just 2 equations, this insight is exact: $\nabla w^j = l_j$ (up to constant multiples, of course), so we may say that the Riemann invariant w is just the potential for the left eigenvector. For $n > 2$ we have to modify our statements to left eigenspaces rather than just single eigenvectors, etc., but the geometric insight is still useful.

3. ** The forerunner for all upwinding schemes for application to gasdynamic schemes is that of Godunov (1959,1962). It can be written in predictor-corrector form

---

*See the book by Courant and Friedrichs cited in Appendix A.1. for an excellent discussion of contact discontinuities in one and more space dimensions. The fact that velocity and pressure are immediately continuous (i.e., *regularized*) across a contact discontinuity, and density is not, should give pause to the reader. As the computational Riemann flows show, in the Test Problem the initial velocities $u_l = u_r = 0$ instantaneously develop a 'smooth delta function spike' to drive the ensuing flow. It should be mentioned that contact discontinuities are the most difficult of the qualitative discontinuities to resolve computationally, as they have tendencies to smear out or develop unwanted small expansion waves. Physically, e.g., with viscous forces allowed to act, these tendencies are natural.

** We had hoped to give a more extensive treatment of all of these schemes, but Pause and Book length do not permit. Accordingly we will be very brief, but will mention some key references for scheme details.

$$\overline{u}_{i+1/2}^{n+1} = \frac{1}{2}(u_{i+1}^n + u_i^n) - (\Delta t/\Delta x)(F_{i+1}^n - F_i^n)$$

$$u_i^{n+1} = u_i^n - (\Delta t/\Delta x)(\overline{F}_{i+1/2}^{n+1} - \overline{F}_{i-1/2}^{n+1})$$

where Δt = uniform time increment, $n\Delta t$ = time of evaluation, Δx uniform space increment, $i\Delta x$ and $(i+1)\Delta x$ bound the ith x cell (finite volume), the $i \pm 1/2$ denoting values placed at cell centers. The $\overline{u}$ denotes the average over the cell (see Pause 5, Exercise 8), the $\overline{F}$ meaning that this predictor average $\overline{u}$ is used to evaluate the flux terms. Clearly the using of piecewise constant approximations u is advantageous conceptually, since then the data propagates ahead along characteristics for a short time at least in piecewise constant segments. The two-step time advances may be numerical or may use the exact analytical solution to the Riemann Problem. For stability one needs $(\Delta t/\Delta x) \le 1$ but in practice smaller Δt are used to assure that only the two adjacent cells interact to produce the next solution values at the left and right cell boundary. As piecewise constant approximations are used, these cell boundaries are all regarded and treated as possible "shocks".

Glimm (1965) and others later have used this numerical procedure for existence proofs for hyperbolic conservation laws. Glimm introduced a random choice for $\overline{u}$ which seems to improve shock resolution at the cost of shock location. Chorin (1976) returned Glimm's method to a numerical setting quite useful for fluid dynamical simulations where one wishes to include some randomness.**

4. Centered schemes without flux splitting, such as the classical Lax-Wendroff two-step scheme

$$u_{i+1/2}^{n+1/2} = \frac{1}{2}(u_{i+1}^n + u_i^n) - (\Delta t/2\Delta x)(F_{i+1}^n - F_i^n)$$

$$u_i^{n+1} = u_i^n - (\Delta t/\Delta x)(F_{i+1/2}^{n+1/2} - F_{i-1/2}^{n+1/2})$$

produce unwanted oscillations and even negative physical values near solution discontinuities. Thus for a while the classical second order centered schemes such as Lax-Wendroff, or more basically, Crank-Nicolson (see Appendix B) lost favor. However, by the addition of conservative dissipative numerical flux terms, these schemes could be stabilized, and are now competitive with the TVD schemes.***

5. TVD (Total Variation Diminishing) schemes are designed to cut down the oscillations at solution discontinuities. Recall that total variation T.. V. $(u^n) = \sum_i |u_i^n - u_{i+1}^n|$ measures the total oscillation over the grid. A scheme is said to be total variation diminishing if T . V .$(u^{n+1}) \le$ T. V .(u^n), i.e., the oscillations diminish with time.

** For other upwind schemes of Van Leer, Steger, Warming, Roe, Osher, Enquist, Fromm, many others, see the cited paper by Sod and the recent literature.

***For an account and numerical comparisons of the symmetric centered schemes of Davis, Roe, Yee, and others, see H. Yee, *J. Computational Physics* **68** (1987). As noted in Pause 6, an upwind scheme can usually be rewritten as a centered scheme. Conversely, a centered scheme can usually be interpreted as creating some upwinding. Thus some care should be taken when distinguishing these schemes.

For application to gasdynamics (most notably, computational aerodynamics), considerable effort went into the development of TVD schemes in the 1980s.[*] An important subclass of TVD schemes are the monotone schemes:if written on a stencil as

$$u_i^{n+1} = H(u_{i-1}^n, u_i^n, u_{i+1}^n)$$

the scheme is said to be *monotone* if H is a monotone nondecreasing function of each of its arguments. Many older schemes such as the Godunov and Lax-Friedrichs schemes are monotone. Monotone schemes produce smooth solutions at discontinuities (smearing) but are only first order accurate there.

An important TVD scheme, second order away from shocks, can be written as follows:
$$u_i^{n+1} = u_i^n + (\Delta t/\Delta x)[C^-\,(a_{i+\frac{1}{2}}^n)\,\Delta_{i+\frac{1}{2}}\,u^n - C^+(a_{i-\frac{1}{2}}^n)\cdot\Delta_{i-\frac{1}{2}}u^n$$
where

$$\Delta_{i+\frac{1}{2}}\,u^n = u_{i+1}^n - u_i^n,$$
$$a_{i+\frac{1}{2}}^n = (F(u_{i+1}^n) - F(u_i^n))/\Delta_{i+\frac{1}{2}}u^n$$

or some other good approximation to $F'(u)$ if $\Delta_{i+j}u^n = 0$, e.g., the appropriate eigenvalue $\lambda_k(u^n)$ of the matrix A, and where

$$C^\pm(z) = \frac{1}{2}\left| Q(z) \pm z \right|$$

where Q is a numerical viscosity term used to assure satisfaction of certain entropy inequalities.

6.   For the backward spatial differencing $\dot{u}_j = \lambda(u_j - u_{j-1})/\Delta x$ the mode $u_j(t) = v(t)e^{ikj\Delta x}$ yields the ODE $\dot{v} + cv = 0$ where $c = \lambda(1 - e^{-ik\Delta x})/\Delta x$. The solution $v(t) = Ce^{-t}$ is bounded for all $t \geq 0$ iff the real part of the coefficient c is non-negative. But Re $c = \lambda(1 - \cos k\Delta x)/\Delta x = \lambda(2\sin^2(k\Delta x/2))\Delta x$. Thus λ must be nonnegative.

Similarly, the forward spatial differencing requires $\lambda \leq 0$ for the solution to remain bounded for all $t > 0$.

7.   For more details on flux splitting, especially as combined with finite volume and TVD schemes for important applications to computational aerodynamics, we defer to the literature, as already cited here,[**] and moreover as it has appeared in the AIAA journals and NASA publications. There are several variations on how to split flux.

Let us comment on the meaning of the Euler invariance property $F(\lambda U) = \lambda F(U)$ so useful in the flux-splitting computations of gasdynamics flows. what this property says is that if $\rho(x, t)$, $u(x, t)$, and $p(x, t)$ are a solution,then so is $\lambda\,\rho(x, t)$, $\lambda u(x, t)$, and $\lambda p(x, t)$,

---

[*] A first paper was that of A. Harten, *J. Computational Physics* 49 (1983). For further references to the work of many other investigators, see the paper by Yee of preceding footnote. There are schemes now called TVB, UNO, ENO, ..., all attempting to effect the best compromises between too much smearing (nonsharp shocks) and too much oscillation (as in the Gibbs' effect).

[**] We add the excellent reference J. Steger and R. Warming, *J. Computational Physics* 40 (1981), as an accessible exposition, with numerical tests on the Shock Tube problem we have considered. It was there that the full force of the homogeneity property $F(\lambda U) = \lambda F(U)$ of the Euler gasdynamic equations was first   exploited computationally within a flux splitting context.

for arbitrary multiple λ. The λ will in an application be pinned down by the initial data $\rho(x, 0)$, $u(x, 0)$, and $p(x,0)$; stated another way, the first order Euler homogeneity ** obtains if one ignores initial conditions. This is typical of group properties applied to PDE's.

8.   A little viscosity goes a long way. Suppose we are deriving a gas law for wave propagation in a continuous medium. We define a density $\rho(x, t)$ per unit length, a flux $F(x, t)$ per unit time. The flow velocity is then defined by $u = F/\rho$. Conservation of material gives us a material balance

$$\frac{d}{dt}\int_a^b \rho(x, t)dx = F(b,t) - F(a, t)$$

and assuming that $\rho(x, t)$ is C^1, we arrive at the differential conservation law

$$\rho_t + F_x = 0$$

If we now assume that F depends only on ρ, $F = F(\rho)$, and smoothly so, we go to the first order quasilinear PDE

$$\rho_t + F'(\rho)\rho_x = 0$$

On the other hand, if we assume that the flux F has *some* dependence on density changes, e.g., on ρ_x, $F = F(\rho, \rho_x)$, then the conservation law implicitly includes a higher order ρ_{xx} effect. For example, suppose $F(\rho, \rho_x) = f(\rho) - \nu\rho_x$, ν a positive constant. This simple linear dependence on ρ_x produces the PDE

$$\rho_t + f'(\rho)\rho_x = \nu\rho_{xx}.$$

No matter how small the (viscosity) coefficient ν may be, the solution qualitative behavior is radically different from the previous (inviscid) law. The Laplacian term $\nu\rho_{xx}$ introduces diffusion, solution smoothing, and very different data propagation paths. For example, when $f'(\rho) = 0$, we have a heat equation, which propagates data infinitely fast instantaneously along characteristics parallel to the x axis.

This example brings out the changed point of view one takes when treating the full viscous Navier-Stokes equations of fluid dynamics, to which we will turn briefly in the next, and last, Appendix C. Although oversimplified, we may say that for the full viscous Navier-Stokes equations, the viscous, elliptic terms dominate the analysis, whereas for the

---

**One needs an equation of state law of the form $p = \rho f(\varepsilon)$, ε the internal energy, in order that this Euler homogeneity holds. The reader should recognize that this amounts to a selection of a special separated similarity solution for the equations. The gas law $p = (\gamma - 1)\rho\varepsilon$ we are using is a nice linear version. Here the gas constant $\gamma = c_p/c_v$, the ratio of specific heat at constant pressure to specific heat at constant volume. As we pointed out in a footnote in Pause 5, and earlier in Section 1.7.1, the need to assume constant specific heats plays a critical role in making otherwise intractable equations more susceptible to analysis.

To go a bit further, let us note (Answer to Exercise 2, above) that the Riemann invariants at the shock and rarefactions are entropy $s = c_v \ln(p/\rho^\gamma)$ and relative enthalpy $u \pm h$, where the enthalpy $h = 2c/(\gamma - 1) = 2(\gamma p/\rho)^{\frac{1}{2}}/(\gamma - 1)$. The relative efficacies of assumptions of similarity solution gas laws and of constant specific heats c_v and c_p in them should be kept in mind when asserting that entropy and enthalpy inequalities pick out the 'correct' physical solution.

truncated Euler equations treated in Pause 5 and Pause 6, the convective, hyperbolic terms dominate the analysis.*

9. *Existence* of solutions for hyperbolic conservation systems, especially the Euler gasdynamics system, has been shown, for nice (eg, piecewise flat) initial data, by *construction* of solutions which are piecewise constant and connected by shocks, fans, or contact discontinuities. *Uniqueness* obtains to the extent of limiting solution construction to these rules. Computational *approximation* methods rely heavily on these piecewise constant (or, more recently, piecewise polynomial) solutions and the rules for interpolating them. The initial data cannot have too much variation if one expects *stability* of analytical and numerical solutions: we must keep ρ away from zero, or else by the $u_t + uu_x = -\rho^{-1}p_x$, we will get velocity "blowup". We have advanced the role of *regularity* in viewing the choice of preferred (physical) solution. The viewpoint was that of pure *Initial Value Problems. Boundary Value Problems* were not considered. *Separation of Variables* is attained along characteristic coordinates. The *Divergence Theorem* was implicit throughout. The *Eigenvalue Problem* for the Jacobian A of the hyperbolic system gave us the principal discontinuity propagation directions, and the properties of solutions across those discontinuities.

In higher space dimensions, correct geometrical intuition for an evolving solution to the Euler equations, even in situations analogous to the one-dimensional shock tube problem, becomes more difficult. To the paramount role of characteristics in the formation of shock waves, are added the effects of diffraction, reflection, and interaction. A key insight has been to develop a theory for a disturbance propagating along a shock surface.**

Adding to one's developing intuition are the results of physical experiments in aerodynamics and acoustics. Also the accompanying computational simulations now aid the development of intuition. The latter permit the inclusion of boundary conditions which are more difficult to include in the pure theory. Eventually a theory based upon the computational methods supersedes the analytical approach and becomes as much of the intuition as the analytical and the physical. This is even more the case when one brings in the viscous effects.

---

*From Lord Rayleigh, *Phil. Magazine* 34 (1892):

As Sir George Stokes has shown, the steady motion is the same whatever be the degree of viscosity; and yet it is entirely different from the flow of an inviscid fluid in which no rotation can be generated. Considerations such as this raise doubts as to the interpretation of much that has been written on the subject of inviscid fluids in the neighbourhood of solid obstacles.

** Such a theory may be found in Z. Y. Han and X. Yin, *Shock Dynamics* (Kluwer, Beijing, 1993).

C.1.(a) The momentum equation becomes

$$u_t = -p_x,$$

because the incompressibility condition $u_x = 0$ knocks out both the advection term uu_x and the diffusion term νu_{xx}. The equations are thus reduced to a first order equation relating $u(x, t)$ and $p_x(x, t)$.

(b) Taking the divergence of both sides of the momentum equation and interchanging the order of differentiation we have

$$\operatorname{div} \frac{\partial u}{\partial t} - \operatorname{div} \frac{1}{Re} \Delta u + \operatorname{div}(u \cdot \nabla)u = -\operatorname{div} \operatorname{grad} p$$

from which

$$-\Delta p = \operatorname{div}(u \cdot \nabla)u \text{ in } \Omega.$$

Keep in mind the vector components of the velocity u and the fact that pressure is a scaler function. Notice how blithely we assume that we can differentiate the given equation. But this "overregularity operation" is physically justified for any smooth solution.

Similarly we are entitled to regard the incompressibility condition on the boundary $\partial\Omega$ as the immiscibility condition $n \cdot u = 0$ there, so that upon dotting the momentum equation by the outer normal n we arrive at*

$$n \cdot \nabla p = -n \cdot ((u \cdot \nabla)u) + \frac{1}{Re} \cdot n \cdot \Delta u \text{ on } \partial\Omega.$$

Thus if we know the velocity u, we have a Neumann-Poisson problem for the pressure. This is the basis for a number of numerical schemes, which step alternately between velocity U_N and pressure P_{N+1} as one advances in time.

(c) The pressure equation of (b) becomes

$$\frac{\partial^2 p}{\partial x^2} = 0, \quad a < x < b.$$

In other words, at each instant in the evolution of the fluid motion, the pressure must be a harmonic function, that is, a linear one

$$p(x, t) = c(t)x + d(t).$$

From the momentum equation this means the acceleration depends, only on t

$$u_t(x, t) = -c'(t)$$

hence $u(x, t) = -c(t) + k(x)$ where $k'(x) = 0$ so that k is a constant.

Up to now we have not invoked initial or boundary conditions. Setting $t = 0$ we have $u(x, 0) = -c(0) + k$ meaning that only a constant initial velocity distribution is permitted. Adding to this a Dirichler boundary condition means thre is no flow.

* This remains sometimes controversial in the literature but can be justified.

C.2. (a) For smooth Ω we know there exists a unique solution ϕ_3. Note that

$$\text{div } u_d = \text{div grad } \phi_2 = \Delta\phi_3 = \text{div } u,$$

and that curl u_d = curl grad ϕ_3 = 0.

(b) Again we know that a solution exists provided the compatibility condition (see Section 1.6.1 Problem 2)

$$0 = \int_\Omega \Delta\phi_1 = \oint_{\partial\Omega} \frac{\partial\phi_1}{\partial n} = \oint_{\partial\Omega} n \cdot u - \oint_{\partial\Omega} \frac{\partial\phi_3}{\partial n}$$

is satisfied. This is so because

$$\oint_{\partial\Omega} \frac{\partial\phi_3}{\partial n} = \int_\Omega \Delta\phi_3 = \int_\Omega \text{div } u = \oint_{\partial\Omega} u \cdot n.$$

Note that although ϕ_1 is only determined up to a constant its gradient u_p is uniquely determined, and that div $u_p = \Delta\phi_1 = 0$, curl u_p = curl grad $p_1 = 0$.

(c) The orthogonality* is an exercise in Green's Identity and the divergence theorem. We have

$$\int_\Omega u_d u_p = \int_\Omega \text{grad } \phi_3 \cdot \text{grad } \phi_1 = \oint_{\partial\Omega} \phi_3 \frac{\partial\phi_1}{\partial n} - \int_\Omega \phi_3\Delta\phi_1 = 0$$

and

$$\int_\Omega u_d u_c = \int_\Omega (u - u_d) \text{ grad } \phi_3$$

$$= \oint_{\partial\Omega} \phi_3(u - u_d) \cdot n - \int_\Omega \phi_3 \text{div}(u - u_d) = 0.$$

C.3. (a) $\int_\Omega (u \cdot \nabla v)w = -\int_\Omega (u \cdot \nabla w)v$ for all u, v, w that are divergence free and vanish on the boundary. Hence the given expression is zero when $v = w$.

(b) The weak form states that $-\Delta u - f$ is orthogonal to all solenoidal vectors v in H_s. Hence it must equal a gradient ∇p in H_d. That one always exists can be shown, albeit delicately.

(c) For the unsteady case a few solutions are also known, in some cases by the methods of Appendix A.3. It can happen that a solution is independent of the Reynolds number.

C.4. (a) For Ω the cube, the number of small cubes within the large one Ω is $(l_0/l_d)^3$. Thus the total number of small eddies to monitor near the turbulent threshold is $0((l_0/l_d)^3)$.

(b) The fractal† dimension of a set is a type of Hausdorff measure obtained as an infimum of a covering of the set by ε-balls. There is a universal attractor

---

* As Green's identities hold as well for L^p spaces, the direct sum decomposition of the Helmholtz Theorem holds more generally beyond the Hilbert space $(L^2(\Omega))^n$, the orthogonality then giving way to the existence of linear functionals or distributions in a dual space.

† B.Mandelbrot *Fractals: Form, Chance, and Dimension* (Freeman, San Francisco, 1977).

set into which all flows of a certain type must eventually go. By tight *a priori* estimates on nonsingular deterministic Navier–Stokes equations it may be shown * that (roughly)

$$\begin{matrix} \text{Attractor} \\ \text{Fractal} \\ \text{Dimension} \end{matrix} \leqq c\left(\frac{l_0}{l_d}\right)^3$$

$$= c\left(\frac{1}{\lambda_1^{3/2}}\right)\left(\frac{\varepsilon^{3/4}}{\nu^{9/4}}\right)$$

$$\sim c\,\frac{Re^{9/4}}{\lambda_1^{3/2}} \cdot \left(\frac{1}{Re}\right)^{3/4} \cdot \left(\lambda_m^M\right)^{3/4}$$

where $\nu = Re^{-1}$ denotes the viscosity, λ_1 is the first eigenvalue of the Stokes operator (note: $\lambda_1 \sim \lambda_0$), and ε denotes the average rate of energy dissipation per unit space and time mentioned above, more precisely here, the limit

$$\varepsilon = \nu \lim_{t \to \infty} \sup \left(\sup_{U_0} \frac{1}{t} \int_0^t \sup_\Omega |\text{grad } u(x,\, t)|^2 d\tau\right)$$

over all flows U_0 which at some time enter the attractor, where we have denoted the lim sup by λ_m^M in the above.

(c) The lim sup expression above may be interpreted as the largest eigenvalue (highest frequency) which may persist forever in the flow. Higher frequencies would break down according to the Kolmogorov postulate.

Because it is known that $\lambda_m \sim c\lambda_1 m^{2/3}$ in three dimensions for large m, the dimension of the universal attracting set can be bounded above by

$$\frac{Re^{3/2}}{\lambda_1^{3/2}} \cdot \left(\lambda_m^M\right)^{3/4} \sim \left(\frac{Re}{\sqrt{\lambda_1}}\right)^{3/2} \cdot m^{1/2}.$$

As Re and m are expected to be rather large for turbulent flows, any realistic simulation would seem to need to monitor many (e.g., 10^9) degrees of freedom. On the other hand the above estimators are upper bounds, and very much dependent on the expression taken for the energy dissipation rate ε.

C.5. (a) Not yet.

(b) Possibly. See the next section.

(c) The Van der Pol equation of Section 1.9.8, letting λ tend to zero. Better examples show a transition from a steady final state to a periodic one as an eigenvalue parameter λ tends to a Hopf bifurcation point.

C.6. (a) The idea† is that in three dimensions, the flow can wander rapidly and erratically between two or three loosely coupled tori. Such attracting sets are called strange attractors.

---

* See P. Constantin, C. Foias, R. Temam, *Attractors representing turbulent flow* (*Mem. Amer. Math. Soc.* **314**, Providence, 1985), where this theory has been put on a rigorous basis (under the assumption that solutions remain regular for all time).

† See D. Ruelle, F. Takens, *Comm. Math. Phys.* **20** (1971).

(b) The original example* was found in an extreme simplification of equations from meteorology, namely, the system

$$\begin{cases} \dot{x}(t) = -\sigma x(t) - \sigma y(t) \\ \dot{y}(t) = rx(t) - y(t) - x(t)z(t) \\ \dot{z}(t) = y(t)x(t) - bz(t) \end{cases}$$

(c) There is evidence both ways.†

C.7. (a) $\omega = v_x - u_y = -\psi_{xx} - \psi_{yy}$, $\Delta\omega = -\Delta^2\psi = 0$ because curl grad $p = 0$. The existence of a stream function ψ follows from the incompressibility condition $u_x = -v_y$, and setting $\psi = 0$ on $\partial\Omega$ renders the stream function unique.

(b) The setting of discrete vorticity boundary conditions is the weak link in using the stream function vorticity formulations, but the suggested assignment is rather well accepted; see the literature for this and alternate formulations.

(c) For stability some relaxation is needed and a correction such as

$$\omega_{corrected} = \gamma\omega_{new} + (1 - \gamma)\omega_{old}$$

works. Here γ can be taken small, for example, $\gamma = 0.1$ or 0.01, good values found by numerical experimentation.

C.8. (a) The problem is linear.

(b) A series of papers produced this. A key step‡ was looking at the transverse velocity $v = -\text{Real}(\partial\psi/\partial r)$ evaluated along the corner bisector. It may be seen that as $r \to 0$, $v = v(r)$ has an infinite number of zeros r_n, $n = 1, 2, \ldots$.

(c) Green's second identity. The plate eigenvalues are all positive, so the only solution at $\lambda = 0$ is the trivial one.

C.9. (a) The flow deep in a corner is very slow and thus may be approximated accurately by a Stokes flow.

(b) For example, the first " $-$ " state, which comes off the left wall, persists to the end, even though it splits and recombines into and from several subordinate " $-$ " states.

(c) Particularly important are the subjects of cyclic reduction techniques, conjugate gradient algorithms, successive subspace correction methods, and the finding of effective preconditioners for specific classes of applications.§ The author has recently found a rather general beautiful new geometric theory of their convergence.**

---

* See E. Lorenz, *J. Atmos. Sciences* **20** (1963).

† See J. Curry, J. Herring, J. Loncaric, S. Orszag, *J. Fluid Mech.* **147** (1984).

‡ H. Moffatt, *J. Fluid Mech.* **18** (1964).

§ See for example R. Sweet, "A generalized cyclic reduction algorithm," *SIAM J Numer. Anal.* **11** (1974), M. Hestenes, *Conjugate Direction Methods in Optimization* (Springer, New York, 1980), and the recent books O. Axelsson, *Iterative Solution Methods* (Cambridge, U.K., 1994), W. Hackbusch, *Iterative Solution of Large Sparse Systems of Equations* (Springer, New York, 1994), Y. Saad, *Iterative Methods for Sparse Linear Systems* (PWS Publishing, Boston, 1996).

** See K. Gustafson, "Operator trigonometry of iterative methods," *Num. Lin. Alg. with Applic.* **4** (1997).

EPILOGUE

When the Second Edition of this book was translated into Japanese,* the Japanese editors[†] noted that the book did not treat aspects of the theory of distributions as they relate to partial differential equations as found, for example, in treatises by Hörmander and Sato.[‡] That is certainly the case, for indeed as was pointed out in the original preface to the book, a goal was to avoid the pitfalls of digressing into related topics such as complex analysis or functional analysis. However, and independently, certain European colleagues have privately expressed to the author the wish for at least some further discussion of distributions in the book. In recognition of these constructive wishes from East and West, some further words about distribution theory as it relates to partial differential equations have been added here in the first part of this epilogue.

The second part of the epilogue contains a brief presentation of the drift-diffusion equations so useful in electrical engineering for the design of semiconductor devices.[§] The author is embarrassed to say he did not even know of these equations until 1988 when, in a small consulting task, he was introduced to them for the first time. Just as the Navier–Stokes equations are the bread and butter of the aerospace industry, these drift-diffusion equations are the bread and butter of the electronic device industry.

Another class of important "modern" partial differential equations came to the author's attention even later, in 1994. These are the Black–Scholes equations, which, since their formulation about twenty years ago,[¶] have revolutionized the financial derivatives industry. We give a very brief treatment of them in the third part of this epilogue.

Many other partial differential equations and their theories and applications have not been treated here: the book is at the point of becoming long.

E.1 DISTRIBUTION EQUATIONS

In our treatment of the Dirichlet principle in Section 2.6 and in the Fourth Pause which follows it, some notions of weak or distributional derivatives entered

---

*K. Gustafson, *Applied Partial Differential Equations* 1 (Kaigai, Tokyo, 1991), *Applied Partial Differential Equations* 2 (Kaigai, Tokyo, 1992).

†See the (translated) Preface to the Japanese Edition, at the beginning of this book.

‡L. Hörmander, *The Analysis of Linear Partial Differential Operators I* (Springer, Berlin, 1983) and M. Sato, "Theory of Hyperfunctions I," *J. Fac. Sci. Univ. Tokyo* I, 8 (1959), "Theory of Hyperfunctions II," *J. Fac. Sci. Univ. Tokyo* I, 8 (1960).

§These are often called the Van Roosbroeck equations, after W. Van Roosbroeck, who first formulated them. See W. Van Roosbroeck, *Bell System Tech. J.* 29 (1950).

¶After F. Black and M. Scholes, *J. Pol. Econ.* 81 (1973). See also R. Merton, *Bell J. Econ. Management Sci.* 4 (1973).

naturally. One approach to these (which the author tends to defend in the never-ending philosophical arguments about the "best" approach) is one that is usually called Sobolev spaces; it is cultivated and generally favored by the Russian school. This is the weak solution* approach in which the partial derivative expressions become closed unbounded operators in strong Hilbert space normed topologies. The weak solutions then must be regularized (fixed up on sets of measure zero) by a technique often called Weyl's lemma so that they become classical solutions. Some further discussion of this closed operator approach was given in Problem 2.9.2. The weak formulation of $\Delta u = 0$ was described and tied to the variational formulation in Problem 1.9.5(3). Weak solutions entered naturally into our discussion of shock dynamics in the Fifth Pause.

The second main approach is usually called Distributions. This was pioneered by the French school[†] and led to the extensive development of a branch of functional analysis called topological vector spaces (t.v.s.). In this approach the partial derivative expressions become continuous operators in weak t.v.s. seminorm topologies.[‡] These spaces, although more complicated than Hilbert space, nonetheless have the property that in them the strong topology equals the weak topology and the latter is quite accessible. Conceptually, one limitation to this approach is its extensive use of Fourier transforms, which tacitly imply that you are treating problems whose domain Ω is the whole space. We briefly treated this aspect of the Fourier transform applied to partial differential equations in Section 2.7.3.

There is a third approach, which we will not examine closely here, which exploits to the full the Lebesgue differentiation theory and local *a priori* estimates. This approach has the advantages of remaining closer to the classical viewpoint and remaining local hence not dependent upon the Fourier transform. It depends on tools such as the mollifiers introduced in Problem 2.9.3, which can be done in the same way in higher dimensions. Our treatment of the heat equation initial value problem in Problem 1.9.7(3) reflects this approach.

All three approaches depend essentially on the fact that the test functions $C_0^\infty(\Omega)$ are dense in practically every space you would want to work in. We proved this in one dimension and for L^p spaces in the fundamental approximation theorem of Problem 2.9.3. This fact, coupled to appropriate integration by parts, which also occurs in all three approaches, is the justification for our statement on p. 128 that the notions of weak derivative, generalized derivative, distributional derivative, and

---

*In lectures the author sometimes advertises this approach as "if you keep weakening the formulation of the problem, eventually you will find the existence of some kind of solution . . . ." Both the weak solution and the distribution solution approaches may be described in this way.

†L. Schwartz, *Théorie des distributions* I, II (Hermann, Paris, 1950).

‡Therein lies the trade-off. The distributions approach adjusts the topology to the partial derivative so that the latter becomes continuous. The Sobolev spaces are a more natural topology but the partial derivative is not continuous. The distinction between the two approaches in practice is not large and indeed L. Schwartz originally set his distribution theory within the context of weak solutions.

strong derivative as they are developed in the literature, usually produce essentially the same results a.e. in partial differential equations.

Let us, then, consider distributional equations

$$P(D)u = f$$

where data and solution may be discontinuous or distributions. To be specific, and because there are already many volumes in which the subject is treated very generally, we will focus on the entities in

$$-\Delta_Q G(P,Q) = \delta(P,Q),$$

the defining equation for the Green's function for a partial differential equation, as already treated in Section 1.5.2 and Section 1.6.1. As we saw there, Green's functions have the form $G(P,Q) = S(P,Q) + g(P,Q)$ where $S(P,Q)$ is the fundamental singularity for the Laplacian Δ for dimension $n = 1,2,3,...$, and where $g(P,Q)$ is a smooth adjusting function depending on Dirichlet, Neumann, Robin, or other boundary condition. We are not interested in the latter so we will focus attention just on $S(P,Q)$ in this equation. How then may one theoretically justify the equation

$$-\Delta_Q S(P,Q) = \delta(P,Q)?$$

First we turn to the delta function $\delta(P,Q)$. This symbol was introduced in the 1920's by the British physicist P. Dirac so that it would have the properties (we consider one dimension here): $\delta(x) = 0$ for $x \neq 0$, $\int_{-\infty}^{\infty} \delta(x)dx = 1$, and for any continuous function $f(x)$ one obtains

$$f(x_0) = \int_{-\infty}^{\infty} f(x)\delta(x_0 - x)dx.$$

We call this the *point evaluation property* and we will concentrate on it as the principal desired property of the delta function. Moreover, Dirac imagined δ to be infinitely differentiable, so that

$$f^{(n)}(x_0) = \int_{-\infty}^{\infty} f(x)\delta^{(n)}(x_0 - x)dx.$$

In particular, the δ function was to be the derivative of the Heaviside step function. Note that these were the properties shown in Problem 2 of Section 1.5.2 although in the guise of the Green's function there:

$$-G'' = \delta, \ -G' = H, \text{ hence } \delta = H'.$$

Using δ and its derivatives formally, Dirac built a beautiful theory of quantum mechanics.*

As we pointed out in Section 1.5.2, the delta function δ is not a function. Its point evaluator property is that of a *linear functional*: a linear mapping ℓ from functions f to the real or complex numbers such that

$$\ell_{\delta(x_0)} = f(x_0).$$

*P.A.M. Dirac, *The Principles of Quantum Mechanics* (Oxford, 1947).

This is why that field of mathematical analysis is called Functional Analysis: it analyzes linear functionals on function spaces. The delta functional is just one of them. In that theory it is easily established that the proper view of the delta function δ is as a measure, not a function.

Specifically, let $C_0^\infty(\Omega)$ be the infinitely differentiable functions f on some domain Ω in R^n and consider all linear functionals ℓ defined on those functions. Require that ℓ in addition be continuous with respect to $C_0^\infty(\Omega)$ equipped with the inductive limit topology under which all partial derivatives are continuous. With this topology, $C_0^\infty(\Omega)$ is commonly denoted $D(\Omega)$. All continuous linear functionals on $D(\Omega)$ are denoted by $D'(\Omega)$. The latter are called generalized functions or distributions. The delta functional is one of them. It is induced by a measure δ concentrated at the point x_0 so that one may represent the point evaluator property as an integral with respect to that measure

$$\ell_{\delta(x_0)} = f(x_0) = \int_{-\infty}^{\infty} f(x)\delta_{(x_0)}(dx)$$

for all $C_0^\infty(\Omega)$ test functions f in $D(\Omega)$. By integration by parts against the test functions in $C_0^\infty(\Omega)$ (note they and all their derivatives always vanish in a strip near $\partial\Omega$ so that all boundary terms in all integration by parts always vanish) one can easily define in the same way all "derivatives" $\delta^{(n)}$ of δ as linear functionals also in $D(\Omega)'$.

All of the resultant theory is widely available in books on functional analysis and topological vector spaces, and a number of these treatments also include how that theory may be applied to partial differential equations.* We may summarize by saying that all distributional equations $P(D)u = f$ are treated by integration by parts against test functions ϕ in $C_0^\infty(\Omega)$ to obtain weak or generalized solutions.

We close our treatment of distribution equations here in two ways: to the general, and to the particular. First to the general, and very briefly. The theory of distributions, once it was recognized as that of linear functionals on smooth test function spaces, has been generalized in several ways. One generalization in particular, as noted at the beginning of this epilogue, is the hyperfunction theory originated by M. Sato.[†] This theory in our opinion is quite interesting and has yet to achieve its full potential. One of its qualities is that it may be constructively viewed from several perspectives, including those of half-plane boundary values of holomorphic or harmonic functions, in terms of a Cech Cohomology, and in terms of Cauchy principal values. Viewed, as here, from the perspective of continuous linear functionals, hyperfunctions are rather general. We may portray this in terms of the following inclusions

$$O \subset A \subset \mathscr{E} \subset C_0^\infty = D \subset L^p \subset M \subset D' \subset \mathscr{E}' \subset B \subset O'.$$

---

*For example, K. Yosida, *Functional Analysis* (Springer, Berlin, 1968), among many others.
†A good recent reference is M. Morimoto, *An Introduction to Sato's Hyperfunctions* (Amer. Math. Soc., Providence, 1993).

Here O = holomorphic functions, A = real analytic functions, $\mathcal{S}$ = Gevrey functions, M = complex measures, B = hyperfunctions. The prime denotes continuous linear functionals on the function space indicated. The hyperfunctions are essentially analytic functionals.

Now we turn to the particular. We continue to like our explanation of the meaning of the delta function as given in our proof of Green's third identity in Section 1.6.1. Accordingly we will take that discussion even further. As shown in Section 1.6.1, the true meaning of the point evaluator property $-\Delta S(P,Q) = \delta(P,Q)$ in three space dimensions when $S(P,Q) = -1/4\pi r(P,Q)$ was placed into Green's second identity

$$\oint_{\partial\Omega} u \frac{\partial v}{\partial n} - \oint_{\partial\Omega} v \frac{\partial u}{\partial n} = \int_{\Omega} u\Delta v - \int_{\Omega} v\Delta u$$

became that of mean value properties of arbitrary test functions u on the surface S_ϵ of the ϵ-ball B_ϵ centered at the point P. For the geometry, see Fig. 1.6d on page 51. What turned out to be needed was that

$$\oint_{S_\epsilon} u \frac{\partial S}{\partial n} = \frac{1}{4\pi\epsilon^2} \oint_{S_\epsilon} u(Q) = u(Q_\epsilon) \rightarrow u(P)$$

$$\oint_{S_\epsilon} S \frac{\partial u}{\partial n} = \frac{1}{4\pi\epsilon} \oint_{S_\epsilon} \frac{\partial u}{\partial n} = \epsilon \frac{\partial u}{\partial n}(Q'_\epsilon) \rightarrow 0$$

as $\epsilon \rightarrow 0$. The points Q_ϵ and Q'_ϵ are points on the S_ϵ surface at which u and $\partial u/\partial n$ attain their mean values, respectively, over the surface. It is entertaining to imagine in one's mind how these points Q_ϵ and Q'_ϵ may rather wildly move over the surfaces S_ϵ as the ball B_ϵ collapses to P as $\epsilon \rightarrow 0$. It is easy to generalize B_ϵ to deformations of it, to classes of collapsing regions, for example, sandwiched between such B_ϵ balls. The point evaluator δ function property must work for all of these collapsing regions. This picture is an interesting three-dimensional version of the way in which the one-dimensional δ function is often defined in terms of equivalence classes of functions converging to it. The above three-dimensional picture is more interesting because it is in physical space, because not only are there equivalence classes of functions but also related equivalence classes of collapsing regions, and because the δ function property clearly needs appropriate collapsing mean value behavior for both u and $\partial u/\partial n$.

The same is true in higher dimensions for $S(P,Q) = 1/(n-2)\,\omega_n r_{PQ}^{n-2}$ and in two dimensions for $S(P,Q) = (-1/2\pi)\ln r_{PQ}$. Let us see how it looks in one dimension. Green's second identity becomes

$$\int_a^b uv'' - \int_a^b vu'' = [uv' - vu']\big|_a^b.$$

Let us extract a small ϵ-ball B_ϵ about P = the origin, taken for simplicity within the interval $a < 0 < b$. Then the entity of interest is $uv' - vu'$ evaluated at $-\epsilon$ and ϵ. We now note that the fundamental singularity $S(0,x)$ for the Laplacian in one dimension is just $\frac{1}{2}|x|$. Letting $v = \frac{1}{2}|x|$ we have

$$uv'\big|_{-\epsilon}^{\epsilon} = \frac{u(\epsilon) + u(-\epsilon)}{2} \to u(0) \text{ as } \epsilon \to 0$$

$$vu'\big|_{-\epsilon}^{\epsilon} = \frac{\epsilon(u'(\epsilon) - u'(-\epsilon))}{2} \to 0 \text{ as } \epsilon \to 0$$

just as in higher dimensions. The only difference is that here, because $S_\epsilon = \partial B_\epsilon$ is two disconnected points, we must stay, conceptually, with the full "integral" of u and u' on S_ϵ, even though those integrals are also the mean values of u and u'. Note that the same phenomenon obtains for B_ϵ replaced with any small collapsing interval, not necessarily symmetric about the origin, since $u \to u(0)$ from each endpoint individually.

E.2 SEMICONDUCTOR EQUATIONS

Many semiconductor devices (e.g., transistors and their descendents) have been and are currently modelled by the following partial differential equations and elaborations and variations thereof.* There are two principal parts of the modelling of such devices: the chemical doping of the silicon or other material being used, and then the electron and hole flow which will result from the doping and the geometry of the device as it is driven by externally applied voltages. Originally this modelling was done separately but current software is now including both in combined models. In recent years quantum level effects are also brought in, and there is current interest in optical input and output (e.g., lasing) effects which may be obtained from such devices.†

Here, we treat only the device simulation, assuming that the chemical doping, which is a diffusion process, has already been done. The partial differential equations describing the electron and hole movement which will result when voltages are applied at key points on the boundary, that is, the equations which describe the operation of the device, are the following:

$$-\epsilon \text{div (grad } \psi) = q(p - n + C) \qquad \text{Potential Equation}$$

$$\frac{\partial n}{\partial t} = \frac{1}{q} \text{div } (q\mu_n(U_T \text{grad} n - n\text{grad}\psi)) - R \qquad \text{Electron Equation}$$

$$\frac{\partial p}{\partial t} = -\frac{1}{q} \text{div } (- q\mu_p(U_T \text{grad} p + p\text{grad}\psi)) - R \qquad \text{Hole Equation}$$

These are commonly called the *drift-diffusion equations*. Here ϵ = device material permittivity, q = electron charge ($\cong 1.6022 \times 10^{-19}$ As), and U_T = thermal voltage ($\cong 0.026V$ at room temperature) are to be regarded as constants, and the doping profile C as given. The unknowns are the electrostatic potential ψ, the electron

---

*Two excellent references are S. Selberherr, *Analysis and Simulation of Semiconductor Devices* (Springer, New York, 1984), and P. Markowich, *The Stationary Semiconductor Device Equations* (Springer, Vienna, 1986).

†For further discussion and references see K. Gustafson, "Semiconductor Device Simulation," *Computer Physics Communications* 65 (1991).

concentration $n(x,t)$, and the hole concentration $p(x,t)$. The electron and hole mobilities μ_n and μ_p, respectively, and the recombination-generation rate R, generally depend in a given manner nonlinearly on the unknowns ψ, n, and p.

Boundary conditions are a combination of Dirichlet- and Neumann-type, and represent insulated or applied voltage boundary portions. When voltages are applied at key points on the boundary of the device, enormous gradients in the electron and hole concentrations are created. Dynamic ranges of 10^{20}, with the electron and hole concentrations forming huge step-function-like offsetting profiles, are commonly seen. This is desired device action, for example, intuitively stated, parts of it are switched on and off as wanted.

As an example of these drift-diffusion equations, and one to which we will wish to refer later, let us consider only the hole concentration equation. Assuming constant positive hole mobilily μ_p and constant electric field $E = \text{grad } \psi$ and only one space dimension, we have

$$p_t = \mu_p(U_T p_{xx} + E p_x) - R.$$

Such equations describe, for example, the changing minority hole concentration $p(x,t)$ caused by the diffusing in silicon of a given initial light pulse under the influence of an electric field which is turned on at time $t = 0$.

In the numerical modelling of the full drift-diffusion equations, the large shock-like transition regions are a formidable challenge. To reduce the size of such transitions, a rescaling by change of dependent variable $u = n\delta^{-2}e^{-\psi}$, $v = p\delta^{-2}e^{-\psi}$ produces the nonlinear coupled elliptic equations

$$\lambda^2 \Delta \psi = \delta^2 e^{\psi} u - \delta^2 e^{-\psi} v - C$$

$$\text{div }(\mu_n e^{\psi}\text{grad } u) = \delta^{-2}R$$

$$\text{div }(\mu_p e^{-\psi}\text{grad } v) = \delta^{-2}R$$

where we have now dropped the time dependent terms $\partial n/\partial t$ and $\partial p/\partial t$. Here λ is the scaled Debye length of the device (e.g., $\lambda \approx 10^{-4}$), and δ^2 is a scaling factor (e.g., $\delta^2 \approx 10^{-10}$) coming from the somewhat complicated recombination-generation term, which is generally a low order rational function of n and p. U_T and ϵ have been absorbed into the rescaling. The solutions to the drift-diffusion equations in this form have smaller dynamic range but this has been paid for by their increasing nonlinearity.

Let us now further simplify these equations so as to detect some kind of expected basic solution behavior that satisfies our intuition. We drop the doping and recombination terms C and R, we absorb (e.g., as constants) the mobilities μ_n and μ_p, and we uncouple the equations by assuming thermal equilibrium so that $u = v = c = $ constant. This produces the nonlinear electrostatic potential equation

$$\Delta \psi = \lambda \sinh \psi$$

where $\lambda = 2c(\delta\lambda^{-1})^2$ has now absorbed some previous constants. This is an equation of the type we considered in Section 1.8 where we treated nonlinear bifurcation theory, for example, equations of the form $\Delta \psi = -\lambda e^{\psi}$.

From Section 1.8 we expect such equations to have more than one solution for given specified boundary conditions, for example, two or three solutions, such as those indicated in the bifurcation diagrams Fig. 1.8j,k of Section 1.8. In the chemical kinetics model of Section 1.8, the lower and upper solutions of Fig. 1.8k were the "cooler" and "hotter" ones. There the λ axis, that of the bifurcation parameter λ, represents a lumped effect of all of the physical constants of the material being studied, including especially the self-activation energy of the substance—for example, its tendency to combust. Beyond a critical λ_0, there are no solutions. This has been sometimes called solution "blowup" but actually is due to a simplifying modelling assumption of infinite activation energy. For true (finite) activation energies, solutions exist for all positive λ and the bifurcation diagram resembles that of Fig. 1.8j, an S-shaped bifurcation curve. The lower and upper right-turning branches of that bifurcation curve represent stable lower and upper temperatures, the left-turning branch an unstable regime.*

From the analogous simplified semiconductor electrostatic potential equation above, one finds similar S-shaped bifurcation curves. The stable lower branch corresponds to the device being "off" and the stable upper branch corresponds to the device being "on." Such bifurcation curves go under the name *current-voltage diagrams* in electrical engineering.†

E.3   FINANCE EQUATIONS

A major technological change occurred in stock market modelling (and resultant behavior), especially that of options and that of other financial derivative instruments, with the formulation of the Black–Scholes equations about twenty years ago. The underlying assumption leading to these equations is that of a no-arbitrage market, one in which any instantaneous deviation in a price which would allow someone to buy or sell for a risk-free profit, also instantaneously vanishes, or at least vanishes before one can act to profit on it.‡ This assumption sets up a random-walk diffusion equation model which leads to large coupled systems of parabolic partial differential equations. Evidently, this system of partial differential equations, treated numerically by the finite difference or finite element methods we have described in this book, is running 24 hours a day in brokerage houses around the world.§

Here we will describe the European Call Option (an easier one) and derive the Black–Scholes equation for it. Then we will show how that equation may be

---

*For these results and further information, see Problem 1.9.9.8 and the references cited in the answer thereto.

†The analogy developed here is taken from K. Gustafson, *Inter. J. for Comp. and Math. in Electrical and Electronic Engineering* 10 (1991).

‡The phraseology is the author's and is overly simplified. Good recent references are R. Merton, *Continuous Time Finance* (Blackwell, Cambridge, 1990) and D. DeRosa, *Options on Foreign Exchange,* (Probus, Chicago, 1992).

§It should be stated that Monte Carlo or probabilistic (e.g., binomial) discretizations are often competitive with the numerical PDE discretizations.

converted by change of variable to a heat equation. One begins with the efficient market hypothesis, which assumes that the market price of an asset reflects accurately all past transactions in its current price, and that the current price will accurately respond instantly to any relevant new information. Mathematically this is an assumption that the asset price S is a Markov innovation process. Then one models the relative change of S according to the stochastic differential equation

$$\frac{dS}{S} = \mu dt + \sigma dz.$$

Here μdt represents a deterministic predictable contribution to the return dS/S in a short time interval, for example, $dS = \mu S dt$ where μ could be an interest rate. The nondeterministic unpredictable part σdz represents an instantaneous standard deviation σ where dz is a stochastic process with stationary mean equal to zero and with variance equal to dt. Usually this z process is assumed to be a Brownian motion modelled by a normal distribution. By recognizing that dS/S is the differential of the log S function and by a Taylor series expansion of differentials truncated to second order, one can show that the asset price S is lognormal distributed, and that any function $V(S,t)$ of S and t has differential

$$dV = \left(\mu S \frac{\partial V}{\partial S} + \frac{1}{2}\sigma^2 S^2 \frac{\partial^2 V}{\partial S^2} + \frac{\partial V}{\partial t} \right) dt + \sigma S \frac{\partial V}{\partial S} dz$$

up to the terms retained in the cut-off Taylor series. This equation adequately represents a number of financial derivative instruments.

To be more specific, let $V = V(S,t)$ be the value of a European future call option. This is the right to buy the asset at a preset price E (the exercise price) at a preset time (the expiry date T) in the future. How much of this asset should we buy? Our augmented asset value would be the present value $V(S,t)$ of our portfolio plus the amount we buy, say, δS. The change in our value during the short time interval dt of the buying would be

$$d(V + \delta S) = dV + \delta dS$$
$$= (\mu S V_S + \frac{\sigma^2}{2} S^2 V_{SS} + V_t)dt + \sigma S V_S dz$$
$$+ \delta S(\mu dt + \sigma dz).$$

Here we have put in the expression for dV given above and the expression for dS from our original ansatz of a combined deterministic and stochastic Markovian description of the asset price S.

At this point the main idea of hedging enters. The random component in the above expression can be eliminated by taking $\delta = -V_S$. Then the two dz terms cancel, leaving the change of value

$$d(V + \delta S) = (V_t + \frac{\sigma^2}{2} S^2 V_{SS})dt.$$

Note that the $V_S dt$ terms also happened to cancel.

Now we invoke the underlying assumption of the first paragraph of this section, that of a no-arbitrage market. At this point we must also bring in simple interest rate theory. If we invested our augmented value $V + \delta S$ in a riskless asset (U.S.A. Treasury Bonds are so considered), we would get a return of $r(V + \delta S)dt$ from interest rate r over time interval dt. If this were smaller than the amount $d(V + \delta S)$ derived above, we could borrow $V + \delta S$ from the bank and buy the increased asset with no risk. On the other hand, if $r(V + \delta S)dt$ were larger than $d(V + \delta S)$, then we could put in a call option and make money by putting $V + \delta S$ in the bank until the expiry date T, when we will buy the asset at a profit. In a hedged no-arbitrage market* we must therefore set

$$d(V + \delta S) = r(V + \delta S)dt = r(V - SV_S).$$

Equating the two expressions we have derived for the change of value of the asset, there results

$$V_t + \frac{\sigma^2}{2}S^2V_{SS} + rSV_S - rV = 0.$$

This is the Black–Scholes equation. As is the case for many physical partial differential equations, it has been obtained as a balance equation, here based upon conservation of future value.

The Black–Scholes equation may be transformed into the heat equation by changes of variable. Before doing this, let us note that as it stands above, it is a backward parabolic equation which is to be solved by taking as initial value at $t = T$ an acceptable payoff value, and then running the equation backward from $t = T$ to $t = 0$ to determine the acceptable option price now ($t = 0$). Let us therefore determine the payoff value at $t = T$. If at the time S is greater than the option exercise price, one should exercise the option, obtaining profit $S - E$. If not, we want no loss and do not exercise the option. Thus once again a no-arbitrage ansatz tells us that the proper payoff value at $t = T$ is

$$V(S,t)\big|_{t = T} = \max(S - E, 0).$$

Since S and E are both positive, our initial condition at $t = T$ is just V equal 0 up to the point $S = E$, and then the 45° line increasing to the right. Spatial boundary conditions $V(0,t) = 0$ and $V(S,t) = S$ as $S \to \infty$ are also imposed.

Now to the variable changes. Because we want to end up with the heat equation, clearly we want to reverse the time direction. Motivated by this, we let

$$\tau = \frac{\sigma^2}{2}(T - t), \quad x = \ln\left(\frac{S}{E}\right), \quad v = \frac{V}{E}$$

from which we obtain

$$v_\tau = v_{xx} + av_x + bv$$

---

*Which assumes of course that there are many, many arbitragers out there keeping the market honest.

where $a = (2r/\sigma^2) - 1$ and $b = -(2r/\sigma^2)$. To eliminate the last two terms we let

$$u = e^{\frac{1}{2}ax+(\frac{1-b}{2})^2\tau}v$$

from which we obtain the heat equation

$$u_\tau = u_{xx}.$$

The reader may verify that the payoff initial condition at $t = T$ has been transformed to the initial condition at $\tau = 0$

$$u(x,0) = \max\left(2e^{\frac{r}{\sigma^2}x}\sinh\left(\frac{x}{2}\right),0\right).$$

The spatial boundary conditions are now $u(-\infty) = 0$ and $u(\infty) = \infty$ although these will of course be cut off by finite realistic upper and lower bounds for S.

*   *   *

The Navier–Stokes equations which we treated in Appendix C, the drift-diffusion equations of Section E.2, and the Black–Scholes equations of Section E.3, constitute three of the most important "modern" partial differential equations. They are used widely and lucratively in the aerospace industry, the electronics industry, the finance industry. It is interesting, therefore, to attempt to bring them "under one roof" for a comparison to bring this epilogue to a close.

In one-dimensional form the Navier–Stokes equation for the velocity $u(x,t)$ becomes

$$u_t = \gamma u_{xx} + uu_x.$$

We have dropped the pressure gradient forcing term for convenience. Also we neglect the additional incompressibility constraint* to enable the comparisons we seek. The one-dimensional drift-diffusion equation of Section E.2 for the hole concentration $p(x,t)$ in constant electric field E becomes

$$p_t = U_T p_{xx} + E p_x$$

where we have dropped the recombination rate forcing term. Also we have taken $\mu_p = 1$, or, more correctly, we have absorbed the hole mobility coefficient μ_p into the time scale. The above one-dimensional Black–Scholes equation for the value of a call option becomes

$$v_t = v_{xx} + av_x + bv$$

where a and b are coefficients depending on the bank interest rate r and the asset variance σ^2.

---

*In two and three dimensions we have called this "imperfect incompressibility" in K. Gustafson, *Lectures on Computational Fluid Dynamics, Mathematical Physics and Linear Algebra* (Kaigai, Tokyo, 1996), (World-Scientific, Singapore, 1997). Allowing this leads to an interesting new concept of *numerical rotational release* which is key to understanding which physical solutions are being captured by implicit numerical schemes in unsteady flow simulations in two- and three-dimensional computational fluid dynamics.

These three equations are all what are commonly called convection-diffusion equations (see Pause 5 and answers for a discussion of most appropriate names: there we favored the term advection-diffusion, but here we yield to the majority). Each came from a conservation law: conservation of momentum, conservation of current, conservation of value, respectively. All three may be converted by change of variable to the heat equation: we did this for the (nonlinear) Navier–Stokes equation (there called Burger's equation) in Problem 2.9.8; it was explicitly carried out above for the Black–Scholes equation, and it is easy to show that it may be done for any equation of the form $\omega_t = \omega_{xx} + a\omega_x + b\omega$ even when a and b are functions of t.

Thus we have brought all three of these currently important equations under the roof of the heat equation, for which we have explicit solution representation in terms of its Gaussian Green's function.[†] Our intuition is pleased that all three flows, those of fluid velocities, electron hole concentrations, and financial derivative values, are principally diffusion controlled evolutions.

Countervailingly, in their details and in higher dimensions, their mathematical properties and solution behaviours become quite different. Fluid equations are essentially those of vortex or rotation development, which can happen only in two and three space dimensions. Semiconductor devices are manufactured in layers of different materials, and that is only meaningful in two and three space dimensions. Financial portfolios in currencies have up to twenty assets and in stocks perhaps three hundred assets, so financial derivative equations must be treated in inherently coupled fashion.

In complexity there is simplicity. In simplicity there is complexity. The game is never over.

[†]The Green's function representation of the solution of the Black–Scholes equation is especially interesting. For the European future call option derived above, the solution becomes $V(S,t) = SN(d_1) - Ee^{-r(T-t)}N(d_2)$ where $N(\cdot)$ is the cumulative probability distribution for a standardized normal variable, $d_1 = [\ln(S/E)+(r+\sigma^2/2)(T-t)]/\sigma(T-t)^{1/2}$, $d_2 = d_1 - \sigma(T-t)^{1/2}$. For their work on the Black–Scholes formula, Robert Merton and Myron Scholes were awarded the Nobel Prize in economics in 1997. (Fischer Black died in 1995. Nobel prizes are not awarded posthumously.)

AUTHOR INDEX

SUBJECT INDEX

A CATALOG OF SELECTED
DOVER BOOKS
IN SCIENCE AND MATHEMATICS

A CATALOG OF SELECTED

DOVER BOOKS

IN SCIENCE AND MATHEMATICS

QUALITATIVE THEORY OF DIFFERENTIAL EQUATIONS, V.V. Nemytskii and V.V. Stepanov. Classic graduate-level text by two prominent Soviet mathematicians covers classical differential equations as well as topological dynamics and ergodic theory. Bibliographies. 523pp. 5⅜ x 8½. 65954-2 Pa. $14.95

MATRICES AND LINEAR ALGEBRA, Hans Schneider and George Phillip Barker. Basic textbook covers theory of matrices and its applications to systems of linear equations and related topics such as determinants, eigenvalues and differential equations. Numerous exercises. 432pp. 5⅜ x 8½. 66014-1 Pa. $12.95

QUANTUM THEORY, David Bohm. This advanced undergraduate-level text presents the quantum theory in terms of qualitative and imaginative concepts, followed by specific applications worked out in mathematical detail. Preface. Index. 655pp. 5⅜ x 8½. 65969-0 Pa. $15.95

ATOMIC PHYSICS (8th edition), Max Born. Nobel laureate's lucid treatment of kinetic theory of gases, elementary particles, nuclear atom, wave-corpuscles, atomic structure and spectral lines, much more. Over 40 appendices, bibliography. 495pp. 5⅜ x 8½. 65984-4 Pa. $13.95

ELECTRONIC STRUCTURE AND THE PROPERTIES OF SOLIDS: The Physics of the Chemical Bond, Walter A. Harrison. Innovative text offers basic understanding of the electronic structure of covalent and ionic solids, simple metals, transition metals and their compounds. Problems. 1980 edition. 582pp. 6⅛ x 9¼. 66021-4 Pa. $19.95

BOUNDARY VALUE PROBLEMS OF HEAT CONDUCTION, M. Necati Özisik. Systematic, comprehensive treatment of modern mathematical methods of solving problems in heat conduction and diffusion. Numerous examples and problems. Selected references. Appendices. 505pp. 5⅜ x 8½. 65990-9 Pa. $12.95

A SHORT HISTORY OF CHEMISTRY (3rd edition), J.R. Partington. Classic exposition explores origins of chemistry, alchemy, early medical chemistry, nature of atmosphere, theory of valency, laws and structure of atomic theory, much more. 428pp. 5⅜ x 8½. (Available in U.S. only) 65977-1 Pa. $11.95

A HISTORY OF ASTRONOMY, A. Pannekoek. Well-balanced, carefully reasoned study covers such topics as Ptolemaic theory, work of Copernicus, Kepler, Newton, Eddington's work on stars, much more. Illustrated. References. 521pp. 5⅜ x 8½. 65994-1 Pa. $12.95

PRINCIPLES OF METEOROLOGICAL ANALYSIS, Walter J. Saucier. Highly respected, abundantly illustrated classic reviews atmospheric variables, hydrostatics, static stability, various analyses (scalar, cross-section, isobaric, isentropic, more). For intermediate meteorology students. 454pp. 6⅛ x 9¼. 65979-8 Pa. $14.95

RELATIVITY, THERMODYNAMICS AND COSMOLOGY, Richard C. Tolman. Landmark study extends thermodynamics to special, general relativity; also applications of relativistic mechanics, thermodynamics to cosmological models. 501pp. 5⅜ x 8½. 65383-8 Pa. $15.95

APPLIED ANALYSIS, Cornelius Lanczos. Classic work on analysis and design of finite processes for approximating solution of analytical problems. Algebraic equations, matrices, harmonic analysis, quadrature methods, much more. 559pp. 5⅜ x 8½. 65656-X Pa. $13.95

INTRODUCTION TO ANALYSIS, Maxwell Rosenlicht. Unusually clear, accessible coverage of set theory, real number system, metric spaces, continuous functions, Riemann integration, multiple integrals, more. Wide range of problems. Undergraduate level. Bibliography. 254pp. 5⅜ x 8½. 65038-3 Pa. $9.95

INTRODUCTION TO QUANTUM MECHANICS With Applications to Chemistry, Linus Pauling & E. Bright Wilson, Jr. Classic undergraduate text by Nobel Prize winner applies quantum mechanics to chemical and physical problems. Numerous tables and figures enhance the text. Chapter bibliographies. Appendices. Index. 468pp. 5⅜ x 8½. 64871-0 Pa. $12.95

ASYMPTOTIC EXPANSIONS OF INTEGRALS, Norman Bleistein & Richard A. Handelsman. Best introduction to important field with applications in a variety of scientific disciplines. New preface. Problems. Diagrams. Tables. Bibliography. Index. 448pp. 5⅜ x 8½. 65082-0 Pa. $13.95

MATHEMATICS APPLIED TO CONTINUUM MECHANICS, Lee A. Segel. Analyzes models of fluid flow and solid deformation. For upper-level math, science and engineering students. 608pp. 5⅜ x 8½. 65369-2 Pa. $14.95

ELEMENTS OF REAL ANALYSIS, David A. Sprecher. Classic text covers fundamental concepts, real number system, point sets, functions of a real variable, Fourier series, much more. Over 500 exercises. 352pp. 5⅜ x 8½. 65385-4 Pa. $11.95

PHYSICAL PRINCIPLES OF THE QUANTUM THEORY, Werner Heisenberg. Nobel Laureate discusses quantum theory, uncertainty, wave mechanics, work of Dirac, Schroedinger, Compton, Wilson, Einstein, etc. 184pp. 5⅜ x 8½. 60113-7 Pa. $7.95

INTRODUCTORY REAL ANALYSIS, A.N. Kolmogorov, S.V. Fomin. Translated by Richard A. Silverman. Self-contained, evenly paced introduction to real and functional analysis. Some 350 problems. 403pp. 5⅜ x 8½. 61226-0 Pa. $11.95

PROBLEMS AND SOLUTIONS IN QUANTUM CHEMISTRY AND PHYSICS, Charles S. Johnson, Jr. and Lee G. Pedersen. Unusually varied problems, detailed solutions in coverage of quantum mechanics, wave mechanics, angular momentum, molecular spectroscopy, scattering theory, more. 280 problems plus 139 supplementary exercises. 430pp. 6½ x 9¼. 65236-X Pa. $14.95

ASYMPTOTIC METHODS IN ANALYSIS, N.G. de Bruijn. An inexpensive, comprehensive guide to asymptotic methods–the pioneering work that teaches by explaining worked examples in detail. Index. 224pp. 5⅜ x 8½. 64221-6 Pa. $7.95

OPTICAL RESONANCE AND TWO-LEVEL ATOMS, L. Allen and J. H. Eberly. Clear, comprehensive introduction to basic principles behind all quantum optical resonance phenomena. 53 illustrations. Preface. Index. 256pp. 5⅜ x 8½.
65533-4 Pa. $8.95

COMPLEX VARIABLES, Francis J. Flanigan. Unusual approach, delaying complex algebra till harmonic functions have been analyzed from real variable viewpoint. Includes problems with answers. 364pp. 5⅜ x 8½. 61388-7 Pa. $10.95

ATOMIC SPECTRA AND ATOMIC STRUCTURE, Gerhard Herzberg. One of best introductions; especially for specialist in other fields. Treatment is physical rather than mathematical. 80 illustrations. 257pp. 5⅜ x 8½. 60115-3 Pa. $7.95

APPLIED COMPLEX VARIABLES, John W. Dettman. Step-by-step coverage of fundamentals of analytic function theory–plus lucid exposition of five important applications: Potential Theory; Ordinary Differential Equations; Fourier Transforms; Laplace Transforms; Asymptotic Expansions. 66 figures. Exercises at chapter ends. 512pp. 5⅜ x 8½. 64670-X Pa. $14.95

ULTRASONIC ABSORPTION: An Introduction to the Theory of Sound Absorption and Dispersion in Gases, Liquids and Solids, A.B. Bhatia. Standard reference in the field provides a clear, systematically organized introductory review of fundamental concepts for advanced graduate students, research workers. Numerous diagrams. Bibliography. 440pp. 5⅜ x 8½. 64917-2 Pa. $11.95

UNBOUNDED LINEAR OPERATORS: Theory and Applications, Seymour Goldberg. Classic presents systematic treatment of the theory of unbounded linear operators in normed linear spaces with applications to differential equations. Bibliography. I99pp. 5⅜ x 8½. 64830-3 Pa. $7.95

LIGHT SCATTERING BY SMALL PARTICLES, H.C. van de Hulst. Comprehensive treatment including full range of useful approximation methods for researchers in chemistry, meteorology and astronomy. 44 illustrations. 470pp. 5⅜ x 8½.
64228-3 Pa. $12.95

CONFORMAL MAPPING ON RIEMANN SURFACES, Harvey Cohn. Lucid, insightful book presents ideal coverage of subject. 334 exercises make book perfect for self-study. 55 figures. 352pp. 5⅜ x 8¼. 64025-6 Pa. $11.95

OPTICKS, Sir Isaac Newton. Newton's own experiments with spectroscopy, colors, lenses, reflection, refraction, etc., in language the layman can follow. Foreword by Albert Einstein. 532pp. 5⅜ x 8½. 60205-2 Pa. $12.95

GENERALIZED INTEGRAL TRANSFORMATIONS, A.H. Zemanian. Graduate-level study of recent generalizations of the Laplace, Mellin, Hankel, K. Weierstrass, convolution and other simple transformations. Bibliography. 320pp. 5⅜ x 8½.
65375-7 Pa. $8.95

THE ELECTROMAGNETIC FIELD, Albert Shadowitz. Comprehensive undergraduate text covers basics of electric and magnetic fields, builds up to electromagnetic theory. Also related topics, including relativity. Over 900 problems. 768pp. 5⅝ x 8¼. 65660-8 Pa. $19.95

FOURIER SERIES, Georgi P. Tolstov. Translated by Richard A. Silverman. A valuable addition to the literature on the subject, moving clearly from subject to subject and theorem to theorem. 107 problems, answers. 336pp. 5⅜ x 8½. 63317-9 Pa. $10.95

THEORY OF ELECTROMAGNETIC WAVE PROPAGATION, Charles Herach Papas. Graduate-level study discusses the Maxwell field equations, radiation from wire antennas, the Doppler effect and more. xiii + 244pp. 5⅜ x 8½. 65678-0 Pa. $9.95

DISTRIBUTION THEORY AND TRANSFORM ANALYSIS: An Introduction to Generalized Functions, with Applications, A.H. Zemanian. Provides basics of distribution theory, describes generalized Fourier and Laplace transformations. Numerous problems. 384pp. 5⅜ x 8½. 65479-6 Pa. $13.95

THE PHYSICS OF WAVES, William C. Elmore and Mark A. Heald. Unique overview of classical wave theory. Acoustics, optics, electromagnetic radiation, more. Ideal as classroom text or for self-study. Problems. 477pp. 5⅜ x 8½. 64926-1 Pa. $13.95

CALCULUS OF VARIATIONS WITH APPLICATIONS, George M. Ewing. Applications-oriented introduction to variational theory develops insight and promotes understanding of specialized books, research papers. Suitable for advanced undergraduate/graduate students as primary, supplementary text. 352pp. 5⅜ x 8½. 64856-7 Pa. $9.95

A TREATISE ON ELECTRICITY AND MAGNETISM, James Clerk Maxwell. Important foundation work of modern physics. Brings to final form Maxwell's theory of electromagnetism and rigorously derives his general equations of field theory. 1,084pp. 5⅜ x 8½. 60636-8, 60637-6 Pa., Two-vol. set $25.90

AN INTRODUCTION TO THE CALCULUS OF VARIATIONS, Charles Fox. Graduate-level text covers variations of an integral, isoperimetrical problems, least action, special relativity, approximations, more. References. 279pp. 5⅜ x 8½. 65499-0 Pa. $8.95

HYDRODYNAMIC AND HYDROMAGNETIC STABILITY, S. Chandrasekhar. Lucid examination of the Rayleigh-Benard problem; clear coverage of the theory of instabilities causing convection. 704pp. 5⅝ x 8¼. 64071-X Pa. $17.95

CALCULUS OF VARIATIONS, Robert Weinstock. Basic introduction covering isoperimetric problems, theory of elasticity, quantum mechanics, electrostatics, etc. Exercises throughout. 326pp. 5⅜ x 8½. 63069-2 Pa. $9.95

DYNAMICS OF FLUIDS IN POROUS MEDIA, Jacob Bear. For advanced students of ground water hydrology, soil mechanics and physics, drainage and irrigation engineering and more. 335 illustrations. Exercises, with answers. 784pp. 6⅛ x 9¼. 65675-6 Pa. $19.95

NUMERICAL METHODS FOR SCIENTISTS AND ENGINEERS, Richard Hamming. Classic text stresses frequency approach in coverage of algorithms, polynomial approximation, Fourier approximation, exponential approximation, other topics. Revised and enlarged 2nd edition. 721pp. 5⅜ x 8½.   65241-6 Pa. $16.95

THEORETICAL SOLID STATE PHYSICS, Vol. 1: Perfect Lattices in Equilibrium; Vol. II: Non-Equilibrium and Disorder, William Jones and Norman H. March. Monumental reference work covers fundamental theory of equilibrium properties of perfect crystalline solids, non-equilibrium properties, defects and disordered systems. Appendices. Problems. Preface. Diagrams. Index. Bibliography. Total of 1,301pp. 5⅜ x 8½. Two volumes.   Vol. I: 65015-4 Pa. $16.95
Vol. II: 65016-2 Pa. $16.95

OPTIMIZATION THEORY WITH APPLICATIONS, Donald A. Pierre. Broad spectrum approach to important topic. Classical theory of minima and maxima, calculus of variations, simplex technique and linear programming, more. Many problems, examples. 640pp. 5⅜ x 8½.   65205-X Pa. $17.95

THE CONTINUUM: A Critical Examination of the Foundation of Analysis, Hermann Weyl. Classic of 20th-century foundational research deals with the conceptual problem posed by the continuum. 156pp. 5⅜ x 8½.   67982-9 Pa. $6.95

ESSAYS ON THE THEORY OF NUMBERS, Richard Dedekind. Two classic essays by great German mathematician: on the theory of irrational numbers; and on transfinite numbers and properties of natural numbers. 115pp. 5⅜ x 8½.
21010-3 Pa. $6.95

THE FUNCTIONS OF MATHEMATICAL PHYSICS, Harry Hochstadt. Comprehensive treatment of orthogonal polynomials, hypergeometric functions, Hill's equation, much more. Bibliography. Index. 322pp. 5⅜ x 8½. 65214-9 Pa. $9.95

NUMBER THEORY AND ITS HISTORY, Oystein Ore. Unusually clear, accessible introduction covers counting, properties of numbers, prime numbers, much more. Bibliography. 380pp. 5⅜ x 8½.   65620-9 Pa. $10.95

THE VARIATIONAL PRINCIPLES OF MECHANICS, Cornelius Lanczos. Graduate level coverage of calculus of variations, equations of motion, relativistic mechanics, more. First inexpensive paperbound edition of classic treatise. Index. Bibliography. 418pp. 5⅜ x 8½.   65067-7 Pa. $14.95

COMBINATORIAL TOPOLOGY, P. S. Alexandrov. Clearly written, well-organized, three-part text begins by dealing with certain classic problems without using the formal techniques of homology theory and advances to the central concept, the Betti groups. Numerous detailed examples. 654pp. 5⅜ x 8½.   40179-0 Pa. $18.95

THEORETICAL PHYSICS, Georg Joos, with Ira M. Freeman. Classic overview covers essential math, mechanics, electromagnetic theory, thermodynamics, quantum mechanics, nuclear physics, other topics. First paperback edition. xxiii + 885pp. 5⅜ x 8½.   65227-0 Pa. $21.95

HANDBOOK OF MATHEMATICAL FUNCTIONS WITH FORMULAS, GRAPHS, AND MATHEMATICAL TABLES, edited by Milton Abramowitz and Irene A. Stegun. Vast compendium: 29 sets of tables, some to as high as 20 places. 1,046pp. 8 x 10½. 61272-4 Pa. $29.95

MATHEMATICAL METHODS IN PHYSICS AND ENGINEERING, John W. Dettman. Algebraically based approach to vectors, mapping, diffraction, other topics in applied math. Also generalized functions, analytic function theory, more. Exercises. 448pp. 5⅜ x 8¼. 65649-7 Pa. $12.95

A SURVEY OF NUMERICAL MATHEMATICS, David M. Young and Robert Todd Gregory. Broad self-contained coverage of computer-oriented numerical algorithms for solving various types of mathematical problems in linear algebra, ordinary and partial, differential equations, much more. Exercises. Total of 1,248pp. 5⅜ x 8½. Two volumes. Vol. I: 65691-8 Pa. $16.95
Vol. II: 65692-6 Pa. $16.95

TENSOR ANALYSIS FOR PHYSICISTS, J.A. Schouten. Concise exposition of the mathematical basis of tensor analysis, integrated with well-chosen physical examples of the theory. Exercises. Index. Bibliography. 289pp. 5⅜ x 8½. 65582-2 Pa. $10.95

INTRODUCTION TO NUMERICAL ANALYSIS (2nd Edition), F.B. Hildebrand. Classic, fundamental treatment covers computation, approximation, interpolation, numerical differentiation and integration, other topics. 150 new problems. 669pp. 5⅜ x 8½. 65363-3 Pa. $16.95

INVESTIGATIONS ON THE THEORY OF THE BROWNIAN MOVEMENT, Albert Einstein. Five papers (1905–8) investigating dynamics of Brownian motion and evolving elementary theory. Notes by R. Fürth. 122pp. 5⅜ x 8½. 60304-0 Pa. $5.95

CATASTROPHE THEORY FOR SCIENTISTS AND ENGINEERS, Robert Gilmore. Advanced-level treatment describes mathematics of theory grounded in the work of Poincaré, R. Thom, other mathematicians. Also important applications to problems in mathematics, physics, chemistry and engineering. 1981 edition. References. 28 tables. 397 black-and-white illustrations. xvii + 666pp. 6⅛ x 9¼. 67539-4 Pa. $17.95

AN INTRODUCTION TO STATISTICAL THERMODYNAMICS, Terrell L. Hill. Excellent basic text offers wide-ranging coverage of quantum statistical mechanics, systems of interacting molecules, quantum statistics, more. 523pp. 5⅜ x 8½. 65242-4 Pa. $13.95

STATISTICAL PHYSICS, Gregory H. Wannier. Classic text combines thermodynamics, statistical mechanics and kinetic theory in one unified presentation of thermal physics. Problems with solutions. Bibliography. 532pp. 5⅜ x 8½. 65401-X Pa. $14.95

ORDINARY DIFFERENTIAL EQUATIONS, Morris Tenenbaum and Harry Pollard. Exhaustive survey of ordinary differential equations for undergraduates in mathematics, engineering, science. Thorough analysis of theorems. Diagrams. Bibliography. Index. 818pp. 5⅜ x 8½. 64940-7 Pa. $19.95

STATISTICAL MECHANICS: Principles and Applications, Terrell L. Hill. Standard text covers fundamentals of statistical mechanics, applications to fluctuation theory, imperfect gases, distribution functions, more. 448pp. 5⅜ x 8½. 65390-0 Pa. $14.95

ORDINARY DIFFERENTIAL EQUATIONS AND STABILITY THEORY: An Introduction, David A. Sánchez. Brief, modern treatment. Linear equation, stability theory for autonomous and nonautonomous systems, etc. 164pp. 5⅜ x 8¼. 63828-6 Pa. $6.95

THIRTY YEARS THAT SHOOK PHYSICS: The Story of Quantum Theory, George Gamow. Lucid, accessible introduction to influential theory of energy and matter. Careful explanations of Dirac's anti-particles, Bohr's model of the atom, much more. 12 plates. Numerous drawings. 240pp. 5⅜ x 8½. 24895-X Pa. $7.95

THEORY OF MATRICES, Sam Perlis. Outstanding text covering rank, nonsingularity and inverses in connection with the development of canonical matrices under the relation of equivalence, and without the intervention of determinants. Includes exercises. 237pp. 5⅜ x 8½. 66810-X Pa. $8.95

GREAT EXPERIMENTS IN PHYSICS: Firsthand Accounts from Galileo to Einstein, edited by Morris H. Shamos. 25 crucial discoveries: Newton's laws of motion, Chadwick's study of the neutron, Hertz on electromagnetic waves, more. Original accounts clearly annotated. 370pp. 5⅜ x 8½. 25346-5 Pa. $11.95

INTRODUCTION TO PARTIAL DIFFERENTIAL EQUATIONS WITH APPLICATIONS, E.C. Zachmanoglou and Dale W. Thoe. Essentials of partial differential equations applied to common problems in engineering and the physical sciences. Problems and answers. 416pp. 5⅜ x 8½. 65251-3 Pa. $11.95

BURNHAM'S CELESTIAL HANDBOOK, Robert Burnham, Jr. Thorough guide to the stars beyond our solar system. Exhaustive treatment. Alphabetical by constellation: Andromeda to Cetus in Vol. 1; Chamaeleon to Orion in Vol. 2; and Pavo to Vulpecula in Vol. 3. Hundreds of illustrations. Index in Vol. 3. 2,000pp. 6⅛ x 9¼. 23567-X, 23568-8, 23673-0 Pa., Three-vol. set $44.85

CHEMICAL MAGIC, Leonard A. Ford. Second Edition, Revised by E. Winston Grundmeier. Over 100 unusual stunts demonstrating cold fire, dust explosions, much more. Text explains scientific principles and stresses safety precautions. 128pp. 5⅜ x 8½. 67628-5 Pa. $5.95

AMATEUR ASTRONOMER'S HANDBOOK, J.B. Sidgwick. Timeless, comprehensive coverage of telescopes, mirrors, lenses, mountings, telescope drives, micrometers, spectroscopes, more. 189 illustrations. 576pp. 5⅝ x 8¼. (Available in U.S. only) 24034-7 Pa. $13.95

SPECIAL FUNCTIONS, N.N. Lebedev. Translated by Richard Silverman. Famous Russian work treating more important special functions, with applications to specific problems of physics and engineering. 38 figures. 308pp. 5⅜ x 8½. 60624-4 Pa. $9.95

OBSERVATIONAL ASTRONOMY FOR AMATEURS, J.B. Sidgwick. Mine of useful data for observation of sun, moon, planets, asteroids, aurorae, meteors, comets, variables, binaries, etc. 39 illustrations. 384pp. 5⅜ x 8¼. (Available in U.S. only) 24033-9 Pa. $8.95

INTEGRAL EQUATIONS, F.G. Tricomi. Authoritative, well-written treatment of extremely useful mathematical tool with wide applications. Volterra Equations, Fredholm Equations, much more. Advanced undergraduate to graduate level. Exercises. Bibliography. 238pp. 5⅜ x 8½. 64828-1 Pa. $8.95

POPULAR LECTURES ON MATHEMATICAL LOGIC, Hao Wang. Noted logician's lucid treatment of historical developments, set theory, model theory, recursion theory and constructivism, proof theory, more. 3 appendixes. Bibliography. 1981 edition. ix + 283pp. 5⅜ x 8½. 67632-3 Pa. $8.95

MODERN NONLINEAR EQUATIONS, Thomas L. Saaty. Emphasizes practical solution of problems; covers seven types of equations. ". . . a welcome contribution to the existing literature...."–*Math Reviews*. 490pp. 5⅜ x 8½. 64232-1 Pa. $13.95

FUNDAMENTALS OF ASTRODYNAMICS, Roger Bate et al. Modern approach developed by U.S. Air Force Academy. Designed as a first course. Problems, exercises. Numerous illustrations. 455pp. 5⅜ x 8½. 60061-0 Pa. $11.95

INTRODUCTION TO LINEAR ALGEBRA AND DIFFERENTIAL EQUATIONS, John W. Dettman. Excellent text covers complex numbers, determinants, orthonormal bases, Laplace transforms, much more. Exercises with solutions. Undergraduate level. 416pp. 5⅜ x 8½. 65191-6 Pa. $11.95

INCOMPRESSIBLE AERODYNAMICS, edited by Bryan Thwaites. Covers theoretical and experimental treatment of the uniform flow of air and viscous fluids past two-dimensional aerofoils and three-dimensional wings; many other topics. 654pp. 5⅜ x 8½. 65465-6 Pa. $16.95

INTRODUCTION TO DIFFERENCE EQUATIONS, Samuel Goldberg. Exceptionally clear exposition of important discipline with applications to sociology, psychology, economics. Many illustrative examples; over 250 problems. 260pp. 5⅜ x 8½. 65084-7 Pa. $8.95

THREE PEARLS OF NUMBER THEORY, A. Y. Khinchin. Three compelling puzzles require proof of a basic law governing the world of numbers. Challenges concern van der Waerden's theorem, the Landau-Schnirelmann hypothesis and Mann's theorem, and a solution to Waring's problem. Solutions included. 64pp. 5⅜ x 8½. 40026-3 Pa. $4.95

LECTURES ON CLASSICAL DIFFERENTIAL GEOMETRY, Second Edition, Dirk J. Struik. Excellent brief introduction covers curves, theory of surfaces, fundamental equations, geometry on a surface, conformal mapping, other topics. Problems. 240pp. 5⅜ x 8½. 65609-8 Pa. $9.95

ROTARY-WING AERODYNAMICS, W.Z. Stepniewski. Clear, concise text covers aerodynamic phenomena of the rotor and offers guidelines for helicopter performance evaluation. Originally prepared for NASA. 537 figures. 640pp. 6⅛ x 9¼.
64647-5 Pa. $16.95

DIFFERENTIAL GEOMETRY, Heinrich W. Guggenheimer. Local differential geometry as an application of advanced calculus and linear algebra. Curvature, transformation groups, surfaces, more. Exercises. 62 figures. 378pp. 5⅜ x 8½.
63433-7 Pa. $11.95

INTRODUCTION TO SPACE DYNAMICS, William Tyrrell Thomson. Comprehensive, classic introduction to space-flight engineering for advanced undergraduate and graduate students. Includes vector algebra, kinematics, transformation of coordinates. Bibliography. Index. 352pp. 5⅜ x 8½.
65113-4 Pa. $9.95

A SURVEY OF MINIMAL SURFACES, Robert Osserman. Up-to-date, in-depth discussion of the field for advanced students. Corrected and enlarged edition covers new developments. Includes numerous problems. 192pp. 5⅜ x 8½.
64998-9 Pa. $8.95

ANALYTICAL MECHANICS OF GEARS, Earle Buckingham. Indispensable reference for modern gear manufacture covers conjugate gear-tooth action, gear-tooth profiles of various gears, many other topics. 263 figures. 102 tables. 546pp. 5⅜ x 8½.
65712-4 Pa. $16.95

SET THEORY AND LOGIC, Robert R. Stoll. Lucid introduction to unified theory of mathematical concepts. Set theory and logic seen as tools for conceptual understanding of real number system. 496pp. 5⅜ x 8¼.
63829-4 Pa. $14.95

A HISTORY OF MECHANICS, René Dugas. Monumental study of mechanical principles from antiquity to quantum mechanics. Contributions of ancient Greeks, Galileo, Leonardo, Kepler, Lagrange, many others. 671pp. 5⅜ x 8½.
65632-2 Pa. $18.95

FAMOUS PROBLEMS OF GEOMETRY AND HOW TO SOLVE THEM, Benjamin Bold. Squaring the circle, trisecting the angle, duplicating the cube: learn their history, why they are impossible to solve, then solve them yourself. 128pp. 5⅜ x 8½.
24297-8 Pa. $5.95

MECHANICAL VIBRATIONS, J.P. Den Hartog. Classic textbook offers lucid explanations and illustrative models, applying theories of vibrations to a variety of practical industrial engineering problems. Numerous figures. 233 problems, solutions. Appendix. Index. Preface. 436pp. 5⅜ x 8½.
64785-4 Pa. $13.95

CURVATURE AND HOMOLOGY: Enlarged Edition, Samuel I. Goldberg. Revised edition examines topology of differentiable manifolds; curvature, homology of Riemannian manifolds; compact Lie groups; complex manifolds; curvature, homology of Kaehler manifolds. New Preface. Four new appendixes. 416pp. 5⅜ x 8½.
40207-X Pa. $14.95

HISTORY OF STRENGTH OF MATERIALS, Stephen P. Timoshenko. Excellent historical survey of the strength of materials with many references to the theories of elasticity and structure. 245 figures. 452pp. 5⅜ x 8½.
61187-6 Pa. $14.95

GEOMETRY OF COMPLEX NUMBERS, Hans Schwerdtfeger. Illuminating, widely praised book on analytic geometry of circles, the Moebius transformation, and two-dimensional non-Euclidean geometries. 200pp. 5⅝ x 8¼. 63830-8 Pa. $8.95

MECHANICS, J.P. Den Hartog. A classic introductory text or refresher. Hundreds of applications and design problems illuminate fundamentals of trusses, loaded beams and cables, etc. 334 answered problems. 462pp. 5⅜ x 8½. 60754-2 Pa. $12.95

TOPOLOGY, John G. Hocking and Gail S. Young. Superb one-year course in classical topology. Topological spaces and functions, point-set topology, much more. Examples and problems. Bibliography. Index. 384pp. 5⅜ x 8¼. 65676-4 Pa. $11.95

STRENGTH OF MATERIALS, J.P. Den Hartog. Full, clear treatment of basic material (tension, torsion, bending, etc.) plus advanced material on engineering methods, applications. 350 answered problems. 323pp. 5⅜ x 8½. 60755-0 Pa. $9.95

ELEMENTARY CONCEPTS OF TOPOLOGY, Paul Alexandroff. Elegant, intuitive approach to topology from set-theoretic topology to Betti groups; how concepts of topology are useful in math and physics. 25 figures. 57pp. 5⅜ x 8½. 60747-X Pa. $4.95

ADVANCED STRENGTH OF MATERIALS, J.P. Den Hartog. Superbly written advanced text covers torsion, rotating disks, membrane stresses in shells, much more. Many problems and answers. 388pp. 5⅜ x 8½. 65407-9 Pa. $11.95

COMPUTABILITY AND UNSOLVABILITY, Martin Davis. Classic graduate-level introduction to theory of computability, usually referred to as theory of recurrent functions. New preface and appendix. 288pp. 5⅜ x 8½. 61471-9 Pa. $8.95

GENERAL CHEMISTRY, Linus Pauling. Revised 3rd edition of classic first-year text by Nobel laureate. Atomic and molecular structure, quantum mechanics, statistical mechanics, thermodynamics correlated with descriptive chemistry. Problems. 992pp. 5⅜ x 8½. 65622-5 Pa. $19.95

AN INTRODUCTION TO MATRICES, SETS AND GROUPS FOR SCIENCE STUDENTS, G. Stephenson. Concise, readable text introduces sets, groups, and most importantly, matrices to undergraduate students of physics, chemistry, and engineering. Problems. 164pp. 5⅜ x 8½. 65077-4 Pa. $7.95

THE HISTORICAL BACKGROUND OF CHEMISTRY, Henry M. Leicester. Evolution of ideas, not individual biography. Concentrates on formulation of a coherent set of chemical laws. 260pp. 5⅜ x 8½. 61053-5 Pa. $8.95

THE PHILOSOPHY OF MATHEMATICS: An Introductory Essay, Stephan Körner. Surveys the views of Plato, Aristotle, Leibniz & Kant concerning propositions and theories of applied and pure mathematics. Introduction. Two appendices. Index. 198pp. 5⅜ x 8½. 25048-2 Pa. $8.95

THE DEVELOPMENT OF MODERN CHEMISTRY, Aaron J. Ihde. Authoritative history of chemistry from ancient Greek theory to 20th-century innovation. Covers major chemists and their discoveries. 209 illustrations. 14 tables. Bibliographies. Indices. Appendices. 851pp. 5⅜ x 8½. 64235-6 Pa. $18.95

DE RE METALLICA, Georgius Agricola. The famous Hoover translation of greatest treatise on technological chemistry, engineering, geology, mining of early modern times (1556). All 289 original woodcuts. 638pp. 6¾ x 11.   60006-8 Pa. $21.95

SOME THEORY OF SAMPLING, William Edwards Deming. Analysis of the problems, theory and design of sampling techniques for social scientists, industrial managers and others who find statistics increasingly important in their work. 61 tables. 90 figures. xvii + 602pp. 5⅜ x 8½.   64684-X Pa. $16.95

THE VARIOUS AND INGENIOUS MACHINES OF AGOSTINO RAMELLI: A Classic Sixteenth-Century Illustrated Treatise on Technology, Agostino Ramelli. One of the most widely known and copied works on machinery in the 16th century. 194 detailed plates of water pumps, grain mills, cranes, more. 608pp. 9 x 12.

28180-9 Pa. $24.95

LINEAR PROGRAMMING AND ECONOMIC ANALYSIS, Robert Dorfman, Paul A. Samuelson and Robert M. Solow. First comprehensive treatment of linear programming in standard economic analysis. Game theory, modern welfare economics, Leontief input-output, more. 525pp. 5⅜ x 8½.   65491-5 Pa. $17.95

ELEMENTARY DECISION THEORY, Herman Chernoff and Lincoln E. Moses. Clear introduction to statistics and statistical theory covers data processing, probability and random variables, testing hypotheses, much more. Exercises. 364pp. 5⅜ x 8½.   65218-1 Pa. $10.95

THE COMPLEAT STRATEGYST: Being a Primer on the Theory of Games of Strategy, J.D. Williams. Highly entertaining classic describes, with many illustrated examples, how to select best strategies in conflict situations. Prefaces. Appendices. 268pp. 5⅜ x 8½.   25101-2 Pa. $8.95

CONSTRUCTIONS AND COMBINATORIAL PROBLEMS IN DESIGN OF EXPERIMENTS, Damaraju Raghavarao. In-depth reference work examines orthogonal Latin squares, incomplete block designs, tactical configuration, partial geometry, much more. Abundant explanations, examples. 416pp. 5⅜ x 8¼.

65685-3 Pa. $10.95

THE ABSOLUTE DIFFERENTIAL CALCULUS (CALCULUS OF TENSORS), Tullio Levi-Civita. Great 20th-century mathematician's classic work on material necessary for mathematical grasp of theory of relativity. 452pp. 5⅜ x 8½.

63401-9 Pa. $11.95

VECTOR AND TENSOR ANALYSIS WITH APPLICATIONS, A.I. Borisenko and I.E. Tarapov. Concise introduction. Worked-out problems, solutions, exercises. 257pp. 5⅜ x 8¼.   63833-2 Pa. $9.95

THE FOUR-COLOR PROBLEM: Assaults and Conquest, Thomas L. Saaty and Paul G. Kainen. Engrossing, comprehensive account of the century-old combinatorial topological problem, its history and solution. Bibliographies. Index. 110 figures. 228pp. 5⅜ x 8½.   65092-8 Pa. $7.95

CATALYSIS IN CHEMISTRY AND ENZYMOLOGY, William P. Jencks. Exceptionally clear coverage of mechanisms for catalysis, forces in aqueous solution, carbonyl- and acyl-group reactions, practical kinetics, more. 864pp. 5⅜ x 8½.
65460-5 Pa. $19.95

PROBABILITY: An Introduction, Samuel Goldberg. Excellent basic text covers set theory, probability theory for finite sample spaces, binomial theorem, much more. 360 problems. Bibliographies. 322pp. 5⅜ x 8½.
65252-1 Pa. $10.95

LIGHTNING, Martin A. Uman. Revised, updated edition of classic work on the physics of lightning. Phenomena, terminology, measurement, photography, spectroscopy, thunder, more. Reviews recent research. Bibliography. Indices. 320pp. 5⅜ x 8¼.
64575-4 Pa. $8.95

PROBABILITY THEORY: A Concise Course, Y.A. Rozanov. Highly readable, self-contained introduction covers combination of events, dependent events, Bernoulli trials, etc. Translation by Richard Silverman. 148pp. 5⅜ x 8¼.
63544-9 Pa. $7.95

AN INTRODUCTION TO HAMILTONIAN OPTICS, H. A. Buchdahl. Detailed account of the Hamiltonian treatment of aberration theory in geometrical optics. Many classes of optical systems defined in terms of the symmetries they possess. Problems with detailed solutions. 1970 edition. xv + 360pp. 5⅜ x 8½.
67597-1 Pa. $10.95

STATISTICS MANUAL, Edwin L. Crow, et al. Comprehensive, practical collection of classical and modern methods prepared by U.S. Naval Ordnance Test Station. Stress on use. Basics of statistics assumed. 288pp. 5⅜ x 8½.
60599-X Pa. $8.95

DICTIONARY/OUTLINE OF BASIC STATISTICS, John E. Freund and Frank J. Williams. A clear concise dictionary of over 1,000 statistical terms and an outline of statistical formulas covering probability, nonparametric tests, much more. 208pp. 5⅜ x 8½.
66796-0 Pa. $7.95

STATISTICAL METHOD FROM THE VIEWPOINT OF QUALITY CONTROL, Walter A. Shewhart. Important text explains regulation of variables, uses of statistical control to achieve quality control in industry, agriculture, other areas. 192pp. 5⅜ x 8½.
65232-7 Pa. $8.95

METHODS OF THERMODYNAMICS, Howard Reiss. Outstanding text focuses on physical technique of thermodynamics, typical problem areas of understanding, and significance and use of thermodynamic potential. 1965 edition. 238pp. 5⅜ x 8½.
69445-3 Pa. $8.95

STATISTICAL ADJUSTMENT OF DATA, W. Edwards Deming. Introduction to basic concepts of statistics, curve fitting, least squares solution, conditions without parameter, conditions containing parameters. 26 exercises worked out. 271pp. 5⅜ x 8½.
64685-8 Pa. $9.95

TENSOR CALCULUS, J.L. Synge and A. Schild. Widely used introductory text covers spaces and tensors, basic operations in Riemannian space, non-Riemannian spaces, etc. 324pp. 5⅜ x 8¼.
63612-7 Pa. $11.95

A CONCISE HISTORY OF MATHEMATICS, Dirk J. Struik. The best brief history of mathematics. Stresses origins and covers every major figure from ancient Near East to 19th century. 41 illustrations. 195pp. 5⅜ x 8½. 60255-9 Pa. $8.95

A SHORT ACCOUNT OF THE HISTORY OF MATHEMATICS, W.W. Rouse Ball. One of clearest, most authoritative surveys from the Egyptians and Phoenicians through 19th-century figures such as Grassman, Galois, Riemann. Fourth edition. 522pp. 5⅜ x 8½. 20630-0 Pa. $13.95

HISTORY OF MATHEMATICS, David E. Smith. Nontechnical survey from ancient Greece and Orient to late 19th century; evolution of arithmetic, geometry, trigonometry, calculating devices, algebra, the calculus. 362 illustrations. 1,355pp. 5⅜ x 8½. 20429-4, 20430-8 Pa., Two-vol. set $27.90

THE GEOMETRY OF RENÉ DESCARTES, René Descartes. The great work founded analytical geometry. Original French text, Descartes' own diagrams, together with definitive Smith-Latham translation. 244pp. 5⅜ x 8½. 60068-8 Pa. $8.95

GAMES, GODS & GAMBLING: A History of Probability and Statistical Ideas, F. N. David. Episodes from the lives of Galileo, Fermat, Pascal, and others illustrate this fascinating account of the roots of mathematics. Features thought-provoking references to classics, archaeology, biography, poetry. 1962 edition. 304pp. 5⅜ x 8½. (USO) 40023-9 Pa. $9.95

THE HISTORY OF THE CALCULUS AND ITS CONCEPTUAL DEVELOPMENT, Carl B. Boyer. Origins in antiquity, medieval contributions, work of Newton, Leibniz, rigorous formulation. Treatment is verbal. 346pp. 5⅜ x 8½. 60509-4 Pa. $9.95

THE THIRTEEN BOOKS OF EUCLID'S ELEMENTS, translated with introduction and commentary by Sir Thomas L. Heath. Definitive edition. Textual and linguistic notes, mathematical analysis. 2,500 years of critical commentary. Not abridged. 1,414pp. 5⅜ x 8½. 60088-2, 60089-0, 60090-4 Pa., Three-vol. set $34.85

GAMES AND DECISIONS: Introduction and Critical Survey, R. Duncan Luce and Howard Raiffa. Superb nontechnical introduction to game theory, primarily applied to social sciences. Utility theory, zero-sum games, n-person games, decision-making, much more. Bibliography. 509pp. 5⅜ x 8½. 65943-7 Pa. $13.95

THE HISTORICAL ROOTS OF ELEMENTARY MATHEMATICS, Lucas N.H. Bunt, Phillip S. Jones, and Jack D. Bedient. Fundamental underpinnings of modern arithmetic, algebra, geometry and number systems derived from ancient civilizations. 320pp. 5⅜ x 8½. 25563-8 Pa. $8.95

CALCULUS REFRESHER FOR TECHNICAL PEOPLE, A. Albert Klaf. Covers important aspects of integral and differential calculus via 756 questions. 566 problems, most answered. 431pp. 5⅜ x 8½. 20370-0 Pa. $8.95

CHALLENGING MATHEMATICAL PROBLEMS WITH ELEMENTARY SOLUTIONS, A.M. Yaglom and I.M. Yaglom. Over 170 challenging problems on probability theory, combinatorial analysis, points and lines, topology, convex polygons, many other topics. Solutions. Total of 445pp. 5⅜ x 8½. Two-vol. set.

Vol. I: 65536-9 Pa. $8.95
Vol. II: 65537-7 Pa. $7.95

FIFTY CHALLENGING PROBLEMS IN PROBABILITY WITH SOLUTIONS, Frederick Mosteller. Remarkable puzzlers, graded in difficulty, illustrate elementary and advanced aspects of probability. Detailed solutions. 88pp. 5⅜ x 8½.

65355-2 Pa. $4.95

EXPERIMENTS IN TOPOLOGY, Stephen Barr. Classic, lively explanation of one of the byways of mathematics. Klein bottles, Moebius strips, projective planes, map coloring, problem of the Koenigsberg bridges, much more, described with clarity and wit. 43 figures. 210pp. 5⅜ x 8½.

25933-1 Pa. $6.95

RELATIVITY IN ILLUSTRATIONS, Jacob T. Schwartz. Clear nontechnical treatment makes relativity more accessible than ever before. Over 60 drawings illustrate concepts more clearly than text alone. Only high school geometry needed. Bibliography. 128pp. 6⅛ x 9¼.

25965-X Pa. $7.95

AN INTRODUCTION TO ORDINARY DIFFERENTIAL EQUATIONS, Earl A. Coddington. A thorough and systematic first course in elementary differential equations for undergraduates in mathematics and science, with many exercises and problems (with answers). Index. 304pp. 5⅜ x 8½.

65942-9 Pa. $8.95

FOURIER SERIES AND ORTHOGONAL FUNCTIONS, Harry F. Davis. An incisive text combining theory and practical example to introduce Fourier series, orthogonal functions and applications of the Fourier method to boundary-value problems. 570 exercises. Answers and notes. 416pp. 5⅜ x 8½.   65973-9 Pa. $13.95

AN INTRODUCTION TO ALGEBRAIC STRUCTURES, Joseph Landin. Superb self-contained text covers "abstract algebra": sets and numbers, theory of groups, theory of rings, much more. Numerous well-chosen examples, exercises. 247pp. 5⅜ x 8½.

65940-2 Pa. $8.95

STARS AND RELATIVITY, Ya. B. Zel'dovich and I. D. Novikov. Vol. 1 of *Relativistic Astrophysics* by famed Russian scientists. General relativity, properties of matter under astrophysical conditions, stars and stellar systems. Deep physical insights, clear presentation. 1971 edition. References. 544pp. 5⅜ x 8½.

69424-0 Pa. $14.95
